# Automorphic Forms and Shimura Varieties of PGSp(2)

# Automorphic Forms and Shimura Varieties of PGSp(2)

*by*

Yuval Z Flicker

*The Ohio State University, USA*

World Scientific

NEW JERSEY • LONDON • SINGAPORE • BEIJING • SHANGHAI • HONG KONG • TAIPEI • CHENNAI

*Published by*

World Scientific Publishing Co. Pte. Ltd.

5 Toh Tuck Link, Singapore 596224

*USA office:* 27 Warren Street, Suite 401-402, Hackensack, NJ 07601

*UK office:* 57 Shelton Street, Covent Garden, London WC2H 9HE

**British Library Cataloguing-in-Publication Data**
A catalogue record for this book is available from the British Library.

**AUTOMORPHIC FORMS AND SHIMURA VARIETIES OF PGSp(2)**



ISBN 981-256-403-9

Printed in Singapore.

# PREFACE

This volume concerns two main topics of interest in the theory of automorphic representations, both are by now classical. The first concerns the question of classification of the automorphic representations of a group, connected and reductive over a number field $F$. We consider here the classical example of the projective symplectic group PGSp(2) of similitudes. It is related to Siegel modular forms in the analytic language. We reduce this question to that for the projective general linear group PGL(4) by means of the theory of liftings with respect to the dual group homomorphism $\mathrm{Sp}(2,\mathbb{C}) \hookrightarrow \mathrm{SL}(4,\mathbb{C})$. To describe this classification we introduce the notion of packets and quasi-packets of representations – admissible and automorphic – of PGSp(2). The lifting implies a rigidity theorem for packets and multiplicity one theorem for the discrete spectrum of PGSp(2). The classification uses the theory of endoscopy, and twisted endoscopy. This leads to a notion of stable and unstable packets of automorphic forms. The stable ones are those which do not come from a proper endoscopic group.

This first topic was developed in part to access the second topic of these notes, which is the decomposition of the étale cohomology with compact supports of the Shimura variety associated with PGSp(2), over an algebraic closure $\overline{F}$, with coefficients in a local system. This is a Hecke-Galois bi-module, and its decomposition into irreducibles associates to each geometric (cohomological components at infinity) automorphic representation (we show they all appear in the cohomology) a Galois representation. They are related at almost all places as the Hecke eigenvalues are the Frobenius eigenvalues, up to a shift. In the stable case we obtain Galois representations of dimension $4^{[F:\mathbb{Q}]}$. In the unstable case the dimension is half that, since endoscopy shows up. The statement, and the definition of stability, is based on the classification and lifting results of the first, main, part. The description of the Zeta function of the Shimura variety, also with coefficients in the local system, follows formally from the decomposition of the cohomology.

The third part – which is written for non-experts in representation theory – consists of a brief introduction to the Principle of Functoriality in the theory of automorphic forms. It puts the first two parts in perspective. Parts 1 and 2 are examples of the general – mainly conjectural – theory described in this last part. Part 3 can be read independently of parts 1 and 2. It can be consulted as needed. It contains many of the definitions used in parts 1 and 2, but is not a prerequisite to them. For this reason this Background part is put at the back and not at the fore. Regrettably, it does not discuss the trace formula. But this would require another book. Part 3 is based on a graduate course at Ohio State in Autumn 2003.

Yuval Flicker
Jerusalem, Tevet 5765

# ACKNOWLEDGMENT

It is my pleasure to acknowledge conversations with J. Bernstein, M. Borovoi, G. Harder, D. Kazhdan, J.G.M. Mars, S. Rallis, G. Savin, R. Schmidt, J.-P. Serre, R. Weissauer, Yuan Yi, E.-W. Zink, Th. Zink, on aspects of part 1 of this work. P.-S. Chan suggested many improvements to an earlier draft, and to part 3. For useful comments I thank audiences in HU Berlin, HU Jerusalem, Princeton, Tel-Aviv, MPI Bonn, MSC Beijing, ICM's Weihai, Saarbrücken, Köln, Lausanne, where segments of this part 1 were discussed since the completion of a first draft in 2001.

This work has been announced in particular at MPI-Bonn Arbeitstagung 2003, ERA-AMS and Proc. Japan Acad. [F6], in 2004.

I wish to thank M. Borovoi for encouraging me to carry out part 2 after I communicated to him the results of part 1, R. Pink, D. Kazhdan, and A. Panchishkin, for inviting me to ETH Zürich, and the Hebrew University in Dec. 2003, and Grenoble in Jan. 2004, and the directors, esp. G. Harder and G. Faltings, for inviting me to MPI, Bonn, in summer 2004, where I talked on part 2 of this work. My indebtedness to the publications of R. Langlands and R. Kottwitz is transparent. I benefited also from talking with P. Deligne, N. Katz, G. Shimura, Y. Tschinkel. The interest of the students who took my Autumn 2003 course at Ohio State prompted me to write-up part 3. The editorial advice of Sim Chee Khian and TeXpertise of Ed Overman are appreciated.

Support of the Humboldt Stiftung at Universität Mannheim, Universität Bielefeld, Universität zu Köln, and the Humboldt Universität Berlin, of the National University of Singapore, of the Max-Planck-Institut für Mathematik at Bonn, and of a Lady Davis Visiting Professorship at the Hebrew University, is gratefully acknowledged.

# CONTENTS

# PART 1. LIFTING AUTOMORPHIC FORMS OF PGSp(2) TO PGL(4)

# I. PRELIMINARIES

## 1. Introduction

According to the "principle of functoriality", "Galois" representations $\rho : L_F \to {}^LG$ of the hypothetical Langlands group $L_F$ of a global field $F$ into the complex dual group ${}^LG$ of a reductive group $\mathbf{G}$ over $F$ should parametrize "packets" of automorphic representations of the adèle group $\mathbf{G}(\mathbb{A})$. Thus a map $\lambda : {}^LH \to {}^LG$ of complex dual groups should give rise to lifting of automorphic representations $\pi_H$ of $\mathbf{H}(\mathbb{A})$ to those $\pi$ of $\mathbf{G}(\mathbb{A})$.

Here we prove the existence of the expected lifting of automorphic representations of the projective symplectic group of similitudes $\mathbf{H} = \mathrm{PGSp}(2)$ to those on $\mathbf{G} = \mathrm{PGL}(4)$. The image is the set of the self-contragredient representations of PGL(4) which are not lifts of representations of the rank two split orthogonal group SO(4).

The global lifting is defined by means of local lifting. We define the local lifting in terms of character relations. This permits us to introduce a definition of packets and quasi-packets of representations of PGSp(2) as the sets of representations that occur in these relations. Our main local result is that packets exist and partition the set of tempered representations. We give a detailed description of the structure of packets.

Our global results include a detailed description of the structure of the global packets and quasi-packets (the latter are almost everywhere nontempered). We obtain a *multiplicity one theorem for the discrete spectrum of* PGSp(2), a *rigidity theorem for packets and quasi-packets*, determine all *counterexamples to the naive Ramanujan conjecture*, compute the *multiplicity of each member in a packet or quasi-packet in the discrete spectrum*, conclude that *in each local tempered packet there is precisely one generic representation*, and that *in each global packet which lifts to a generic representation of* PGL(4) *there is precisely one representation which is generic everywhere.* The latter representation is generic if it lifts to a properly induced representation of $\mathrm{PGL}(4, \mathbb{A})$.

We also prove the lifting from SO(4) to PGL(4). This amounts to establishing a product of two representations of GL(2) with central characters whose product is 1. Our rigidity theorem for SO(4) amounts to a strong rigidity statement for a pair of representations of $\mathrm{GL}(2,\mathbb{A})$.

Our method is based on an interplay of global and local tools, e.g. the trace formula and the fundamental lemma. We deal with all, not only generic or tempered, representations.

# 2. Statement of Results

## 2a. Homomorphisms of Dual Groups

Let $\mathbf{G}$ be the projective general linear group $\mathrm{PGL}(4) = \mathrm{PSL}(4)$ over a number field $F$. Our initial purpose is to determine the automorphic representations $\pi$ (Borel-Jacquet [BJ], Langlands [L4]) of $\mathbf{G}(\mathbb{A})$, $\mathbb{A}$ is the ring of adèles of $F$, which are self-contragredient: $\pi \simeq \check{\pi}$, equivalently (Bernstein-Zelevinski [BZ1]), $\theta$-invariant: $\pi \simeq {}^\theta\pi$. Here $\theta$, $\theta(g) = J^{-1}{}^t g^{-1} J$, is the involution defined by

$$J = \begin{pmatrix} 0 & w \\ -w & 0 \end{pmatrix}, \qquad w = \begin{pmatrix} 0 & 1 \\ 1 & 0 \end{pmatrix},$$

where ${}^t g$ denotes the transpose of $g \in \mathbf{G}$, and ${}^\theta\pi(g) = \pi(\theta(g))$. According to the principle of functoriality (Borel [Bo1], Arthur [A2]) these automorphic representations are essentially described by representations of the Weil group $W_F$ of $F$ into the dual group $\widehat{G} = \mathrm{SL}(4,\mathbb{C})$ of $\mathbf{G}$ which are $\hat{\theta}$-invariant, namely representations of $W_F$ into centralizers $Z_{\widehat{G}}(\hat{s}\hat{\theta})$ of $\mathrm{Int}(\hat{s})\hat{\theta}$ in $\widehat{G}$. Here $\hat{\theta}$ is the dual involution $\hat{\theta}(\hat{g}) = J^{-1}{}^t\hat{g}^{-1}J$, and $\hat{s}$ is a semisimple element in $\widehat{G}$. These centralizers are the duals of the twisted (by $\hat{s}\hat{\theta}$) endoscopic groups (Kottwitz-Shelstad [KS]). In fact these are the connected components of the identity of the duals of the twisted endoscopic groups $Z_{\widehat{G}}(\hat{s}\hat{\theta}) \times W_F$. But in our case the endoscopic groups are split so the product of $Z_{\widehat{G}}(\hat{s}\hat{\theta})$ with the Weil group $W_F$ is direct. Hence it suffices for us to work here with the connected component of the identity.

A twisted endoscopic group is called *elliptic* if its dual is not contained in a proper parabolic subgroup of $\widehat{G}$. Representations of nonelliptic endoscopic groups can be reduced by parabolic induction to known ones of smaller rank groups. For our $\widehat{G}$, up to conjugacy the elliptic twisted endoscopic groups have as duals the symplectic group $\widehat{H} = Z_{\widehat{G}}(\hat{\theta}) = \mathrm{Sp}(2,\mathbb{C})$ and the special orthogonal group $\widehat{C} = Z_{\widehat{G}}(\hat{s}\hat{\theta}) =$ "$\mathrm{SO}(4,\mathbb{C})$"

$$= \mathrm{SO}\left(\left(\begin{smallmatrix} 0 & \omega \\ \omega^{-1} & 0 \end{smallmatrix}\right), \mathbb{C}\right) = \left\{g \in \mathrm{SL}(4,\mathbb{C}); g\hat{s}J^t g = \hat{s}J = \left(\begin{smallmatrix} 0 & \omega \\ \omega^{-1} & 0 \end{smallmatrix}\right)\right\},$$

which consists of all $A \otimes B = \left(\begin{smallmatrix} aB & bB \\ cB & dB \end{smallmatrix}\right)$, where

$$\left(A = \left(\begin{smallmatrix} a & b \\ c & d \end{smallmatrix}\right), B\right) \in [\mathrm{GL}(2,\mathbb{C}) \times \mathrm{GL}(2,\mathbb{C})]/\mathbb{C}^\times$$

satisfy $\det A \cdot \det B = 1$. Here $z \in \mathbb{C}^\times$ embeds as the central element $(z, z^{-1})$, $\hat{s} = \mathrm{diag}(-1,1,-1,1)$ and $\omega = \left(\begin{smallmatrix} 0 & -1 \\ 1 & 0 \end{smallmatrix}\right)$.

The group $\widehat{H}$ is the dual group of the simple $F$-group $\mathbf{H} = \mathrm{PSp}(2) = \mathrm{PGSp}(2)$, the projective group of symplectic similitudes, which can also be denoted by the shorter symbol $\mathrm{PGp}(2)$. It is the quotient of

$$\mathrm{GSp}(2) = \{(g, \boldsymbol{\lambda}) \in \mathrm{GL}(4) \times \mathbb{G}_m;\ {}^t g J g = \boldsymbol{\lambda} J\}$$

by its center $\{(\boldsymbol{\lambda}, \boldsymbol{\lambda}^2)\} \simeq \mathbb{G}_m$. Since $\boldsymbol{\lambda}$ is uniquely determined by $g$ (we write $\boldsymbol{\lambda} = \boldsymbol{\lambda}(g)$), we view $\mathrm{GSp}(2)$ as a subgroup of $\mathrm{GL}(4)$ and $\mathrm{PGSp}(2)$ of $\mathrm{PGL}(4)$.

The group $\widehat{C}$ is the dual group of the special orthogonal group ("SO(4)")

$$\mathbf{C} = \{(g_1, g_2) \in \mathrm{GL}(2) \times \mathrm{GL}(2); \det g_1 = \det g_2\}/\mathbb{G}_m.$$

Here $z \in \mathbb{G}_m$ embeds as the central element $(z, z)$. Also we write

$$[\mathrm{GL}(2) \times \mathrm{GL}(2)]'/\,\mathrm{GL}(1)$$

for $\mathbf{C}$, where the prime indicates that the two factors in $\mathrm{GL}(2)$ have equal determinants.

The principle of functoriality suggests that automorphic discrete spectrum representations of $\mathbf{H}(\mathbb{A})$ and $\mathbf{C}(\mathbb{A})$ parametrize (or lift to) the $\theta$-invariant automorphic discrete spectrum representations of the group of $\mathbb{A}$-valued points, $\mathbf{G}(\mathbb{A})$, of $\mathbf{G}$. Our main purpose is to describe this lifting, or parametrization. In particular we define tensor products of two automorphic forms of $\mathrm{GL}(2, \mathbb{A})$ the product of whose central characters is 1. Moreover we describe the automorphic representations of the projective symplectic group of similitudes of rank two, $\mathrm{PGSp}(2, \mathbb{A})$, in terms of $\theta$-invariant representations of $\mathrm{PGL}(4, \mathbb{A})$.

Motivation for the theory of automorphic forms is attractively explained in some articles by S. Gelbart, see, e.g. [G]. For a more technical introduction see part 3, "Background", of this volume. It is based on a course I gave at the Ohio State University in 2003. It gives most definitions used in this work, from adèles to Weil and L-groups, to twisted endoscopy, and a proof of (Emil) Artin's conjecture for two dimensional Galois representations with image $A_4$, $S_4$ in $\mathrm{PGL}(2, \mathbb{C})$.

## 2b. Unramified Liftings

We proceed to explain how the liftings are defined, first for unramified representations.

An irreducible admissible representation $\pi$ of an adèle group $\mathbf{G}(\mathbb{A})$ is the restricted tensor product $\otimes\pi_v$ of irreducible admissible ([BZ1]) representations $\pi_v$ of the groups $\mathbf{G}(F_v)$ of $F_v$-points of $\mathbf{G}$, where $F_v$ is the completion of $F$ at the place $v$ of $F$. Almost all the local components $\pi_v$ are unramified, that is contain a (necessarily unique up to a scalar multiple) nonzero $K_v$-fixed vector. Here $K_v$ is the standard maximal compact subgroup of $\mathbf{G}(F_v)$, namely the group $\mathbf{G}(R_v)$ of $R_v$-points, $R_v$ being the ring of integers of the nonarchimedean local field $F_v$; $\mathbf{G}$ is defined over $R_v$ at almost all nonarchimedean places $v$. For such $v$, an irreducible unramified $\mathbf{G}(F_v)$-module $\pi_v$ is the unique unramified irreducible constituent in an unramified principal series representation $I(\eta_v)$, normalizedly induced (thus induced in the normalized way of [BZ2]) from an unramified character $\eta_v$ of the maximal torus $\mathbf{T}(F_v)$ of a Borel subgroup $\mathbf{B}(F_v)$ of $\mathbf{G}(F_v)$ (extended trivially to the unipotent radical $\mathbf{N}(F_v)$ of $\mathbf{B}(F_v)$). The space of $I(\eta_v)$ consists of the smooth functions $\phi : \mathbf{G}(F_v) \to \mathbb{C}$ with

$$\phi(ank) = (\delta_v^{1/2}\eta_v)(a)\phi(k), \quad k \in K_v, \quad n \in \mathbf{N}(F_v), \quad a \in \mathbf{T}(F_v),$$

$\delta_v(a) = \det[\mathrm{Ad}(a)|\,\mathrm{Lie}\,\mathbf{N}(F_v)]$, and the $\mathbf{G}(F_v)$-action is $(g \cdot \phi)(h) = \phi(hg)$, $g, h \in \mathbf{G}(F_v)$.

The character $\eta_v$ is unramified, thus it factors as $\eta_v : \mathbf{T}(F_v)/\mathbf{T}(R_v) \to$

$\mathbb{C}^\times$. As $X_*(\mathbf{T}) = \mathrm{Hom}(\mathbb{G}_m, \mathbf{T}) \simeq \mathbf{T}(F_v)/\mathbf{T}(R_v)$, $\eta_v$ lies in

$$\mathrm{Hom}(X_*(\mathbf{T}), \mathbb{C}^\times) = \mathrm{Hom}(X^*(\widehat{T}), \mathbb{C}^\times),$$

where $\widehat{T}$ is the maximal torus in the Borel subgroup $\widehat{B}$ of $\widehat{G}$, both fixed in the definition of the (complex) dual group $\widehat{G}$ (Borel [Bo1], Kottwitz [Ko2]). Now

$$\mathrm{Hom}(X^*(\widehat{T}), \mathbb{C}^\times) = X_*(\widehat{T}) \otimes \mathbb{C}^\times = \widehat{T} \subset \widehat{G}.$$

Thus the unramified irreducible $\mathbf{G}(F_v)$-module $\pi_v$ determines a conjugacy class $t(\pi_v) = t(I(\eta_v))$ in $\widehat{G}$ represented by the image of $\eta_v$ in $\widehat{T}$. This class $t(\pi_v)$ is called the Langlands parameter of the unramified $\pi_v$.

In the case of $\mathbf{G} = \mathrm{GL}(n)$, take $\mathbf{B}$ to be the group of upper triangular matrices, $\mathbf{T}$ the diagonal subgroup, and $\eta_v(a_1, \dots, a_n) = \prod \eta_i(a_i)$ $(1 \le i \le n)$. If $\boldsymbol{\pi}_v$ is a generator of the maximal ideal of $R_v$ then $t(I(\eta_v))$ is the class of $\mathrm{diag}(\eta_1(\boldsymbol{\pi}_v), \dots, \eta_n(\boldsymbol{\pi}_v))$ in $\widehat{G} = \mathrm{GL}(n, \mathbb{C})$. If $\mathbf{G} = \mathrm{PGL}(n)$ then $\eta_1 \dots \eta_n = 1$ and $t(I(\eta_v))$ is a class in $\widehat{G} = \mathrm{SL}(n, \mathbb{C})$.

We make the following notational conventions: If the components of $\eta$ are $\eta_{1v}, \eta_{2v}, \dots$, we write $I(\eta_{1v}, \eta_{2v}, \dots)$ for $I(\eta_v)$. For a representation $\pi$ and a character $\chi$ we write $\chi\pi$ for $g \mapsto \chi(g)\pi(g)$, and not $\chi \otimes \pi$, reserving the notation $\pi_1 \otimes \pi_2$, or $\pi_1 \times \pi_2$, for products on different groups: $(h, g) \mapsto \pi_1(h) \otimes \pi_2(g)$ (for example, if $(h, g)$ ranges over a Levi subgroup, the representation normalizedly induced from the representation $\pi_1 \otimes \pi_2$ on the Levi will be denoted by $I(\pi_1, \pi_2)$ or $\pi_1 \times \pi_2$, depending on the context). We prefer the notation $\pi_1 \times \pi_2$ for a representation of a group which is a product of two groups, such as our $C = \mathrm{SO}(4, F)$. By a representation we mean an irreducible one, unless otherwise is specified.

## 2c. The Lifting from SO(4) to PGL(4)

We next describe our results on our secondary lifting $\lambda_1$, from $\mathbf{C} = \mathrm{SO}(4)$ to $\mathbf{G} = \mathrm{PGL}(4)$.

We now return to $\mathbf{G} = \mathrm{PGL}(4)$, $\theta$ and $\mathbf{C} = [\mathrm{GL}(2) \times \mathrm{GL}(2)]'/\mathrm{GL}(1)$. Note that an irreducible unramified $\mathrm{GL}(2, F_v)$-module $\pi_{1v}$ is parametrized by a conjugacy class $t(\pi_{1v})$ in $\mathrm{GL}(2, \mathbb{C})$ (the Langlands parameter of the representation; its eigenvalues are called the Hecke eigenvalues of the representation). An unramified irreducible representation $\pi_{1v} \times \pi_{2v}$ of $\mathbf{C}(F_v)$ is parametrized by a class $t(\pi_{1v}) \times t(\pi_{2v})$ in

$$[\mathrm{GL}(2, \mathbb{C}) \times \mathrm{GL}(2, \mathbb{C})]''/\mathbb{C}^\times \simeq \mathrm{SO}\left(\left(\begin{smallmatrix} 0 & \omega \\ \omega^{-1} & 0 \end{smallmatrix}\right), \mathbb{C}\right) = \widehat{C} \subset \widehat{G}.$$

(Double prime means $\det g_1 \cdot \det g_2 = 1$). If $\pi_{iv}$ is the unramified constituent of

$$I(\eta_{iv}), \quad t(\pi_{iv}) = \operatorname{diag}(\eta_{i1}, \eta_{i2}), \quad \eta_{ij} = \eta_{ijv}(\boldsymbol{\pi}_v), \quad \eta_{11}\eta_{12}\eta_{21}\eta_{22} = 1,$$

we define the "lift" $\pi_{1v} \boxtimes \pi_{2v} = \lambda_1(\pi_{1v} \times \pi_{2v})$ of $\pi_{1v} \times \pi_{2v}$ with respect to the dual group homomorphism $\lambda_1 : \widehat{C} = \mathrm{SO}(4, \mathbb{C}) \hookrightarrow \widehat{G} = \mathrm{SL}(4, \mathbb{C})$ (the natural embedding) to be the unramified irreducible constituent $\pi_v$ of the $\mathrm{PGL}(4, F_v)$-module $I(\eta_v)$ parametrized by the class

$$t(\pi_v) = \operatorname{diag}(\eta_{11}\eta_{21}, \eta_{11}\eta_{22}, \eta_{12}\eta_{21}, \eta_{12}\eta_{22})$$

in $\widehat{G} = \mathrm{SL}(4, \mathbb{C})$. In different notations,

$$\lambda_1(I(a_1, a_2) \times I(b_1, b_2)) = I(a_1b_1, a_1b_2, a_2b_1, a_2b_2) \qquad (a_i, b_i \in \mathbb{C}^\times),$$

provided that $a_1a_2b_1b_2 = 1$. Note that the inverse image under $\lambda_1$ of $I(a_1b_1, a_1b_2, b_1a_2, a_2b_2)$ consists only of

$$\chi I(a_1, a_2) \times \chi^{-1} I(b_1, b_2) \quad \text{and} \quad \chi I(b_1, b_2) \times \chi^{-1} I(a_1, a_2)$$

where $\chi$ is any character of $F_v^\times$. Thus, $\lambda_1$ is two-to-one unless $\pi_{1v} = \check{\pi}_{2v}$ (the contragredient of $\pi_{2v}$), where $\lambda_1$ is injective on the set of orbits of multiplication by $\chi$ in $\mathrm{Hom}(F_v^\times, \mathbb{C}^\times)$.

The rigidity theorem for the discrete spectrum automorphic representations of $\mathrm{GL}(n, \mathbb{A})$ asserts that discrete spectrum automorphic representations $\pi_1 = \otimes \pi_{1v}$ and $\pi_2 = \otimes \pi_{2v}$ which have $\pi_{1v} \simeq \pi_{2v}$ for almost all places $v$ of $F$ are equivalent (Jacquet-Shalika [JS], Moeglin-Waldspurger [MW1]). Moreover they are even equal, by the multiplicity one theorem for $\mathrm{GL}(n)$ (Shalika [Shal]). Representations of $\mathrm{PGL}(n, \mathbb{A})$ (or $\mathrm{PGL}(n, F_v)$) are simply representations of $\mathrm{GL}(n, \mathbb{A})$ (or $\mathrm{GL}(n, F_v)$) with trivial central character (since $H^1(F, \mathbb{G}_m) = \{0\}$), and the rigidity theorem applies then to $\mathrm{PGL}(n)$. Both multiplicity one theorem, and the rigidity theorem for packets (the latter asserts that $\pi = \otimes \pi_v$ and $\pi' = \otimes \pi'_v$ must lie in the same packet if $\pi_v \simeq \pi'_v$ for almost all $v$) hold for $\mathrm{SL}(2)$ ([F3]) and fail for $\mathrm{SL}(n)$, $n \geq 3$ (Blasius [Bla]).

The rigidity theorem holds for $\mathbf{C} = \mathrm{SO}(4)$; this is the content of the assertion that the lifting $\lambda_1$ is injective, made in the second paragraph of the following theorem. The first paragraph asserts that the lifting exists. By an *elliptic* representation we mean one whose character (Harish-Chandra [H]) is not identically zero on the set of elliptic elements.

2.1 THEOREM (SO(4) TO PGL(4)). *Let $\pi_1 = \otimes\pi_{1v}$, $\pi_2 = \otimes\pi_{2v}$ be discrete spectrum automorphic representations of* $\mathrm{GL}(2,\mathbb{A})$ *whose central characters $\omega_1$, $\omega_2$ are equal, and whose components at two places $v_1$, $v_2$ are elliptic. Then there* **exists** *an automorphic representation $\pi = \lambda_1(\pi_1 \times \check{\pi}_2)$ of* $\mathrm{PGL}(4,\mathbb{A})$ *with $\pi_v = \lambda_1(\pi_{1v} \times \check{\pi}_{2v})$ for almost all $v$.*

*We have $\lambda_1(\chi_1\pi_1 \times \chi_2\pi_2) = \chi_1\chi_2\lambda_1(\pi_1 \times \pi_2)$ for $\chi_i : \mathbb{A}^\times/F^\times \to \mathbb{C}^\times$ with $(\chi_1\chi_2)^2 = 1$.*

*If $\pi_1 = \pi_E(\mu_1)$, $\pi_2 = \pi_E(\mu_2)$ are cuspidal monomial representations of* $\mathrm{GL}(2,\mathbb{A})$ *associated with characters $\mu_1$, $\mu_2$ of $\mathbb{A}_E^\times/E^\times$ where $E$ is a quadratic extension of $F$ such that the restriction of $\mu_1\mu_2$ to $\mathbb{A}^\times$ is 1, then $\lambda_1(\pi_E(\mu_1) \times \pi_E(\mu_2)) = I_{(2,2)}(\pi_E(\mu_1\overline{\mu}_2), \pi_E(\mu_1\mu_2))$.*

*If $\{\pi_1, \pi_2\}$ are cuspidal but not of the form $\{\pi_E(\mu_1), \pi_E(\mu_2)\}$, and $\pi_1 \neq \chi\pi_2$ for any quadratic character $\chi$ of $\mathbb{A}^\times/F^\times$, then $\pi_1 \boxtimes \pi_2$ is cuspidal.*

*If $\pi_1$ is the trivial representation $\mathbf{1}_2$ and $\pi_2$ is a cuspidal representation of* $\mathrm{PGL}(2,\mathbb{A})$*, then $\lambda_1(\mathbf{1}_2 \times \pi_2)$ is the discrete spectrum noncuspidal* $\mathrm{PGL}(4,\mathbb{A})$*-module $J(\nu^{1/2}\pi_2, \nu^{-1/2}\pi_2)$. Here $\nu(x) = |x|$, and $J$ is the quotient of the representation $I(\nu^{1/2}\pi_2, \nu^{-1/2}\pi_2)$ normalizedly induced from the parabolic subgroup of type $(2,2)$ of* $\mathrm{PGL}(4)$.

*The global map $\lambda_1$ is* **injective** *on the set of pairs $\pi_1 \times \check{\pi}_2$ with $\omega_1 = \omega_2$ up to the equivalence $\pi_1 \times \check{\pi}_2 \simeq \chi\pi_1 \times \chi^{-1}\check{\pi}_2$, $\chi$ a character of $\mathbb{A}^\times/F^\times$, and $\pi_1 \times \check{\pi}_2 \simeq \check{\pi}_2 \times \pi_1$.*

The injectivity means that if $\pi_1, \pi_2, \pi_1^0, \pi_2^0$ are discrete spectrum automorphic representations of $\mathrm{GL}(2,\mathbb{A})$ with central characters $\omega_1, \omega_2, \omega_1^0, \omega_2^0$ satisfying $\omega_1\omega_2 = 1 = \omega_1^0\omega_2^0$, each of which has elliptic components at least at the three places $v_1$, $v_2$, $v_3$, and if for each $v$ outside a fixed finite set of places of $F$ there is a character $\chi_v$ of $F_v^\times$ such that the set $\{\pi_{1v}\chi_v, \pi_{2v}\chi_v^{-1}\}$ is equal to the set $\{\pi_{1v}^0, \pi_{2v}^0\}$ (up to equivalence of representations), then there is a character $\chi$ of $\mathbb{A}^\times/F^\times$ such that the set $\{\pi_1\chi, \pi_2\chi^{-1}\}$ is equal to the set $\{\pi_1^0, \pi_2^0\}$. In particular, starting with a pair $\pi_1, \pi_2$ of automorphic discrete spectrum representations of $\mathrm{GL}(2,\mathbb{A})$ with $\omega_1\omega_2 = 1$, we cannot get another such pair by interchanging a set of their components $\pi_{1v}, \pi_{2v}$ and multiplying $\pi_{1v}$ by a local character and $\pi_{2v}$ by its inverse, unless we interchange $\pi_1, \pi_2$ and multiply $\pi_1$ by a global character and $\pi_2$ by its inverse.

A considerably weaker result, where the notion of equivalence is generated only by $\pi_{1v} \times \check{\pi}_{2v} \simeq \check{\pi}_{2v} \times \pi_{1v}$ but not by $\pi_{1v} \times \check{\pi}_{2v} \simeq \chi_v\pi_{1v} \times \chi_v^{-1}\check{\pi}_{2v}$, follows also on using the Jacquet-Shalika [JS] theory of $L$-functions, comparing the poles at $s=1$ of the partial, product $L$-functions

$$L^V(s,\pi_1^0\times\check{\pi}_1)L^V(s,\pi_2^0\times\check{\pi}_1)=L^V(s,\pi_1\times\check{\pi}_1)L^V(s,\pi_2\times\check{\pi}_1).$$

Our global results are complemented and strengthened by very precise local results. If $\pi\simeq{}^\theta\pi$ there is an intertwining operator $A$ with $A\pi(g)=\pi(\theta(g))A$ for all $g$. By Schur's lemma we may assume that $A^2=1$. Then $A$ is unique up to a sign. We put $\pi(\theta)=A$ and $\pi(f\times\theta)=\pi(f)A$. We define $\lambda_1$-lifting locally by means of character relations:

$$\lambda_1(\pi_1\times\check{\pi}_2)=\pi\quad\text{if}\quad \operatorname{tr}\pi(f\times\theta)=\operatorname{tr}(\pi_1\times\check{\pi}_2)(f_C)$$

for all matching functions $f$, $f_C$ (and a suitable choice of $A$). This definition is compatible with the one given above for purely induced $\pi_1$ and $\pi_2$ and unramified representations. We have $\lambda_1(I_2(\mu,\mu')\times\check{\pi}_2)=I_4(\mu\check{\pi}_2,\mu'\check{\pi}_2)$ (the central character of the $\mathrm{GL}(2,F)$-module $\pi_2$ is $\mu\mu'$). The local and global results are closely analogous.

## 2d. Special Cases of the Lifting from SO(4)

Let us describe some special cases of the lifting $\lambda_1$. When $\pi_2=\check{\pi}_1$ is the contragredient of $\pi_1$, $\lambda_1(\pi_1\times\check{\pi}_1)$ is the $\mathrm{PGL}(4,\mathbb{A})$-module normalizedly induced from the maximal parabolic of type (3,1) and the $\mathrm{PGL}(3,\mathbb{A})$-module $\mathrm{Sym}^2(\pi_1)$ on the GL(3)-factor of the Levi subgroup (extended trivially to the GL(1)-factor of the Levi, and to the unipotent radical). Here $\mathrm{Sym}^2(\pi_1)$ is the symmetric square lifting from GL(2) to PGL(3) ([F3]). Indeed, if the local component $\pi_{1v}$ of $\pi_1$ at $v$ is unramified then $t(\pi_{1v})=\operatorname{diag}(a,b)$ (thus $\pi_{1v}$ is a constituent of $I_2(a,b)$), $\pi_v=\lambda_1(\pi_{1v}\times\check{\pi}_{1v})$ has $t(\pi_v)=\operatorname{diag}(a/b,1,1,b/a)$ (thus $\pi_v$ is a constituent of $I_4(I_3(a/b,1,b/a),1)$, and $I_3(a/b,1,b/a)$ is the symmetric square lifting of $I_2(a,b)$). We write $I_n$ to emphasize that the representation is of the group $\mathrm{GL}(n)$, and e.g. $I_{(3,1)}(\pi_3,\pi_1)$ to indicate the representation of GL(4) induced from its maximal parabolic subgroup of type (3,1). However, the results of [F3] are stronger, in lifting representations of $\mathrm{SL}(2,\mathbb{A})$ to $\mathrm{PGL}(3,\mathbb{A})$ and consequently providing new results such as multiplicity one for SL(2).

Although we do not obtain here a new proof of the existence of the symmetric square lift of discrete spectrum representations of $\mathrm{PGL}(2,\mathbb{A})$, we do obtain new character identities, relating the $\theta$-twisted character of $I_{(3,1)}(\mathrm{Sym}^2\,\pi_2, 1)$ with that of $\pi_2 \times \check{\pi}_2$. Clearly in this case the lift $\lambda_1$ is injective: if

$$\lambda_1(\pi_1 \times \check{\pi}_2) = \lambda_1(\pi_0 \times \check{\pi}_0) \quad (= I_{(3,1)}(\mathrm{Sym}^2(\pi_0), 1))$$

then $\pi_1 = \pi_2 = \pi_0\chi$ for some character $\chi$ of $\mathbb{A}^\times/F^\times$.

In particular, if $\pi_1$ is a one dimensional representation $g \mapsto \chi(\det g)$ of $\mathrm{GL}(2,\mathbb{A})$, then $\lambda_1(\pi_1 \times \check{\pi}_1) = I_{(3,1)}(\mathbf{1}_3, 1)$ is the representation of $\mathrm{PGL}(4,\mathbb{A})$ normalizedly induced from the trivial representation of the maximal parabolic subgroup of type (3,1). An alternative purely local computation of this twisted character is developed in [FZ].

Let $\pi_1 = \pi(\mu)$ be a cuspidal monomial representation of $\mathrm{GL}(2,\mathbb{A})$ associated with a character $\mu$ of $\mathbb{A}_E^\times/E^\times$ where $E$ is a quadratic extension of $F$ (denote by $\sigma$ the nontrivial element of $\mathrm{Gal}(E/F)$). Then

$$\mathrm{Sym}^2\,\pi_1 = I_{(2,1)}(\pi(\mu/{}^\sigma\mu), \chi_{E/F}),$$

where $\chi_{E/F}$ is the quadratic character of $\mathbb{A}^\times/F^\times N_{E/F}\mathbb{A}_E^\times$ ($N_{E/F}$ is the norm map from $E$ to $F$). Moreover,

$$\lambda_1(\pi(\mu) \times \check{\pi}(\mu)) = I_{(2,1,1)}(\pi(\mu/{}^\sigma\mu), \chi_{E/F}, 1)$$

is an induced representation from the parabolic subgroup of type (2,1,1) of $\mathrm{PGL}(4)$. Note that the central character of the $\mathrm{GL}(2,\mathbb{A})$-module $\pi(\mu)$ is $\chi_{E/F} \cdot \mu|\mathbb{A}^\times$, for any character $\mu$ of $\mathbb{A}_E^\times/E^\times$. If $\pi(\mu)$ is a $\mathrm{PGL}(2,\mathbb{A})$-module we have that the restriction of $\mu$ to $\mathbb{A}_F^\times/F^\times$ is $\chi_{E/F}$, nontrivial but trivial on $F^\times N_{E/F}\mathbb{A}_E^\times$.

If $\pi_1 = \pi_E(\mu_1)$, $\pi_2 = \pi_E(\mu_2)$, cuspidal monomial representations of $\mathrm{GL}(2,\mathbb{A})$ associated with characters $\mu_1$, $\mu_2$ of $\mathbb{A}_E^\times/E^\times$ where $E$ is a quadratic extension of $F$ such that the restriction of $\mu_1\mu_2$ to $\mathbb{A}^\times$ is 1, then

$$\lambda_1(\pi_E(\mu_1) \times \pi_E(\mu_2)) = I_{(2,2)}(\pi_E(\mu_1\overline{\mu}_2), \pi_E(\mu_1\mu_2)).$$

Indeed

$$W_{E/F} = \langle z, \sigma; z \in C_E, \sigma z \sigma^{-1} = \overline{z}, \sigma^2 \in C_F - N_{E/F}C_E\rangle$$

where $C_E = \mathbb{A}_E^\times/E^\times$ (globally, and $E^\times$ locally), and the representation corresponds to

$$\begin{aligned} z &\mapsto \begin{pmatrix} \mu_1(z) & 0 \\ 0 & \mu_1(\overline{z}) \end{pmatrix} \times \begin{pmatrix} \mu_2(z) & 0 \\ 0 & \mu_2(\overline{z}) \end{pmatrix} \\ &\overset{\lambda_1}{\mapsto} \begin{pmatrix} \mu_1\mu_2 & & & 0 \\ & \mu_1\overline{\mu}_2 & & \\ & & \mu_2\overline{\mu}_1 & \\ 0 & & & \overline{\mu}_1\overline{\mu}_2 \end{pmatrix} \overset{(13)}{\mapsto} \begin{pmatrix} \mu_2\overline{\mu}_1 & & & 0 \\ & \mu_1\overline{\mu}_2 & & \\ & & \mu_1\mu_2 & \\ 0 & & & \overline{\mu}_1\overline{\mu}_2 \end{pmatrix}, \\ \sigma &\mapsto \begin{pmatrix} 0 & 1 \\ \mu_1(\sigma^2) & 0 \end{pmatrix} \times \begin{pmatrix} 0 & 1 \\ \mu_2(\sigma^2) & 0 \end{pmatrix} \\ &\overset{\lambda_1}{\mapsto} \begin{pmatrix} 0 & & & 1 \\ & & \mu_2(\sigma^2) & \\ & \mu_1(\sigma^2) & & \\ 1 & & & 0 \end{pmatrix} \overset{(13)}{\mapsto} \begin{pmatrix} 0 & \mu_1(\sigma^2) & & \\ \mu_2(\sigma^2) & 0 & & \\ & & 0 & 1 \\ & & 1 & 0 \end{pmatrix}, \end{aligned}$$

where $\mu_1\mu_2(\sigma^2) = 1$ and $\overline{\mu}_i(z) = \mu_i(\overline{z})$, $\mu_1\overline{\mu}_1\mu_2\overline{\mu}_2 = 1$ and $\mu_i(z)$ are abbreviated to $\mu_i$ in the line of $z$. When $\mu_1 = \mu_2^{-1}$ we have $\pi(\mu_1\overline{\mu}_2) = \pi(\mu_1/\overline{\mu}_1)$ and $\pi(\mu_1\mu_2) = I(\chi_{E/F}, 1)$. Thus

$$\lambda_1(\pi(\mu_1) \times \check{\pi}(\mu_1)) = I_{(2,1,1)}(\pi(\mu_1/\overline{\mu}_1), \chi_{E/F}, 1) = I_{(3,1)}(\mathrm{Sym}^2(\pi(\mu_1)), 1).$$

Note that if $\mu : \mathbb{A}_E^\times \to \mathbb{C}^\times$ has $(\mu/\overline{\mu})^2 = 1 \neq \overline{\mu}/\mu$ then there are quadratic extensions $E_2$, $E_3$ and characters $\mu_i : \mathbb{A}_{E_i}^\times/E_i^\times \to \mathbb{C}^\times$ with $\pi_{E_i}(\mu_i) = \pi_E(\mu)$.

Another interesting special case is when $\pi_1$ is taken to be the trivial representation $\mathbf{1}_2$ of $\mathrm{PGL}(2, \mathbb{A})$ while $\pi_2$ is a cuspidal representation of $\mathrm{PGL}(2, \mathbb{A})$. Then $\lambda_1(\mathbf{1}_2 \times \pi_2)$ is the discrete spectrum noncuspidal representation $J(\nu^{1/2}\pi_2, \nu^{-1/2}\pi_2)$ of $\mathrm{PGL}(4, \mathbb{A})$, the quotient of the normalizedly induced $I(\nu^{1/2}\pi_2, \nu^{-1/2}\pi_2)$ from the parabolic of type (2,2) of PGL(4). Here $\nu(x) = |x|$. Indeed, $\mathbf{1}_2$ is the quotient of the induced $I(\nu^{1/2}, \nu^{-1/2})$. Hence

$$t(\lambda_1(\mathbf{1}_{2v} \times \pi_{2v})) \quad \text{is} \quad (t(\nu_v^{1/2}\pi_{2v}), t(\nu_v^{-1/2}\pi_{2v})). \quad \text{Then} \quad \lambda_1(\mathbf{1}_{2v} \times \pi_{2v})$$

is the quotient $J(\nu_v^{1/2}\pi_{2v}, \nu_v^{-1/2}\pi_{2v})$ of the induced $I(\nu_v^{1/2}\pi_{2v}, \nu_v^{-1/2}\pi_{2v})$ for all $v$ where $\pi_{2v}$ is unramified. Hence it is $J(\nu^{1/2}\pi_2, \nu^{-1/2}\pi_2)$ globally by the rigidity theorem for this noncuspidal discrete spectrum ([MW1]).

On the set of pairs $\pi_1 \times \pi_2$ such that at least one of $\pi_1$ or $\pi_2$ is one dimensional, the lifting $\lambda_1$ is injective. Indeed, a discrete spectrum representation of $\mathrm{GL}(2, \mathbb{A})$ with a one-dimensional component is necessarily one-dimensional. If $\pi_2$ is not cuspidal but rather trivial, then the quotient $J(\nu^{1/2}\mathbf{1}_2, \nu^{-1/2}\mathbf{1}_2)$ of $I_4(\nu^{1/2}\mathbf{1}_2, \nu^{-1/2}\mathbf{1}_2)$ is not discrete spectrum, but the induced $I_4(\mathbf{1}_3)$ from the trivial representation of the (3,1)-parabolic; this is $\lambda_1(\mathbf{1}_2 \times \mathbf{1}_2)$.

## 2e. The Lifting from PGSp(2) to PGL(4)

We now turn to the study of our main lifting $\lambda$, and of the automorphic representations of the $F$-group $\mathbf{H} = (\mathrm{PSp}(2) =) \,\mathrm{PGSp}(2) = \mathrm{GSp}(2)/\mathbb{G}_m$, where the center $\mathbb{G}_m$ of

$$\mathrm{GSp}(2) = \{g \in \mathrm{GL}(4);\ {}^tgJg = \boldsymbol{\lambda} J, \quad \exists \boldsymbol{\lambda} = \boldsymbol{\lambda}(g) \in \mathbb{G}_m\}$$

consists of the scalar matrices. Its dual group is $\widehat{H} = \mathrm{Sp}(2,\mathbb{C}) = Z_{\widehat{G}}(\widehat{\theta}) \subset \widehat{G} = \mathrm{SL}(4,\mathbb{C})$, where $\widehat{\theta}(g) = J^{-1}{}^tg^{-1}J$. It has a single elliptic endoscopic group $\mathbf{C}_0$ different than $\mathbf{H}$ itself. Thus

$$\widehat{C}_0 = Z_{\widehat{H}}(\widehat{s}_0) = \left\{ \begin{pmatrix} a & 0 & 0 & b \\ 0 & \alpha & \beta & 0 \\ 0 & \gamma & \delta & 0 \\ c & 0 & 0 & d \end{pmatrix} \in \widehat{H} \right\} \simeq \mathrm{SL}(2,\mathbb{C}) \times \mathrm{SL}(2,\mathbb{C}),$$

where $\widehat{s}_0 = \mathrm{diag}(-1,1,1,-1)$, and $\mathbf{C}_0 = \mathrm{PGL}(2) \times \mathrm{PGL}(2)$. Write $\lambda_0$ for the embedding $\widehat{C}_0 \hookrightarrow \widehat{H}$, and $\lambda$ for the embedding $\widehat{H} \hookrightarrow \widehat{G}$.

The embedding $\lambda_0 : \widehat{C}_0 = \mathrm{SL}(2,\mathbb{C}) \times \mathrm{SL}(2,\mathbb{C}) \hookrightarrow \widehat{H} = \mathrm{Sp}(2,\mathbb{C})$ defines the "endoscopic" lifting

$$\lambda_0 : \pi_2(\mu_1, \mu_1^{-1}) \times \pi_2(\mu_2, \mu_2^{-1}) \mapsto \pi_{\mathrm{PGSp}(2)}(\mu_1, \mu_2).$$

Here $\pi_2(\mu_i, \mu_i^{-1})$ is the unramified irreducible constituent of the normalizedly induced representation $I(\mu_i, \mu_i^{-1})$ of $\mathrm{PGL}(2, F_v)$ ($\mu_i$ are unramified characters of $F_v^\times$, $i = 1,\ 2$); $\pi_{\mathrm{PGSp}(2)}(\mu_1,\mu_2)$ is the unramified irreducible constituent of the $\mathrm{PGSp}(2,F_v)$-module $I_{\mathrm{PGSp}(2)}(\mu_1,\mu_2)$ normalizedly induced from the character $n \cdot \mathrm{diag}(\alpha,\beta,\gamma,\delta) \mapsto \mu_1(\alpha/\gamma)\mu_2(\alpha/\beta)$ of the upper triangular subgroup of $\mathrm{PGSp}(2,F_v)$ ($n$ is in the unipotent radical, $\alpha\delta = \beta\gamma$).

The embedding $\lambda : \widehat{H} = \mathrm{Sp}(2,\mathbb{C}) \hookrightarrow \mathrm{SL}(4,\mathbb{C}) = \widehat{G}$ defines the lifting $\lambda$ which maps the unramified irreducible representation $\pi_{\mathrm{PGSp}(2)}(\mu_1,\mu_2)$ of $\mathrm{PGSp}(2,F_v)$ to $\pi_4(\mu_1,\mu_2,\mu_2^{-1},\mu_1^{-1})$, an unramified irreducible representation of $\mathrm{PGL}(4,F_v)$.

The composition $\lambda \circ \lambda_0 : \widehat{C}_0 = \mathrm{SL}(2,\mathbb{C}) \times \mathrm{SL}(2,\mathbb{C}) \to \widehat{G} = \mathrm{SL}(4,\mathbb{C})$ takes $\pi_2(\mu_1,\mu_1^{-1}) \times \pi_2(\mu_2,\mu_2^{-1})$ to

$$\pi_4(\mu_1,\mu_2,\mu_2^{-1},\mu_1^{-1}) = \pi_4(\mu_1,\mu_1^{-1},\mu_2,\mu_2^{-1}),$$

namely the unramified irreducible $\mathrm{PGL}(2,F_v)\times\mathrm{PGL}(2,F_v)$-module $\pi_2\times\pi_2'$ to the unramified irreducible constituent $\pi_4(\pi_2,\pi_2')$ of the $\mathrm{PGL}(4,F_v)$-module $I_4(\pi_2,\pi_2')$ normalizedly induced from the representation $\pi_2\otimes\pi_2'$ of the parabolic of type (2,2) of $\mathrm{PGL}(4,F_v)$ (extended trivially on the unipotent radical). For example $\lambda\circ\lambda_0$ takes the trivial $\mathrm{PGL}(2,F_v)\times\mathrm{PGL}(2,F_v)$-module $\mathbf{1}_2 \times \mathbf{1}_2$ to the unramified irreducible constituent $\pi_4(\mathbf{1}_2,\mathbf{1}_2)$ of $I_4(\mathbf{1}_2,\mathbf{1}_2)$, and $\mathbf{1}_2\times\pi_2$ to $\pi_4(\mathbf{1}_2,\pi_2) = \pi_4(\nu^{1/2}\pi_2,\nu^{-1/2}\pi_2)$. Note that this last $\pi_4$ is traditionally denoted by $J$.

The definition of lifting is extended from the case of unramified representations to that of any admissible representations. For this purpose we define below norm maps from the set of $\theta$-stable $\theta$-regular conjugacy classes in $G = \mathbf{G}(F)$ to the set of stable conjugacy classes in $H = \mathbf{H}(F)$, and from this to the set of conjugacy classes in $\mathbf{C}_0(F)$, extending the norm maps on the split tori in these groups which are dual to the dual groups homomorphisms $\lambda$ and $\lambda_0$. This is used to define a relation of matching functions $f$, $f_H$ and $f_{C_0}$ (they have suitably defined matching orbital integrals) and a dual relation of liftings of representations.

To express the lifting results we use the following notations for induced representations of $H = \mathrm{PGSp}(2,F)$. For characters $\mu_1$, $\mu_2$, $\sigma$ of $F^\times$ with $\mu_1\mu_2\sigma^2 = 1$ we write $\mu_1 \times \mu_2 \rtimes \sigma$ for the $H$-module normalizedly induced from the character

$$p = mu \mapsto \mu_1(a)\mu_2(b)\sigma(\boldsymbol{\lambda}), \quad m = \mathrm{diag}(a,b,\boldsymbol{\lambda}/b,\boldsymbol{\lambda}/a), \quad u \in U,$$

$a,b,\boldsymbol{\lambda} \in F^\times$, of the upper triangular minimal parabolic of $H$.

For a $\mathrm{GL}(2,F)$-module $\pi_2$ and character $\mu$ we write $\pi_2 \rtimes \mu$ for the $\mathrm{PGSp}(2,F)$-module normalizedly induced from the representation

$$p = mu \mapsto \pi_2(g)\mu(\boldsymbol{\lambda}), \quad m = \mathrm{diag}(g, \boldsymbol{\lambda} w\, {}^t g^{-1} w), \quad u \in U_{(2)}, \quad \boldsymbol{\lambda} \in F^\times$$

(here the product of the central character $\omega$ of $\pi_2$ with $\mu^2$ is 1) of the Siegel parabolic subgroup (whose unipotent radical $U_{(2)}$ is abelian).

We write $\mu \rtimes \pi_2$, if $\omega\mu = 1$, for the representation of $\mathrm{PGSp}(2,F)$ normalizedly induced from the representation

$$p = mu \mapsto \mu(a)\pi_2(g), \qquad m = \mathrm{diag}(a, g, \boldsymbol{\lambda}(g)/a), \qquad u \in U_{(1)},$$

$\boldsymbol{\lambda}(g) = \det g$, of the Heisenberg parabolic subgroup (whose unipotent radical $U_{(1)}$ is a Heisenberg group).

These inductions are normalized by multiplying the inducing representation by the character $p \mapsto |\det(\mathrm{Ad}(p))|\,\mathrm{Lie}\,U|^{1/2}$, as usual. For example,

$$I_H(\mu_1, \mu_2) = \mu_1\mu_2 \times \mu_1/\mu_2 \rtimes \mu_1^{-1}.$$

Note that $\pi \rtimes \sigma \simeq \check{\pi} \rtimes \omega\sigma$ and $\mu(\pi \rtimes \sigma) = \pi \rtimes \mu\sigma$.

Complete results describing reducibility of these induced representations, stated in Sally-Tadic [ST] following earlier work of Rodier [Ro2], Shahidi [Sh2,3], Waldspurger [W1], are recorded in chapter V, section 1, Propositions 2.1-2.3, below. For notations see chapter II, section 4.

For properly induced representations, defining $\lambda$- and $\lambda_0$-liftings by character relations ($\lambda(\pi_H) = \pi_4$ if $\mathrm{tr}\,\pi_4(f \times \theta) = \mathrm{tr}\,\pi_H(f_H)$ for all matching $f$, $f_H$, and $\lambda_0(\pi_1 \times \pi_2) = \pi_H$ if $\mathrm{tr}\,\pi_H(f_H) = \mathrm{tr}(\pi_1 \times \pi_2)(f_{C_0})$ for all matching $f_H$, $f_{C_0}$), our preliminary results (obtained by local character evaluations), are that $\omega^{-1} \rtimes \pi_2$ $\lambda$-lifts to $\pi_4 = I_G(\pi_2, \check{\pi}_2)$, that $\mu\pi_2 \rtimes \mu^{-1}$ (here $\omega = 1$) $\lambda$-lifts to $\pi_4 = I_G(\mu, \pi_2, \mu^{-1})$, and that $I_2(\mu, \mu^{-1}) \times \pi_2$ $\lambda_0$-lifts from $C_0$ to $\mu\pi_2 \rtimes \mu^{-1}$ on $H = \mathrm{PGSp}(2,F)$.

Let $\chi$ be a character of $F^\times/F^{\times 2}$. It defines a one-dimensional representation $\chi_H(h) = \chi(\boldsymbol{\lambda}(h))$ of $H$, which $\lambda$-lifts to the one-dimensional representation $\chi(g) = \chi(\det g)$ of $G$ (if $h = Ng$ then $\boldsymbol{\lambda}(h) = \det g$; on diagonal matrices $N(\mathrm{diag}(a,b,c,d)) = \mathrm{diag}(ab, ac, db, dc)$). The Steinberg representation of $H$ $\lambda$-lifts to the Steinberg representation of $G$, and for any character $\chi$ of $F^\times$ with $\chi^2 = 1$ we have $\lambda(\chi_H\,\mathrm{St}_H) = \chi\,\mathrm{St}_G$.

## 2f. Elliptic Representations

Our finer local lifting results concern elliptic representations (whose characters are nonzero on the elliptic set). They follow on using global techniques. Elliptic representations include the cuspidal ones (terminology of [BZ]. These are called "supercuspidal" by Harish-Chandra, who used the word "cuspidal" for what is currently named "discrete series" or "square integrable" representations).

2.2 LOCAL THEOREM (PGSP(2) TO PGL(4)). (1) *For any unordered pair* $\pi_1$, $\pi_2$ *of square integrable irreducible representations of* PGL(2, $F$) *there exists a unique pair* $\pi_H^+$, $\pi_H^-$ *of tempered (square integrable if* $\pi_1 \neq \pi_2$, *cuspidal if* $\pi_1 \neq \pi_2$ *are cuspidal) representations of* $H$ *with*

$$\operatorname{tr}(\pi_1 \times \pi_2)(f_{C_0}) = \operatorname{tr}\pi_H^+(f_H) - \operatorname{tr}\pi_H^-(f_H),$$
$$\operatorname{tr} I_G(\pi_1, \pi_2; f \times \theta) = \operatorname{tr}\pi_H^+(f_H) + \operatorname{tr}\pi_H^-(f_H)$$

*for all matching functions* $f$, $f_H$, $f_{C_0}$.

*If* $\pi_1 = \pi_2$ *is cuspidal,* $\pi_H^+$ *and* $\pi_H^-$ *are the two inequivalent constituents of* $1 \rtimes \pi_1$.

*If* $\pi_1 = \pi_2 = \sigma\,\mathrm{sp}_2$ *where* $\sigma$ *is a character of* $F^\times$ *with* $\sigma^2 = 1$, *then* $\pi_H^+$ *and* $\pi_H^-$ *are the two tempered inequivalent constituents* $\tau(\nu^{1/2}\,\mathrm{sp}_2, \sigma\nu^{-1/2})$, $\tau(\nu^{1/2}\mathbf{1}_2, \sigma\nu^{-1/2})$ *of* $1 \rtimes \sigma\,\mathrm{sp}_2$.

*If* $\pi_1 = \sigma\,\mathrm{sp}_2$, $\sigma^2 = 1$, *and* $\pi_2$ *is cuspidal, then* $\pi_H^+$ *is the square integrable constituent* $\delta(\sigma\nu^{1/2}\pi_2, \sigma\nu^{-1/2})$ *of the induced* $\sigma\nu^{1/2}\pi_2 \rtimes \sigma\nu^{-1/2}$; $\pi_H^-$ *is cuspidal, denoted here by*

$$\delta^-(\sigma\nu^{1/2}\pi_2, \sigma\nu^{-1/2}).$$

*If* $\pi_1 = \sigma\,\mathrm{sp}_2$ *and* $\pi_2 = \xi\sigma\,\mathrm{sp}_2$, $\xi$ $(\neq 1 = \xi^2)$ *and* $\sigma$ $(\sigma^2 = 1)$ *are characters of* $F^\times$, *then* $\pi_H^+$ *is the square integrable constituent*

$$\delta(\xi\nu^{1/2}\,\mathrm{sp}_2, \sigma\nu^{-1/2})$$

*of the induced* $\xi\nu^{1/2}\,\mathrm{sp}_2 \rtimes \sigma\nu^{-1/2}$; $\pi_H^-$ *is cuspidal, denoted here by*

$$\delta^-(\xi\nu^{1/2}\,\mathrm{sp}_2, \sigma\nu^{-1/2}).$$

(2) *For every character* $\sigma$ *of* $F^\times/F^{\times 2}$ *and square integrable* $\pi_2$ *there exists a nontempered representation* $\pi_H^\times$ *of* $H$ *such that*

$$\operatorname{tr}(\sigma\mathbf{1}_2 \times \pi_2)(f_{C_0}) = \operatorname{tr}\pi_H^\times(f_H) + \operatorname{tr}\pi_H^-(f_H),$$
$$\operatorname{tr} I_G(\sigma\mathbf{1}_2, \pi_2; f \times \theta) = \operatorname{tr}\pi_H^\times(f_H) - \operatorname{tr}\pi_H^-(f_H),$$

*for all matching* $f$, $f_H$, $f_{C_0}$. *Here*

$$\pi_H^- = \pi_H^-(\sigma\,\mathrm{sp}_2 \times \pi_2), \qquad \pi_H^\times = L(\sigma\nu^{1/2}\pi_2, \sigma\nu^{-1/2}).$$

(3) *For any characters* $\xi$, $\sigma$ *of* $F^\times/F^{\times 2}$ *and matching* $f$, $f_H$, $f_{C_0}$ *we have that* $\operatorname{tr}(\sigma\xi\mathbf{1}_2 \times \sigma\mathbf{1}_2)(f_{C_0})$ *is*

$$= \operatorname{tr} L(\nu\xi, \xi \rtimes \sigma\nu^{-1/2})(f_H) - \operatorname{tr} X(\xi\nu^{1/2}\operatorname{sp}_2, \xi\sigma\nu^{-1/2})(f_H),$$

*and* $\operatorname{tr} I_G(\sigma\xi\mathbf{1}_2, \sigma\mathbf{1}_2; f \times \theta)$ *is*

$$= \operatorname{tr} L(\nu\xi, \xi \rtimes \sigma\nu^{-1/2})(f_H) + \operatorname{tr} X(\xi\nu^{1/2}\operatorname{sp}_2, \xi\sigma\nu^{-1/2})(f_H).$$

*Here* $X = \delta^-$ *if* $\xi \neq 1$ *and* $X = L$ *if* $\xi = 1$.

(4) *Any* $\theta$*-invariant irreducible square integrable representation* $\pi$ *of* $G$ *which is not a* $\lambda_1$*-lift is a* $\lambda$*-lift of an irreducible square integrable representation* $\pi_H$ *of* $H$, *thus* $\operatorname{tr}\pi(f \times \theta) = \operatorname{tr}\pi_H(f_H)$ *for all matching* $f$, $f_H$. *In particular, the square integrable (resp. nontempered) constituent* $\delta(\xi\nu, \nu^{-1/2}\pi_2)$ *(resp.* $L(\xi\nu, \nu^{-1/2}\pi_2)$*) of the induced representation* $\xi\nu \rtimes \nu^{-1/2}\pi_2$ *of* $H$, *where* $\pi_2$ *is a cuspidal (irreducible) representation of* $\mathrm{GL}(2, F)$ *with central character* $\xi \neq 1 = \xi^2$ *and* $\xi\pi_2 = \pi_2$, $\lambda$*-lifts to the square integrable (resp. nontempered) constituent*

$$S(\nu^{1/2}\pi_2, \nu^{-1/2}\pi_2) \qquad (\textit{resp.} \quad J(\nu^{1/2}\pi_2, \nu^{-1/2}\pi_2))$$

*of the induced representation* $I_G(\nu^{1/2}\pi_2, \nu^{-1/2}\pi_2)$ *of* $G = \mathrm{PGL}(4, F)$.

These character relations permit us to introduce the notion of a packet of an irreducible representation, and of a quasi-packet, over a local field. Thus we say that the *packet* of a representation $\pi_H$ of $H$ consists of $\pi_H$ alone unless it is tempered of the form $\pi_H^+$ or $\pi_H^-$ for some pair $\pi_1$, $\pi_2$ of (irreducible) square integrable representations of $\mathrm{PGL}(2, F)$, in which case the packet $\{\pi_H\}$ is defined to be $\{\pi_H^+, \pi_H^-\}$, and we write $\lambda_0(\pi_1 \times \pi_2) = \{\pi_H^+, \pi_H^-\}$ and $\lambda(\{\pi_H^+, \pi_H^-\}) = I_G(\pi_1, \pi_2)$. Further, we define a *quasi-packet* only for the nontempered (irreducible) representations $\pi_H^\times$ and $L = L(\nu\xi, \xi \rtimes \sigma\nu^{-1/2})$, to consist of $\{\pi_H^\times, \pi_H^-\}$ and $\{L, X\}$, $X = X(\xi\nu^{1/2}\operatorname{sp}_2, \xi\sigma\nu^{-1/2})$. We say that $\sigma\mathbf{1}_2 \times \pi_2$ $\lambda_0$-lifts to the quasi-packet $\lambda_0(\sigma\mathbf{1}_2 \times \pi_2) = \{\pi_H^\times, \pi_H^-\}$, which in turn $\lambda$-lifts to $I_G(\sigma\mathbf{1}_2, \pi_2)$, and similarly, $\sigma\xi\mathbf{1}_2 \times \sigma\mathbf{1}_2$ $\lambda_0$-lifts to $\lambda_0(\sigma\xi\mathbf{1}_2 \times \sigma\mathbf{1}_2) = \{L, X\}$ which $\lambda$-lifts to $I_G(\sigma\xi\mathbf{1}_2, \sigma\mathbf{1}_2)$.

Conjecturally our packets and quasi-packets coincide with the L-packets and A-packets conjectured to exist by Langlands and Arthur [A2-3].

Using the notations of section V.11 below, we state the analogue of these results in the real case: $F = \mathbb{R}$. For clarity, denote $\pi_1$ and $\pi_2$

above by $\pi^1$ and $\pi^2$. In (1), $\pi^1 = \pi_{k_1}$ and $\pi^2 = \pi_{k_2}$, $k_1 \geq k_2 > 0$ and $k_1$, $k_2$ are odd, are discrete series representations of $\mathrm{PGL}(2,\mathbb{R})$, and $\pi_H^+$ is the generic $\pi_{k_1,k_2}^{\mathrm{Wh}}$, $\pi_H^-$ is the holomorphic $\pi_{k_1,k_2}^{\mathrm{hol}}$, which are discrete series representations of $\mathrm{PGSp}(2,\mathbb{R})$ when $k_1 > k_2$. When $k_1 = k_2$, $\pi_H^+$ is the generic and $\pi_H^-$ is the nongeneric (tempered) constituents of the induced $1 \rtimes \pi_{2k_1+1}$. There is no special or Steinberg representation of $\mathrm{GL}(2,\mathbb{R})$; the analogue is the lowest discrete series $\pi_1$. The $\pi_k$ are self invariant under twist with sgn. In (2) with $\pi^2 = \pi_{2k+3}$ $(k \geq 0)$, $\pi_H^\times$ is $L(\sigma\nu^{1/2}\pi_{2k+3}, \sigma\nu^{-1/2})$, $\pi_H^-$ is $\pi_{2k+3,1}^{\mathrm{hol}}$, $\pi_H^+$ is $\pi_{2k+3,1}^{\mathrm{Wh}}$. In (3), if $\xi = \mathrm{sgn}$ then $X$ is the tempered $\pi_H^- \subset 1 \rtimes \pi_1$, if $\xi = 1$ then $X$ is $L(\nu^{1/2}\pi_1, \sigma\nu^{-1/2})$. Both of these $X$, as well as $L(\nu\xi, \xi \rtimes \sigma\nu^{-1/2})$, are not cohomological. In (4), $\pi^2$ is $\pi_{2k+2}$, $L(\xi\nu, \nu^{-1/2}\pi^2)$ is $L(\mathrm{sgn}\nu, \nu^{-1/2}\pi_{2k+2})$, $\delta(\xi\nu, \nu^{-1/2}\pi^2)$ is $\pi_{2k+3,2k+1}^{\mathrm{hol}} \oplus \pi_{2k+3,2k+1}^{\mathrm{Wh}}$.

## 2g. Automorphic Representations

With these local definitions we can state our *global results.* These global results *are partial, since we work with test functions whose components are elliptic at least at three places,* and consequently we cannot detect automorphic representations which do not have at least three components whose ($\theta$-) characters are nonzero on the ($\theta$-) elliptic set. Thus we fix three places $\{v_1, v_2, v_3\}$ and discuss only $\pi_1 \times \pi_2$, $\pi_H$ and $\pi = \pi_G$ whose components there are ($\theta$-) elliptic.

Let us explain the reason for this restriction. The (noninvariant) trace formula, as developed by Arthur, involves weighted orbital integrals and logarithmic derivatives of induced representations. Arthur's splitting formula shows that these can be expressed as products of local distributions, which are all invariant (orbital integrals or traces of induced representations) except at most at $\mathrm{rank}(H)$ places. Working with test functions $f_H = \otimes f_{Hv}$ with $\mathrm{rank}(H)+1$ components $f_{Hv}$ with $\mathrm{tr}\,\pi_{Hv}(f_{Hv}) = 0$ for every tempered properly induced representation $\pi_{Hv}$ of $H_v$ (equivalently: $f_{Hv}$ whose orbital integrals vanish on the regular nonelliptic set of $H_v$), all non elliptic terms vanish. We call such $f_{Hv}$ elliptic. At an additional place we use a regular Iwahori biinvariant component (see [FK1], [FK2], [F2] or [F3;VI]) to annihilate the singular orbital integrals. For the twisted trace formula we use the twisted rank, which is equal to $\mathrm{rank}(H)$, to obtain the same vanishing. This removes all complicated terms in the trace formulae comparison. Here *rank* means the $F$-dimension of a maximal split torus in the derived group, or in the derived group of the group of fixed points of the involution in the twisted case.

For very little effort we can reduce the number of restrictions to two, rather than three. Using elliptic components $f_{Hv_1}$, $f_{Hv_2}$, implies that the local factors at each $v \neq v_1$, $v_2$, in the terms in the trace formula, are invariant. We then use at a third, nonarchimedean, place $v_3$ a regular-Iwahori function (as in [FK1], [FK2], [F2], [F3;VI]). Similar choice is made for the twisted formula. The geometric sides of the trace formulae consist now of elliptic terms only. As the distributions at $v_3$ which occur in the trace formula are invariant, such $f_{Hv_3}$ can also be taken to be a spherical function with the same orbital integrals as the Iwahori-regular component. The resulting equality of discrete and continuous measures (the continuous measure comes from the spectral sides), which are invariant distributions in $f_{Hv_3}$, implies their vanishing by the (standard) argument of "generalized linear independence of characters" (using the Stone-Weierstrass theorem) employed in this context in [FK1], [FK2], [F2], [F3]. To simplify our exposition we do not record this argument here, but our global results can safely be used with two restriction, at $v_1$, $v_2$, rather than three.

One can do better, and require that only one component, $f_{Hv}$, be elliptic, at a single real place $v$. This argument, explained in Laumon [Lau], requires very extensive referencing to much of Arthur's deep analysis of the distributions appearing in the trace formula. Inclusion of these arguments here would have made this work more complicated than the relatively elementary exposition I wish to present. However, our results are provable for global representations with a single elliptic component at a real place. This suffices for all purposes of studying the decomposition of the $\ell$-adic cohomology with compact supports of the Shimura variety associated with our group, and any coefficients, as a Galois-Hecke module ([F7]).

These constraints will be removed once the trace formulae identity is established for general test functions. This is being developed by Arthur. A simpler method, based on regular functions, has been introduced when the rank is one (see [F2;I], [F3;VI], [F4;III]) to establish unconditional comparison of trace formulae. But it has not yet been extended to the higher rank cases.

*With this reservation*, emphasized by a $*$-superscript in the following

Global Theorem, the discrete spectrum representations of $\mathrm{PGSp}(2,\mathbb{A})$, i.e. $\mathbf{H}(\mathbb{A})$, can now be described by means of the liftings. They consist of two types, stable and unstable. Global packets and quasi-packets define a partition of the spectrum. To define a (global) [quasi-] packet $\{\pi_H\}$, fix a local [quasi-] packet $\{\pi_{Hv}\}$ at each place $v$ of $F$, such that $\{\pi_{Hv}\}$ contains an unramified member $\pi^0_{Hv}$ (and then $\{\pi_{Hv}\}$ consists only of $\pi^0_{Hv}$ in case it is a packet) for almost all $v$. The [quasi-] *packet* $\{\pi_H\}$ is then defined to consist of all products $\otimes_v \pi'_{Hv}$ with $\pi'_{Hv}$ in $\{\pi_{Hv}\}$ for all $v$, and $\pi'_{Hv} = \pi^0_{Hv}$ for almost all $v$. The [quasi-] packet $\{\pi_H\}$ of an automorphic representation $\pi_H$ is defined by the local [quasi-] packets $\{\pi_{Hv}\}$ of the components $\pi_{Hv}$ of $\pi_H$ at almost all places.

The discrete spectrum of $\mathrm{PGSp}(2,\mathbb{A})$ will be described by means of the $\lambda_0$- and $\lambda$-liftings. We say that the discrete spectrum $\pi_1 \times \pi_2$ $\lambda_0$-*lifts* to a packet $\{\pi_H\}$ (or to a member thereof) if $\{\pi_{Hv}\} = \lambda_0(\pi_{1v} \times \pi_{2v})$ for almost all $v$, and that a packet $\{\pi_H\}$ (or a member of it) $\lambda$-*lifts* to an irreducible self-contragredient automorphic representation $\pi$ if $\lambda(\{\pi_{Hv}\}) = \pi_v$ for almost all $v$. The *unstable* spectrum of $\mathrm{PGSp}(2,\mathbb{A})$ is the set of discrete spectrum representations which are $\lambda_0$-lifts; its complement is the *stable* spectrum. A [quasi-] packet whose automorphic members lie in the (un)stable spectrum is called a(n un)stable [quasi-] packet.

2.3 Global Theorem* (PGSp(2) to PGL(4)). *The packets and quasi-packets partition the discrete spectrum of the group* $\mathrm{PGSp}(2,\mathbb{A})$, *thus they satisfy the rigidity theorem: if* $\pi_H$ *and* $\pi'_H$ *are discrete spectrum representations locally equivalent at almost all places then their packets or quasi-packets are equal.*

*The* $\lambda$*-lifting is a bijection between the set of packets (resp. quasi-packets) of discrete spectrum representations in the stable spectrum (of* $\mathrm{PGSp}(2,\mathbb{A})$*) and the set of self contragredient discrete spectrum representations of* $\mathrm{PGL}(4,\mathbb{A})$ *which are one dimensional, or cuspidal and not a* $\lambda_1$*-lift from* $\mathbf{C}(\mathbb{A})$ *(or residual* $J(\nu^{1/2}\pi_2, \nu^{-1/2}\pi_2)$ *where* $\pi_2$ *is a cuspidal representation of* $\mathrm{GL}(2,\mathbb{A})$ *with central character* $\xi \neq 1 = \xi^2$ *and* $\xi\pi_2 = \pi_2$*).*

*The* $\lambda_0$*-lifting is a bijection between the set of pairs of discrete spectrum representations*

$$\{\pi_1 \times \pi_2, \pi_2 \times \pi_1; \pi_1 \neq \pi_2\} \quad \text{of} \quad \mathrm{PGL}(2,\mathbb{A}) \times \mathrm{PGL}(2,\mathbb{A}),$$

*and the set of packets and quasi-packets in the unstable spectrum of the group* PGSp(2, $\mathbb{A}$). *The* $\lambda$*-lifting is a bijection from this last set to the set of automorphic representations* $I_G(\pi_1, \pi_2)$ *of* PGL(4, $\mathbb{A}$), *normalizedly induced from discrete spectrum* $\pi_1 \times \pi_2$ ($\pi_1 \neq \pi_2$) *on the parabolic subgroup with Levi factor of type* (2, 2). *If* $\pi_1 \times \pi_2$ *is cuspidal, its* $\lambda_0$*-lift is a packet, otherwise: quasi-packet.*

*Each member of a stable packet occurs in the discrete spectrum of the group* PGSp(2, $\mathbb{A}$) *with multiplicity one. The multiplicity* $m(\pi_H)$ *of a member* $\pi_H = \otimes \pi_{Hv}$ *of an unstable [quasi-]packet* $\lambda_0(\pi_1 \times \pi_2)$ ($\pi_1 \neq \pi_2$) *is not ("stable", or) constant over the [quasi-]packet. If* $\pi_1 \times \pi_2$ *is cuspidal, it is*

$$m(\pi_H) = \frac{1}{2}(1 + (-1)^{n(\pi_H)}) \quad (\in \{0, 1\}).$$

*Here* $n(\pi_H)$ *is the number of components* $\pi_{Hv}^-$ *of* $\pi_H$ *(it is bounded by the number of places* $v$ *where both* $\pi_{1v}$ *and* $\pi_{2v}$ *are square integrable). Each* $\pi_H$ *with* $m(\pi_H) = 1$ *is cuspidal.*

*The multiplicity* $m(\pi_H)$ *(in the discrete spectrum of* PGSp(2, $\mathbb{A}$)) *of* $\pi_H = \otimes \pi_{Hv}$ *from a quasi-packet* $\lambda_0(\sigma \mathbf{1}_2 \times \pi_2)$, *where* $\pi_2$ *is a cuspidal representation of* PGL(2, $\mathbb{A}$) *and* $\sigma$ *is a character of* $\mathbb{A}^\times / F^\times \mathbb{A}^{\times 2}$, *is*

$$\frac{1}{2}(1 + \varepsilon(\sigma\pi_2, \frac{1}{2})(-1)^{n(\pi_H)}) \qquad (= 0 \quad \textit{or} \quad 1),$$

*where* $n(\pi_H)$ *is the number of components* $\pi_{Hv}^-$ *of* $\pi_H$, *and* $\varepsilon(\pi_2, s)$ *is the usual* $\varepsilon$*-factor which appears in the functional equation of the* $L$*-function* $L(\pi_2, s)$. *In particular* $\pi_H^\times = \otimes \pi_{Hv}^\times$ ($n(\pi_H) = 0$) *is in the discrete spectrum if and only if* $\varepsilon(\sigma\pi_2, \frac{1}{2}) = 1$.

*Finally we have* $m(\pi_H) = \frac{1}{2}(1 + (-1)^{n(\pi_H)})$ *for* $\pi_H = \otimes \pi_{Hv}$ *in* $\lambda_0(\sigma\xi \mathbf{1}_2 \times \sigma \mathbf{1}_2)$ *with* $n(\pi_H)$ *components* $\pi_{Hv} = X_v$. *Here* $\pi_H = \otimes L_v$ ($n(\pi_H) = 0$) *is residual.*

## 2h. Unstable Spectrum

Note that the quasi-packet $\lambda_0(\sigma\xi\mathbf{1}_2 \times \sigma\mathbf{1}_2)$ is defined by the local quasi-packets

$$\{L_v = L(\nu_v\xi_v, \xi_v \rtimes \sigma_v\nu_v^{-1/2}), \qquad X_v = X(\xi_v\nu_v^{1/2}\operatorname{sp}_{2v}, \xi_v\sigma_v\nu_v^{-1/2})\}$$

for every $v$, where $\xi$ $(\neq 1)$, $\sigma$ are characters of $\mathbb{A}^\times/F^\times$ with $\xi^2 = 1 = \sigma^2$ and $\xi_v$, $\sigma_v$ are their components. When $\xi_v$, $\sigma_v$ are unramified, this quasi-packet contains the unramified representation $\pi_{Hv}^0 = L_v$. Members of this quasi-packet have been studied by means of the theta correspondence by Howe and Piatetski-Shapiro, see, e.g., [PS1], Theorem 2.5. They attracted interest since they violate the naive generalization of the Ramanujan conjecture, which expects the components of a cuspidal representation to be tempered. (The form of the Ramanujan conjecture which is expected to be true asserts that the components of a cuspidal representation of $\mathrm{PGSp}(2,\mathbb{A})$ which $\lambda$-lifts to a representation of $\mathrm{PGL}(4,\mathbb{A})$ induced from a cuspidal representation of a Levi subgroup, are tempered.) Members of this quasi-packet are equivalent at almost all places to the quotient of the properly induced representation $\nu\xi \times \xi \rtimes \sigma\nu^{-1/2}$.

Let $\pi_2$ be a cuspidal representation of $\mathrm{PGL}(2,\mathbb{A})$, $\sigma$ a character of $\mathbb{A}^\times/F^\times\mathbb{A}^{\times 2}$. The packet $\lambda_0(\sigma\mathbf{1}_2 \times \pi_2)$ contains the constituent $\pi_H^\times = \otimes_v\pi_{Hv}^\times$ of the representation $\sigma\nu^{1/2}\pi_2 \rtimes \sigma\nu^{-1/2} \simeq \sigma\nu^{-1/2}\pi_2 \rtimes \sigma\nu^{1/2}$ properly induced from an automorphic representation, hence it is automorphic by [L4]. It is known that $\pi_H^\times$ is residual precisely when $L(\sigma\pi_2, \frac{1}{2}) \neq 0$; hence $\varepsilon(\sigma\pi_2, \frac{1}{2}) = 1$ in this case.

Let $n(\pi_2)$ denote the number of square integrable components of $\pi_2$. The quasi-packet $\lambda_0(\sigma\mathbf{1}_2 \times \pi_2)$ thus consists of $2^{n(\pi_2)}$ (irreducible) representations. If $n(\pi_2) \geq 1$, half of them in the discrete spectrum, all cuspidal if $L(\sigma\pi_2, \frac{1}{2}) = 0$, all but one: $\pi_H^\times = \otimes_v\pi_{Hv}^\times$, are cuspidal if $L(\sigma\pi_2, \frac{1}{2}) \neq 0$. If $n(\pi_2) \geq 1$ and $L(\sigma\pi_2, \frac{1}{2}) = 0$, the automorphic nonresidual $\pi_H^\times$ is cuspidal when $\varepsilon(\sigma\pi_2, \frac{1}{2}) = 1$.

If $\pi_2$ has no square integrable components ($n(\pi_2) = 0$), the packet $\lambda_0(\sigma\mathbf{1}_2 \times \pi_2)$ consists only of $\pi_H^\times$. This $\pi_H^\times$ is residual if $L(\sigma\pi_2, \frac{1}{2}) \neq 0$; cuspidal (by [PS1], Theorem 2.6 and [PS2], Theorem A.2) if $L(\sigma\pi_2, \frac{1}{2}) = 0$ and $\varepsilon(\sigma\pi_2, \frac{1}{2}) = 1$; or (automorphic but) not in the discrete spectrum otherwise: $L(\sigma\pi_2, \frac{1}{2}) = 0$ and $\varepsilon(\sigma\pi_2, \frac{1}{2}) = -1$. In this last case the $\lambda_0$-

lift of $\sigma\mathbf{1}_2 \times \pi_2$ is not in the discrete spectrum, and there is no discrete specrtum representation $\lambda$-lifting to $I_G(\sigma\mathbf{1}_2, \pi_2)$.

At a place $v$ where $\pi_{2v}$ is induced $I(\mu_v, \mu_v^{-1})$, the packet

$$\pi_{Hv} = \lambda_0(\sigma_v\mathbf{1}_2 \times \pi_{2v})$$

is the irreducible induced $\mu_v\sigma_v\mathbf{1}_2 \rtimes \mu_v^{-1}$, which $\lambda$-lifts to the induced $I_G(\mu_v, \sigma_v\mathbf{1}_2, \mu_v^{-1})$, and *not* the irreducible induced

$$\sigma_v\mu_v\nu_v^{1/2} \times \sigma_v\mu_v^{-1}\nu_v^{1/2} \rtimes \sigma_v\nu_v^{-1/2} = \sigma_v\mu_v\nu_v^{1/2} \rtimes I(\mu_v^{-1}, \sigma_v\nu_v^{-1/2}),$$

which $\lambda$-lifts to the reducible induced $I_G(\mu_v, \sigma_v I(\nu_v^{1/2}, \nu_v^{-1/2}), \mu_v^{-1})$, which has the constituent $I_G(\mu_v, \sigma_v\mathbf{1}_2, \mu_v^{-1})$.

Members of the quasi-packet $\lambda_0(\sigma\mathbf{1}_2 \times \pi_2)$ were studied numerically by H. Saito and N. Kurokawa, and using the theta correspondence by Piatetski-Shapiro and others, see [PS1], Theorem 2.6. They attracted interest since they violate the naive generalization of the Ramanujan conjecture. They are equivalent at almost all places to the quotient of the properly induced representation $\sigma\nu^{1/2}\pi_2 \rtimes \sigma\nu^{-1/2}$.

A discrete specrtum representation $\pi_H$ with a local component

$$L(\nu_v\xi_v, \nu_v^{-1/2}\pi_{2v})$$

(whose packet consists of itself), where $\pi_{2v}$ is a cuspidal representation with central character $\xi_v \neq 1 = \xi_v^2$ and $\xi_v\pi_{2v} = \pi_{2v}$, is in the packet of $L(\nu\xi, \nu^{-1/2}\pi_2)$. Here $\pi_2$ is cuspidal with central character $\xi \neq 1 = \xi^2$, hence $\xi\pi_2 = \pi_2$, whose components at $v$ are $\pi_{2v}$ and $\xi_v$. It $\lambda$-lifts to $J_G(\nu^{1/2}\pi_2, \nu^{-1/2}\pi_2)$. At $v$ with $\xi_v = 1$ the component $\pi_{2v}$ is induced. If $\pi_{2v} = I(\mu_v, \mu_v\xi_v)$, $\xi_v^2 = 1$ and $\mu_v^2 = 1$ (in particular whenever $\xi_v \neq 1$ and $\pi_{2v}$ is not cuspidal), then $L(\nu_v\xi_v, \nu_v^{-1/2}\pi_{2v})$ is $L_v = L(\nu_v\xi_v, \xi_v \rtimes \mu_v\nu_v^{-1/2})$, which $\lambda$-lifts to $I_G(\mu_v\mathbf{1}_2, \mu_v\xi_v\mathbf{1}_2)$, and its packet contains also $X_v = X(\nu_v^{1/2}\xi_v\,\mathrm{sp}_{2v}, \xi_v\mu_v\nu_v^{-1/2})$. Thus the packet of $\pi_H$ is determined by $\{L_v, X_v\}$ at all $v$ where $\pi_{2v} = I(\mu_v, \mu_v\xi_v)$, $\mu_v^2 = 1 = \xi_v^2$, and by the singleton $\{L_v = L(\nu_v\xi_v, \nu_v^{-1/2}\pi_{2v})\}$ at all other $v$, where $\pi_{2v}$ is cuspidal, or $\xi_v = 1$ and $\pi_{2v} = I(\mu_v, \mu_v^{-1})$, $\mu_v^2 \neq 1$. Each member of this infinite packet occurs in the discrete specrtum with multiplicity one, and is cuspidal, with the exception of $L(\nu\xi, \nu^{-1/2}\pi_2) = \otimes_v L(\nu_v\xi_v, \nu_v^{-1/2}\pi_{2v})$, which is residual ([Kim], Theorem 7.2). Members of the packet $L(\nu\xi, \nu^{-1/2}\pi_2)$ are considered in the Appendix of [PS1] and its corrigendum.

If $\pi_1$ and $\pi_2$ are cuspidal but there is no place $v$ where both are square integrable, $\lambda_0(\pi_1 \times \pi_2)$ consists of a single irreducible cuspidal representation. This instance of the lifting $\lambda_0$ – where $\pi_i$ are cuspidal – can also be studied ([Rb]) using the theta correspondence for suitable dual reductive pairs (SO(4), PGSp(2)) for the isotropic and anisotropic forms of the orthogonal group, to describe further properties of the packets, such as their periods.

## 2i. Generic Representations

Our proof of the existence of the lifting $\lambda$ uses only the trace formula, orbital integrals and character relations. However, for cuspidal representations $\pi_1$, $\pi_2$ of $\mathrm{PGL}(2,F)$, $F$ local, we get only the character relation

$$\operatorname{tr} I_G(\pi_1, \pi_2; f \times \theta) = (2m+1)[\operatorname{tr} \pi_H^+(f_H) + \operatorname{tr} \pi_H^-(f_H)].$$

Here $f$ on $G = \mathrm{PGL}(4,F)$ and $f_H$ on $H = \mathrm{PGSp}(2,F)$ are any matching functions, and $m = m(\pi_1, \pi_2)$ is a nonnegative integer. To prove multiplicity one theorem for $\mathrm{PGSp}(2,\mathbb{A})$ we need the fact that $m = 0$.

Our proof is global. It uses the following results from the theory of the theta correspondence, Whittaker models and Eisenstein series. (1) Ginzburg-Rallis-Soudry [GRS], Theorem A: Each representation $I(\pi_1, \pi_2)$ of $\mathrm{PGL}(4,\mathbb{A})$ normalizedly induced from a cuspidal representation $\pi_1 \times \pi_2$ of its $(2,2)$-parabolic, where $\pi_1 \neq \pi_2$ are cuspidal representations of $\mathrm{PGL}(2,\mathbb{A})$, is a $\lambda$-lift of a unique *generic* cuspidal representation $\pi_H$ of $\mathrm{SO}(5,\mathbb{A}) = \mathrm{PGSp}(2,\mathbb{A})$. (2) Kudla-Rallis-Soudry [KRS], Theorem 8.1: If $\pi_0$ is a locally generic cuspidal representation of $\mathrm{Sp}(2,\mathbb{A})$ and the partial degree 5 $L$-function $L(S, \pi_0, \mathrm{id}_5, s)$ is $\neq 0$ at $s = 1$ then $\pi_0$ is (globally) generic. (3) Shahidi [Sh1], Theorem 5.1: If $\pi_0$ is a generic cuspidal representation of $\mathrm{Sp}(2,\mathbb{A})$, then $L(S, \pi_0, \mathrm{id}_5, s)$ is $\neq 0$ at $s = 1$. See chapter V, section 7, and the final remark in section 6, for further comments. We do not use the assertion (attributed to "a yet to be published result of Jacquet and Shalika") in the Remark following the statement of Theorem 8.1 in [KRS], p. 535 (that a cuspidal representation of GSp(2) is generic iff it lifts to GL(4)), which contradicts – at least as stated – our result that all representations but one in a packet of PGSp(2) are nongeneric, yet they all lift to PGL(4).

Our global proof resembles (but is strictly different from) the second proof of [F4;II], Proposition 3.5, p. 48, which is also based on the theory of generic representations. This Proposition claims the multiplicity one theorem for the discrete spectrum of $\mathrm{U}(3, E/F)$. However, the proof of [F4;II], p. 48, is not complete. Indeed, the claim in Proposition 2.4(i) in reference [GP] to [F4;II], that "$L^2_{0,1}$ has multiplicity 1", is interpreted in [F4;II] as asserting that generic representations of U(3) occur in the discrete spectrum with multiplicity one. But it should be interpreted as asserting that irreducible $\pi$ in $L^2_{0,1}$ have multiplicity one *only in the subspace* $L^2_{0,1}$ of the discrete spectrum. This claim does not exclude the possibility of having a cuspidal $\pi'$ perpendicular and equivalent to $\pi \subset L^2_{0,1}$.

Multiplicity one for the generic spectrum would follow via this global argument from the statement that a locally generic cuspidal representation is globally generic (multiplicity one implies this statement too). In our case of PGSp(2) we deduce from [KRS], [GRS], [Sh1] that a locally generic cuspidal representation which is equivalent at almost all places to a generic cuspidal representation is globally generic. A proof for U(3) still needs to be written down.

The usage of the theory of generic representations in the proof described above is not natural. A purely local proof of multiplicity one theorem for the discrete spectrum of U(3) based only on character relations is proposed in [F4;II], Proof of Proposition 3.5, p. 47. It is based on Rodier's result [Ro1] that the number of Whittaker models is encoded in the character of the representation near the origin. Details of this proof are given in [F4;IV] in odd residual characteristic in the case of basechange for U(3). It implies that in a tempered packet of representations of $\mathrm{U}(3, E/F)$ there is precisely one generic representation, and that each generic packet of discrete spectrum representations of $\mathrm{U}(3, \mathbb{A}_E/\mathbb{A}_F)$ – where a generic packet means one which lifts to a generic representation of $\mathrm{GL}(3, \mathbb{A}_E)$ – would contain precisely one generic member. Moreover, a locally generic cuspidal representation of $\mathrm{U}(3, \mathbb{A}_E/\mathbb{A}_F)$ is generic.

This type of a local argument was introduced in [FK1] in the proof of the metaplectic correspondence and the multiplicity one theorem for the discrete spectrum of the metaplectic group of $\mathrm{GL}(n, \mathbb{A})$. We have not carried out this local proof in the case of PGSp(2) as yet.

In the case of PGSp(2) our global proof implies that a local tempered packet contains precisely one generic representation, and that a global packet which lifts to a generic representation of $\mathrm{PGL}(4,\mathbb{A})$ contains precisely one everywhere generic representation. The latter is generic if the packet is unstable (in the image of the lifting $\lambda_0$). We do not show that a locally generic cuspidal representation of $\mathrm{PGSp}(2,\mathbb{A})$ which is stable ($\lambda$-lifts to a cuspidal representation of $\mathrm{PGL}(4,\mathbb{A})$) is generic.

There is some overlap between our results on the existence of the $\lambda$-lifting and the work of [GRS] which asserts that the weak (i.e., in terms of almost all places) lifting establishes a bijection from the set of equivalence classes of (irreducible automorphic) cuspidal *generic* representations of the split group $\mathrm{SO}(2n+1,\mathbb{A})$, to the set of representations of $\mathrm{PGL}(2n,\mathbb{A})$ of the form $\pi = I(\pi_1,\dots,\pi_r)$, normalized induction from the standard parabolic subgroup of type $(2n_1,\dots,2n_r)$, $n = n_1 + \dots + n_r$, where $\pi_i$ are cuspidal representations of $\mathrm{GL}(2n_i,\mathbb{A})$ such that $L(S,\pi_i,\Lambda^2,s)$ has a pole at $s=1$ and $\pi_i \neq \pi_j$ for all $i \neq j$, and the partial $L$-function is defined as a product outside a finite set $S$ where all $\pi_i$ are unramified. Of course we are concerned only with the case $n=2$, where $\mathrm{PGSp}(2) \simeq \mathrm{SO}(5)$.

Our characterization of the lifting $\lambda$ is (as in [GRS]) that $I(\pi_1,\pi_2)$, cuspidal representations $\pi_1 \neq \pi_2$ of $\mathrm{PGL}(2,\mathbb{A})$, are in the image; and that self contragredient cuspidal representations $\pi$ of $\mathrm{PGL}(4,\mathbb{A})$ are in the image of the lifting $\lambda$ from $\mathrm{PGSp}(2,\mathbb{A})$ ($= \mathrm{SO}(5,\mathbb{A})$) precisely if they are not in the image of the lifting $\lambda_1$ from $\mathrm{SO}(4,\mathbb{A})$. The cuspidal $\pi = \lambda(\pi_H)$, generic $\pi_H$, are characterized in [GRS] as the $\pi \simeq \check{\pi}$ such that $L(S,\pi,\Lambda^2,s)^{-1}$ is 0 at $s=1$. Thus the characterization of the cuspidal image of $\lambda$ here is complementary to but different than that of [GRS].

However, the methods of [GRS] apply only to generic representations, while our methods apply to all representations of PGSp(2). In particular, we can define packets, describe their structure, establish multiplicity one theorem and rigidity theorem for packets of PGSp(2), specify which member in a packet or a quasi-packet is in the discrete spectrum, and we can also $\lambda$-lift the nongeneric nontempered (at almost all places) packets to residual self-contragredient representations of $\mathrm{PGL}(4,\mathbb{A})$. Our liftings are proven in terms of all places, not only almost all places. In addition we establish the lifting $\lambda_1$ from SO(4) to PGL(4), determine its fibers (that is, prove multiplicity one theorem for SO(4) and rigidity in the sense explained above), and show that any self-contragredient discrete spectrum representation of $\mathrm{PGL}(4,\mathbb{A})$ which is not a $\lambda$-lift from $\mathrm{PGSp}(2,\mathbb{A})$ is a $\lambda_1$-lift from $\mathrm{SO}(4,\mathbb{A})$.

## 2j. Orientation

This work is an analogue for (SO(4), PGSp(2), PGL(4)) of [F3], which dealt with (PGL(2), SL(2), PGL(3)), thus with the symmetric square lifting, and of [F4], which dealt with quadratic basechange for the unitary group $\mathrm{U}(3,E/F)$, thus with $(\mathrm{U}(2,E/F)$, $\mathrm{U}(3,E/F)$, $\mathrm{GL}(3,E))$. These works use the twisted – by transpose-inverse (and the Galois action in the unitary groups case) – trace formulae on PGL(4), PGL(3), $\mathrm{GL}(3,E)$. They are based on the fundamental lemma: [F5] in our case, [F3;V] and [F4;I] in the other cases. The technique employed in these last works benefited from work of Weissauer [W] and Kazhdan [K1]. The present work, which deals with the applications of the fundamental lemma and the trace formula to character relations, liftings and the definition of packets, is analogous to [F3;IV] and [F4;II].

The trace formula identity is proven in [F3;VI] and [F4;III] for all test functions. Here we deal only with test functions which have at least three elliptic components. The trace formulae identity for a general test function has not yet been proven in our case. Perhaps the method of [AC] could be used for that, as it has been applied in a general rank case. It would be interesting to pursue the elementary techniques of [F3;VI] and [F4;III], and [F2;I], which establish the trace formulae identity for basechange for GL(2) by elementary means, based on the usage of regular, Iwahori test functions. In particular the present work does not develop the trace formula. It only uses a form of it.

Our approach uses the trace formula, developed by Arthur (see [A1]), as envisaged by Langlands e.g. in his work on basechange for GL(2).

Of course Siegel modular forms have been extensively studied by many authors (e.g., Siegel, Maass, Shimura, Andrianov, Freitag, Klingen...) over a long period of time, and several textbooks are available.

As noted above, an important representation theoretic approach alternative to the trace formula, based on the theta correspondence, Weil representation, Howe's dual reductive pairs, $L$-functions and converse theorems, has been fruitfully developed in our context of the symplectic group by Piatetski-Shapiro, Howe, Kudla, Rallis, Ginzburg, Roberts, Schmidt, Soudry, and others, see, e.g., [PS], [KRS], [GRS], [Rb], [Sch].

A purely local approach to character computations is developed in [FZ].

Our results are used by P.-S. Chan [Ch] to determine the representations of GSp(2) which are invariant under twisting by a quadratic character.

The classification of the automorphic representations of PGSp(2) has applications to the decomposition of the étale cohomology with compact supports and twisted coefficients of the Shimura varieties associated with GSp(2), see [F7]. Our techniques extend to deal with admissible and automorphic representations of GSp(2), but this we do not do here.

The present part is divided into five chapters: I. Introduction, II. Basic Facts, III. Trace Formulae, IV. The Lifting $\lambda_1$, V. The Lifting $\lambda$. Each is divided into sections. Definitions or propositions are numbered together in each section.

## 3. Conjectural Compatibility

Our local results are analogous to those of Arthur [A2], who verified them in the real case, and are consistent with his conjectures. We shall assume in this section, not to be used anywhere else in this work, familiarity with [A2], [A3], and briefly highlight some of the definitions and conjectures of [A2] in our context, in our notations ($H, C_0$ in place of Arthur's $G, H$). For brevity we write $W_F$ for the Weil group of the local field, but as in [A2], 2.1, this group has to be the motivic Galois group of the conjecturally Tannakian category of tempered representations of all GL($n$)'s in the global case, a complex pro-reductive group, or an extension of $W_F$ by a connected compact group ($W_F \times \mathrm{SU}(2,\mathbb{R})$ in the p-adic case).

Thus $\Phi(H/F)$ denotes the set of $\widehat{H}$-conjugacy classes of admissible (in particular, $\mathrm{pr}_2 \circ \phi = \mathrm{id}_{W_F}$) maps

$$\phi : W_F \to {}^L H = \widehat{H} \times W_F \qquad (\widehat{H} = {}^L H^0).$$

It contains the subset $\Phi_{\mathrm{temp}}(H/F)$ defined using the $\phi$ with bounded $\mathrm{Im}(\mathrm{pr}_1 \circ \phi)$. Note that for a split adjoint group $H$ over $F$, $\widehat{H}$ is simply connected, and for any semisimple $s$ in $\widehat{H}$, the centralizer $\widehat{C}_0 = Z_{\widehat{H}}(s)$ of $s$ in $\widehat{H}$ specifies the endoscopic group $H$ uniquely (up to isomorphism). Write $S_\phi = S_\phi^H = Z_{\widehat{H}}(\phi(W_F))$ (centralizer in the connected group $\widehat{H}$ of the image of $\phi$), $\widehat{Z} = Z_{\widehat{H}}({}^L H) \subset \widehat{H}$, and note that $\mathcal{S}_\phi = S_\phi/S_\phi^0\widehat{Z}$ is a finite abelian group, conjecturally in duality with the packet $\Pi_\phi$ to be associated with $\phi \in \Pi_{\text{temp}}(H/F)$ (this is the case when $F = \mathbb{R}$, see [A2]). Arthur [A2] defines a further set $\Psi(H/F)$ of $\widehat{H}$-conjugacy classes of maps $\psi : W_F \times \mathrm{SL}(2,\mathbb{C}) \to {}^L H$ such that $\psi|W_F \in \Phi_{\text{temp}}(H/F)$, and a map

$$\psi \mapsto \phi_\psi, \qquad \phi_\psi(w) = \psi(w, \begin{pmatrix} |w|^{1/2} & 0 \\ 0 & |w|^{-1/2} \end{pmatrix}),$$

which embeds $\Psi(H/F)$ in $\Phi(H/F)$. Each $\psi$ can be viewed as a pair

$$(\phi, \rho) \in (\Phi_{\text{temp}}(H/F) \times \mathrm{Hom}(\mathrm{SL}(2,\mathbb{C}), S_\phi))/\,\mathrm{Int}(S_\phi)$$

(quotient by $S_\phi$-conjugacy). Then $\Phi_{\text{temp}}(H/F)$ embeds in $\Psi(H/F)$ as the $(\phi, 1)$. Put

$$S_\psi = S_\psi^H = Z_{\widehat{H}}(\psi(W_F \times \mathrm{SL}(2,\mathbb{C}))).$$

It is equal to

$$Z_{S_{\phi_\psi}}(\rho(\mathrm{SL}(2,\mathbb{C}))),$$

a subgroup of $S_{\phi_\psi}$, and there is a surjection $\mathcal{S}_\psi = S_\psi/S_\psi^0\widehat{Z} \twoheadrightarrow \mathcal{S}_{\phi_\psi}$. The group $\mathcal{S}_\psi$ is in duality with the quasi-packet $\Pi_\psi$ conjecturally associated with $\psi$. Globally, the quasi-packet $\Pi_\psi$ contains no discrete spectrum representations of $H$ unless $S_\psi$ is finite.

Let us review the examples of [A2], where $\widehat{H} = \mathrm{Sp}(2,\mathbb{C}) \supset \widehat{C}_0 = \mathrm{SL}(2,\mathbb{C}) \times \mathrm{SL}(2,\mathbb{C}) = \left\{ \begin{pmatrix} a & 0 & 0 & b \\ 0 & \alpha & \beta & 0 \\ 0 & \gamma & \delta & 0 \\ c & 0 & 0 & d \end{pmatrix} \right\}$. The parameter $\psi$ can be described by the maps

$$(\phi = \phi_1 \times \phi_2, \rho = \rho_1 \times \rho_2): \ W_F \times \mathrm{SL}(2,\mathbb{C}) \to \mathrm{SL}(2,\mathbb{C}) \times \mathrm{SL}(2,\mathbb{C}).$$

If $\phi_i : W_F \to \mathrm{SL}(2,\mathbb{C})$ are irreducible and inequivalent, $\rho = 1$,

$$Z_{\mathrm{SL}(2,\mathbb{C})}(\mathrm{Im}\,\phi_i) = \{\pm I\}, \quad S_{\phi_\psi} = \mathbb{Z}/2 \times \mathbb{Z}/2, \quad \mathcal{S}_{\phi_\psi} = \mathbb{Z}/2,$$

$S_\psi = \mathbb{Z}/2 \times \mathbb{Z}/2$, $\mathcal{S}_\psi = \mathbb{Z}/2$. This is a "classical" tempered case, as $\mathrm{Im}\,\phi_i$ are bounded.

If $\phi_1 = \phi_2$ is irreducible, $\rho = 1$, $S_{\phi_\psi} = \mathrm{O}(2,\mathbb{C}) = S_\psi$ (this group consists of the $\mathrm{diag}(g, g^*)$, $g^* = w^t g^{-1} w$, $w = \begin{pmatrix} 0 & 1 \\ 1 & 0 \end{pmatrix}$, which commute with $\begin{pmatrix} 0 & w \\ w & 0 \end{pmatrix}$, thus $g^t g = I$), $S_\psi^0 = \mathrm{SO}(2,\mathbb{C})$ and $\mathcal{S}_{\phi_\psi} = \mathcal{S}_\psi = \mathbb{Z}/2$ $(= \langle \mathrm{diag}(w,w) \rangle)$.

These cases correspond to $\lambda_0(\pi_1 \times \pi_2)$, where $\pi_1$, $\pi_2$ are in the discrete spectrum; a local packet consists of $2 = [\mathbb{Z}/2]$ elements. A global packet in the second case consists of no discrete spectrum representations since $S_\psi = \mathrm{O}(2,\mathbb{C})$ is not finite. In the first case, where $\pi_2 \neq \pi_1$, the packet consists of $2^n$ irreducibles, where $n$ is the number of places where both $\pi_1$ and $\pi_2$ are square integrable; half of the members in the packet are in the discrete spectrum (one, if $n = 0$).

If $\phi_1$ is irreducible and $\mathrm{Im}(\phi_2) \subset \{\pm I\}$, and $\rho = 1 \times \mathrm{id}$, we have $S_{\phi_\psi} = \mathbb{Z}/2 \times \mathbb{C}^\times$ $(= \{\mathrm{diag}(\iota, z, z^{-1}, \iota);\ z \in \mathbb{C}^\times,\ \iota \in \{\pm 1\}\})$, $\mathcal{S}_{\phi_\psi} = \{1\}$, $S_\psi = \mathbb{Z}/2 \times \mathbb{Z}/2$, $\mathcal{S}_\psi = \mathbb{Z}/2$. This is the case of $\lambda_0(\pi_1 \times \phi_2 \mathbf{1}_2)$, where $\phi_2$ is a character.

If $\mathrm{Im}\,\phi_i \subset \{\pm I\}$ but $\phi_1 \neq \phi_2$, and $\rho_i = \mathrm{id}$, $S_{\phi_\psi} = \mathbb{C}^\times \times \mathbb{C}^\times$ $(= \{\mathrm{diag}(z, t, t^{-1}, z^{-1});\ z, t \in \mathbb{C}^\times\})$, $\mathcal{S}_{\phi_\psi} = \{1\}$, $S_\psi = \mathbb{Z}/2 \times \mathbb{Z}/2$, $\mathcal{S}_\psi = \mathbb{Z}/2$. This is the case of $\lambda_0(\phi_1 \mathbf{1}_2 \times \phi_2 \mathbf{1}_2)$, where $\phi_1 \neq \phi_2$ are characters of $F^\times/F^{\times 2}$ or $\mathbb{A}^\times/F^\times \mathbb{A}^{\times 2}$.

If $\phi_1 = \phi_2$ with image in $\{\pm I\}$, and $\rho_i = \mathrm{id}$, $S_{\phi_\psi} = \mathrm{GL}(2,\mathbb{C})$ $(= \{\mathrm{diag}(g, g^*);\ g \in \mathrm{GL}(2,\mathbb{C})\})$, $\mathcal{S}_{\phi_\psi} = \{1\}$, $S_\psi = \mathrm{O}(2,\mathbb{C})$, $\mathcal{S}_\psi = \mathbb{Z}/2$. This is the case of $\lambda_0(\phi_1 \mathbf{1}_2 \times \phi_1 \mathbf{1}_2)$, whose packet contains no discrete spectrum representations, and indeed $S_\psi = \mathrm{O}(2,\mathbb{C})$ is not finite.

In addition we determine that the multiplicity $d_\psi$ of [A2], p. 28, is one.

# 4. Conjectural Rigidity

This section explains the rigidity theorem for SO(4) via the principle of functoriality. It is based on conversations with J.-P. Serre at Singapore.

4.1 Proposition. *Let $\eta_1$, $\eta_2$, $\eta_1'$, $\eta_2'$: $W_F \to \mathrm{GL}(2,\mathbb{C})$ be (irreducible continuous) representations of the Weil group $W_F$ of $F$ which are unramified at almost all places $v$ (so they depend there only on the Frobenius element) with $\eta_1 \otimes \eta_2 | W_{F_v} \simeq \eta_1' \otimes \eta_2' | W_{F_v}$ for almost all $v$ and with $\det \eta_1 \cdot \det \eta_2 = \det \eta_1' \cdot \det \eta_2'$. Then there exists a homomorphism $\chi : W_F \to \mathbb{C}^\times$ such that $\eta_1' = \chi \eta_1$ and $\eta_2' = \chi^{-1} \eta_2$, or $\eta_2' = \chi \eta_2$ and $\eta_2' = \chi^{-1} \eta_1$.*

Since the subgroup of $W_F$ generated by the Frobenii is dense, we may consider instead a group $\Gamma$ (instead of $W_F$), and two representations $\rho_i$

(instead of $\eta_1 \otimes \eta_2$) which are *locally conjugate*, which means that $\rho_1(\gamma)$ is conjugate to $\rho_2(\gamma)$ for each $\gamma$ in $\Gamma$, or alternatively that the restrictions of $\rho_1, \rho_2$ to any cyclic subgroup are conjugate. We wish to know whether they are conjugate as representations.

We say that a group $G$ over $\mathbb{C}$ has the *rigidity-property* if for any group $\Gamma$, any two locally conjugate representations $\rho_1, \rho_2 : \Gamma \to G(\mathbb{C})$ are conjugate. Variants are naturally defined (for special $\Gamma$ and $\rho$). For example, if $\Gamma$ is finite and $G = \mathrm{GL}(n)$, character theory asserts that locally conjugate $\rho_1, \rho_2 : \Gamma \to \mathrm{GL}(n, \mathbb{C})$ are conjugate. The group $G = \mathrm{GL}(n)$ has the rigidity-property for any semisimple continuous representations $\rho_1, \rho_2$ of the Weil group. On the other hand, the group $\mathrm{PGL}(n, \mathbb{C})$ does not have the rigidity-property since it is the dual group of $\mathrm{SL}(n)$, for which rigidity does not hold.

In our case we wish to know whether locally conjugate $\rho_1$, $\rho_2$ into $\mathrm{SO}(4, \mathbb{C})$ are conjugate. They are not, but almost are: they are conjugate in $\mathrm{O}(4, \mathbb{C})$, which is the semidirect product of $\mathrm{SO}(4, \mathbb{C})$ with an element which maps $\eta_1 \otimes \eta_2$ to $\eta_2 \otimes \eta_1$. We proceed to explain this via the group theoretical notion of fusion control.

4.2 Definition. Given groups $G \supset H' \supset H$ we say that $H'$ *controls the fusion* of $H$ in $G$ if for any sets $A$, $B$ in $H$ and $g$ in $G$ with $gAg^{-1} = B$ there is $h$ in $H'$ with $hah^{-1} = gag^{-1}$ for every $a$ in $A$, namely $h^{-1}g$ lies in the centralizer $C_G(A)$ of $A$ in $G$.

4.3 Example. Let $S$ be an abelian p-Sylow subgroup in a finite group $G$, and $N = N_G(S)$ the normalizer of $S$ in $G$. Then $S \subset N \subset G$ and $N$ controls the fusion of $S$ in $G$.

Proof. Since $S$ is abelian and $A$ is a subset of $S$ we have that $S$ is contained in the centralizer $C_G(A)$ of $A$ in $G$. Hence $S$ is a p-Sylow subgroup of $C_G(A)$. Now the abelian $S$ commutes with any subset $B$ of $S$, hence $g^{-1}Sg$ commutes with $g^{-1}Bg = A$, and so $g^{-1}Sg$ is a p-Sylow subgroup of $C_G(A)$ for any $g$ in $G$. Since p-Sylow subgroups are conjugate, there is $u$ in $C_G(A)$ with $g^{-1}Sg = uSu^{-1}$; take $h = gu \in N_G(S)$. Then $hah^{-1} = guau^{-1}g^{-1} = gag^{-1}$ for any $a$ in $A$. □

4.4 Example. Let $G$ be an algebraic reductive group, $T$ a maximal torus and $N = N_G(T)$ the normalizer of $T$ in $G$. Then $T \subset N \subset G$ and $N$ controls the fusion of $T$ in $G$.

PROOF. If $A$ is any subset of the abelian $T$, we have that $T$ lies in the centralizer $C_G(A)$ of $A$ in $G$. Hence $T$ is a maximal torus in $C_G(A)$. Now $T$ commutes with any of its subsets $B$, hence $g^{-1}Tg$ commutes with $g^{-1}Bg = A$, and so $g^{-1}Tg$ is a maximal torus in $C_G(A)$. Since maximal tori of a reductive group are conjugate, there exists $u$ in $C_G(A)$ such that $g^{-1}Tg = uTu^{-1}$. Hence $h = gu$ lies in $N_G(T)$ and satisfies $hah^{-1} = guau^{-1} = gag^{-1}$ for any $a$ in $A$. □

4.5 PROPOSITION. *Let $S = {}^tS$ be a symmetric matrix in $\mathrm{GL}(n,\mathbb{C})$. Put $g^* = S^tg^{-1}S^{-1}$. Then the orthogonal group $\mathrm{O}(S,\mathbb{C}) = \{g \in \mathrm{GL}(n,\mathbb{C}); g = g^*\}$ controls its own fusion in $\mathrm{GL}(n,\mathbb{C})$.*

PROOF. Suppose that $A$, $B$ are subsets of $\mathrm{O}(S,\mathbb{C})$ and $g \in \mathrm{GL}(n,\mathbb{C})$ satisfies $gAg^{-1} = B$. For each $a$ in $A$ we have $a^* = a$, hence $g^*ag^{*-1} = (gag^{-1})^* = gag^{-1}$ (as $b = b^*$ for $b = gag^{-1}$). Then $c = g^{-1}g^*$ commutes with each $a$ in $A$, and $c^{*-1} = S^tcS^{-1} = S^tg^{*t}g^{-1}S^{-1} = g^{-1}S^tg^{-1}S^{-1} = g^{-1}g^* = c$. Let $d$ be a square root of $c$, thus $c = d^2$. Using the binomial expansion $u^{1/2} = \sum_{n=0}^{\infty}\binom{\frac{1}{2}}{n}(u-1)^n$ for a unipotent matrix $u$ and $(re^{i\theta})^{1/2} = r^{1/2}e^{i\theta/2}$ $(0 \leqslant \theta < 2\pi,\ r > 0)$, the Jordan decomposition $c = su = us$ and diagonalization, we express $d$ as a function $f(c)$ in $c$, where $f$ satisfies $f(xyx^{-1}) = xf(y)x^{-1}$ and $f({}^tx) = {}^tf(x)$. Then

$$d^{*-1} = S\,{}^tdS^{-1} = Sf({}^tc)S^{-1} = f(S^tcS^{-1}) = f(c^{*-1}) = f(c) = d$$

and $h = (gd)^* = g^*d^* = gcd^{-1} = gd$ satisfies $(gd)a(gd)^{-1} = gag^{-1}$, for all $a$ in $A$. □

4.6 COROLLARY. *If $\rho_1$, $\rho_2 : \Gamma \to \mathrm{O}(S,\mathbb{C})$ are representations of a group $\Gamma$ into the orthogonal group $\mathrm{O}(S,\mathbb{C})$, and there is $g$ in $\mathrm{GL}(n,\mathbb{C})$ with $\rho_2 = g\rho_1g^{-1}$, then there is $h$ in $\mathrm{O}(S,\mathbb{C})$ with $\rho_2 = h\rho_1h^{-1}$.* □

REMARK. The last Proposition and its Corollary hold (with the same proof) for the symplectic group $\mathrm{Sp}(S,\mathbb{C})$, defined using $S = -{}^tS$.

4.7 PROPOSITION. *Let $\eta_1, \eta_2, \eta_1', \eta_2'$: $\Gamma \to \mathrm{GL}(2,\mathbb{C})$ be representations of a group $\Gamma$ with $\eta_1 \otimes \eta_2 \simeq \eta_1' \otimes \eta_2'$ in $\mathrm{GL}(4,\mathbb{C})$ and $\det\eta_1 \cdot \det\eta_2 = \det\eta_1' \cdot \det\eta_2'$. Then there exists a homomorphism $\chi : \Gamma \to \mathbb{C}^\times$ such that $\eta_1' = \chi\eta_1$ and $\eta_2' = \chi^{-1}\eta_2$ or $\eta_1' = \chi\eta_2$ and $\eta_2' = \chi^{-1}\eta_1$.*

PROOF. The tensor products $\rho = \eta_1 \otimes \eta_2$ and $\rho' = \eta_1' \otimes \eta_2'$ have images in $SO(S,\mathbb{C}) \subset O(S,\mathbb{C})$ where $S = \hat{s}J = \operatorname{antidiag}(-1,1,1,-1)$,

$$\hat{s} = \operatorname{diag}(-1,1,-1,1), \quad J = \begin{pmatrix} 0 & w \\ -w & 0 \end{pmatrix}, \quad w = \begin{pmatrix} 0 & 1 \\ 1 & 0 \end{pmatrix}.$$

Hence $\rho$ and $\rho'$ are equivalent in $O(S,\mathbb{C}) = SO(S,\mathbb{C}) \times \langle \iota \rangle$, where $\iota = \operatorname{diag}(1,w,1)$ acts on $a \otimes b$ in $SO(S,\mathbb{C})$ ($a, b$ in $GL(2,\mathbb{C})$, $\det ab = 1$) by $\iota : a \otimes b \mapsto b \otimes a$. So $\rho$ is equivalent under $SO(S,\mathbb{C})$ to $\rho'$ or to ${}^{\iota}\rho' = \eta_2' \otimes \eta_1'$, and $(\eta_1, \eta_2)$ is equivalent to $(\chi\eta_1', \chi^{-1}\eta_2')$ or to $(\chi\eta_2', \chi^{-1}\eta_1')$. The map $\chi : \Gamma \to \mathbb{C}^\times$ is a homomorphism since so are the $\eta_i$, $\eta_i'$, $i = 1, 2$. □

We also note the following analogue for the group of similitudes.

4.8 PROPOSITION. *If the representations $\rho$, $\rho' : \Gamma \to GO(S,\mathbb{C})$ (of a group $\Gamma$ into the group of orthogonal similitudes) are conjugate in $GL(n,\mathbb{C})$ ($\ni S = {}^tS$) and have the same factor $\boldsymbol{\lambda}$ of similitudes, then they are conjugate in $O(S,\mathbb{C})$.*

PROOF. Replacing $\Gamma$ by the 2-fold cover $\tilde{\Gamma} = \Gamma_{\boldsymbol{\lambda}} \times_{\mathbb{C}^\times, \square} \mathbb{C}^\times$ (fiber product of $\boldsymbol{\lambda} : \Gamma \to \mathbb{C}^\times$ with $\mathbb{C}^\times \to \mathbb{C}^\times, \square : z \mapsto z^2$), there is a character $\mu : \tilde{\Gamma} \to \mathbb{C}^\times$ with $\boldsymbol{\lambda} = \mu^2$:

$$\begin{array}{ccc} \tilde{\Gamma} & \xrightarrow{\mu} & \mathbb{C}^\times \\ \downarrow & & \downarrow \square \\ \Gamma & \xrightarrow{\boldsymbol{\lambda}} & \mathbb{C}^\times \end{array}$$

Then $\mu^{-1}\rho$, $\mu^{-1}\rho' : \tilde{\Gamma} \to O(S,\mathbb{C})$ are conjugate in $GL(n,\mathbb{C})$ hence also in $O(S,\mathbb{C})$, and so $\rho$, $\rho' : \tilde{\Gamma} \to O(S,\mathbb{C})$ are conjugate in $O(S,\mathbb{C})$ and they factorize via $\mathrm{pr}:\tilde{\Gamma} \to \Gamma$. □

We can now return to our initial Proposition 4.1. If the irreducible continuous representations $\eta_1$, $\eta_2$, $\eta_1'$, $\eta_2' : W_F \to GL(2,\mathbb{C})$ are unramified and satisfy $\eta_1 \otimes \eta_2(\mathrm{Fr}_v) \simeq \eta_1' \otimes \eta_2'(\mathrm{Fr}_v)$ for almost all places $v$, then $\rho = \eta_1 \otimes \eta_2$ and

$$\rho' = \eta_1' \otimes \eta_2' : W_F \to SO(S,\mathbb{C}) \subset O(S,\mathbb{C}) \subset GL(4,\mathbb{C})$$

are conjugate in $GL(4,\mathbb{C})$ (since the Frobenii are dense in $W_F$ and $\rho$, $\rho'$ are semisimple). Hence they are conjugate in $O(S,\mathbb{C})$ and there is a homomorphism $\chi : W_F \to \mathbb{C}^\times$ with $\eta_1' = \chi\eta_1, \eta_2' = \chi^{-1}\eta_2$, or $\eta_1' = \chi\eta_2$, $\eta_2' = \chi^{-1}\eta_1$.

Had we known the Principle of Functoriality, namely that discrete spectrum representations $\pi_i$ of $\mathrm{GL}(2,\mathbb{A})$ are parametrized by two dimensional representations $\eta_i : \Gamma \to \mathrm{GL}(2,\mathbb{C})$ of a suitable Weil group $\Gamma(= W_F)$, we could conclude the rigidity theorem part of our global theorem about the lifting $\lambda_1$ from $\mathbf{C} = \mathrm{SO}(4)$ to $\mathrm{PGL}(4)$. However, this Principle is known only for monomial representations $\eta_i = \mathrm{Ind}(\mu_i; W_{E_i/E_i}, W_{E_i/F})$, induced from characters $\mu_i$ of $W_{E_i/E_i} = \mathbb{A}_{E_i}^\times/E_i^\times$, where $E_i$ is a quadratic extension of $F$. Thus we get an alternative proof – based only on class field theory and the basic group theoretic consideration above – of the special case for monomial representations $\pi_i = \pi(\mu_i)$ stated after that theorem.

Note that the rigidity property, that any locally conjugate $\rho$, $\rho' : \Gamma \to G(\mathbb{C})$ are conjugate, holds for $G = \mathrm{GL}(n)$, $\mathrm{O}(n)$, $\mathrm{Sp}(n)$ and $G_2$, and for any connected, simply connected, complex Lie group precisely if it has no direct factors of type $B_n (n \geq 4)$, $D_n (n \geq 4)$, $E_n$ or $F_4$. For this and related results see Larsen ([Lar]).

# II. BASIC FACTS

## 1. Norm Maps

The norm maps are formally defined by the dual group maps, as we proceed to explain. Denote by $\widehat{T}_0$ the diagonal torus in $\widehat{C}_0$, and by $\widehat{T}_H$ the diagonal torus in $\widehat{H}$, $\mathbf{T}_0^*$ in $\mathbf{C}_0$ and $\mathbf{T}_H^*$ in $\mathbf{H}$. Then

$$X_*(\widehat{T}_0) = X_*(\widehat{T}_H) = \{(a, b, -b, -a); a, b \in \mathbb{Z}\}$$

is the lattice of 1-parameter subgroups, while the lattices of characters are

$$X^*(\widehat{T}_0) = X^*(\widehat{T}_H) = \{(x, y, z, t) \operatorname{mod}(n, m, m, n); x, y, z, t \in \mathbb{Z}\};$$

here $(x, y, z, t)$ takes $\operatorname{diag}(a, b, b^{-1}, a^{-1})$ in $\widehat{T}_H$ or $\operatorname{diag}(a, a^{-1}) \times \operatorname{diag}(b, b^{-1})$ in $\widehat{T}_0$ to $a^{x-t}b^{y-z}$. Further we have $X^*(\widehat{T}_0) = X_*(\mathbf{T}_0^*)$, while the isomorphism $X^*(\widehat{T}_H) \xrightarrow{\sim} X_*(\mathbf{T}_H^*)$

$$= \{(\alpha, \beta, \gamma, \delta) \operatorname{mod}(\varepsilon, \varepsilon, \varepsilon, \varepsilon); \alpha, \beta, \gamma, \delta, \varepsilon \in \mathbb{Z}, \alpha + \delta = \beta + \gamma\}$$

is given by

$$(x, y, z, t) \mapsto (x + y, x + z, y + t, z + t),$$

with inverse

$$(\alpha, \beta, \gamma, \delta) \mapsto (\alpha - \gamma, \alpha - \beta, 0, 0).$$

In particular the map

$$X_*(\mathbf{T}_H^*) \xrightarrow{\sim} X_*(\mathbf{T}_0^*) \quad \text{is} \quad (\alpha, \beta, \gamma, \delta) \mapsto (\alpha - \gamma, \alpha - \beta, 0, 0),$$

and we make

1.1 Definition. The *norm map* $N : \mathbf{T}_H^* \xrightarrow{\sim} \mathbf{T}_0^*$ is defined by

$$\operatorname{diag}(\alpha, \beta, \gamma, \delta) \mapsto \begin{pmatrix} \alpha/\gamma & 0 \\ 0 & 1 \end{pmatrix} \times \begin{pmatrix} \alpha/\beta & 0 \\ 0 & 1 \end{pmatrix}.$$

The elements $(a,b,-b,-a)$ of $X_*(\widehat{T}_0) = X^*(\mathbf{T}_0^*)$ can be viewed as characters of $\mathbf{T}_0^*$:

$$(a,b,-b,-a) : \left(\begin{pmatrix} \alpha_1 & 0 \\ 0 & \alpha_2 \end{pmatrix}, \begin{pmatrix} \beta_1 & 0 \\ 0 & \beta_2 \end{pmatrix}\right) \mapsto (\alpha_1/\alpha_2)^a (\beta_1/\beta_2)^b.$$

Under the isomorphism $N : \mathbf{T}_H^* \xrightarrow{\sim} \mathbf{T}_0^*$,

$$\operatorname{diag}(\alpha,\beta,\gamma,\delta) \bmod zI_4 \mapsto \left(\begin{pmatrix} \alpha/\gamma & 0 \\ 0 & 1 \end{pmatrix}, \begin{pmatrix} \alpha/\beta & 0 \\ 0 & 1 \end{pmatrix}\right), \qquad \alpha\delta = \beta\gamma,$$

the elements $(a,b,-b,-a)$ of $X_*(\widehat{T}_H) \simeq X^*(\mathbf{T}_H^*)$ can be viewed as characters of $T_H^*$:

$$(a,b,-b,-a) : \operatorname{diag}(\alpha,\beta,\gamma,\delta) \mapsto (\alpha/\gamma)^a (\alpha/\beta)^b.$$

Hence corresponding to $\lambda_0 : \widehat{T}_0 \xrightarrow{\sim} \widehat{T}_H$ induced by $\lambda_0 : \widehat{C}_0 \hookrightarrow \widehat{H}$ we have the "endoscopic" lifting

$$\lambda_0 : \pi_2(\mu_1, \mu_1^{-1}) \times \pi_2(\mu_2, \mu_2^{-1}) \mapsto \pi_{\mathrm{PGSp}(2)}(\mu_1, \mu_2).$$

Here $\pi_2(\mu_i, \mu_i^{-1})$ is the unramified irreducible constituent of the normalizedly induced representation $I(\mu_i, \mu_i^{-1})$ of $\mathrm{PGL}(2, F_v)$ ($\mu_i$ are unramified characters of $F_v^\times$, $i = 1, 2$); $\pi_{\mathrm{PGSp}(2)}(\mu_1, \mu_2)$ is the unramified irreducible constituent of the $\mathrm{PGSp}(2, F_v)$-module $I_{\mathrm{PGSp}(2)}(\mu_1, \mu_2)$ normalizedly induced from the character $n \cdot \operatorname{diag}(\alpha,\beta,\gamma,\delta) \mapsto \mu_1(\alpha/\gamma)\mu_2(\alpha/\beta)$ of the upper triangular subgroup of $\mathrm{PGSp}(2, F_v)$ ($n$ is in the unipotent radical, $\alpha\delta = \beta\gamma$).

Corresponding to the embedding $\lambda : \widehat{H} = \mathrm{Sp}(2,\mathbb{C}) \hookrightarrow \mathrm{SL}(4,\mathbb{C}) = \widehat{G}$ we have the natural embedding

$$\begin{aligned} &X^*(\mathbf{T}_H^*) = X_*(\widehat{T}_H) = \{(x,y,-y,-x); x,y \in \mathbb{Z}\} \\ &\hookrightarrow X_*(\widehat{T}) = \{(x,y,z,t) \in \mathbb{Z}^4; x+y+z+t = 0\} = X^*(\mathbf{T}^*). \end{aligned}$$

The torus $\mathbf{T}_H^*$ consists of $\operatorname{diag}(\alpha,\beta,\gamma,\delta) \bmod(zI_4)$, $\alpha\delta = \beta\gamma$, and the character $(x,y,-y,-x)$ maps this element to $(\alpha/\gamma)^x(\alpha/\beta)^y$ $(\bullet)$. The torus $\mathbf{T}^*$ consists of $\operatorname{diag}(\alpha,\beta,\gamma,\delta)$ in $\mathrm{PGL}(4)$.

Dual to the embedding

$$\lambda : \widehat{T}_H = \{\operatorname{diag}(a,b,b^{-1},a^{-1})\} \hookrightarrow \widehat{T} = \{\operatorname{diag}(a,b,c,d); abcd = 1\}$$

there is the map of the character lattices

$$\begin{aligned}(X_*(\mathbf{T}^*) =)X^*(\widehat{T}) &= \{(x,y,z,t) \operatorname{mod}(z,z,z,z) \in \mathbb{Z}^4/\mathbb{Z}\}\\ \to X^*(\widehat{T}_H) &= \{(x,y,z,t)/(\alpha,\beta,\beta,\alpha); x,y,z,t,\alpha,\beta \in \mathbb{Z}\}.\end{aligned}$$

The isomorphism

$$X^*(\widehat{T}_H) \xrightarrow{\sim} X_*(\mathbf{T}_H^*), \qquad (x,y,z,t) \mapsto (x+y, x+z, y+t, z+t),$$

leads us to make the

1.2 Definition. The norm map $N : \mathbf{T}^* \to \mathbf{T}_H^*$ is given by

$$N(\operatorname{diag}(a,b,c,d)) = \operatorname{diag}(ab, ac, bd, cd).$$

The dual map of characters

$$X^*(\mathbf{T}_H^*) \quad (= X_*(\widehat{T}_H)) \quad \xrightarrow{\lambda} \quad (X_*(\widehat{T}) =) \quad X^*(\mathbf{T}^*), \qquad \chi \mapsto \lambda(\chi),$$

is given by

$$\lambda(\chi)(\operatorname{diag}(a,b,c,d)) = \chi(N(\operatorname{diag}(a,b,c,d)) = \chi(\operatorname{diag}(ab,ac,bd,cd)).$$

If $\chi = (x, y, -y, -x)$ then

$$\lambda(\chi)(\operatorname{diag}(a,b,c,d)) = (ab/bd)^x(ab/ac)^y = a^x b^y c^{-y} d^{-x}$$

(by (•) 18 lines above) as expected. In other words the lifting $\lambda$ maps the unramified irreducible $\mathrm{PGSp}(2, F_v)$-module $\pi_{\mathrm{PGSp}(2)}(\mu_1, \mu_2)$ to the unramified irreducible $\mathrm{PGL}(4, F_v)$-module $\pi_4(\mu_1, \mu_2, \mu_2^{-1}, \mu_1^{-1})$.

Note that the norm map extends to the Levi $M_{(2,2)}$ of PGL(4) of type (2,2) by $N\begin{pmatrix} A & 0 \\ 0 & B \end{pmatrix} = \begin{pmatrix} \det A & & 0 \\ & \varepsilon B \varepsilon A & \\ 0 & & \det B \end{pmatrix}$, where $\varepsilon = \begin{pmatrix} -1 & 0 \\ 0 & 1 \end{pmatrix}$. It takes $\theta$-conjugacy classes in $M_{(2,2)}$ to conjugacy classes in the Levi of type (1,2,1) in PGSp(2). Indeed,

$$\theta\begin{pmatrix} C & 0 \\ 0 & D \end{pmatrix}^{-1} \begin{pmatrix} A & 0 \\ 0 & B \end{pmatrix} \begin{pmatrix} C & 0 \\ 0 & D \end{pmatrix} = \begin{pmatrix} w^t D w A C & 0 \\ 0 & w^t C w B D \end{pmatrix}$$

$$\overset{N}{\mapsto} \begin{pmatrix} cd \det A & & 0 \\ & X & \\ 0 & & cd \det B \end{pmatrix} = cd \begin{pmatrix} \det A & & 0 \\ & C^{-1} \varepsilon B \varepsilon A C & \\ 0 & & \det B \end{pmatrix}$$

where $c = \det C$, $d = \det D$, and

$$X = \varepsilon w^t C w B D \varepsilon w^t D w A C = c d C^{-1} \varepsilon B \varepsilon A C$$

is conjugate to $\varepsilon B \varepsilon A$ times $cd$.

Moreover, it extends to the Levi of PGL(4) of type (1,2,1) by

$$N\begin{pmatrix} a & & 0 \\ & A & \\ 0 & & d \end{pmatrix} = \begin{pmatrix} aA & 0 \\ 0 & d\varepsilon A\varepsilon \end{pmatrix}.$$

It takes

$$\theta\begin{pmatrix} u & & 0 \\ & B & \\ 0 & & v \end{pmatrix}^{-1}\begin{pmatrix} a & & 0 \\ & A & \\ 0 & & d \end{pmatrix}\begin{pmatrix} u & & 0 \\ & B & \\ 0 & & v \end{pmatrix}$$

to

$$uv \det B \begin{pmatrix} aB^{-1}AB & 0 \\ 0 & d\varepsilon B^{-1}AB\varepsilon \end{pmatrix}.$$

The composition

$$\lambda \circ \lambda_0 : \widehat{C}_0 = \mathrm{SL}(2,\mathbb{C}) \times \mathrm{SL}(2,\mathbb{C}) \to \widehat{G} = \mathrm{SL}(4,\mathbb{C})$$

takes $\quad \pi_2(\mu_1, \mu_1^{-1}) \times \pi_2(\mu_2, \mu_2^{-1}) \quad$ to

$$\pi_4(\mu_1, \mu_2, \mu_2^{-1}, \mu_1^{-1}) = \pi_4(\mu_1, \mu_1^{-1}, \mu_2, \mu_2^{-1}),$$

namely the unramified irreducible $\mathrm{PGL}(2, F_v) \times \mathrm{PGL}(2, F_v)$-module $\pi_2 \times \pi_2'$ to the unramified irreducible constituent $\pi_4(\pi_2, \pi_2')$ of the $\mathrm{PGL}(4, F_v)$-module $I_4(\pi_2, \pi_2')$ normalizedly induced from the representation $\pi_2 \otimes \pi_2'$ of the parabolic of type (2,2) of $\mathrm{PGL}(4, F_v)$ (extended trivially on the unipotent radical). For example $\lambda \circ \lambda_0$ takes the trivial $\mathrm{PGL}(2, F_v) \times \mathrm{PGL}(2, F_v)$-module $\mathbf{1}_2 \times \mathbf{1}_2$ to the unramified irreducible constituent $\pi_4(\mathbf{1}_2, \mathbf{1}_2)$ of $I_4(\mathbf{1}_2, \mathbf{1}_2)$, and $\mathbf{1}_2 \times \pi_2$ to $\pi_4(\mathbf{1}_2, \pi_2) = \pi_4(\nu^{1/2}\pi_2, \nu^{-1/2}\pi_2)$. Note that this last $\pi_4$ is traditionally denoted by $J$.

The embedding

$$\lambda_1 : \widehat{C} = [\mathrm{GL}(2,\mathbb{C}) \times \mathrm{GL}(2,\mathbb{C})]'/\mathbb{C}^\times \xrightarrow{\sim} \mathrm{SO}\left(\begin{pmatrix} 0 & \omega \\ \omega^{-1} & 0 \end{pmatrix}\right) \hookrightarrow \widehat{G} = \mathrm{SL}(4,\mathbb{C})$$

defines an embedding of diagonal subgroups

$$\widehat{T}_C = \{(\begin{pmatrix} a_1 & 0 \\ 0 & a_2 \end{pmatrix}, \begin{pmatrix} b_1 & 0 \\ 0 & b_2 \end{pmatrix}) \equiv \mathrm{diag}(a_1 b_1, a_1 b_2, b_1 a_2, a_2 b_2); a_1 a_2 b_1 b_2 = 1\},$$

$$\hookrightarrow \widehat{T} = \{\mathrm{diag}(a, b, c, d); abcd = 1\},$$

and lattices

$$(X^*(\mathbf{T}_C^*) =) \quad X_*(\widehat{T}_C) \hookrightarrow X_*(\widehat{T}) \quad (= X^*(\mathbf{T}^*)),$$

$(x_1, x_2; y_1, y_2) \mapsto (x_1+y_1, x_1+y_2; y_1+x_2, x_2+y_2)$, $x_1+x_2+y_1+y_2 = 0$. Here $\mathbf{T}_C^*$ and $\mathbf{T}^*$ are the diagonal subgroups of $\mathbf{C} = [\mathrm{GL}(2)\times\mathrm{GL}(2)]'/\,\mathrm{GL}(1)$ and $\mathbf{G} = \mathrm{PGL}(4)$. The dual map $N : X_*(\mathbf{T}^*) = X^*(\widehat{T}) \to X^*(\widehat{T}_C) = X_*(\mathbf{T}_C^*)$, or

$$N : \mathbf{T}^* \to \mathbf{T}_C^*, \qquad N(\mathrm{diag}(\alpha,\beta,\gamma,\delta)) = \left(\left(\begin{smallmatrix} a_1 & 0 \\ 0 & a_2 \end{smallmatrix}\right), \left(\begin{smallmatrix} b_1 & 0 \\ 0 & b_2 \end{smallmatrix}\right)\right),$$

satisfies

$$\begin{aligned} a_1^{x_1} a_2^{x_2} b_1^{y_1} b_2^{y_2} &= (x_1, x_2; y_1, y_2)\left(\left(\begin{smallmatrix} a_1 & 0 \\ 0 & a_2 \end{smallmatrix}\right), \left(\begin{smallmatrix} b_1 & 0 \\ 0 & b_2 \end{smallmatrix}\right)\right) = \chi(N(\alpha,\beta,\gamma,\delta)) \\ &= (x_1+y_1, x_1+y_2; y_1+x_2, x_2+y_2)(\mathrm{diag}(\alpha,\beta,\gamma,\delta)) \\ &= \alpha^{x_1+y_1}\beta^{x_1+y_2}\gamma^{y_1+x_2}\delta^{x_2+y_2} \\ &= (\alpha\beta)^{x_1}(\gamma\delta)^{x_2}(\alpha\gamma)^{y_1}(\beta\delta)^{y_2} \end{aligned}$$

for all $\chi = (x_1, x_2; y_1, y_2)$ in $X^*(\mathbf{T}_C^*)$, hence we are led to make the

1.3 Definition. The norm map $N : \mathbf{T}^* \to \mathbf{T}_C^*$ is defined to be

$$N(\mathrm{diag}(\alpha,\beta,\gamma,\delta)) = \left(\left(\begin{smallmatrix} \alpha\beta & 0 \\ 0 & \gamma\delta \end{smallmatrix}\right), \left(\begin{smallmatrix} \alpha\gamma & 0 \\ 0 & \beta\delta \end{smallmatrix}\right)\right).$$

## 2. Induced Representations

Let us recall the computation of the character of a representation $\pi = I(\eta)$ of $G = \mathbf{G}(F_v)$ normalizedly induced from the character $\eta$ of the Borel subgroup $B = AN$, $N$ the unipotent radical and $A$ the maximal torus in $B$. If $K$ is the maximal compact subgroup with $G = BK = NAK$, the space of $\pi$ consists of the smooth $\phi : G \to \mathbb{C}$ with $\phi(nak) = (\delta^{1/2}\eta)(a)\phi(k)$, where

$$\delta(a) = |\det(\mathrm{Ad}\, a|\,\mathrm{Lie}\, N)|$$

and $\pi$ acts by right translation; of course $a \in A$, $k \in K$, $n \in N$. In Lemma 2.1 $\mathbf{G}$ can be any quasi-split connected reductive group.

Recall that the *character* ([H]) of an admissible representation $\pi$ is a conjugacy invariant locally integrable function $\chi_\pi$ satisfying $\mathrm{tr}\,\pi(fdg) = \int_G \chi_\pi(g)f(g)dg$ for any test function $f \in C_c^\infty(G)$. It characterizes the representation up to isomorphism.

2.1 LEMMA. *The character $\chi_\pi$ of the induced representation $\pi = I(\eta)$ is supported on the split set and we have for regular $a \in A$*

$$(\Delta\chi_\pi)(t) = \sum_{w\in W} \eta(w(a)).$$

PROOF. There is a measure decomposition $dg = \delta^{-1}(a)dndadk$ corresponding to $g = nak$, $G = NAK$. For a test function $f \in C_c^\infty(G)$ the convolution operator $\pi(fdg) = \int_G \pi(g)f(g)dg$ maps $\phi \in \pi$ to

$$\begin{aligned}(\pi(fdg)\phi)(h) &= \int_G f(g)\phi(hg)dg = \int_G f(h^{-1}g)\phi(g)dg \\ &= \int_N\int_A\int_K f(h^{-1}n_1ak)(\delta^{1/2}\eta)(a)\phi(k)\delta^{-1}(a)dn_1dadk.\end{aligned}$$

The change of variables $n_1 \mapsto n$, where $n$ is defined by $n^{-1}ana^{-1} = n_1$, has the Jacobian

$$|\det(1 - \mathrm{Ad}\, a)|\,\mathrm{Lie}\, N|.$$

The trace of $\pi(fdg)$ is obtained on integrating the kernel of the convolution operator – viewed as a trivial vector bundle over $K$ – on the diagonal $h = k \in K$. Hence, writing

$$\Delta(a) = \delta^{-1/2}(a)|\det(1 - \mathrm{Ad}\, a)|\,\mathrm{Lie}\, N|,$$

we have

$$\begin{aligned}\mathrm{tr}\,\pi(fdg) &= \int_K\int_N\int_A \Delta\eta(a)f(k^{-1}n^{-1}ank)dndadk \\ &= w(A)^{-1}\int_A [\sum_{w\in W}\eta(w(a))](\Delta(a)\int_{G/A} f(gag^{-1})d\dot{g})da,\end{aligned}$$

where $w(A)$ is the cardinality of the Weyl group $W$. Here $W$ is the quotient of the normalizer of $A$ by the centralizer of $A$ in $G$.

To conclude the proof of the lemma we now use the Weyl integration formula

$$\int_G \chi(g)f(g)dg = \sum_T w(T)^{-1}\int_T \Delta(t)\chi(t)[\Delta(t)\int_{T\backslash G} f(g^{-1}tg)d\dot{g}]dt.$$

Here $T$ ranges over the conjugacy classes of tori, $\chi(g)$ is a conjugacy class function, $\Delta(t)^2$ is the Jacobian

$$|\det(1-\mathrm{Ad}(t))|(\mathrm{Lie}\,N \oplus \mathrm{Lie}\,N^-)|$$

(over an algebraic closure $\overline{F}$ of $F$ the torus $T$ splits). $N$ is the unipotent radical of a Borel subgroup containing $T$ and $N^-$ is the opposite unipotent group:

$$\mathrm{Lie}(G/T) = \mathrm{Lie}\,N \oplus \mathrm{Lie}\,N^-$$

and

$$|\det(1-\mathrm{Ad}(t))|\,\mathrm{Lie}\,N^-| = \delta^{-1}(t)|\det(1-\mathrm{Ad}(t))|\,\mathrm{Lie}\,N|.$$

□

Similar analysis applies in the twisted case, where $\Delta(t\theta)$ is defined in the course of the following proof.

2.2 Lemma. *The twisted character $\chi_\pi(t\theta)$ of the induced $\theta$-invariant representation $\pi = I(\eta)$ with $\eta = \eta \circ \theta$ vanishes outside the $\theta$-split set (the set of $\theta$-conjugacy classes of $A$), and is given by*

$$\Delta(t\theta)\chi_\pi(t\theta) = \sum_{w \in W^\theta} \eta(w(a))$$

*on the $\theta$-regular $a \in A$.*

Proof. Let $\theta$ be an involution of $G$ preserving $B$ and $K$, for example $\theta(g) = J^{-1}{}^t g^{-1} J$ where $G = \mathrm{GL}(n,F)$ (or $\mathrm{PGL}(n,F)$, etc.) and $J$ an anti-diagonal matrix. Then $\mathrm{tr}(\pi(f\theta))$ is zero unless $\pi$ is equivalent to ${}^\theta\pi(: g \mapsto \pi(\theta(g)))$, in which case, for $\pi = I(\eta)$, we have

$$\begin{aligned}(\pi(\theta f dg)\phi)(h) &= \int_G f(g)\phi(\theta(h)g)dg = \int_G f(\theta(h)^{-1}g)\phi(g)dg \\ &= \iiint f(\theta(h^{-1})nak)(\delta^{1/2}\eta)(a)\phi(k)\delta^{-1}(a)dndadk,\end{aligned}$$

hence

$$\mathrm{tr}\,\pi(\theta f dg) = \iiint f(\theta(k)^{-1}n_1 ak)(\delta^{-1/2}\eta)(a)dn_1 dadk.$$

We change variables $n_1 \mapsto n$, where $\theta(n)^{-1}ana^{-1} = n_1$, which has the same Jacobian as if $na\theta(n)^{-1}a^{-1} = n_1$, which is

$$|\det(1 - \mathrm{Ad}(a\theta))|\,\mathrm{Lie}\,N|,$$

to get

$$\operatorname{tr}\pi(\theta f dg) = \int_{A/A^{1-\theta}} \eta(a)\Delta(a\theta) \int_{A^\theta\backslash G} f(\theta(g)^{-1}ag)d\dot{g}\, da.$$

Here we put

$$\Delta(a\theta) = \delta^{-1/2}(a)|\det(1 - \mathrm{Ad}(a\theta))|\,\mathrm{Lie}\,N|,$$

$$A^\theta = \{a \in A; a = \theta(a)\}, \qquad A^{1-\theta} = \{a\theta(a)^{-1}; a \in A\}.$$

We may choose a set of representatives $T$ for the $\theta$-conjugacy classes of tori in $G$ with $T = \theta(T)$ ([KS]), such that on the regular set

$$G = \bigcup_T \bigcup_{t\in T/T^{1-\theta}} \bigcup_{g\in T^\theta\backslash G} \theta(g^{-1})tg.$$

The corresponding Weyl integration formula is $\int_G \chi(g)f(g)dg$

$$= \sum_T w^\theta(T)^{-1} \int_{T/T^{1-\theta}} \Delta(t\theta)\chi(t)\cdot\Delta(t\theta) \int_{T^\theta\backslash G} f(\theta(g^{-1})tg)d\dot{g}\, dt,$$

where

$$\Delta(t\theta)^2 = |\det(1 - \mathrm{Ad}(t\theta))|\,\mathrm{Lie}(G/T)|$$

and $w^\theta(T)$ is the cardinality of the group $W^\theta(T)$ of $\theta$-fixed elements in the Weyl group $W(T)$ of $T$. The lemma follows. □

2.3 Lemma. *For* $t = \mathrm{diag}(a,b,c,d)$ *we have*

$$\Delta(t\theta) = \left|\frac{(ac-bd)^2(ab-cd)^2(a-d)^2(b-c)^2}{(abcd)^3}\right|^{1/2}.$$

Proof. Note that $\mathrm{Lie}(G/T) = \mathrm{Lie}\,N \oplus \mathrm{Lie}\,N^-$ and $N$, $N$ are $\theta$-invariant. We have,

$$|\det(1 - \mathrm{Ad}(t\theta))|\,\mathrm{Lie}\,N| = |\prod_\Theta(1 - \sum_{\alpha\in\Theta}\alpha(t))|$$

where the product ranges over the $\theta$-orbits $\Theta$ of the positive roots $\alpha > 0$, and the sum over the roots in the $\theta$-orbit. Thus for $t = \mathrm{diag}(a, b, c, d)$ we obtain

$$\left|\left(1-\frac{a}{b}\frac{c}{d}\right)\left(1-\frac{a}{c}\frac{b}{d}\right)\left(1-\frac{a}{d}\right)\left(1-\frac{b}{c}\right)\right|.$$

Further,

$$|\det(1-\mathrm{Ad}(t\theta))|\,\mathrm{Lie}\,N^-| = \delta(t\theta)^{-1}|\det(1-\mathrm{Ad}(t\theta))|\,\mathrm{Lie}\,N|$$

where $\delta(t\theta)$ is

$$= \Big|\prod_{\Theta}\Big(\big(\sum_{\alpha\in\Theta}\alpha\big)(t)\Big)\Big| = \left|\left(\frac{a}{b}\frac{c}{d}\right)\left(\frac{a}{c}\frac{b}{d}\right)\left(\frac{b}{c}\right)\left(\frac{a}{d}\right)\right| = \Big|\prod_{\alpha>0}\alpha(t)\Big| = \delta(t).$$

The lemma follows. □

Recall now that the map

$$\lambda_1^* : \mathbf{G} = \mathrm{PGL}(4) \to \mathbf{C} = [\mathrm{GL}(2)\times\mathrm{GL}(2)]'/\,\mathrm{GL}(1)$$

dual to

$$\lambda_1 : \widehat{C} = [\mathrm{GL}(2,\mathbb{C})\times\mathrm{GL}(2,\mathbb{C})]'/\mathbb{C}^\times \hookrightarrow \widehat{G} = \mathrm{SL}(4,\mathbb{C})$$

maps $\mathrm{diag}(a,b,c,d)$ to $\left(\begin{pmatrix} ab & 0 \\ 0 & cd\end{pmatrix}, \begin{pmatrix} ac & 0 \\ 0 & bd\end{pmatrix}\right)$.

2.4 DEFINITION. Let $F$ be a local field. We say that $f \in C_c^\infty(G(F))$ *weakly matches* $f_C \in C_c^\infty(\mathbf{C}(F))$ if

$$F_f(t\theta) = \Delta(t\theta)\int_{T^\theta\backslash G} f(\theta(g)^{-1}tg)d\dot g \qquad (t\in T(F))$$

and

$$F_{f_C}(t) = \Delta_C(t)\int_{T\backslash C} f_C(g^{-1}tg)d\dot g \qquad (t\subset T(F))$$

are related by $F_f(t\theta) = F_{f_C}(\lambda_1^*(t))$ for $t \in \mathbf{A}(F)^{\mathrm{reg}}$.

This is a temporary definition, sufficient for the study of induced representations; it will be completed below.

2.5 DEFINITION. We say that the induced $\mathbf{C}(F)$-module $\pi_1 \times \pi_2$ *lifts* to the induced $\mathbf{G}(F)$-module $\pi$ if $\operatorname{tr}(\pi_1 \times \pi_2)(f_C) = \operatorname{tr} \pi(\theta f)$ for all weakly matching $f$ and $f_C$.

Note that the characters of the induced $\mathbf{C}(F)$-modules and the twisted characters of the induced $\mathbf{G}(F)$-modules are supported on the split set, hence our temporary definition of weakly matching is sufficient. We conclude

2.6 PROPOSITION. *The induced representation*

$$\pi_C = I_2(\mu_1, \mu_1') \times I_2(\mu_2, \mu_2')$$

*of* $\mathbf{C}(F)$ $\lambda_1$*-lifts to the induced representation*

$$\pi = I_4(\mu_1\mu_2, \mu_1\mu_2', \mu_2\mu_1', \mu_1'\mu_2')$$

*of* $\mathbf{G}(F)$. *Here* $\mu_i, \mu_i' : F^\times \to \mathbb{C}^\times$ *are any characters with* $\mu_1\mu_1'\mu_2\mu_2' = 1$.

PROOF. It suffices to observe that

$$\begin{aligned}&(\mu_1\mu_2, \mu_1\mu_2', \mu_2\mu_1', \mu_1'\mu_2')(\operatorname{diag}(a,b,c,d))\\ &= (\mu_1\mu_2)(a)(\mu_1'\mu_2')(d)(\mu_1\mu_2')(b)(\mu_2\mu_1')(c)\end{aligned}$$

on the $G$-side is equal to $\mu_1(ab)\mu_1'(cd)\mu_2(ac)\mu_2'(bd)$ on the $C$-side, and use the computation of the character of the induced and $\theta$-induced representations. □

Similarly the map $\lambda^* : \mathbf{G} = \operatorname{PGL}(4) \to \mathbf{H} = \operatorname{PGSp}(2)$ dual to

$$\lambda : \widehat{H} = \operatorname{Sp}(2, \mathbb{C}) \hookrightarrow \widehat{G} = \operatorname{SL}(4, \mathbb{C})$$

maps $\operatorname{diag}(a,b,c,d)$ to $\operatorname{diag}(ab, ac, bd, cd)$.

2.7 DEFINITION. Let $F$ be a local field. We say that $f \in C_c^\infty(\mathbf{G}(F))$ *weakly matches* $f_H \in C_c^\infty(\mathbf{H}(F))$ if $F_f(t\theta)$ and

$$F_{f_H}(t) = \Delta_H(t) \int_{T\backslash H} f_H(g^{-1}tg)\,d\dot{g}$$

are related by $F_f(t\theta) = F_{f_H}(\lambda^*(t))$ for $t \in \mathbf{A}(F)^{\text{reg}}$.

We say that the induced $H(F)$-module $\pi_H$ *lifts* to the induced $G(F)$-module $\pi$ if for all weakly matching $f$ and $f_H$ we have $\operatorname{tr} \pi_H(f_H) = \operatorname{tr} \pi(\theta f)$.

2.8 PROPOSITION. *The induced* $\mathbf{H}(F)$*-module* $I_H(\mu_1, \mu_2)$ *lifts – via* $\lambda$ *– to the induced* $\mathbf{G}(F)$*-module* $\pi = I_4(\mu_1, \mu_2, \mu_2^{-1}, \mu_1^{-1})$.

PROOF. It suffices to observe that

$$(\mu_1, \mu_2, \mu_2^{-1}, \mu_1^{-1})(\operatorname{diag}(a, b, c, d)) = \mu_1(a/d)\mu_2(b/c);$$

and that $\pi_H(\mu_1, \mu_2)$ is induced from the character

$$\operatorname{diag}(\alpha, \beta, \gamma, \delta) \mapsto \mu_1(\alpha/\gamma)\mu_2(\alpha/\beta)$$

of the diagonal subgroup of $\mathrm{PGSp}(2, F)$, and that the value of this last character at

$$\lambda^*(\operatorname{diag}(a, b, c, d)) = \operatorname{diag}(ab, ac, bd, cd) \quad \text{is} \quad \mu_1(ab/bd)\mu_2(ab/ac).$$

□

Finally note that the map

$$\lambda_0^* : \mathbf{H} = \mathrm{PGSp}(2) \to \mathbf{C}_0 = \mathrm{PGL}(2) \times \mathrm{PGL}(2)$$

dual to $\lambda_0 : \widehat{C}_0 \hookrightarrow \widehat{H}$ takes

$$\operatorname{diag}(\alpha, \beta, \gamma, \delta) \quad \text{to} \quad \left( \begin{pmatrix} \alpha & 0 \\ 0 & \gamma \end{pmatrix}, \begin{pmatrix} \alpha & 0 \\ 0 & \beta \end{pmatrix} \right).$$

We define

2.9 DEFINITION. The functions $f_H \in C_c^\infty(\mathbf{H}(F))$ and $f_0 \in C_c^\infty(\mathbf{C}_0(F))$ are *weakly matching* if $F_{f_H}(t) = F_{f_0}(t)$ on $t \in \mathbf{A}(F)^{\mathrm{reg}}$. The induced $\mathbf{C}_0(F)$-module $\pi_0 = \pi_1 \times \pi_2$ *lifts* to the induced $\mathbf{H}(F)$-module $\pi_H$ if $\operatorname{tr} \pi_H(f_H) = \operatorname{tr} \pi_0(f_0)$ for all weakly matching $f_H$ and $f_0$.

2.10 PROPOSITION. *The induced representation*

$$I_2(\mu_1, \mu_1^{-1}) \times I_2(\mu_2, \mu_2^{-1})$$

*of* $\mathbf{C}_0(F)$ $\lambda_0$*-lifts to the induced representation* $I_H(\mu_1, \mu_2)$ *of* $\mathbf{H}(F)$.

PROOF. On the $H$-side, we induce from

$$\operatorname{diag}(\alpha, \beta, \gamma, \delta) \mapsto \mu_1(\alpha/\gamma)\mu_2(\alpha/\beta).$$

This matrix is mapped by $\lambda_0^*$ to

$$\left( \begin{pmatrix} \alpha & 0 \\ 0 & \gamma \end{pmatrix}, \begin{pmatrix} \alpha & 0 \\ 0 & \beta \end{pmatrix} \right),$$

and the $C_0$-module is induced from the character whose value at this last pair of matrices is $\mu_1(\alpha/\gamma)\mu_2(\alpha/\beta)$. □

# 3. Satake Isomorphism

Our liftings are summarized in the following diagram ($X = \mathrm{GL}(2,\mathbb{C})$)

$$\begin{array}{ccc} \widehat{C}_0 = \mathrm{SL}(2,\mathbb{C}) \times \mathrm{SL}(2,\mathbb{C}) & & \widehat{C} = \mathrm{SO}(4,\mathbb{C}) \simeq [X \times X]'/\mathbb{C}^\times \\ \lambda_0 \searrow & & \swarrow \lambda_1 \\ & \widehat{H} = \mathrm{Sp}(2,\mathbb{C}) \overset{\lambda}{\hookrightarrow} \widehat{G} = \mathrm{SL}(4,\mathbb{C}) & \end{array}$$

The dual group homomorphisms define liftings of unramified (local) representation. These representations are uniquely determined by the semisimple conjugacy classes that they define in the dual group. Thus the $\lambda_0$, $\lambda$, $\lambda_1$ define liftings as follows.

$$\begin{aligned} \lambda_0 : \pi_2(\mu_1, \mu_1^{-1}) \times \pi_2(\mu_2, \mu_2^{-1}) \mapsto \; & \pi_{\mathrm{PGSp}(2)}(\mu_1, \mu_2) \\ \in \; & JH(\mu_1\mu_2 \times \mu_1/\mu_2 \rtimes \mu_1^{-1}), \end{aligned}$$

$$\lambda : \pi_{\mathrm{PGSp}(2)}(\mu_1, \mu_2) \mapsto \pi_4(\mu_1, \mu_2, \mu_2^{-1}, \mu_1^{-1}) \in JH(I_4(\mu_1, \mu_2, \mu_2^{-1}, \mu_1^{-1})),$$

$$\lambda_1 : \pi_2(\mu_1, \mu_1') \times \pi_2(\mu_2, \mu_2') \mapsto \pi_4(\mu_1\mu_2, \mu_1\mu_2', \mu_2\mu_1', \mu_1'\mu_2'), \; \mu_1\mu_1'\mu_2\mu_2' = 1.$$

We write $JH(\pi)$ for the set of irreducible constituents of a representation $\pi$, for example $\pi = \pi_1 \times \cdots \times \pi_r \rtimes \sigma$ on $\mathrm{GSp}(n,F)$ or $\pi = \pi_1 \times \cdots \times \pi_r = I(\pi_1, \ldots, \pi_r)$ on $\mathrm{GL}(|\mathbf{n}|, F)$. The subscript indicates that $\pi_2$ is a representation of $\mathrm{GL}(2, F_v)$ and $\pi_4$ of $\mathrm{PGL}(4, F_v)$.

The $\mu_1$, $\mu_2$ are unramified characters of the local nonarchimedean field $F_v^\times$; write $\mu_i^\bullet$ for their values $\mu_i(\boldsymbol{\pi})$ at a uniformizer. Then the class $t(\pi_2(\mu_1, \mu_1'))$ associated to the unramified irreducible $\pi_2(\mu_1, \mu_1')$ is that of $\mathrm{diag}(\mu_1^\bullet, \mu_1'^\bullet)$ in $\mathrm{GL}(2,\mathbb{C})$, $t(\pi_{\mathrm{PGSp}(2)}(\mu_1, \mu_2))$ is the class of

$$\mathrm{diag}(\mu_1^\bullet, \mu_2^\bullet, \mu_2^{\bullet-1}, \mu_1^{\bullet-1})$$

in $\mathrm{Sp}(2,\mathbb{C})$,

$$t(\pi_4(\mu_1\mu_2, \mu_1\mu_2', \mu_2\mu_1', \mu_1'\mu_2'))$$

is that of

$$\mathrm{diag}(\mu_1^\bullet\mu_2^\bullet, \mu_1^\bullet\mu_2'^\bullet, \mu_2^\bullet\mu_1'^\bullet, \mu_1'^\bullet\mu_2'^\bullet)$$

in $\mathrm{SL}(4,\mathbb{C})$. Note that the homomorphisms $\lambda$, $\lambda_0$, $\lambda_1$ define dual homomorphisms of Hecke algebras, e.g., with $G = G(F_v)$, $K = G(R_v), \ldots$,

$$\lambda^* : \mathbb{H}_G = C_c^\infty(K\backslash G/K) \to \mathbb{H}_H = C_c^\infty(K_H\backslash H/K_H),$$

by $\lambda^* : f \mapsto f_H$, $f_H^\vee(t(\pi_H)) = f^\vee(t(\pi) \times \theta)$, where the Satake isomorphism $f \mapsto f^\vee$, from $\mathbb{H}_G$ to $\mathbb{C}[(A_0(\mathbb{C}) \times \theta)^W]$ is given by $f^\vee(t \times \theta) = \operatorname{tr} \pi(t)(f \times \theta)$, and $f_H^\vee(t_H) = \operatorname{tr} \pi_H(t_H)(f_H)$. In particular, by definition of corresponding functions $f \mapsto \lambda^*(f) = f_H$, we have that

$$\operatorname{tr} I_{\mathrm{PGSp}(2)}(\mu_1, \mu_2)(\lambda^*(f)) = \operatorname{tr} \pi_{\mathrm{PGSp}(2)}(\mu_1, \mu_2)(\lambda^*(f)) = \operatorname{tr} \pi_4(f \times \theta)$$

$= \operatorname{tr} I_4(f \times \theta)$, where the traces of the full induced representation

$$I_4 = I_4(\mu_1\mu_2, \mu_1\mu_2', \mu_2\mu_1', \mu_1'\mu_2')$$

at a spherical function $f$ is equal to that at its unramified constituent.

# 4. Induced Representations of PGSp(2, $F$)

We use results recorded in Sally-Tadic [ST] – using those of Rodier [Ro2], Shahidi [Sh2,3] and Waldspurger [W1] – on reducibility of induced representations of $\mathbf{H}(F) = \mathrm{PGSp}(2, F)$, and unitarizability. Let us recall some notations. Denote by GSp($n$) (or GSp($n$)) the group of symplectic similitudes

$$\left\{ g \in \mathrm{GL}(2n); \quad {}^t g \begin{pmatrix} 0 & w \\ -w & 0 \end{pmatrix} g = \boldsymbol{\lambda}(g) \begin{pmatrix} 0 & w \\ -w & 0 \end{pmatrix} \right\}.$$

Here $w = w_n = (\delta_{i,n-j+1})$ in GL($n$). Its standard parabolic subgroups are the upper triangular subgroups $\mathbf{P_n} = \mathbf{P_n^{\mathit{n}}}$ with Levi subgroups

$$\mathbf{M_n} = \mathbf{M_n^{\mathit{n}}} = \{m = \operatorname{diag}(g_1, \dots, g_r, h, \boldsymbol{\lambda}(h)^\tau g_r^{-1}, \dots, \boldsymbol{\lambda}(h)^\tau g_1^{-1})\};$$

$g_i \in \mathrm{GL}(n_i)$, $h \in \mathrm{GSp}(n - |\mathbf{n}|)$. Here $\mathbf{n} = (n_1, \dots, n_r)$, $\quad n_i \geqslant 1, \quad r \geqslant 0$,

$$|\mathbf{n}| = n_1 + \cdots + n_r \leq n, \quad {}^\tau g_i = w_i^t g_i w_i, \quad w_i = w_{n_i}.$$

Put $\mathrm{GSp}(0) = \mathbb{G}_m = \{\boldsymbol{\lambda}(h)\}$. These groups are in bijection with the set of subsets of the set of simple roots of GSp($n$); to a subset we associate the Levi subgroup generated by the root subgroups of the simple roots in the subset and their negatives. For GSp(2) the standard parabolic subgroups are $\mathbf{P}_{(0)} = \mathbf{P}_{\{\alpha,\beta\}} = \mathrm{GSp}(2)$, the Siegel parabolic $\mathbf{P}_{(2)} = \mathbf{P}_{\{\alpha\}}$ which has Levi

$$\mathbf{M}_{(2)} = \mathbf{M}_{\{\alpha\}} = \{\mathrm{diag}(g, \boldsymbol{\lambda}{}^{\tau}g^{-1}); \quad g \in \mathrm{GL}(2), \quad \boldsymbol{\lambda} \in \mathbb{G}_m\},$$

the Heisenberg parabolic $\mathbf{P}_{(1)} = \mathbf{P}_{\{\beta\}}$ which has Levi $\mathbf{M}_{(1)} = \mathbf{M}_{\{\beta\}}$

$$= \{\mathrm{diag}(a, h, \boldsymbol{\lambda}(h)/a);\ a \in \mathbb{G}_m,\ h \in \mathrm{GSp}(1) = \mathrm{GL}(2),\ \boldsymbol{\lambda}(h) = \det h\},$$

and $\mathbf{P}_{(1,1)} = \mathbf{P}_{\emptyset}$ is the minimal standard parabolic subgroup with Levi subgroup $\mathbf{M}_{(1,1)} = \mathbf{M}_{\emptyset}$ that we usually denote by $\mathbf{A}_0$, consisting of

$$\{\mathrm{diag}(a, b, \boldsymbol{\lambda}/b, \boldsymbol{\lambda}/a);\ a, b, \boldsymbol{\lambda} \in \mathbb{G}_m\}.$$

If $\pi_1, \ldots, \pi_r$ are representations of $\mathrm{GL}(n_i, F)$, and $\sigma$ of $\mathrm{GSp}(n - |\mathbf{n}|, F)$, $F$ a local field, as in [ST] denote by $\pi_1 \times \cdots \times \pi_r \rtimes \sigma$ the representation $I(\pi_1, \ldots, \pi_r, \sigma)$ of $\mathrm{GSp}(n, F)$ normalizedly induced from the representation

$$p = mu \mapsto \pi_1(g_1) \otimes \cdots \otimes \pi_r(g_r) \otimes \sigma(h) \qquad \text{of} \quad \mathbf{P_n} = \mathbf{M_n}\mathbf{U_n}.$$

Here $\mathbf{U_n}$ denotes the unipotent radical of $\mathbf{P_n}$. Note that $\sigma$ is a character if $|\mathbf{n}| = n$ (thus $h \in \mathrm{GSp}(0, F) = F^\times$). The induction is normalized by multiplying the inducing representation by the character $\delta_{\mathbf{n}}^{1/2}(p)$, where $\delta_{\mathbf{n}}(p) = |\det(\mathrm{Ad}(p)|\,\mathrm{Lie}(\mathbf{U_n}))|$. Normalized induction takes unitarizable representations to unitarizable representations.

NOTATION. As in [BZ2], 4.2, we write $\nu(x) = |x|$ for $x \in F^\times$.

EXAMPLE. The simplest example is where GSp(1)=GL(2). Here $\mathbf{M}^1_{(1)}$ is the diagonal subgroup, $\delta(\mathrm{diag}(a, b)) = |a/b|$, and $\mu \rtimes \sigma$ is the representation usually denoted by $I(\mu\sigma, \sigma)$, normalizedly induced from the character $\mathrm{diag}(a, b) \cdot u \mapsto (\mu\sigma)(a)\sigma(b)$ (if $b = \boldsymbol{\lambda}/a$, this is $= (\mu\sigma)(a)\sigma(\boldsymbol{\lambda}/a) = \mu(a)\sigma(\boldsymbol{\lambda})$). The trivial representation $\mathbf{1}_2$ of $\mathrm{GL}(2, F)$ is a subrepresentation of $I(\nu^{-1/2}, \nu^{1/2}) = \nu^{-1} \rtimes \nu^{1/2}$ and a quotient of $I(\nu^{1/2}, \nu^{-1/2}) = \nu \rtimes \nu^{-1/2}$.

EXAMPLE. In the case of $\mathrm{GSp}(2, F)$ and $\mathbf{P}_{(1,1)}$, the representation denoted $I_H(\mu_1, \mu_2)$ normalizedly induced from the character

$$p = u\,\mathrm{diag}(a, b, \boldsymbol{\lambda}/b, \boldsymbol{\lambda}/a) \mapsto \mu_1(ab/\boldsymbol{\lambda})\mu_2(a/b) = \mu_1\mu_2(a)(\mu_1/\mu_2)(b)\mu_1^{-1}(\boldsymbol{\lambda})$$

is the same as $\mu_1\mu_2 \times \mu_1/\mu_2 \rtimes \mu_1^{-1}$. Its central character is trivial, namely it is a representation of $\mathbf{H}(F) = \mathrm{PGSp}(2, F)$. If $\xi^2 = 1$ then $I_H(\xi\mu_1, \xi\mu_2) = \mu_1\mu_2 \times \mu_1/\mu_2 \rtimes \xi/\mu_1$.

4.1 LEMMA. *(i) The central character of* $\pi_1 \times \cdots \times \pi_r \rtimes \sigma$ *is* $\omega_\sigma \omega_{\pi_1} \ldots \omega_{\pi_r}$ *if* $|\mathbf{n}| < n$*; here* $\omega_{\pi_i}$ *are the central characters of* $\pi_i$ *(*$\omega_\sigma$ *of* $\sigma$*,* $\sigma$ *being a representation of* $\mathrm{GSp}(n - |\mathbf{n}|, F)$*). It is* $\sigma^2 \omega_{\pi_1} \ldots \omega_{\pi_r}$ *if* $|\mathbf{n}| = n$*. (ii) For a character* $\mu$ *we have* $\mu(\pi_1 \times \cdots \times \pi_r \rtimes \sigma) = \pi_1 \times \cdots \times \pi_r \rtimes \mu\sigma$*. In particular* $\mu(\pi \rtimes \sigma) = \pi \rtimes \mu\sigma$*. (iii) We have* $\pi \rtimes \sigma = \check{\pi} \rtimes \omega_\pi \sigma$ □

Recall that two parabolic subgroups of a reductive group $\mathbf{G}$ over $F$ are called *associate* if their Levi subgroups are conjugate. This is an equivalence relation. An irreducible representation $\pi$ of $G = \mathbf{G}(F)$, $F$ a p-adic field, is *supported in an associate class* if there is a parabolic subgroup $P$ in this class such that $\pi$ is a composition factor of a representation of $G$ induced from an irreducible cuspidal representation of the Levi factor $M$ of $P$ extended trivially to the unipotent radical $U$ of $P$. In our case an irreducible representation $\pi$ of $H = \mathrm{PGSp}(2, F)$ is supported in $P_{(1,1)}$, $P_{(1)}$, $P_{(2)}$ or it is cuspidal. An unramified representation is supported on $P_{(1,1)}$. It is a subquotient $\pi_H(\mu_1, \mu_2)$ of a fully induced $I_H(\mu_1, \mu_2) = \mu_1\mu_2 \times \mu_1/\mu_2 \rtimes \mu_1^{-1}$, where the $\mu_i$ are unramified characters of $F^\times$.

An irreducible representation $\pi$ is called *essentially tempered* if $\nu^e \pi$ is tempered for some real number $e$, where $(\nu^e\pi)(g) = \nu(\det g)^e \pi(g)$.

The following is the Langlands classification for $\mathrm{GSp}(n, F)$.

4.2 PROPOSITION. *Each representation* $\nu^{e_1}\pi_1 \times \cdots \times \nu^{e_r}\pi_r \rtimes \sigma$*, where* $e_1 \geqslant \cdots \geqslant e_r > 0$*,* $\pi_i$ *are irreducible square integrable representations of* $\mathrm{GL}(n_i, F)$*, and* $\sigma$ *is an irreducible essentially tempered representation of* $\mathrm{GSp}(n - |\mathbf{n}|, F)$*, has a unique irreducible quotient:* $L(\nu^{e_1}\pi_1, \ldots, \nu^{e_r}\pi_r, \sigma)$*. Each irreducible representation of* $\mathrm{GSp}(n, F)$ *is of this form.* □

With these notations we shall use the results stated in [ST]. These concern the reducibility of the induced representations, and description of their properties. In particular [ST], Lemma 3.1 asserts that for characters $\chi_1, \chi_2, \sigma$ of $F^\times$ the representation $\chi_1 \times \chi_2 \rtimes \sigma$ is irreducible if and only if $\chi_i \neq \nu^{\pm 1}$ and $\chi_1 \neq \nu^{\pm 1}\chi_2^{\pm 1}$. In case of reducibility the composition series are described in [ST], together with their properties. The list is recorded in chapter V, section 2, 2.1-2.3 below. Moreover, we shall use [ST], Theorem 4.4, which classifies the irreducible unitarizable representations of $\mathrm{GSp}(2, F)$ supported in minimal parabolic subgroups. It shows that

4.3 LEMMA. *The representation* $L(\nu \times \nu \rtimes \nu^{-1}) = \pi_H(\nu, 1)$ *is not unitarizable.* □

# 5. Twisted Conjugacy Classes

The geometric part of the trace formula is expressed in terms of stable conjugacy classes, whose definition we now recall. We shall need only *strongly regular semisimple* (we abbreviate this to "regular") elements $t$ in $H = \mathbf{H}(F)$, those whose centralizer $Z_{\mathbf{H}}(t)$ in $\mathbf{H}$ is a maximal $F$-torus $\mathbf{T_H}$. The elements $t$, $t'$ of $H$ are *conjugate* if there is $g$ in $H$ with $t'$ equal to $\operatorname{Int}(g^{-1})t$ $(= g^{-1}tg)$. Such $t$, $t'$ in $H$ are *stably conjugate* if there is $g$ in $\mathbf{H}(= \mathbf{H}(\overline{F}))$ with $t' = \operatorname{Int}(g^{-1})t$. Then $g_\sigma = g\sigma(g)^{-1}$ lies in $\mathbf{T_H}$ for every $\sigma$ in the Galois group $\Gamma = \operatorname{Gal}(\overline{F}/F)$, and $g \mapsto \{\sigma \mapsto g_\sigma\}$ defines an isomorphism from the set of conjugacy classes within the stable conjugacy class of $t$ to the pointed set $D(\mathbf{T_H}/F) = \ker[H^1(F, \mathbf{T_H}) \to H^1(F, \mathbf{H})]$.

Using the commutative diagram with exact rows

$$\begin{array}{ccccc} & & H^1(F, \mathbf{T}^{\mathrm{sc}}_{\mathbf{H}}) & \to & H^1(F, \mathbf{H}^{\mathrm{sc}}) \\ & & \downarrow & & \downarrow \\ 1 \to D(\mathbf{T_H}/F) \to & & H^1(F, \mathbf{T_H}) & \to & H^1(F, \mathbf{H}), \end{array}$$

where $\mathbf{H}^{\mathrm{sc}} \xrightarrow{\pi} \mathbf{H}^{\mathrm{der}} \hookrightarrow \mathbf{H}$ is the simply connected covering group of the derived (commutator $[\mathbf{H}, \mathbf{H}]$) group $\mathbf{H}^{\mathrm{der}}$ of $\mathbf{H}$, and noting that for a p-adic field $F$ one has $H^1(F, \mathbf{H}^{\mathrm{sc}}) = \{1\}$, one concludes that $D(\mathbf{T_H}/F) = \operatorname{Im}[H^1(F, \mathbf{T}^{\mathrm{sc}}) \to H^1(F, \mathbf{T})]$ for such $F$, in particular it is a group. Here $\mathbf{T}^{\mathrm{sc}}_{\mathbf{H}} = \pi^{-1}(\mathbf{T}^{\mathrm{der}}_{\mathbf{H}})$, $\mathbf{T}^{\mathrm{der}}_{\mathbf{H}} = \mathbf{H}^{\mathrm{der}} \cap \mathbf{T_H}$. Indeed, if $\{\sigma \mapsto g_\sigma\}$ is in $D(\mathbf{T_H}/F)$, thus $g_\sigma = g\sigma(g)^{-1}$, write $g = g_1 z$, using $\mathbf{H} = \mathbf{H}^{\mathrm{der}}\mathbf{Z_H}$ ($\mathbf{Z_H}$ denotes the center of $\mathbf{H}$), with $z$ in $\mathbf{Z_H}$ and $g_1$ in $\mathbf{H}^{\mathrm{der}}$. Then $g_\sigma = g_{1\sigma} z_\sigma$, and $z_\sigma = z\sigma(z)^{-1}$ is a coboundary, as $\mathbf{Z_H} \subset \mathbf{T_H}$, and $H^1(F, \mathbf{T}^{\mathrm{sc}}_{\mathbf{H}})$ surjects on $D(\mathbf{T_H}/F)$.

It is convenient to compute $H^1(F, \mathbf{T_H})$ using the Tate-Nakayama isomorphism which identifies this group with

$$H^{-1}(F, X_*(\mathbf{T_H})) = \{X \in X_*(\mathbf{T_H}); NX = 0\}/\langle X - \tau X; \tau \in \operatorname{Gal}(L/F)\rangle.$$

Here $L$ is a sufficiently large Galois extension of the local field $F$ which splits $\mathbf{T_H}$, $N$ denotes the norm from $L$ to $F$, and $X_*(\mathbf{T}_H)$ is the lattice $\operatorname{Hom}(\mathbb{G}_m, \mathbf{T_H})$.

In our case of $\mathbf{H} = \mathrm{PSp}(2) = \mathrm{PGSp}(2)$, $H^{\mathrm{sc}}$ is $\mathrm{Sp}(2)$, and $H^1(F, \mathbf{H}) = \{0\}$, hence $D(\mathbf{T_H}/F) = H^1(F, \mathbf{T_H})$ is a group.

Denote by $\mathbf{N}$ the normalizer $\mathrm{Norm}(\mathbf{T}^*_\mathbf{H}, \mathbf{H})$ of $\mathbf{T}^*_\mathbf{H}$ in $\mathbf{H}$, and let $W = \mathbf{N}/\mathbf{T}^*_\mathbf{H}$ be the Weyl group of $\mathbf{T}^*_\mathbf{H}$ in $\mathbf{H}$. Signify by $H^1(F, W)$ the group of continuous homomorphisms $\delta : \Gamma \to W$, where $\Gamma$ acts trivially on $W$.

5.1 LEMMA. *The set of stable conjugacy classes of $F$-tori in $\mathbf{H}$ injects naturally in the image of* $\ker[H^1(F, \mathbf{N}) \to H^1(F, \mathbf{H})]$ *in $H^1(F, W)$. When $\mathbf{H}$ is quasi-split this map is an isomorphism.* □

This is proven in Section I.B of [F5] where it is used to list the (stable) conjugacy classes in GSp(2). Our case of $\mathbf{H} = \mathrm{PSp}(2)$ is similar but simpler. The Weyl group $W$ is the dihedral group $D_4$, generated by the reflections $s_1 = (12)(34)$ and $s_2 = (23)$. Its other elements are $1, (12)(34)(23) = (3421)$ (taking 1 to 2, 2 to 4, 4 to 3, 3 to 1), $(23)(12)(34) = (2431)$, $(23)(3421) = (42)(31)$, $(3421)^2 = (32)(41)$, $(23)(23)(41) = (41)$.

Our list of $F$-tori follows that of *loc. cit.* The list of $F$-tori $\mathbf{T_H}$ is parametrized by the subgroup of $W$. If $\mathbf{T_H}$ splits over the Galois extension $E$ of $F$ then $H^1(F, \mathbf{T_H}) = H^1(\mathrm{Gal}(E/F), \mathbf{T}^*_\mathbf{H}(E))$ where $\mathbf{T}^*_\mathbf{H}(E) = \{t = \mathrm{diag}(a, b, \boldsymbol{\lambda}/b, \boldsymbol{\lambda}/a) \bmod \mathbf{Z_H}; a, b, \boldsymbol{\lambda} \in E^\times\}$ and $\mathrm{Gal}(E/F)$ acts via $\rho : \Gamma \to W$. If $\rho(\sigma) = \mathrm{Int}(g_\sigma)$ then $\Gamma$ acts on $\mathbf{T}^*_\mathbf{H}$ by $\sigma^*(t) = g_\sigma \cdot \sigma t \cdot g_\sigma^{-1}$, and $\sigma t = (\sigma a, \sigma b, \sigma\boldsymbol{\lambda}/\sigma b, \sigma\boldsymbol{\lambda}/\sigma a) \bmod \mathbf{Z_H}$. The split torus corresponds to the subgroup $\{1\}$ of $W$, its stable conjugacy class consists of a single class. There are nonelliptic tori $T_H$, with trivial $H^1(F, \mathbf{T_H})$, corresponding to $\rho(\Gamma)$ being $\langle(23)\rangle$, $\langle(14)\rangle$, $\langle(12)(34)\rangle$, $\langle(13)(24)\rangle$. The elliptic tori are:

(I) $\rho(\Gamma) = \langle(14)(23)\rangle$, $T_H \simeq \{\mathrm{diag}(a, b, \boldsymbol{\lambda}/b = \sigma b, \boldsymbol{\lambda}/a = \sigma a); a, b \in E^\times, \boldsymbol{\lambda} \in N_{E/F}E^\times\}$, [E:F]=2. To compute $D(\mathbf{T_H}/F)$ we take the quotient of $X_*(\mathbf{T}^{\mathrm{sc}}_\mathbf{H}) = \langle(x, y, -y, -x); x, y \in \mathbb{Z}\rangle$ (note that the generator $\sigma$ of $\mathrm{Gal}(E/F)$ maps $(x, y, z-y, z-x)$ to $(z-x, z-y, y, x)$ and the norm $N_{E/F} = N$ is the sum of the two) by the span $\langle X - \sigma X = (x, y, z-y, z-x) - (z-x, z-y, y, x) = (2x-z, 2y-z, z-2y, z-2x)\rangle$ ($X$ ranges over $X_*(\mathbf{T_H})$); it is $\mathbb{Z}/2$.

(II) $\rho(\Gamma) = \langle(14)(23), (12)(34), (13)(24)\rangle$, then $\mathbf{T_H}$ splits over an extension $E = E_1E_2$, biquadratic over $F$, $\mathrm{Gal}(E/F) = \mathbb{Z}/2 \times \mathbb{Z}/2$ is generated by $\sigma$ and $\tau$ whose fixed fields are $E_3 = E^{\langle\sigma\rangle}$, $E_2 = E^{\langle\sigma\tau\rangle}$, $E_1 = E^{\langle\tau\rangle}$, say $\rho(\sigma) = (14)(23)$, $\rho(\tau) = (12)(34)$. Then $H^{-1}$ is the quotient of $\langle(x, y, -y, -x)\rangle$ by $\langle(2x-z, 2y-z, z-2y, z-2x)\rangle$, namely $\mathbb{Z}/2$.

(III) $\rho(\Gamma) = \langle(14),(23)\rangle$, again $E = E_1E_2$, $\mathrm{Gal}(E/F) = \mathbb{Z}/2 \times \mathbb{Z}/2$ is generated by $\sigma$ and $\tau$ with $E_3 = E^{\langle\sigma\rangle}$, $E_2 = E^{\langle\sigma\tau\rangle}$, $E_1 = E^{\langle\tau\rangle}$, and $\rho(\tau) = (23)$, $\rho(\tau\sigma) = (14)$, and $H^{-1}$ is $\{0\}$, being the quotient of $\langle(x,y,-y,-x)\rangle$ by $\langle(2x-z,0,0,z-2x),\,(0,2y-z,z-2y,0)\rangle = \langle(x,0,0,-x),(0,y,-y,0)\rangle$.

(IV) When $\rho(\Gamma)$ contains an element of order four, $H^{-1}$ is $\{0\}$, as explained in [F5], I.B, (IV).

Next we describe the (stable) $\theta$-conjugacy classes of a strongly $\theta$-regular element $t$ in $G$. We fix a $\theta$-invariant $F$-torus $\mathbf{T}^*$, to wit: the diagonal subgroup. The stable $\theta$-conjugacy class of $t$ in $G$ intersects $\mathbf{T}^*$ ([KS], Lemma 3.2.A). Hence there is $h \in \mathbf{G}(=\overline{G} = \mathbf{G}(\overline{F}))$ and $t^* \in \mathbf{T}^*$ such that $t = h^{-1}t^*\theta(h)$. The centralizers are related by $Z_{\mathbf{G}}(t\theta) = h^{-1}Z_{\mathbf{G}}(t^*\theta)h$, and $Z_{\mathbf{G}}(t^*\theta) = \mathbf{T}^{*\theta}$. The centralizer of $Z_{\mathbf{G}}(t\theta)$ in $\mathbf{G}$ is an $F$-torus $\mathbf{T}_t$: it is $Z_{\mathbf{G}}(Z_{\mathbf{G}}(t\theta))$

$$= \{g \in \mathbf{G}; g^{-1}t_1g = t_1 \forall t_1 \in Z_{\mathbf{G}}(t\theta) = h^{-1}Z_{\mathbf{G}}(t^*\theta)h = h^{-1}\mathbf{T}^{*\theta}h\}$$

$= h^{-1}\mathbf{T}^*h = \mathbf{T}_t$. The torus $\mathbf{T}_t$ is $\theta_t = \mathrm{Int}(t) \circ \theta$-invariant:

$$\mathrm{Int}(t)\theta(h^{-1}t_1^*h) = h^{-1}t^*\theta(h)\cdot\theta(h)^{-1}\theta(t_1^*)\theta(h)\cdot\theta(h)^{-1}t^{*-1}h = h^{-1}\theta(t_1^*)h.$$

We have $Z_{\mathbf{G}}(t\theta) = \mathbf{T}_t^{\theta_t}$: if $t_1 \in Z_{\mathbf{G}}(t\theta) = h^{-1}\mathbf{T}^{*\theta}h \subset h^{-1}\mathbf{T}^*h = \mathbf{T}_t$ then $t_1^{-1}\cdot t\theta\cdot t_1 = t\theta$, thus $\theta_t(t_1) = t\theta(t_1)t^{-1} = t_1$.

The $\theta$-conjugacy classes within the stable $\theta$-conjugacy class of $t$ can be classified as follows.

If $t_1 = g^{-1}t\theta(g)$ and $t$ are stably $\theta$-conjugate in $G$ then $g_\sigma = g\sigma(g)^{-1} \in Z_{\mathbf{G}}(t\theta) = \mathbf{T}_t^{\theta_t}$. The set

$$D(F,\theta,t) = \ker[H^1(F,\mathbf{T}_t^{\theta_t}) \to H^1(F,\mathbf{G})]$$

parametrizes, via $(t_1,t) \mapsto \{\sigma \mapsto g_\sigma\}$, the $\theta$-conjugacy classes within the stable $\theta$-conjugacy class of $t$. The Galois action on $\mathbf{T}_t$:

$$\sigma(t) = \sigma(h^{-1}t^*\theta(h)) = h^{-1}\cdot h\sigma(h)^{-1}\cdot\sigma(t^*)\cdot\theta(\sigma(h)h^{-1})\theta(h),$$

induces the Galois action $\sigma^*$ on $\mathbf{T}^*$, given by $\sigma^*(t^*) = h\sigma(h)^{-1}\cdot\sigma(t^*)\cdot\theta(\sigma(h)h^{-1})$, and

$$H^1(F,\mathbf{T}_t^{\theta_t}) = H^1(F,\mathbf{T}^{*\theta}).$$

The norm map $N : \mathbf{T}^* \to \mathbf{T}^*_{\mathbf{H}}$ is defined to be the composition of the projection $\mathbf{T}^* \to \mathbf{T}^*_\theta = \mathbf{T}^*/(1-\theta)\mathbf{T}^*$ and the isomorphism $\mathbf{T}^*_\theta \xrightarrow{\sim} \mathbf{T}^*_{\mathbf{H}}$. If the norm $Nt^*$ of $t^* \in \mathbf{T}^*$ is defined over $F$ then for each $\sigma \in \Gamma$ there is $\ell \in \mathbf{T}^*$ such that $\sigma^*(t^*) = \ell t^* \theta(\ell)^{-1}$. Then

$$h^{-1}t^*\theta(h) = t = \sigma(t) = \sigma(h)^{-1} \cdot \sigma t^* \cdot \theta(\sigma h) = \sigma(h)^{-1}\ell t^*\theta(\ell^{-1}\sigma(h)),$$

hence

$$t^* = h_\sigma \ell \cdot t^* \cdot \theta(h_\sigma \ell)^{-1}, \quad h_\sigma = h\sigma(h)^{-1},$$

and $h_\sigma \ell \in Z_{\mathbf{G}}(t^*\theta) = \mathbf{T}^{*\theta}$, so that $h_\sigma \in \mathbf{T}^*$. Moreover, $(1-\theta)(h_\sigma) = t^*\sigma^*(t^*)^{-1}$, hence $(h_\sigma, t^*)$ lies in the subset

$$H^1(F, \mathbf{T}^* \xrightarrow{1-\theta} \mathbf{V}^*_t) \quad \text{of} \quad H^1(F, \mathbf{T}^* \xrightarrow{1-\theta} \mathbf{T}^*),$$

and it parametrizes the $\theta$-conjugacy classes of strongly $\theta$-regular elements which have the same norm. We put $\mathbf{V}_t = (1-\theta_t)\mathbf{T}_t$.

While not necessary in our case, recall that the *first hypercohomology group* $H^1(G, A \xrightarrow{f} B)$ of the short complex $A \xrightarrow{f} B$ of $G$-modules placed in degrees 0 and 1 is the group of 1-hypercocycles, quotient by the subgroup of 1-hypercoboundaries. A 1-*hypercocycle* is a pair $(\alpha, \beta)$ with $\alpha$ being a 1-cocycle of $G$ in $A$ and $\beta \in B$ such that $f(\alpha) = \partial\beta$; $\partial\beta$ is the 1-cocycle $\sigma \mapsto \beta^{-1}\sigma(\beta)$ of $G$ in $B$. A 1-*hypercoboundary* is a pair $(\partial\alpha, f(\beta))$, $\alpha \in A$.

This hypercohomology group fits in an exact sequence

$$H^0(G,A) \xrightarrow{f} H^0(G,B) \to H^1(G, A \xrightarrow{f} B) \to H^1(G,A) \xrightarrow{f} H^1(G,B).$$

We need only the case where $A = \mathbf{T}_t$, $B = \mathbf{V}_t = (1-\theta_t)\mathbf{T}_t$, $f = 1-\theta_t$, $G = \mathrm{Gal}(\overline{F}/F)$. The exact sequence $1 \to \mathbf{T}_t^{\theta_t} \to \mathbf{T}_t \xrightarrow{1-\theta_t} \mathbf{V}_t \to 1$ induces the exact sequence $H^0(F, \mathbf{T}_t) = \mathbf{T}_t^\Gamma = T_t \to H^0(F, \mathbf{V}_t) = V_t$

$$\to H^1(F, \mathbf{T}_t^{\theta_t}) \to H^1(F, \mathbf{T}_t) \xrightarrow{1-\theta_t} H^1(F, \mathbf{V}_t).$$

Hence $H^1(F, \mathbf{T}_t^{\theta_t}) = H^1(F, \mathbf{T}_t \xrightarrow{1\theta_t} \mathbf{V}_t)$.

If $t$ is a strongly $\theta$-regular element in $G$, then $\mathbf{T}_t = Z_{\mathbf{G}}(Z_{\mathbf{G}}(t\theta)^0)$ is a maximal torus in $\mathbf{G}$. Denote by $\mathbf{T}_t^{\mathrm{sc}}$ the inverse image of $\mathbf{T}_t$ under the natural homomorphism $\pi : \mathbf{G}^{\mathrm{sc}} \to \mathbf{G}^{\mathrm{der}} \hookrightarrow \mathbf{G}$. Note that $\mathbf{G} = \pi(\mathbf{G}^{\mathrm{sc}})Z(\mathbf{G})$.

If $t_1 = g^{-1}t\theta(g) \in G$ is stably $\theta$-conjugate to $t \in G$ then $g = \pi(g_1)z$ for some $g_1 \in \mathbf{G}^{\mathrm{sc}}$ and $z \in Z(\mathbf{G})$. Then $\sigma(g_1)g_1^{-1}$ lies in $\mathbf{T}_t^{\mathrm{sc}}$, and

$$(1-\theta_t)\pi(\sigma(g_1)g_1^{-1}) = \sigma(b)b^{-1} \quad \text{where} \quad b = \theta(z)z^{-1} = (1-\theta_t)(z^{-1}) \in \mathbf{V}_t;$$

$(\sigma \mapsto \sigma(g_1)g_1^{-1}, b)$ defines an element $\mathrm{inv}(t, t_1)$ of $H^1(F, \mathbf{T}_t^{\mathrm{sc}} \xrightarrow{(1-\theta_t)\pi} \mathbf{V}_t)$. This element parametrizes the $\theta$-conjugacy classes under $\mathbf{G}^{\mathrm{sc}}$ within the stable $\theta$-conjugacy class of $t$. The image in $H^1(F, \mathbf{T}_t \xrightarrow{1-\theta_t} \mathbf{V}_t)$ under the map $[\mathbf{T}_t^{\mathrm{sc}} \to \mathbf{V}_t] \to [\mathbf{T}_t \to \mathbf{V}_t]$ induced by $\pi : \mathbf{T}_t^{\mathrm{sc}} \to \mathbf{T}_t$ is denoted by $\mathrm{inv}'(t, t_1)$. The set of $\theta$-conjugacy classes within the stable $\theta$-conjugacy class of $t$, $D(F, \theta, t)$

$$= \ker[H^1(F, \mathbf{T}_t^{\theta_t}) \to H^1(F, \mathbf{G})] = \ker[H^1(F, \mathbf{T}_t \xrightarrow{1-\theta_t} \mathbf{V}_t) \to H^1(F, \mathbf{G})],$$

is the image under

$$H^1(F, \mathbf{T}_t^{\mathrm{sc}} \xrightarrow{(1-\theta_t)\pi} \mathbf{V}_t) \to H^1(F, \mathbf{T}_t \xrightarrow{1-\theta_t} \mathbf{V}_t)$$

of

$$\ker[H^1(F, \mathbf{T}_t^{\mathrm{sc}} \xrightarrow{(1-\theta_t)\pi} \mathbf{V}_t) \to H^1(F, \mathbf{G}^{\mathrm{sc}})],$$

hence a subset of the abelian group

$$\mathrm{Im}[H^1(F, \mathbf{T}_t^{\mathrm{sc}} \xrightarrow{(1-\theta_t)\pi} \mathbf{V}_t) \to H^1(F, \mathbf{T}_t \xrightarrow{1-\theta_t} \mathbf{V}_t)].$$

In our case of $\mathbf{G} = \mathrm{PGL}(4)$, the pointed set $H^1(F, \mathbf{G})$ is trivial, hence $D(F, \theta, t) = H^1(F, \mathbf{T}_t^{\theta_t}) = H^1(F, \mathbf{T}_t \xrightarrow{1-\theta_t} \mathbf{V}_t)$. Since $H^1(F, \mathbf{T})$ is trivial for every maximal torus $\mathbf{T}$, we have that $H^1(F, \mathbf{T}_t \xrightarrow{1-\theta} \mathbf{V}_t)$ is $V_t/(1-\theta_t)T_t$.

We list the stable $\theta$-conjugacy classes of strongly $\theta$-regular elements $t$ in $\mathbf{G} = \mathrm{PGL}(4)$ as in [F5]. Thus we describe the F-tori $\mathbf{T}$, as $Z_{\mathbf{G}}(t\theta) = \mathbf{T}_t^{\theta_t}$ and $\mathbf{T} = \mathbf{T}_t = Z_{\mathbf{G}}(Z_{\mathbf{G}}(t\theta))$. The conjugacy classes of $F$-tori $\mathbf{T}$ are determined by the homomorphisms $\rho : \Gamma \to W = W(\mathbf{T}^{*\theta}, \mathbf{G}^\theta) = W(\mathbf{T}^*, \mathbf{G})^\theta$. We list only the $\theta$-elliptic, or $\theta$-anisotropic ($T^\theta$ does not contain a split torus) as the other tori can be dealt with using parabolic induction.

(I) $\rho(\Gamma) = \langle(14)(23)\rangle$, $[E : F] = 2$,

$$T^* = \{(a, b, \sigma b, \sigma a); a, b \in E^\times\}/Z; \qquad \mathbf{V} = \{(a, b, b, a)\}/\mathbf{Z}.$$

Hence $V = \{(a,b,b,a) = (z\sigma a, z\sigma b, z\sigma b, z\sigma a);\ z,a,b \in E^\times\}$. Then $a/\sigma a = b/\sigma b$, or $a/b = \sigma(a/b)$, and $(a,b,b,a) \equiv (1, b/a, b/a, 1)$ with $b/a$ in $F^\times$. Finally $(1-\theta)T^* = \{(a\sigma a, b\sigma b, b\sigma b, a\sigma a); a,b \in E^\times\}/Z$,

$$V/(1-\theta)T^* = F^\times/NE^\times = \mathbb{Z}/2.$$

(II) $\rho(\Gamma) = \langle \rho(\sigma\tau) = (14), \rho(\tau) = (23)\rangle$. The splitting field of $T$ is $E = E_1E_2$, where $E_1 = E(\sqrt{D}) = E^{\langle\tau\rangle}$,

$$E_2 = E(\sqrt{AD}) = E^{\langle\sigma\tau\rangle}, \quad E_3 = E(\sqrt{A}) = E^{\langle\sigma\rangle}$$

are the quadratic extensions of $F$ in $E$. Then

$$T^* = \{(a,b,\tau b,\sigma a); a \in E_1^\times, b \in E_2^\times\}/Z, \quad V = \{(a,b,b,a); a,b \in F^\times\}/\mathbf{Z}$$

(since $(a,b,b,a) \equiv (\tau a, \tau b, \tau b, \tau a) \equiv (\sigma a, \sigma b, \sigma b, \sigma a) \bmod Z$ implies $a/b \in F^\times$),

$$(1-\theta)T^* = \{(a\sigma a, b\tau b, b\tau b, a\sigma a); a \in E_1^\times, b \in E_2^\times\}/Z,$$

hence

$$V/(1-\theta)T^* = F^\times/N_{E_1/F}E_1^\times = F^\times/N_{E_2/F}E_2^\times = \mathbb{Z}/2.$$

(III) $\rho(\Gamma) = \langle\rho(\tau) = (12)(34), \rho(\sigma) = (14)(23)\rangle$, the splitting field of $T$ is $E = E_1E_2$, a biquadratic extension of $F$, $\mathrm{Gal}(E/F) = \langle 1, \sigma, \tau, \sigma\tau\rangle$, $E_1 = E^{\langle\tau\rangle} = F(\sqrt{D})$, $E_3 = E^{\langle\sigma\rangle} = F(\sqrt{A})$, $E_2 = E^{\langle\sigma\tau\rangle} = F(\sqrt{AD})$ are the quadratic subextensions, and so $T^* = \{(a, \tau a, \tau\sigma a, \sigma a); a \in E^\times\}/Z$, and

$$(1-\theta)T^* = \{(a\sigma a, \tau(a\sigma a), \tau(a\sigma a), a\sigma a); a \in E^\times\}/Z.$$

Now $V$ consists of $(a,b,b,a)$ which equal $(\sigma a, \sigma b, \sigma b, \sigma a) \bmod Z$. Thus $a/b = \sigma(a/b)$ lies in $E^\sigma = E_3$, and also $(a,b,b,a) = (\tau b, \tau a, \tau a, \tau b) \bmod Z$. Hence $b/a = \tau(a/b)$, and so $\frac{a}{b} = u/\tau u$, $u \in E_3^\times$. Then

$$(a,b,b,a) = (bu/\tau u, b, b, bu/\tau u) = (u, \tau u, \tau u, u).$$

Hence $V/(1-\theta)T^*$ is $E_3^\times/N_{E/E_3}E^\times = \mathbb{Z}/2$.

(IV) If $\rho(\Gamma)$ contains $\rho(\sigma) = (3421)$, $T$ is isomorphic to the multiplication group $E^\times$ of an extension $E = F(\sqrt{D}) = E_3(\sqrt{D})$ of $F$ of degree

4, where $E_3 = F(\sqrt{A})$ is a quadratic extension of $F$ ($A \in F - F^2, D = \alpha + \beta\sqrt{A} \in E_3$). The Galois closure $\tilde{E}/F$ of $F(\sqrt{D})/F$ is $E = F(\sqrt{D})$ when $F(\sqrt{D})/F$ is cyclic, and $\tilde{E} = F(\sqrt{D}, \zeta)$ when $F(\sqrt{D})/F$ is not Galois. Here $\zeta^2 = -1$ and $\mathrm{Gal}(\tilde{E}/F)$ is the dihedral group $D_4$. In either case

$$T^* = \{(a, \sigma a, \sigma^3 a, \sigma^2 a); \quad a \in E^\times\}/Z,$$

$$(1-\theta)T^* = \{(a\sigma^2 a, \sigma(a\sigma^2 a), \sigma(a\sigma^2 a), a\sigma^2 a); a \in E^\times\}/Z,$$

and $V$ consists of $(a,b,b,a)$ with $(\sigma b, \sigma a, \sigma a, \sigma b) = (a,b,b,a)z$, thus $a/\sigma b = b/\sigma a$, or $a/b = \sigma b/\sigma a$ so $a/b = \sigma^2(a/b)$ lies in $E_3$, and $a/b = u/\sigma u$ for some $u \in E_3^\times$. Thus $(a,b,b,a) = (bu/\sigma u, b, b, bu/\sigma u) \equiv (u, \sigma u, \sigma u, u), u \in E_3^\times$, and $V/(1-\theta)T^*$ is $E_3^\times/N_{E/E_3}E^\times = \mathbb{Z}/2$.

We recall some results of [F5] concerning representatives of (stable) $\theta$-twisted regular conjugacy classes. These are listed according to the four types of $\theta$-elliptic classes: I, II, III, IV.

A set of representatives for the $\theta$-conjugacy classes within a stable semi simple $\theta$-conjugacy class of type I in $\mathrm{GL}(4,F)$ which splits over a quadratic extension $E = F(\sqrt{D})$ of $F$, $D \in F - F^2$, is parametrized by $(\mathbf{r}, \mathbf{s}) \in F^\times/N_{E/F}E^\times \times F^\times/N_{E/F}E^\times$ ([F5], p. 16). Representatives for the $\theta$-regular (thus $t\theta(t)$ is regular) stable $\theta$-conjugacy classes of type (I) in $\mathrm{GL}(4,F)$ which split over $E$ can be found in a torus $T = \mathbf{T}(F)$, $\mathbf{T} = h^{-1}\mathbf{T}^*h$, $\mathbf{T}^*$ denoting the diagonal subgroup in $\mathbf{G}$, $h = \theta(h)$, and

$$T = \left\{ t = \begin{pmatrix} a_1 & 0 & 0 & a_2 D \\ 0 & b_1 & b_2 D & 0 \\ 0 & b_2 & b_1 & 0 \\ a_2 & 0 & 0 & a_1 \end{pmatrix} = h^{-1}t^*h; \quad t^* = \mathrm{diag}(a, b, \sigma b, \sigma a) \in T^* \right\}.$$

Here $a = a_1 + a_2\sqrt{D}$, $b = b_1 + b_2\sqrt{D} \in E^\times$, and $t$ is regular if $a/\sigma a$ and $b/\sigma b$ are distinct and not equal to $\pm 1$. Note that here $T^* = \mathbf{T}^*(F)$ where the Galois action is that obtained from the Galois action on $T$.

A set of representatives for the $\theta$-conjugacy classes within a stable $\theta$-conjugacy class can be chosen in $T$. Indeed, if $t = h^{-1}t^*h$ and $t_1 = h^{-1}t_1^*h$ in $T$ are stably $\theta$-conjugate, then there is $g = h^{-1}\mu h$ with $t_1 = gt\theta(g)^{-1}$, thus $t_1^* = \mu t^*\theta(\mu)^{-1}$ and $t_1^*\theta(t_1^*) = \mu t^*\theta(t^*)\mu^{-1}$. Since $t$ is $\theta$-regular, $\mu$ lies in the $\theta$-normalizer of $\mathbf{T}^*(\overline{F})$ in $\mathbf{G}(\overline{F})$. Since the group $W^\theta(\mathbf{T}^*, \mathbf{G}) = N^\theta(\mathbf{T}^*, \mathbf{G})/\mathbf{T}^*$, quotient by $\mathbf{T}^*(\overline{F})$ of the $\theta$-normalizer of

$\mathbf{T}^*(\overline{F})$ in $\mathbf{G}(\overline{F})$, is represented by the group $W^\theta(T^*,G) = N^\theta(T^*,G)/T^*$, quotient by $T^*$ of the $\theta$-normalizer of $T^*$ in $G$, we may modify $\mu$ by an element of $W^\theta(T^*,G)$, that is replace $t_1$ by a $\theta$-conjugate element, and assume that $\mu$ lies in $\mathbf{T}^*(\overline{F})$. In this case $\mu\theta(\mu)^{-1} = \mathrm{diag}(u,u',\sigma u',\sigma u)$ (since $t$, $t_1$ lie in $T^*$), with $u = \sigma u$, $u' = \sigma u'$ in $F^\times$. Such $t$, $t_1$ are $\theta$-conjugate if $g \in G$, thus $g \in T$, so $\mu = \mathrm{diag}(v,v',\sigma v',\sigma v) \in T^*$ and $\mu\theta(\mu)^{-1} = \mathrm{diag}(v\sigma v, v'\sigma v', v'\sigma v', v\sigma v)$. Hence a set of representatives for the $\theta$-conjugacy classes within the stable $\theta$-conjugacy class of the $\theta$-regular $t$ in $T$ is given by $t\cdot\mathrm{diag}(\mathbf{r},\mathbf{s},\mathbf{s},\mathbf{r})$, where $\mathbf{r}$, $\mathbf{s} \in F^\times/N_{E/F}E^\times$. Clearly in $\mathrm{PGL}(4,F)$ the $\theta$-classes within a stable class are parametrized only by $\mathbf{r}$, or equivalently only by $\mathbf{s}$.

A set of representatives for the $\theta$-conjugacy classes within a stable semi simple $\theta$-conjugacy class of type II in $\mathrm{GL}(4,F)$ which splits over the biquadratic extension $E = E_1E_2$ of $F$ with Galois group $\langle\sigma,\tau\rangle$, where $E_1 = F(\sqrt{D}) = E^\tau$, $E_2 = F(\sqrt{AD}) = E^{\sigma\tau}$, $E_3 = F(\sqrt{A}) = E^\sigma$ are quadratic extensions of $F$, thus $A, D \in F - F^2$, is parametrized by $\mathbf{r} \in F^\times/N_{E_1/F}E_1^\times$, $\mathbf{s} \in F^\times/N_{E_2/F}E_2^\times$ ([F5], p. 16). It is given by

$$\begin{pmatrix} a_1\mathbf{r} & 0 & 0 & a_2D\mathbf{r} \\ 0 & b_1\mathbf{s} & b_2AD\mathbf{s} & 0 \\ 0 & b_2\mathbf{s} & b_1\mathbf{s} & 0 \\ a_2\mathbf{r} & 0 & 0 & a_1\mathbf{r} \end{pmatrix} = h^{-1}t^*h\cdot\mathrm{diag}(\mathbf{r},\mathbf{s},\mathbf{s},\mathbf{r}), \qquad t^* = \mathrm{diag}(a,b,\tau b,\sigma a).$$

Here $a = a_1 + a_2\sqrt{D} \in E_1^\times$, $b = b_1 + b_2\sqrt{AD} \in E_2^\times$, $\theta(h) = h$. In $\mathrm{PGL}(4,F)$ the $\theta$-classes within a stable class are parametrized only by $\mathbf{r}$, or equivalently only by $\mathbf{s}$.

A set of representatives for the $\theta$-conjugacy classes within a stable semi simple $\theta$-conjugacy class of type III in $\mathrm{GL}(4,F)$ which splits over the biquadratic extension $E = E_1E_2$ of $F$ with Galois group $\langle\sigma,\tau\rangle$, where $E_1 = F(\sqrt{D}) = E^\tau$, $E_2 = F(\sqrt{AD}) = E^{\sigma\tau}$, $E_3 = F(\sqrt{A}) = E^\sigma$ are quadratic extensions of $F$, thus $A, D \in F - F^2$, is parametrized by $\boldsymbol{r}(= \boldsymbol{r}_1 + \boldsymbol{r}_2\sqrt{A}) \in E_3^\times/N_{E/E_3}E^\times$ ([F5], p. 16). Representatives for the stable regular $\theta$-conjugacy classes can be taken in the torus $T = h^{-1}T^*h$, consisting of

$$t = \begin{pmatrix} \boldsymbol{a} & \boldsymbol{b}D \\ \boldsymbol{b} & \boldsymbol{a} \end{pmatrix} = h^{-1}t^*h, \qquad t^* = \mathrm{diag}(\alpha,\tau\alpha,\sigma\tau\alpha,\sigma\alpha),$$

where $h = \theta(h)$ is described in [F5], p. 16. This $t$ is $\theta$-regular when $\alpha/\sigma\alpha$, $\tau(\alpha/\sigma\alpha)$ are distinct and $\neq \pm 1$. Here

$$\boldsymbol{a} = \begin{pmatrix} a_1 & a_2A \\ a_2 & a_1 \end{pmatrix}, \qquad \boldsymbol{b} = \begin{pmatrix} b_1 & b_2A \\ b_2 & b_1 \end{pmatrix}; \qquad \text{put also} \qquad \boldsymbol{r} = \begin{pmatrix} r_1 & r_2A \\ r_2 & r_1 \end{pmatrix}.$$

Further $\alpha = a + b\sqrt{D} \in E^\times$, $a = a_1 + a_2\sqrt{A} \in E_3^\times$, $b = b_1 + b_2\sqrt{A} \in E_3^\times$, $\sigma\alpha = a - b\sqrt{D}$, $\tau\alpha = \tau a + \tau b\sqrt{D}$. Representatives for all $\theta$-conjugacy classes within the stable $\theta$-conjugacy class of $t$ can be taken in $T$. In fact if $t' = gt\theta(g)^{-1}$ lies in $T$ and $g = h^{-1}\mu h$, $\mu \in \mathbf{T}^*(\overline{F})$, then $\mu\theta(\mu)^{-1} = \operatorname{diag}(u, \tau u, \sigma\tau u, \sigma u)$ has $u = \sigma u$, thus $u \in E_3^\times$. If $g \in T$, thus $\mu \in T^*$, then

$$\mu = \operatorname{diag}(v, \tau v, \sigma\tau v, \sigma v) \quad \text{and} \quad \mu\theta(\mu)^{-1} = \operatorname{diag}(v\sigma v, \tau v\sigma\tau v, \tau v\sigma\tau v, v\sigma v),$$

with $v\sigma v \in N_{E/E_3}E^\times$. We conclude that a set of representatives for the $\theta$-conjugacy classes within the stable $\theta$-conjugacy class of $t$ is given by $t \cdot \operatorname{diag}(\boldsymbol{r}, \boldsymbol{r})$, as $r$ ranges over $E_3^\times / N_{E/E_3}E^\times$.

Representatives for the stable regular $\theta$-conjugacy classes of type (IV) can be taken in the torus $T = h^{-1}T^*h$, consisting of

$$t = \begin{pmatrix} \boldsymbol{a} & \boldsymbol{bD} \\ \boldsymbol{b} & \boldsymbol{a} \end{pmatrix} = h^{-1}t^*h, \qquad t^* = \operatorname{diag}(\alpha, \sigma\alpha, \sigma^3\alpha, \sigma^2\alpha),$$

where $h = \theta(h)$ is described in [F5], p. 18. Here $\alpha$ ranges over a quadratic extension $E = F(\sqrt{D}) = E_3(\sqrt{D})$ of a quadratic extension $E_3 = F(\sqrt{A})$ of $F$. Thus $A \in F - F^2$, $D = d_1 + d_2\sqrt{A}$ lies in $E_3 - E_3^2$ where $d_i \in F$. The normal closure $E'$ of $E$ over $F$ is $E$ if $E/F$ is cyclic with Galois group $\mathbb{Z}/4$, or a quadratic extension of $E$, generated by a fourth root of unity $\zeta$, in which case the Galois group is the dihedral group $D_4$. In both cases the Galois group contains an element $\sigma$ with $\sigma\sqrt{A} = -\sqrt{A}$, $\sigma\sqrt{D} = \sqrt{\sigma D}$, $\sigma^2\sqrt{D} = -\sqrt{D}$. In the $D_4$ case $\operatorname{Gal}(E'/F)$ contains also $\tau$ with $\tau\zeta = -\zeta$, we may choose $D = \sqrt{A}$, $\tau D = D$ and $\sigma\sqrt{D} = \zeta\sqrt{D}$.

In any case, $t$ is $\theta$-regular when $\alpha \neq \sigma^2\alpha$. We write $\alpha = a + b\sqrt{D} \in E^\times$, $a = a_1 + a_2\sqrt{A} \in E_3^\times$, $b = b_1 + b_2\sqrt{A} \in E_3^\times$, $\sigma\alpha = \sigma a + \sigma b\sqrt{\sigma D}$, $\sigma^2\alpha = a - b\sqrt{D}$. Also

$$\boldsymbol{a} = \begin{pmatrix} a_1 & a_2A \\ a_2 & a_1 \end{pmatrix}, \qquad \boldsymbol{b} = \begin{pmatrix} b_1 & b_2A \\ b_2 & b_1 \end{pmatrix}, \qquad \boldsymbol{D} = \begin{pmatrix} d_1 & d_2A \\ d_2 & d_1 \end{pmatrix}.$$

Representatives for all $\theta$-conjugacy classes within the stable $\theta$-conjugacy class of $t$ can be taken in $T$. In fact if $t' = gt\theta(g)^{-1}$ lies in $T$ and $g = h^{-1}\mu h$, $\mu \in \mathbf{T}^*(\overline{F})$, then $\mu\theta(\mu)^{-1} = \operatorname{diag}(u, \sigma u, \sigma^3 u, \sigma^2 u)$ has $u = \sigma^2 u$, thus $u \in E_3^\times$. If $g \in T$, thus $\mu \in T^*$, then $\mu = \operatorname{diag}(v, \sigma v, \sigma^3 v, \sigma^2 v)$ and

$$\mu\theta(\mu)^{-1} = \operatorname{diag}(v\sigma^2 v, \sigma(v\sigma^2 v), \sigma(v\sigma^2 v), v\sigma^2 v),$$

with $v\sigma v \in N_{E/E_3}E^\times$. It follows that a set of representatives for the $\theta$-conjugacy classes within the stable $\theta$-conjugacy class of

$$t = h^{-1}t^*h = \begin{pmatrix} \boldsymbol{a} & \boldsymbol{b}D \\ \boldsymbol{b} & \boldsymbol{a} \end{pmatrix}, \qquad \text{where} \qquad t^* = \mathrm{diag}(\alpha, \sigma\alpha, \sigma^3\alpha, \sigma^2\alpha),$$

is given by multiplying $\alpha$ by $r$, that is $t^*$ by $t_0^* = \mathrm{diag}(r, \sigma r, \sigma^3 r, \sigma^2 r)$, where $r = \sigma^2 r$ ranges over a set of representatives for $E_3^\times / N_{E/E_3}E^\times$. Now $t_0 = h^{-1}t_0^*h = \begin{pmatrix} \boldsymbol{r} & 0 \\ 0 & \boldsymbol{r} \end{pmatrix}$. Hence a set of representatives is given by $t \cdot \mathrm{diag}(\boldsymbol{r}, \boldsymbol{r})$, $\boldsymbol{r} \in E_3^\times / N_{E/E_3}E^\times$.

# III. TRACE FORMULAE

## 1. Twisted Trace Formula: Geometric Side

The comparison of representations is based on comparison of trace formulae, which are equalities of geometric and spectral sides. In this section we first state the spectral side of the $\theta$-twisted trace formula on the discrete spectrum, and then we record the geometric side of the $\theta$-twisted trace formula for $\mathbf{G}$, in fact only its $\theta$-elliptic strongly $\theta$-regular part, stabilized according to its $\theta$-elliptic endoscopic groups, as in [KS]. This geometric side is a linear form, with complex values, on the space $C_c^\infty(\mathbf{G}(\mathbb{A}))$ of smooth compactly supported complex valued functions on $\mathbf{G}(\mathbb{A})$. This space is spanned by products $\otimes f_v$, where $f_v \in C_c^\infty(\mathbf{G}(F_v))$ for all $v$ and $f_v = f_v^0$ is the characteristic function of $K_v = \mathbf{G}(R_v)$ for almost all $v$. In fact we need the measure $f dg$ where $dg = \otimes dg_v$ is a Haar measure on $\mathbf{G}(\mathbb{A})$, but we suppress the $dg$ from the notations. We later compare trace formulae for test measures $f dg$, $f_H dh$, $f_{C_0} dc_0$, etc., with matching orbital integrals. The dependence on measures is implicit.

The trace formula is obtained on integrating over the diagonal $g = h$ in $\mathbf{G}(F)\backslash\mathbf{G}(\mathbb{A})$ the kernel $K_f(h,g)$ of the convolution operator $r(f)r(\theta)$ on $L^2 = L^2(\mathbf{G}(F)\backslash\mathbf{G}(\mathbb{A}))$, defined by

$$(r(f)\phi)(h) = \int_{\mathbf{G}(\mathbb{A})} f(g)\phi(hg)dg \quad \text{and} \quad (r(\theta)\phi)(h) = \phi(\theta^{-1}(h))$$

for $\phi \in L^2$. The discrete part $L_d$ of $L^2$ splits as a direct sum $\oplus_\pi L_\pi$ of subspaces transforming according to inequivalent irreducible representations $\pi$ of $\mathbf{G}(\mathbb{A})$. Thus $L_\pi = m(\pi)\pi$ is a multiple of an irreducible $\pi$, occurring with finite multiplicity $m(\pi)$ in $L^2$, and the sum is over inequivalent $\pi$.

If $\{\phi_i^\pi\}$ is an orthonormal basis of $L_\pi$ then the kernel of $r(f)r(\theta)$ on $L_d$ is

$$K_d(k,g) = \sum_\pi \sum_{\phi_i^\pi \in L_\pi} \int_h f(h^{-1}\theta(k))\overline{\phi_i^\pi}(h)dh \cdot \phi_i^\pi(g), \quad h \,\text{in}\, \mathbf{G}(F)\backslash\mathbf{G}(\mathbb{A}).$$

Indeed,

$$r(f)r(\theta)\phi(g) = \sum_{\pi}\sum_{\phi_i^\pi}\langle r(f)r(\theta)\phi, \phi_i^\pi\rangle\phi_i^\pi(g)$$

$$= \sum\sum\int_{h\in\mathbf{G}(F)\backslash\mathbf{G}(\mathbb{A})} r(f)r(\theta)\phi(h)\overline{\phi_i^\pi}(h)dh\cdot\phi_i^\pi(g)$$

$$= \sum\sum\int_h\int_{k\in G(\mathbb{A})} f(k)(r(\theta)\phi)(hk)dk\cdot\overline{\phi_i^\pi}(h)dh\cdot\phi_i^\pi(g)$$

$$= \sum\sum\int_h\int_k f(h^{-1}\theta(k))\overline{\phi_i^\pi}(h)dh\cdot\phi_i^\pi(g)\phi(k)dk.$$

The trace of $r(f\times\theta) = r(f)r(\theta)$ over the discrete spectrum is the integral of $K_d$ over the diagonal $k = g$ in $\mathbf{G}(\mathbb{A})$:

$$\sum_{\pi}\sum_{\phi_i^\pi}\int_{g\in\mathbf{G}(\mathbb{A})}\int_{h\in\mathbf{G}(F)\backslash\mathbf{G}(\mathbb{A})}\overline{\phi_i^\pi}(h)f(h^{-1}\theta(g))\phi_i^\pi(g)dhdg$$

$$= \sum\sum\int_h\int_g\overline{\phi_i^\pi}(h)f(g)\phi_i^\pi(\theta^{-1}(hg))dgdh$$

$$= \sum\sum\int_h[r(f)(r(\theta)\phi_i^\pi)](h)\overline{\phi_i^\pi}(h)dh$$

$$= \sum_{\pi}\sum_{\phi_i^\pi}\langle\pi(f)\pi(\theta)\phi_i^\pi,\phi_i^\pi\rangle = \sum_{\pi}m(\pi)\operatorname{tr}\pi(f\times\theta),$$

where $\pi(f)$ and $\pi(\theta)$ denote the restriction of $r(f)$ and $r(\theta)$ to $\pi$, and $\pi(f\times\theta) = \pi(f)\pi(\theta)$. One can see that the sum $\sum_\pi m(\pi)|\operatorname{tr}\pi(f\times\theta)|$ is convergent.

The $\pi$ in $L_d$ which contribute a nonzero term to this sum are those which are $\theta$-invariant: ${}^\theta\pi\simeq\pi$. The contribution to the trace formula from the complement of $L_d$ in $L^2$ is described using Eisenstein series; we describe this spectral side below. This side will be used to study the representations $\pi$ whose traces occur in the sum.

We now turn to the geometric side of the trace formula.

The geometric side of the trace formula is obtained on integrating over the diagonal $g = h\in\mathbf{G}(F)\backslash\mathbf{G}(\mathbb{A})$ the kernel of the convolution operator $r(f)r(\theta)$ on $L^2$: here $(r(f)r(\theta)\phi)(h)$ is

$$= \int_{\mathbf{G}(\mathbb{A})} f(h^{-1}\theta(g))\phi(g)dg = \int_{\mathbf{G}(F)\backslash\mathbf{G}(\mathbb{A})}\sum_{\gamma\in\mathbf{G}(F)} f(h^{-1}\gamma\theta(g))\phi(g)dg,$$

and we consider only the subsum

$$K_e(h,g) = \sum_{\delta \in \mathbf{G}(F)_e} f(h^{-1}\delta\theta(g))$$

over the set $\mathbf{G}(F)_e$ of $\theta$-semisimple, strongly $\theta$-regular and $\theta$-elliptic elements $\delta$ in $\mathbf{G}(F)$.

An element $\delta$ of $\mathbf{G}(F)$ is called *$\theta$-semisimple* if the automorphism $\mathrm{Int}(\delta)\circ\theta = \mathrm{Int}(\delta\theta)$ is quasi-semisimple, by which we mean that its restriction to the derived group is semisimple (thus there is a pair $(\mathbf{B},\mathbf{T})$ in $\mathbf{G}$ fixed by the automorphism). As for $\theta$-regularity, denote by $I_\delta = Z_{\mathbf{G}}(\delta\theta)$ the centralizer of $\delta\theta$ in $\mathbf{G}$ (this is the group $\{g \in G; \delta\theta(g)\delta^{-1} = g\}$ of fixed points of $\mathrm{Int}(\delta)\cdot\theta$). A $\theta$-semisimple $\delta$ in $\mathbf{G}$ is called *$\theta$-regular* if $Z_{\mathbf{G}}(\delta\theta)^0$ is a torus, and *strongly $\theta$-regular* if $Z_{\mathbf{G}}(\delta\theta)$ is abelian. If $\delta$ is strongly $\theta$-regular then $\mathbf{T}_\delta = Z_{\mathbf{G}}(Z_{\mathbf{G}}(\delta\theta)^0))$ (centralizer in $\mathbf{G}$ of $Z_{\mathbf{G}}(\delta\theta)^0$) is a maximal torus in $\mathbf{G}$ fixed under $\mathrm{Int}(\delta\theta)$, and $Z_{\mathbf{G}}(\delta\theta) = \mathbf{T}_\delta^{\mathrm{Int}(\delta\theta)}$. A $\theta$-semisimple element $\delta$ of $\mathbf{G}(F)$ is called *$\theta$-elliptic* if $(Z_{\mathbf{G}}(\delta\theta)/Z(\mathbf{G})^\theta)^0$ is anisotropic over $F$.

The integral $\mathrm{T}_e(f,\mathbf{G},\theta)$ over $h = g$ in $\mathbf{G}(F)\backslash\mathbf{G}(\mathbb{A})$ of $K_e(g,g)$ is the sum over a set of representatives $\delta$ for the $\theta$-conjugacy classes in $\mathbf{G}(F)_e$ of orbital integrals:

$$\begin{aligned}&\sum_\delta \int_{Z_{\mathbf{G}}(\delta\theta)(F)\backslash\mathbf{G}(\mathbb{A})} f(g^{-1}\delta\theta(g))dg\\ &= \sum_\delta \mathrm{vol}_{dt}(Z_{\mathbf{G}}(\delta\theta)(F)\backslash Z_{\mathbf{G}}(\delta\theta)(\mathbb{A})) \int_{Z_{\mathbf{G}}(\delta\theta)(\mathbb{A})\backslash\mathbf{G}(\mathbb{A})} f(g^{-1}\delta\theta(g))dg/dt.\end{aligned}$$

It is rewritten in [KS], (7.4.2) as a sum over a set of representatives $(\mathbf{H},\mathcal{H},s,\xi)$ for the isomorphism classes of elliptic endoscopic data for $(\mathbf{G},\theta)$ ([KS], (2.1)) and over a set of representatives for the $\mathbf{H}(\overline{F})$-conjugacy classes of elliptic strongly $\mathbf{G}$-regular $\gamma$ in $\mathbf{H}(F)$ ($\gamma \in \mathbf{H}$ is called *strongly G-regular* if the image under the norm map $A_{\mathbf{H}/\mathbf{G}}$ ([KS], (3.3)) of the conjugacy class of $\gamma$ consists of (strongly) $\theta$-regular elements):

$$\sum_{(\mathbf{H},\mathcal{H},s,\xi)} a_{\mathbf{G}} \cdot |\mathrm{Out}(\mathbf{H},\mathcal{H},s,\xi)|^{-1} \sum_\gamma \Phi_\gamma^\kappa(f).$$

Here $\mathrm{Out}(\mathbf{H},\mathcal{H},s,\xi)$ is the group defined in [KS], (2.1.8); $a_{\mathbf{G}}$ is the number defined in [KS], (6.4.B); and the twisted $\kappa$-orbital integral $\Phi_\gamma^\kappa(f)$ is defined in [KS], 3 lines above (6.4.10) and 3 lines above (6.4.16).

If $f_H = \otimes f_{vH}$, $f_{vH} \in C_c^\infty(\mathbf{H}(F_v))$ has matching orbital integrals with $f_v$ for all $v$ ([KS], (5.5)), then $\Phi_\gamma^\kappa(f)$ can be replaced by the stable orbital integral $\Phi_\gamma^{\mathrm{st}}(f_H)$, and the stabilized trace formula takes the form ([KS], (7.4.4))

$$\sum_{(\mathbf{H},\mathcal{H},s,\xi)} \iota(\mathbf{G},\theta,\mathbf{H})\,\mathrm{ST}_e(f_H),$$

where

$$\mathrm{ST}_e(f_H) = a_H \sum_\gamma \Phi_\gamma^{\mathrm{st}}(f_H), \qquad a_H = |\pi_0(Z(\widehat{H})^\Gamma)||\ker^1(F, Z(\widehat{H}))|^{-1},$$

and

$$\iota(\mathbf{G},\theta,\mathbf{H}) = a_{\mathbf{G}}|\mathrm{Out}(\mathbf{H},\mathcal{H},s,\xi)|^{-1}a_{\mathbf{H}}^{-1},$$

$$a_{\mathbf{G}} = \frac{|\pi_0(Z(\widehat{G})^\Gamma)|}{|\ker^1(F, Z(\widehat{G}))|}\cdot\frac{|\pi_0((Z(\widehat{G})^\Gamma)^0 \cap (\widehat{T}^{\hat\theta})^0)|}{|\pi_0((Z(\widehat{G})/Z(\widehat{G}) \cap (\widehat{T}^{\hat\theta})^0)^\Gamma)|}.$$

In our case $\mathbf{G} = \mathrm{PGL}(4)$, $\theta(g) = J^{-1}{}^tg^{-1}J$, there are two elliptic $\theta$-endoscopic groups $\mathbf{H} = \mathrm{PGSp}(2)$ and $\mathbf{C} = [\mathrm{GL}(2)\times\mathrm{GL}(2)]'/\mathbb{G}_m'$, with $\widehat{H} = \mathrm{Sp}(2,\mathbb{C})$ and

$$\widehat{C} = \{(A,B) \in [\mathrm{GL}(2,\mathbb{C})\times\mathrm{GL}(2,\mathbb{C})]/\mathbb{C}^\times;\ \det A \det B = 1\},$$

$Z(\widehat{G}) = \mu_4$, $Z(\widehat{H}) = Z(\widehat{C}) = \mu_2 = \{\pm I\}$, the Galois group $\Gamma = \mathrm{Gal}(\overline{F}/F)$ acts trivially as $\mathbf{G}$, $\mathbf{H}$ and $\mathbf{C}$ are split, $Z_1 = Z(\widehat{G}) \cap (\widehat{T}^\theta)^0$ of [KS], Lemma 6.4.B, is $\{\pm I\}$, hence $\overline{Z}$ there $(= Z(\widehat{G})/Z_1)$ is $\mu_2$, $Z_1 \cap (Z(\widehat{G})^\Gamma)^0$ is trivial, $\ker^1(F, Z(\widehat{G})) = 1$, hence $a_{\mathbf{G}} = 2$. For $\mathbf{H}$ and $\mathbf{C}$ there is no $\theta$, $Z_1 = Z(\widehat{H})$ and $\overline{Z} = 1$, $\ker^1(F,\mu_2) = 1$, hence $a_{\mathbf{H}} = 2 = a_{\mathbf{C}}$. In particular $\iota(\mathbf{G},\theta,\mathbf{H}) = |\mathrm{Out}(\mathbf{H},\mathcal{H},s,\xi)|^{-1}$ is 1 and $\iota(\mathbf{G},\theta,\mathbf{C}) = \frac{1}{2}$.

Similarly we consider the elliptic regular part of the geometric side of the trace formula of $\mathbf{H} = \mathrm{PGSp}(2)$ and stabilize it, to obtain (here $\theta$ is trivial and is omitted from the notations)

$$\mathrm{T}_e(f_H,\mathbf{H}) = \mathrm{ST}_e(f_H) + \iota(\mathbf{H},1,\mathbf{C}_0)\,\mathrm{ST}_e(f_{C_0})$$

where $f_{C_0}$ is a function on $\mathbf{C}_0(\mathbb{A})$ matching $f_H$. Here $\mathbf{C}_0 = \mathrm{PGL}(2)\times\mathrm{PGL}(2)$ is the only elliptic endoscopic group of $\mathbf{H}$ other than $\mathbf{H}$, $\widehat{C}_0 =$

$\mathrm{SL}(2,\mathbb{C}) \times \mathrm{SL}(2,\mathbb{C})$ has center $\{\pm I\} \times \{\pm I\}$ of order 4, hence $a_{\mathbf{C}_0} = 4$. Also $\mathrm{Out}(\mathbf{C}_0, \dots)$ has order 2 and $a_{\mathbf{H}} = 2$, hence $\iota(\mathbf{H}, 1, \mathbf{C}_0) = \frac{1}{4}$.

The $\theta$-endoscopic group $\mathbf{C}$ of $\mathbf{G}$ has a proper endoscopic subgroup $\mathbf{C}_E$ for each quadratic extension $E$ of $F$. Its connected dual group $\widehat{C}_E = Z_{\widehat{C}}(\hat{s}_E)^0$, $\hat{s}_E = (\mathrm{diag}(1,-1), \mathrm{diag}(1,-1))$, is

$$\{(\mathrm{diag}(a_1,a_2), \mathrm{diag}(b_1,b_2)) \bmod \mathbb{C}^\times;\ a_1a_2b_1b_2 = 1\},$$

and $\mathrm{Gal}(E/F)$ acts via $\left(\begin{pmatrix} 0 & 1 \\ -1 & 0 \end{pmatrix}, \begin{pmatrix} 0 & 1 \\ -1 & 0 \end{pmatrix}\right)$. Thus

$$\mathbf{C}_E = \{(z_1,z_2) \in (\mathrm{R}_{E/F}\mathbb{G}_m \times \mathrm{R}_{E/F}\mathbb{G}_m)/\mathbb{G}_m; z_1\overline{z}_1 = z_2\overline{z}_2\}$$

and $\mathbf{C}_E(F) = \{(z_1,z_2) \in (E^\times \times E^\times)/F^\times; z_1\overline{z}_1 = z_2\overline{z}_2\}$, and $\widehat{C}_E^{\mathrm{Gal}(E/F)}$ is $\mathbb{Z}/2$, generated by (−I,I). Since $\mathrm{Out}(\mathbf{C}_E, \dots)$ has order 2, $a_{\mathbf{C}_E} = 4$ and $a_{\mathbf{C}} = 2$, we get $\iota(\mathbf{C}, 1, \mathbf{C}_E) = \frac{1}{4}$ and

$$\mathrm{T}_e(f_C, \mathbf{C}) = \mathrm{ST}_e(f_C) + \frac{1}{4}\sum_E \mathrm{ST}_E(f_{C_E}).$$

This identity can be used to associate to any pair $\mu_1, \mu_2$ of characters of $\mathbb{A}_E^\times/E^\times$ whose restriction to $\mathbb{A}^\times/F^\times$ is $\chi_{E/F}$ the pair $\pi(\mu_1) \times \pi(\mu_2)$ of representations of $\mathbf{C}(\mathbb{A}) = [\mathrm{GL}(2,\mathbb{A}) \times \mathrm{GL}(2,\mathbb{A})]'/\mathbb{A}^\times$. This lifting is well-known. Whenever possible we shall work with $f_C = \otimes f_{C,v}$ whose component at a relevant place has orbital integrals which are stable, so that there won't be a contribution from $\mathrm{T}_E(f_{C_E}) = \mathrm{ST}_E(f_{C_E})$.

In summary, the $\theta$-elliptic $\theta$-semisimple strongly $\theta$-regular part of the geometric side of the $\theta$-twisted trace formula for $\mathbf{G}$, $\mathrm{T}_e(f, \mathbf{G}, \theta)$, takes the form

$$\mathrm{T}_e(f_H, \mathbf{H}) - \frac{1}{4}\mathrm{T}_e(f_{C_0}, \mathbf{C}_0) + \frac{1}{2}[\mathrm{T}_e(f_C, \mathbf{C}) - \frac{1}{4}\sum_{[\mathrm{E:F}]=2} \mathrm{T}_E(f_{C_E})].$$

Here $f_H = \otimes f_{Hv}$ and $f_C = \otimes f_{Cv}$ have orbital integrals matching the (stable and unstable) $\theta$-twisted orbital integrals of $f = \otimes f_v$ for each place $v$, those of $f_{C_0} = \otimes f_{C_0 v}$ match those of $f_H$ and those of $f_{C_E} = \otimes f_{C_E v}$ match those of $f_C$. Of course by this we mean that the measures $f dg$, $f_H dh$, $f_C dc$, are matching, and $f_{C_0} dc_0$ and $f_H dh$ are matching, and so are

$f_{C_E} dc_E$ and $f_C dc$. The identities of trace formulae hold for such matching measures. We suppress the measures from the notations.

Complete analysis of the geometric sides of the trace formulae would include terms related to singular and to nonelliptic orbital integrals. In order to not deal with these in this work, we take a component of all global functions at a fixed place $v_0$ of $F$ to vanish on the singular set, and then the integrals over the singular classes vanish a-priori and need not be computed. This mild restriction does not restrict the uses for lifting applications of the identity of trace formulae. For example we may take these functions to be biinvariant under the Iwahori subgroup, and supported on double cosets of elements in the maximal split torus (diagonal, in our case) on which the absolute values of the roots are big (the eigenvalues have distinct absolute values, in our case).

To avoid dealing with the non-($\theta$-)elliptic conjugacy classes, we observe that using the process of truncation, integration over these orbits leads to ($\theta$-) orbital integrals weighted by a factor which can be expressed as a sum of local products involving number of factors bounded by the (twisted) rank. Thus these weighted ($\theta$-) orbital integrals are sums of products of local factors which are all – except for at most rank-$(\mathbf{G}, \theta)$ factors – orbital integrals on the non-$\theta$-elliptic class. In our case the $\theta$-twisted rank of $\mathbf{G}$ is two, and the ranks of $\mathbf{H}$, $\mathbf{C}$ and $\mathbf{C}_0$ are two too. The restrictive assumption that we make is that we fix three places: $v_1$, $v_2$, $v_3$, of $F$, and work with functions $f$ whose components $f_v$ at $v = v_i$ $(i = 1, 2, 3)$ have $\theta$-orbital integrals equal to 0 on the strongly $\theta$-regular orbits which are not $\theta$-elliptic. In this case the geometric side of the twisted trace formula is equal to the $\theta$-elliptic part $\mathrm{T}_e(f, \mathbf{G}, \theta)$.

The matching functions on $\mathbf{H}$, $\mathbf{C}$ and $\mathbf{C}_0$ can also be chosen now to have components at $v_1$, $v_2$, $v_3$ whose orbital integrals vanish on the regular nonelliptic sets of these groups, and the component at $v_0$ vanishes on the non regular set. The geometric sides of the trace formulae are then concentrated on the elliptic regular sets, and are equal for such test functions to $\quad \mathrm{T}_e(f_H, \mathbf{H}), \quad \mathrm{T}_e(f_{C_0}, \mathbf{C}_0), \quad \mathrm{T}_e(f_C, \mathbf{C})$.

The requirement that the orbital integrals of $f_{v_i}$ $(i = 1, 2, 3)$ be zero on the strongly $\theta$-regular non $\theta$-elliptic set is weaker than an assumption that the functions themselves be zero there. The requirement that we make permits applying the trace formula with coefficients of elliptic representations at the places $v_i$ $(i = 1, 2, 3)$.

We compare the geometric sides of the trace formulae with the spectral sides, which include, in addition to the contribution $\sum_\pi m(\pi) \operatorname{tr} \pi(f \times \theta)$ (in the case of the $\theta$-twisted trace formula for $f$ on $G$) from the discrete spectrum $L_d$, also contributions from the continuous spectrum. These contributions are described in terms of Eisenstein series, and lead to a sum of discrete terms and integrals of continuous series of representations, involving logarithmic derivatives. The weight factor splits as sum of local products whose number of terms is bounded by the ($\theta$-)rank. In our case the rank is 2, and assuming as we do the vanishing of the orbital integrals on the regular nonelliptic set at 3 places leads to the vanishing of all continuous sums, or integrals, of traces of representations which contribute to the ($\theta$-)trace formula. We proceed to describe only the discrete sums contributions to the spectral sides of the trace formulae.

## 2. Twisted Trace Formula: Analytic Side

We now record the analytic side of the twisted trace formula; it involves twisted traces of representations. The expression is taken from [CLL], XV, p. 15. Fix a minimal $\theta$-invariant $F$-parabolic subgroup $\mathbf{P}_0$ of $\mathbf{G}$, and its Levi subgroup $\mathbf{M}_0$. Denote by $\mathbf{P}$ any standard (containing $\mathbf{P}_0$) $F$-parabolic subgroup of $\mathbf{G}$, by $\mathbf{M}$ its Levi subgroup which contains $\mathbf{M}_0$, and by $\mathbf{A_M}$ the split component of the center of $\mathbf{M}$. Then $\mathbf{A_M} \subset \mathbf{A}_0 = \mathbf{A}_{\mathbf{M}_0}$. Let $X^*(\mathbf{A_M})$ be the lattice of rational characters of $\mathbf{A_M}$, $\mathcal{A}_\mathbf{M}$ the vector space $X_*(\mathbf{A_M}) \otimes \mathbb{R} = \operatorname{Hom}(X^*(\mathbf{A_M}), \mathbb{R})$, and $\mathcal{A}_\mathbf{M}^*$ the vector space dual to $\mathcal{A}_\mathbf{M}$. Let $W_0 = W(A_0, G)$ be the Weyl group of $A_0$ in $G$. Both $\theta$ and every $s$ in $W_0$ act on $\mathcal{A}_0$. The truncation and the general expression to be recorded depend on a vector $T$ in $\mathcal{A}_0 = \mathcal{A}_{\mathbf{M}_0}$. In our specific case of $\mathbf{G} = \mathrm{PGL}(4)$ we shall use only the constant term, or value at $T = 0$, and in fact only the discrete part, of terms where $\mathcal{A}^* = \{0\}$, below.

2.1 Proposition ([CLL]). *The spectral, or analytic side of the trace formula is equal to a sum over*

(1) *the set of all Levi subgroups* $\mathbf{M}$ *containing* $\mathbf{M}_0$ *of the* $F$*-parabolic subgroups of* $\mathbf{G}$;

*(2) the set of subspaces $\mathcal{A}$ of $\mathcal{A}_0$ such that for some $s$ in $W_0$ we have $\mathcal{A} = \mathcal{A}_{\mathbf{M}}^{s\times\theta}$, where $\mathcal{A}_{\mathbf{M}}^{s\times\theta}$ is the space of $s\times\theta$-invariant elements in the space $\mathcal{A}_{\mathbf{M}}$ associated with a $\theta$-invariant $F$-parabolic subgroup $\mathbf{P}$ of $\mathbf{G}$;*
*(3) the set $W^{\mathcal{A}}(\mathcal{A}_{\mathbf{M}})$ of distinct maps on $\mathcal{A}_{\mathbf{M}}$ obtained as restrictions of the maps $s\times\theta$ ($s$ in $W_0$) on $\mathcal{A}_0$ whose space of fixed vectors is precisely $\mathcal{A}$; and*
*(4) the set of discrete spectrum representations $\tau$ of $\mathbf{M}(\mathbb{A})$ with $(s\times\theta)\tau\simeq\tau$, $s\times\theta$ as in (3).*

*The terms in the sum are equal to the product of*

$$\frac{[W_0^M]}{[W_0]}\left(\det(1-s\times\theta)|_{\mathcal{A}_{\mathbf{M}}/\mathcal{A}}\right)^{-1}$$

*and*

$$\int_{i\mathcal{A}^*}\operatorname{tr}\left[M_{\mathcal{A}}^T(P,\zeta)M_{\mathbf{P}|\theta(\mathbf{P})}(s,0)I_{\mathbf{P},\tau}(\zeta,f\times\theta)\right]|d\zeta|.$$

Here $[W_0^M]$ is the cardinality of the Weyl group $W_0^M = W(A_0,M)$ of $A_0$ in $M$. Also $\mathbf{P}$ is an $F$-parabolic subgroup of $\mathbf{G}$ with Levi component $\mathbf{M}$; $M_{\mathbf{P}|\theta(\mathbf{P})}$ is an intertwining operator; $\mathcal{M}_{\mathcal{A}}^T(\mathbf{P},\lambda)$ is a logarithmic derivative of intertwining operators, and $I_{\mathbf{P},\tau}(\zeta)$ is the $\mathbf{G}(\mathbb{A})$-module normalizedly induced from the $\mathbf{M}(\mathbb{A})$-module $m\mapsto\tau(m)e^{\langle\zeta,H(m)\rangle}$ in standard notations.

The sum of the terms corresponding to $\mathbf{M}=\mathbf{G}$ in the formula is equal to the sum $I=\sum\operatorname{tr}\pi(f\times\theta)$ over all discrete spectrum representations $\pi$ of $\mathbf{G}(\mathbb{A})$.

We proceed to describe, in our case of $\mathbf{G}=\mathrm{PGL}(4)$ and the involution $\theta$, the terms corresponding to $\mathbf{M}\neq\mathbf{G}$ and $\mathcal{A}=\{0\}$ in the formula. Let $\mathbf{M}_0$ be the diagonal subgroup of $\mathbf{G}$.

There are $[W_0]/[W_0^M]=4$ Levi subgroups $\mathbf{M}\supset\mathbf{A}_0$ of maximal parabolic subgroups $\mathbf{P}$ of $\mathbf{G}$ (of type (3,1)) isomorphic to GL(3), that is to the image of $\mathrm{GL}(3)\times\mathrm{GL}(1)$ in PGL(4). The space $\mathcal{A}_{\mathbf{M}}=\{(a,a,a,b)^*;\ a,b\in\mathbb{R}\}$ (the superscript $*$ means image in $\mathbb{R}^4/\mathbb{R}$, where $\mathbb{R}$ is embedded diagonally), has $\mathcal{A}=\mathcal{A}_{\mathbf{M}}^{s\times\theta}=\{0\}$ for any $s\in W$ (for which $s\times\theta$ maps $\mathcal{A}_{\mathbf{M}}$ back to $\mathcal{A}_{\mathbf{M}}$), and the contribution is

$$\sum_{\mathbf{M}}\frac{3!}{4!}\cdot\frac{1}{2}\sum_{\tau}\operatorname{tr}M(s,0)I_{P,\tau}(0;f\times\theta)$$
$$=\frac{1}{2}\sum_{\chi}\sum_{\tau}\operatorname{tr}M(\alpha_3\alpha_2\alpha_1,0)I_{\mathbf{P}_1}(\tau,\chi;f\times\theta).$$

Here $\mathbf{P}_1$ denotes the upper triangular parabolic subgroup of $\mathbf{G}$ of type (3,1). We write $\alpha_1 = (12)$, $\alpha_2 = (23)$, $\alpha_3 = (34)$, $J = (14)(23)$ for the transpositions in the Weyl group $W_0$. In the last sum, $\chi$ ranges over the characters of $\mathbb{A}^\times/F^\times$ of order at most two, while $\tau$ ranges over the discrete spectrum representations of $\mathrm{GL}(3,\mathbb{A})$ whose central character is $\chi$ and $\tau^\theta \simeq \tau$.

There are $[W_0]/2[W_0^M] = 3$ Levi subgroups $\mathbf{M} \supset \mathbf{A}_0$ of maximal parabolic subgroups $\mathbf{P}$ of $\mathbf{G}$ (of type (2,2)) isomorphic to the image of $\mathrm{GL}(2) \times \mathrm{GL}(2)$ in $\mathrm{PGL}(4)$. The space

$$\mathcal{A}_{\mathbf{M}} = \{(a,a,b,b)^*; a,b \in \mathbb{R}^2\}$$

has $\mathcal{A} = \mathcal{A}_{\mathbf{M}}^{s\times\theta}$ equal to $\{(0,0,a,a)^*\}$ for $s \in W_0^M$, and $\mathcal{A} = \{0\}$ for all $s \neq 1$ in $W/W_0^M$. Consider only the case of $\mathcal{A} = \{0\}$, and choose $s = J$ to be a representative. Then

$$1 - J \times \theta : (a,a,b,b)^* \mapsto (-a,-a,-b,-b)^*$$

has determinant 2 on the one dimensional space $\mathcal{A}_{\mathbf{M}}$, and the contribution is

$$\sum_{\mathbf{M}} \frac{2\cdot 2}{4!} \cdot \frac{1}{2} \sum_{\tau} \operatorname{tr} M(J,0) I_{\mathbf{P},\tau}(0; f \times \theta)$$

$$= \frac{1}{4} \sum_{\chi} \sum_{\tau_1 \times \tau_2} \operatorname{tr} M(J,0) I_{\mathbf{P}_2}(\tau_1, \tau_2; f \times \theta).$$

Here $\mathbf{P}_2$ denotes the upper triangular parabolic subgroup of $\mathbf{G}$ of type (2,2). The last sum ranges over the ordered pairs $(\tau_1, \tau_2)$ of discrete spectrum representations $\tau_1$, $\tau_2$ of $\mathrm{GL}(2,\mathbb{A})$ with central characters $\omega_{\tau_1}$ and $\omega_{\tau_2}$ with $\omega_{\tau_1}\omega_{\tau_2} = 1$ and $\tau_i^\theta \simeq \tau_i$, thus $\tau_1$ and $\tau_2$ are discrete spectrum representations of $\mathrm{PGL}(2,\mathbb{A})$ (then we write $\chi = 1$) or $\tau_i = \pi(\mu_i)$, $\mu_i$ characters of $\mathbb{A}_E^\times/E^\times\mathbb{A}^\times$, and $E/F$ is the quadratic extension determined by $\chi = \omega_{\tau_1} = \omega_{\tau_2}$. Thus the sum over $\chi$ ranges over all characters of $\mathbb{A}^\times/F^\times$ of order at most two.

There are $[W_0]/2[W_0^M] = 6$ Levi subgroups $\mathbf{M} \supset \mathbf{A}_0$ of parabolic subgroups $\mathbf{P}$ of $\mathbf{G}$ (of type (2,1,1)) isomorphic to the image of $\mathrm{GL}(2)\times\mathrm{GL}(1)\times \mathrm{GL}(1)$ in $\mathrm{PGL}(4)$. If $s \in W_0$ is such that $s \times \theta$ maps $\mathcal{A}_{\mathbf{M}} = \{(a,a,b,c)^*\}$ to itself, then (up to multiplication by $\langle \alpha_1 = (12)\rangle = W_0^M$), $s$ can be (1) $s = (14)(23)$, in which case

$$s \times \theta : (a,a,b,c)^* \mapsto (-a,-a,-b,-c)^*,$$

$\mathcal{A} = \{0\}$ and $\det(1 - s \times \theta)|_{\mathcal{A}_M} = 4$, or (2) $s = (13)(24)$, then

$$s \times \theta : (a, a, b, c)^* \mapsto (-a, -a, -c, -b)^*$$

and $\mathcal{A} \neq \{0\}$. The term with $\mathcal{A} = \{0\}$ is

$$\frac{1}{3 \cdot 4} \cdot \frac{1}{4} \sum_{\mathbf{M}, \tau} \operatorname{tr} M((14)(23), 0) I_{\mathbf{P}_3, \tau}(0; f \times \theta).$$

Here $\mathbf{P}_3$ is the upper triangular parabolic subgroup of type $(2, 1, 1)$, and $\tau = (\tau_1, \chi_1, \chi_2)$ is equivalent to $(\tau_1^\theta, \chi_1^{-1}, \chi_2^{-1})$. If $\chi$ denotes the central character of $\tau_1$ then $\chi = \chi_1\chi_2$, and $\tau_1 \simeq \tau_1^\theta = \tau_1\chi$. We can write the induced representation as $\chi_2 I(\tau_1, \chi_1, 1)$. If $\chi = \chi_1 \neq 1$ then $\tau_1 = \pi(\mu_1)$ where $\mu_1$ is a character of $\mathbb{A}_E^\times/E^\times$, where $E/F$ is determined by $\chi$. The central character of $\pi(\mu_1)$ is $\chi \cdot \mu_1|\mathbb{A}^\times$; if this is equal to $\chi$, then $\mu_1|\mathbb{A}^\times = 1$, hence there is a character $\mu_0$ of $\mathbb{A}_E^\times/E^\times$ with $\mu_1(z) = \mu_0(z/\overline{z})$. Put $\overline{\mu}_0(z) = \mu_0(\overline{z})$, $z \in \mathbb{A}_E^\times$, then $\tau_1 = \pi(\mu_0/\overline{\mu}_0)$. If on the contrary $\chi = \chi_1 = 1$ then $\tau_1$ is a discrete spectrum representation of $\mathrm{PGL}(2, \mathbb{A})$. We then obtain from the terms with $\mathcal{A} = \{0\}$ the sum

$$\frac{1}{8} \sum_{\chi, \tau_1} \operatorname{tr} M((14)(23), 0)(\chi I_{\mathbf{P}_3}(\tau_1, 1, 1))(f \times \theta)$$

$$+ \frac{1}{4} \sum_{\chi_1 \neq 1, \mu_0, \chi} \operatorname{tr} M((14)(23), 0)(\chi I_{\mathbf{P}_3}(\pi(\mu_0/\overline{\mu}_0), \chi_1, 1))(f \times \theta)$$

where $\chi$ is any quadratic character, $\chi_1$ is a quadratic character $\neq 1$, $\tau_1$ ranges over the discrete spectrum of $\mathrm{PGL}(2, \mathbb{A})$ and $\mu_1 = \mu_0/\overline{\mu}_0$ over the characters of $\mathbb{A}_E^\times/\mathbb{A}^\times E^\times$, or $\mu_0$ over the characters of $\mathbb{A}_E^1/E^1$, where $\mathbb{A}_E^1 = \{z \in \mathbb{A}_E^\times; z\overline{z} = 1\}$. Note that $I(\tau_1, \chi, 1)$ and $I(\tau_1, 1, \chi)$ contribute two equivalent contributions when $\chi \neq 1$.

Let $\pi$ be an irreducible $\theta$-invariant representation of $\mathbf{G} = \mathrm{PGL}(4)$, which is properly induced from a parabolic subgroup. We proceed to list these.

If $\pi$ is induced from the (standard) parabolic of type (3,1) the $\pi = I(\tau, \chi)$, where $\tau$ is a representation of $\mathrm{GL}(3)$ and $\chi$ is a character (of $\mathrm{GL}(1)$), and $\omega\chi = 1$ where $\omega = \omega_\tau$ is the central character of $\tau$. From $\pi^\theta \simeq \pi$ we conclude that

$$\tau^\theta \simeq \tau \qquad (\tau^\theta(g) = \tau(J^t g^{-1} J), \quad J = \text{antidiagonal}(1, -1, 1))$$

and $\chi^2 = 1$. Then $\pi = \chi I(\tau\chi, 1)$, where $\tau\chi$ is a representation of PGL(3) with $(\tau\chi)^\theta \simeq \tau\chi$, hence the image of the symmetric square lifting from GL(2) (or rather SL(2), see [F3]) to PGL(3). Globally we have that the lifting

$$\lambda_1 : \mathrm{SO}(4) = [\mathrm{GL}(2) \times \mathrm{GL}(2)]'/\,\mathrm{GL}(1) \to \mathrm{PGL}(4)$$

takes $\tau_1 \times \tau_1\chi$ to $I(\chi\,\mathrm{Sym}^2(\tau_1), \chi)$.

If $\pi$ is induced from the (standard) parabolic of type (2,2) then it is

$$\pi = I(\tau_1, \tau_2) \simeq \pi^\theta = I(\tau_1^\theta, \tau_2^\theta), \qquad \tau^\theta(g) = \tau(w\,{}^tg^{-1}w^{-1}).$$

If $\tau_1^\theta \simeq \tau_2$, then $\pi$ lies in a continuous family $I(\tau_1\nu^s, \tau_1^\theta\nu^{-s})$, $\nu(g) = |\det g|$, of $\theta$-invariant representations. Otherwise $\tau_1^\theta \simeq \tau_1$ and $\tau_2^\theta \simeq \tau_2$, thus $\omega_{\tau_i}^2 = 1$, and $\omega_{\tau_1}\omega_{\tau_2} = 1$. If $\omega_{\tau_i} = 1$ then $\tau_i$ is a representation of PGL(2), and if $\omega_{\tau_i} \neq 1$ then $\tau_i = \pi(\mu_i)$, where $\mu_i$ is a character of $C_E(= E^\times$ in the local case, $\mathbb{A}_{E_i}^\times/E_i$ in the global case) which is trivial on $C_F$, where $E/F$ is the quadratic extension determined by $\omega_{\tau_1} = \omega_{\tau_2}$.

*Interlude about* GL(2): if $E/F$ is the quadratic extension determined by a quadratic character $\omega$ of $F^\times$ ($F$ local), and $\mu$ is a complex valued character of $E^\times$, there is a two dimensional representation $\rho(\mu)$ of the extension

$$W_{E/F} = \langle z \in E^\times, \sigma; \sigma^2 \in F - N_{E/F}E, \sigma z = \overline{z}\sigma\rangle$$

of $\mathrm{Gal}(E/F) = \langle\sigma\rangle$ by $W_{E/E} = C_E$, given by

$$z \mapsto \begin{pmatrix} \mu(z) & 0 \\ 0 & \mu(\overline{z}) \end{pmatrix}, \quad \sigma \mapsto \begin{pmatrix} 0 & 1 \\ \mu(\sigma^2) & 0 \end{pmatrix}.$$

Then $\det\rho(\mu)(z) = \mu(z\overline{z})$, $\det\rho(\mu)(\sigma) = -\mu(\sigma^2)$. The corresponding admissible (globally: automorphic) representation of GL(2) is denoted by $\pi(\mu)$, and its central character is $\omega(x) = \chi_{E/F}(x)\mu(x)$.

In the case of the parabolic of type (2,2) above, $\omega_{\tau_i} \neq 1$ then implies that $\mu_i|F^\times = 1$, hence there is a character $\mu_i' : E^\times \to \mathbb{C}^\times$ with $\mu_i(z) = \mu_i'(z/\overline{z})$ so that $\tau_i = \pi(\mu_i'/\overline{\mu}_i')$. Choose square roots of

$$a(z)^2 = (\mu_1'\mu_2')(z/\overline{z}), \qquad b(z)^2 = (\mu_1'/\mu_2')(z/\overline{z}),$$

then

$$\begin{pmatrix} a & \\ & a^{-1} \end{pmatrix} \times \begin{pmatrix} b & \\ & b^{-1} \end{pmatrix} \mapsto \begin{pmatrix} ab & & & \\ & a/b & & \\ & & b/a & \\ & & & 1/ab \end{pmatrix} = \begin{pmatrix} \mu_1'/\overline{\mu}_1' & & & \\ & \mu_2'/\overline{\mu}_2' & & \\ & & \overline{\mu}_2'/\mu_2' & \\ & & & \overline{\mu}_1'/\mu_1' \end{pmatrix}$$

and

$$\pi(a) \times \pi(b) \xrightarrow{\lambda_1} I(\pi(\mu_1'/\overline{\mu}_1'), \pi(\mu_2'/\overline{\mu}_2')).$$

If $\pi$ is induced from the standard parabolic of type (2,1,1) then

$$\pi = I(\tau, \chi_1, \chi_2) \simeq \pi^\theta = I(\tau^\theta, \chi_1^{-1}, \chi_2^{-1}), \qquad \chi_i^2 = 1,$$

and $\pi = \chi_2 I(\tau\chi_2, \chi_1\chi_2, 1)$. Further $\tau^\theta \simeq \tau$ (it is a representation of GL(2)), and $\tau_0 = \tau\chi_2 \simeq \tau_0^\theta$, and $\chi_0 = \chi_1\chi_2$ has order two. If $\chi_0 = 1$ then $\tau_0$ is a representation of PGL(2), while if $\chi_0 \neq 1$ then $\tau_0 = \pi(\mu_0/\overline{\mu}_0)$, where $\mu_0$ is a character of the quadratic extension $E$ of $F$ determined by $\chi_0$. In this case

$$\pi(\mu_0) \times \chi_2\pi(\mu_0) \to \chi_2 I(\pi(\mu_0/\overline{\mu}_0), \chi_0, 1).$$

If $\pi$ is induced from the minimal parabolic, of type (1,1,1,1), and

$$\pi = I(\chi_1, \chi_2, \chi_3, \chi_4) \simeq \pi^\theta = I(\chi_1^{-1}, \chi_2^{-1}, \chi_3^{-1}, \chi_4^{-1})$$

is not in a continuous family of $\theta$-invariant representations, then $\chi_i^2 = 1$. If two $\chi_i's$ are equal then $\pi$ is $\chi_0 I(\chi, \chi^{-1}, 1, 1)$. Otherwise $\pi$ is the twist by $\chi_0$ of $I(\chi_1, \chi_2, \chi_1\chi_2, 1)$. Denote by $E_2$ the extension of $F$ determined by $\chi_2$, put $\mu(z) = \chi_1(z\overline{z})$ ($\overline{z}$ is the image of $z \in E_2^\times$ under the Galois action over $F$). Then $\mu = \overline{\mu}$ ($\overline{\mu}(z)$ is $\mu(\overline{z})$) and $\mu = \mu^{-1}$ since $\chi_1^2 = 1$. Then there is $\mu_1$ on $E_2^\times$ with $\mu(z) = \mu_1(z/\overline{z}) (= \mu_1(\overline{z}/z) \not\equiv 1)$, and

$$\pi(\mu_1) \times \chi_0\pi(\mu_1) \to \chi_0 I(\pi(\mu_1/\overline{\mu}_1), \chi_2, 1) = \chi_0 I(\chi_1, \chi_1\chi_2, \chi_2, 1).$$

We now take the Levi subgroup $\mathbf{A}_0$ and list the different types of maps $s \times \theta$. The involution $\theta$ maps an element $(a, b, c, d)^*$ of $\mathcal{A}_0$ to $(-d, -c, -b, -a)^*$, and it is convenient to write $s$ as $sJ$ ($J = (14)(23)$). In these notations, there are 1 (resp. 8, 6, 6, 3) distinct maps $sJ \times \theta$ where $s$ is 1 (resp. has order 3, is a transposition, has order 4, is a product of two transpositions of disjoint support). Representatives are given by $s = 1$ (resp. (321), (12), (4321), (12)(34)). The subspace $\mathcal{A}$ of vectors in $\mathcal{A}_0$ fixed by $sJ \times \theta$ is $\{0\}$ (resp. $\{0\}$, $\{(a,-a,0,0)^*\}$, $\{(a,b,a,b)^*\}$, $\{(a,b,c,a+b-c)^*\}$) and $\det(1 - sJ \times \theta)$ is 8 (resp. 2, 2, 1, 1). We record only the discrete part, where $\mathcal{A} = \{0\}$.

$$\frac{1}{4!} \cdot \frac{1}{8} \sum_{\tau} \operatorname{tr} M(J,0) I_{\mathbf{P}_0,\tau}(0, f \times \theta)$$
$$+ \frac{1}{4!} \cdot \frac{8}{2} \sum_{\tau} \operatorname{tr} M((321)J,0) I_{\mathbf{P}_0,\tau}(0, f \times \theta).$$

The $\tau$ in the first sum are the characters $(\chi_1, \chi_2, \chi_3, \chi_4)$ of $A_0$ fixed by $J \times \theta$, that is $\chi_i^2 = 1$. There are 4! such ordered 4-tuples of distinct $\chi_i'$s, $3! \cdot 2$ ordered 4-tuples where $\{\chi_i\}$ has 3 distinct elements, $3 \times 2$ ordered 4-tuples where each $\chi_i$ occurs twice in $(\chi_1, \chi_2, \chi_3, \chi_4)$, 4 ordered 4-tuples where exactly 3 of the 4 $\chi_i'$s are equal. The first sum becomes

$$\frac{1}{8} \sum_{\chi_i \neq \chi_j, \chi_i^2 = 1, \chi_1\chi_2\chi_3\chi_4 = 1} \operatorname{tr} MI((\chi_1, \chi_2, \chi_3, \chi_4); f \times \theta)$$
$$+ \frac{1}{4 \cdot 8} \sum_{\chi_1 \neq \chi_2, \chi_i^2 = 1} \operatorname{tr} MI((\chi_1, \chi_1, \chi_2, \chi_2); f \times \theta)$$
$$+ \frac{1}{4!8} \sum_{\chi^2 = 1} \operatorname{tr} MI((\chi, \chi, \chi, \chi); f \times \theta).$$

Since $(321)J \times \theta$ maps $\tau$ to $(\chi_3^{-1}, \chi_1^{-1}, \chi_2^{-1}, \chi_4^{-1})$, the fixed $\tau$ have $\chi_4^2 = 1$ and $\chi_1\chi_3 = \chi_1\chi_2 = \chi_2\chi_3 = 1$, thus $\chi_1 = \chi_2 = \chi_3$ has $\chi_1^2 = 1$. Since $\chi_1\chi_2\chi_3\chi_4 = 1$, we get $\chi_4 = \chi$. The contribution is then

$$\frac{1}{6} \sum_{\chi^2 = 1} \operatorname{tr} M((321)J, 0) I((\chi, \chi, \chi, \chi); f \times \theta).$$

# 3. Trace Formula of H: Spectral Side

The spectral side of the trace formula for $\mathbf{H} = \mathrm{PGSp}(2)$ can be written out too as the case where $\theta$ is trivial. We proceed to specify the objects involved. As usual, a superscript $*$ indicates image in the projective group. We choose $\mathbf{P}_0$ to be the upper triangular subgroup in $\mathbf{H}$. Its fixed Levi subgroup is chosen to be $\mathbf{A}_0 = \{t = \mathrm{diag}(a, b, \boldsymbol{\lambda}/b, \boldsymbol{\lambda}/a)^*\}$. A basis of the root system is $\Delta = \Delta(\mathbf{H}, \mathbf{P}_0, \mathbf{A}_0) = \{\alpha, \beta\}$, $\alpha(t) = a/b$, $\beta(t) = b^2/\boldsymbol{\lambda}$, and the root system is $R = R^+ \cup -R^+$, where $R^+ = R^+(\mathbf{H}, \mathbf{P}_0, \mathbf{A}_0) = \{\alpha, \beta, \alpha+\beta, 2\alpha+\beta\}$ is the set of distinct homomorphisms in the action (Int) of $\mathbf{A}_0$ on $\mathrm{Lie}(\mathbf{P}_0/\mathbf{A}_0)$. The group $X_*(\mathbf{A}_0) = \mathrm{Hom}(\mathbb{G}_m, \mathbf{A}_0)$ is a lattice in the vector space $\mathcal{A}_0 = X_*(\mathbf{A}_0) \otimes \mathbb{R}$ which we identify with $\mathbb{R}^2$ via the map $\log : X_*(\mathbf{A}_0) \to \mathcal{A}_0 = \mathbb{R}^2$,

$$\log(a, b, \boldsymbol{\lambda}/b, \boldsymbol{\lambda}/a)^* = (\log_q |a| - \frac{1}{2} \log_q |\boldsymbol{\lambda}|, \log_q |b| - \frac{1}{2} \log_q |\boldsymbol{\lambda}|).$$

The roots, characters of $\mathbf{A}_0(F) = X_*(\mathbf{A}_0) \otimes F^\times$, lie in the group $X^*(\mathbf{A}_0) = \mathrm{Hom}(\mathbf{A}_0, \mathbb{G}_m)$, which is a lattice in the dual space $\mathcal{A}_0^* = X^*(\mathbf{A}_0) \otimes \mathbb{R}$ to $\mathcal{A}_0 = \mathrm{Hom}(X^*(\mathbf{A}_0), \mathbb{R})$. Identifying $\mathcal{A}_0^*$ with $\mathbb{R}^2$ with the usual inner product: $\mathcal{A}_0^* \times \mathcal{A}_0 \to \mathbb{R}$, $((x, y), (u, v)) = xu + yv$, the roots can be identified with the vectors $\alpha = (1, -1)$, $\beta = (0, 2)$, $\alpha + \beta = (1, 1)$, $2\alpha + \beta = (2, 0)$ in $\mathcal{A}_0^* = \mathbb{R}^2$. The coroots $\alpha^\vee = 2\alpha/(\alpha, \alpha), \ldots$ are in $\mathcal{A}_0$ identified with $\alpha^\vee = (1, -1)$, $\beta^\vee = (0, 1)$, $(\alpha + \beta)^\vee = (1, 1)$, $(2\alpha + \beta)^\vee = (1, 0)$.

The Weyl group $W_0 = W(H, A_0)$ of $A_0$ in $H$, which is the quotient by (the centralizer in $H$ of) $A_0$ of the normalizer of $A_0$ in $H$, viewed as a group of permutations in the symmetric group $S_4$ on 4 letters, is generated by the reflections $s_\alpha = (12)(34)$, $s_\beta = (23)$, $s_{\alpha+\beta} = (13)(24)$, $s_{2\alpha+\beta} = (14)$ in $S_4$. Put $\sigma = s_\beta s_\alpha = s_\alpha s_{2\alpha+\beta} (= (23)(12)(34) = (12)(34)(14))$. Then

$$W_0 = \langle \sigma, s_\beta; \sigma^4 = s_\beta^2 = 1, s_\beta \sigma s_\beta = \sigma^{-1} \rangle = \{s_\beta^i \sigma^j; i = 0, 1; j = 0, 1, 2, 3\}$$

is the dihedral group $D_4$. Note that $s_\beta \sigma = s_\alpha$, $s_\beta \sigma^2 = s_{2\alpha+\beta}$, $s_\beta \sigma^3 = s_{\alpha+\beta}$. Under the identification of $X_*(\mathbf{A}_0)$ with a lattice in $\mathcal{A}_0 = \mathbb{R}^2$, the Weyl group can be identified with a group of automorphisms of $\mathcal{A}_0$: $s_\alpha(x, y) = (y, x)$,

$$s_\beta(x, y) = (x, -y), \quad s_{\alpha+\beta}(x, y) = (-y, -x), \quad s_{2\alpha+\beta}(x, y) = (-x, y),$$

$\sigma(x,y)=(y,-x)$, $\sigma^2=-1$. Note that for each root $\gamma$ in $\{\alpha,\beta,\alpha+\beta,2\alpha+\beta\}$ and for $\delta\in\mathcal{A}_0^*$ perpendicular to $\gamma$, we have $s_\gamma\gamma=-\gamma$ and $s_\gamma\delta=\delta$. Then the $s_\gamma$ are reflections, and $\sigma$ is a rotation of $\pi/2$, clockwise.

The Levi (components of parabolic) subgroups of $\mathbf{H}$ containing $\mathbf{A}_0$ other than $\mathbf{H}$ and $\mathbf{A}_0$ are $\mathbf{M}'_\alpha=s_\beta\mathbf{M}_\alpha s_\beta$, $\mathbf{M}'_\beta=s_\alpha\mathbf{M}_\beta s_\alpha$,

$$\mathbf{M}_\alpha=\mathbf{M}_{(2,2)}=\{\mathrm{diag}(A,\ \boldsymbol{\lambda}w^tA^{-1}w)^*;\ A \text{ in } \mathrm{GL}(2),\boldsymbol{\lambda}\text{ in }\mathbb{G}_m\},$$

$$\mathbf{M}_\beta=\{\mathrm{diag}(a,A,\det A/a)^*;\ A\text{ in }\mathrm{GL}(2),\ a\text{ in }\mathrm{GL}(1)\}.$$

We determine the subspaces of $\mathcal{A}_0$ associated with these. The (split component of the) center $A_\alpha=A_{\mathbf{M}_\alpha}$ of $\mathbf{M}_\alpha$ consists of $t=(a,a,\boldsymbol{\lambda}/a,\boldsymbol{\lambda}/a)^*$, thus $X_*(\mathbf{A}_\alpha)=\mathrm{Hom}(\mathbb{G}_m,\mathbf{A}_\alpha)$ is $\mathbb{Z}(1,1)$ and $\mathcal{A}_\alpha=X_*(\mathbf{A}_\alpha)\otimes\mathbb{R}$ is $\mathbb{R}(\alpha+\beta)^\vee$ in $\mathcal{A}_0$. Since $X_*(\mathbf{A}_{\mathbf{M}'_\alpha})=s_\beta X_*(\mathbf{A}_{\mathbf{M}_\alpha})$ we have $\mathcal{A}_{\mathbf{M}'_\alpha}=\mathbb{R}\alpha^\vee$. From $X_*(A_{\mathbf{M}_\beta})=\mathbb{Z}(1,0)$ we obtain $\mathcal{A}_{M_\beta}=\mathbb{R}(2\alpha+\beta)^\vee$ in $\mathcal{A}_0$, and since $X_*(\mathbf{A}_{M'_\beta})=s_\alpha X_*(\mathbf{A}_{\mathbf{M}_\beta})$ we have $\mathcal{A}_{\mathbf{M}'_\beta}=\mathbb{R}\beta^\vee$. Hence

$$\mathcal{A}_0^{s_\alpha}=\mathcal{A}_{\mathbf{M}_\alpha},\quad \mathcal{A}_0^{s_\beta}=\mathcal{A}_{\mathbf{M}_\beta},\quad \mathcal{A}_0^{s_{\alpha+\beta}}=\mathcal{A}_{\mathbf{M}'_\alpha},\quad \mathcal{A}_0^{s_{2\alpha+\beta}}=\mathcal{A}_{\mathbf{M}'_\beta}.$$

Here is the diagram (where $\alpha_1=\varepsilon_1-\varepsilon_2$):

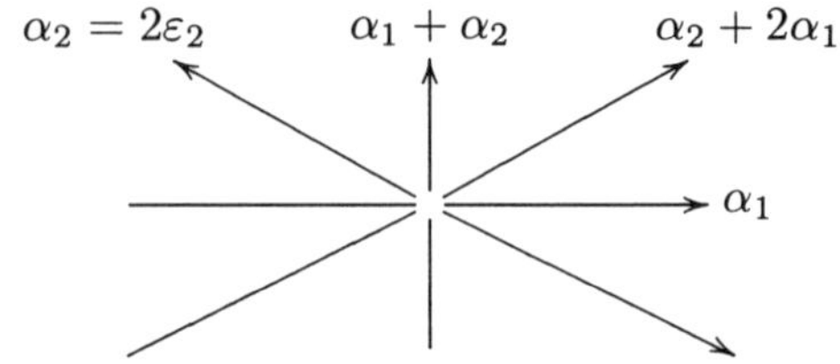


To list the contributions to the trace formula, note that the $w$ in $W_0$ with $\mathcal{A}_0^w=\{0\}$ are $\sigma,\sigma^2,\sigma^3$. Recall that $\sigma(a,b,\boldsymbol{\lambda}/b,\boldsymbol{\lambda}/a)=(b,\boldsymbol{\lambda}/a,a,\boldsymbol{\lambda}/b)$ and the character $(\mu_1,\mu_2)$ from which $I(\mu_1,\mu_2)=\mu_1\mu_2\times\mu_2/\mu_1\rtimes\mu_1^{-1}$ is induced takes at $t=(a,b,\boldsymbol{\lambda}/b,\boldsymbol{\lambda}/a)$ the value

$$\mu_1(a/b)\mu_2(ab/\boldsymbol{\lambda})=\mu_2^{-1}(\boldsymbol{\lambda})(\mu_1\mu_2)(a)(\mu_2/\mu_1)(b).$$

Then ${}^{\sigma^{-1}}(\mu_1,\mu_2)(g)=(\mu_1,\mu_2)(\sigma g)$ is the character

$$t\mapsto(\sigma t\mapsto)\mu_1(ab/\boldsymbol{\lambda})\mu_2(b/a)=\mu_1^{-1}(\boldsymbol{\lambda})(\mu_1/\mu_2)(a)(\mu_1\mu_2)(b).$$

We have ${}^{\sigma}(\mu_1, \mu_2) = (\mu_1, \mu_2)$ if $\mu_1 = \mu_2 = \mu_2^{-1}$. Since

$$\sigma^2 t = (\boldsymbol{\lambda}/a, \boldsymbol{\lambda}/b, b, a), \quad {}^{\sigma^{-2}}(\mu_1, \mu_2)(t) = (\mu_1, \mu_2)(\sigma^2 t) = \mu_1(b/a)\mu_2(\boldsymbol{\lambda}/ab)$$

is equal to $(\mu_1, \mu_2)(t)$ if $\mu_1^2 = 1 = \mu_2^2$. Note also that $1 - \sigma : (x, y) \mapsto (x, y) - (y, -x) = (x - y, y + x)$ has determinant 2, while $\det(1 - \sigma^2) = 4$. Since $[W_0] = 8$ and $W_0^{A_0} = \{1\}$, the contribution to the trace formula from $\mathbf{M} = \mathbf{A}_0$ and $\mathcal{A} = \{0\}$, thus $W^{\{0\}}(\mathcal{A}_0) = \{\sigma, \sigma^2, \sigma^3\}$, is

$$\frac{1}{8} \cdot \frac{1}{2} \sum_{\mu_1 = \mu_2 = \mu_2^{-1}} \operatorname{tr} M(\sigma, 0) I_{\mathbf{P}_0}(\mu_1, \mu_2; f_H)$$
$$+ \frac{1}{8} \cdot \frac{1}{2} \sum_{\mu_1 = \mu_2 = \mu_2^{-1}} \operatorname{tr} M(\sigma^3, 0) I_{\mathbf{P}_0}(\mu_1, \mu_2; f_H)$$
$$+ \frac{1}{8} \cdot \frac{1}{4} \sum_{\mu_1^2 = 1 = \mu_2^2} \operatorname{tr} M(\sigma^2, 0) I_{\mathbf{P}_0}(\mu_1, \mu_2; f_H).$$

Note that the representations $I_{\mathbf{P}_0}(\mu_1, \mu_2) = \mu_1\mu_2 \times \mu_1/\mu_2 \rtimes \mu_1^{-1}$ with $\mu_1^2 = 1 = \mu_2^2$ are irreducible (by [ST]), hence the operators $M$ are scalars which in fact are equal to 1.

Next we consider the $\mathcal{A}_{\mathbf{M}}$ for Levi subgroups other than $\mathbf{H}$ and $\mathbf{A}_0$, and the $w$ in the Weyl group which map $\mathcal{A}_{\mathbf{M}}$ to itself with fixed points $\{0\}$ only. These are $\mathcal{A}_{\mathbf{M}_\alpha}^{s_{\alpha+\beta}} = \{0\}$, $\mathcal{A}_{\mathbf{M}'_\alpha}^{s_\alpha} = \{0\}$, $\mathcal{A}_{\mathbf{M}_\beta}^{s_{2\alpha+\beta}} = \{0\}$ and $\mathcal{A}_{\mathbf{M}'_\beta}^{s_\beta} = \{0\}$. The reflection $s_{\alpha+\beta}$ acts on $\mathbf{M}_\alpha$ by mapping $\operatorname{diag}(A, \boldsymbol{\lambda}A^*)$, $A^* = w^t A^{-1} w$, to $(\boldsymbol{\lambda}A^*, A)$. The representation $\pi_2 \otimes \mu$, from which $I_{\mathbf{M}_\alpha}(\pi_2, \mu) = \pi_2 \rtimes \mu$ is induced, takes $\operatorname{diag}(A, \boldsymbol{\lambda}A^*)$ to $\mu(\boldsymbol{\lambda})\pi_2(A)$. Since $I_{\mathbf{M}_\alpha}(\pi_2, \mu)$ is a representation of PGSp(2), we have $\mu^2\omega = 1$, where $\omega$ is the central character of $\pi$. The representation $\pi_2 \otimes \mu$ takes $(\boldsymbol{\lambda}A^*, A)$ to $\mu(\boldsymbol{\lambda})\pi_2(\boldsymbol{\lambda}A^*) = \mu(\boldsymbol{\lambda})\omega(\boldsymbol{\lambda})\omega^{-1}(\det A)\pi_2(A)$. Then we have $s_{\alpha+\beta}(\pi_2 \otimes \mu) = \pi_2 \otimes \mu$ when $\omega = 1$, thus $\pi_2$ is a representation of PGL(2). Since $[W_0] = 8$, $[W_0^{M_\alpha}] = 2$, $\det(1 - s)|_{\mathcal{A}_{\mathbf{M}_\alpha}} = 2$ and $\mathbf{M}'_\alpha$ contributes a term equal to that contributed by $\mathbf{M}_\alpha$, the contribution to the trace formula from the $W^{\mathcal{A}}(\mathcal{A}_{\mathbf{M}})$ with $\mathcal{A} = \{0\}$ and $\mathbf{M} = \mathbf{M}_\alpha$ and $\mathbf{M}'_\alpha$ is ($\pi_2$ is a representation of $\mathrm{PGL}(2, \mathbb{A})$)

$$2 \cdot \frac{2}{8} \cdot \frac{1}{2} \sum_{\{\mu; \mu^2 = 1\}} \sum_{\pi_2} \operatorname{tr} M(s_{\alpha+\beta}, 0) I_{\mathbf{P}_\alpha}(\pi_2, \mu; f_H).$$

The representations $I_{\mathbf{P}_\alpha}(\pi_2,\mu)=\pi_2\rtimes\mu$ are irreducible (by [ST]), hence the operators $M$ are constants, in fact equal 1.

The only element of $W_0$ which maps $\mathcal{A}_{\mathbf{M}_\beta}=\mathbb{R}(2\alpha+\beta)^\vee$ to itself with $\{0\}$ as its only fixed point is $s_{(\beta+2\alpha)}=(14)$. It takes $t=\operatorname{diag}(a,A,\det A/a)$ to $\operatorname{diag}(\det A/a,A,a)$. The representation $\mu\otimes\pi_2$ of $\mathrm{GL}(1,\mathbb{A})\times\mathrm{GL}(2,\mathbb{A})$ from which the representation $I_{\mathbf{P}_\beta}(\mu,\pi_2)=\mu\rtimes\pi_2$ of $\mathrm{PGSp}(2,\mathbb{A})$ is induced takes the value $\mu(a)\pi_2(A)$ at $t$, and $\mu(\det A/a)\pi_2(A)$ at $s_{2\alpha+\beta}t$. Since it is a representation of the projective group we have $\mu\omega=1$, where $\omega$ denotes the central character of $\pi$. From $\mu\otimes\pi_2\simeq s_{2\alpha+\beta}(\mu\otimes\pi_2)$ we conclude that $\mu^2=1$ and $\pi_2\simeq\omega\pi_2$, $\omega=\mu$ is 1 or has order 2. We have $\det(1-s_{2\alpha+\beta})|\mathcal{A}_{\mathbf{M}_\beta}=2$, $[W_0]=8$, $[W_0^{M_\beta}]=2$, and the contribution from $\mathbf{M}'_\beta$ is the same as that from $\mathbf{M}_\beta$, hence the contribution to the trace formula of $\mathbf{H}$ from $\mathbf{M}_\beta$ and $\mathbf{M}'_\beta$ and the unique element in $W^{\mathcal{A}}(A_{\mathbf{M}_\beta})$ when $\mathcal{A}=\{0\}$ is

$$2\cdot\frac{2}{8}\cdot\frac{1}{2}\sum_{\{\mu;\mu^2=1\}}\ \sum_{\{\pi_2;\pi_2=\mu\pi_2\}}\operatorname{tr}M(s_{2\alpha+\beta},0)I_{\mathbf{P}_\beta}(\mu,\pi_2;f_H).$$

The representations $I_{\mathbf{P}_\beta}(\mu,\pi_2)=\mu\rtimes\pi_2$ are irreducible when $\mu\neq1$, in which case the operator $M$ is a constant, equal to 1. When $\mu=1$, the representation $1\rtimes\pi_2$ is a product of local representations $1\rtimes\pi_{2v}$, which are irreducible unless $\pi_{2v}$ is square integrable or one dimensional. The operator $M$ can be written as a product $m\otimes_v R_v$ of a scalar valued function $m$ and local normalized operators $R_v$ (they map the $K_v$-fixed vector in an unramified $1\rtimes\pi_{2v}$ to itself). When $1\rtimes\pi_{2v}$ is irreducible, $R_v$ acts trivially. When $\pi_{2v}$ is square integrable $1\rtimes\pi_{2v}$ decomposes as a direct sum of two tempered constituents, $\pi^+_{Hv}$ and $\pi^-_{Hv}$, and $R_v$ acts on one constituent trivially, and by multiplication by $-1$ on the other. When $\pi_{2v}$ is $\xi\mathbf{1}_2$, $\xi^2=1$, $1\rtimes\xi\mathbf{1}_2$ has two irreducible (nontempered) constituents: $L(\nu,1\rtimes\nu^{-1/2}\xi)$ and $L(\nu^{1/2}\operatorname{sp}_2,\nu^{-1/2}\xi)$, and $R_v$ acts on the first trivially and by multiplication with $-1$ on the second. The scalar $m$ is 1.

Similarly we describe the spectral discrete contributions to the trace formula of the endoscopic group $\mathbf{C}_0=\mathrm{PGL}(2)\times\mathrm{PGL}(2)$ of $\mathbf{H}=\mathrm{PGSp}(2)$. The terms corresponding to the parabolic group $\mathbf{C}_0$ itself is as usual a sum over the discrete spectrum representations $\pi_1,\pi_2$ of $\mathrm{PGL}(2,\mathbb{A})$:

$$\sum_{\pi_1,\pi_2}\operatorname{tr}(\pi_1\times\pi_2)(f_{C_0}).$$

The proper parabolic subgroups are $\mathbf{M}_\beta = \mathbf{A}_0 \times \mathrm{PGL}(2)$, $\mathbf{M}_\alpha = \mathrm{PGL}(2) \times \mathbf{A}_0$, $\mathbf{M}_0 = \mathbf{A}_0 \times \mathbf{A}_0$, where $\mathbf{A}_0$ denotes here the diagonal subgroup in PGL(2). Thus $\mathbf{M}_0$ consists of $t = \mathrm{diag}(a,1)^* \times \mathrm{diag}(b,1)^*$. The roots are $\alpha(t) = a$, $\beta(t) = b$. They can be viewed as $\alpha = (1,0)$, $\beta = (0,1)$, in the lattice $X^*(\mathbf{M}_0) = \mathbb{Z} \times \mathbb{Z}$ in $\mathcal{A}_0^* = \mathbb{R} \times \mathbb{R}$. The coroots $\alpha^\vee = 2\alpha/(\alpha,\alpha) = (2,0)$, $\beta^\vee = (0,2)$ lie in the lattice $X_*(\mathbf{M}_0) = \mathbb{Z} \times \mathbb{Z}$ in $\mathcal{A}_0 = \mathbb{R} \times \mathbb{R}$. Since $X^*(\mathbf{M}_\beta) = X_*(\mathbf{A}_0 \times \{0\}) = \mathbb{Z} \times \{0\}$ and $X_*(\mathbf{M}_\alpha) = \{0\} \times \mathbb{Z}$, we have $\mathcal{A}_\alpha = \mathcal{A}_{\mathbf{M}_\alpha} = \mathbb{R}\beta^\vee$ and $\mathcal{A}_\beta = \mathcal{A}_{\mathbf{M}_\beta} = \mathbb{R}\alpha^\vee$. The Weyl group $W_0 = W(M_0)$ is generated by the commuting reflections $s_\alpha$ and $s_\beta$, where $s_\alpha(t) = \mathrm{diag}(1,a)^* \times \mathrm{diag}(b,1)^*$ and $s_\beta(t) = \mathrm{diag}(a,1)^* \times \mathrm{diag}(1,b)^*$. Identifying $X^*(\mathbf{M}_0)$ with a lattice in $\mathbb{R}\times\mathbb{R}$ these reflections become $s_\alpha(x,y) = (-x,y)$, $s_\beta(x,y) = (x,-y)$. The other nontrivial element in $W_0$ is $s_\alpha s_\beta = -1$. For $\mathcal{A} = \{0\}$ we have $W^{\mathcal{A}}(\mathcal{A}_0) = \{s_\alpha s_\beta\}$. Since $1 - s_\alpha s_\beta = 2$ and $\dim \mathcal{A}_0 = 2$, $\det(1 - s_\alpha s_\beta)|\mathcal{A}_0 = 4$. Further, $W^{\{0\}}(\mathcal{A}_\alpha) = \{s_\beta\}$ and $W^{\{0\}}(\mathcal{A}_\beta) = \{s_\alpha\}$, $(1-s_\alpha)(0,y) = (0,2y)$ hence $\det(1-s_\alpha)|\mathcal{A}_\beta = 2$ and $\det(1-s_\beta)|\mathcal{A}_\alpha = 2$. The representation $\mu_1 \otimes \mu_2$ of $\mathbf{M}_0$, taking $t$ to $\mu_1(a)\mu_2(b)$, is equal to $s_\alpha s_\beta(\mu_1 \otimes \mu_2)$, whose value at $t$ is $\mu_1^{-1}(a)\mu_2^{-1}(b)$, precisely when the characters $\mu_i$ are of order at most 2. The representation $\mu_1 \otimes \pi_2$ of $\mathbf{M}_\beta$ is equal to $s_\alpha(\mu_1 \otimes \pi_2)$ precisely when $\mu_1^2 = 1$. We obtain

$$\frac{1}{4}\cdot\frac{1}{4}\sum_{\mu_1^2=1=\mu_2^2} \mathrm{tr}\, M(s_\alpha s_\beta, 0) I_{\mathbf{P}_0}(\mu_1,\mu_2; f_{C_0})$$

$$+\frac{2}{4}\cdot\frac{1}{2}\sum_{\mu_1^2=1,\pi_2} \mathrm{tr}\, M(s_\alpha, 0) I_{\mathbf{P}_\beta}(\mu_1,\pi_2; f_{C_0})$$

$$+\frac{2}{4}\cdot\frac{1}{2}\sum_{\mu_2^2=1,\pi_2} \mathrm{tr}\, M(s_\beta, 0) I_{\mathbf{P}_\alpha}(\pi_2,\mu_2, f_{C_0}).$$

Note that the representations which occur in these three sums are well-known to be irreducible, from the theory of GL(2). Hence the operators $M$ are scalars, equal 1.

Similar analysis applies to the $\theta$-twisted endoscopic group

$$\mathbf{C} = [\mathrm{GL}(2) \times \mathrm{GL}(2)]'/\mathbb{G}_m,$$

whose group of $F$-points consists of $(g_1, g_2)$, $g_i$ in $\mathrm{GL}(2,F)$, $\det g_1 = \det g_2$, with $(g_1,g_2) \equiv (zg_1, zg_2)$, $z \in F^\times$. A character

$$(\mu_1,\mu_2';\mu_2,\mu_2') \bmod (\mu,\mu^{-1}), \quad \mu_1\mu_1'\mu_2\mu_2' = 1,$$

of the diagonal subgroup $\mathbf{M}_0$ of

$$t = \operatorname{diag}(a_1, a_2) \times \operatorname{diag}(b_1, b_2) \operatorname{mod}(z, z), \quad a_1 a_2 = b_1 b_2,$$

invariant under $s_\alpha s_\beta$ satisfies

$$\mu_1(a_1)\mu_1'(a_2)\mu_2(b_1)\mu_2'(b_2) = \mu_1(a_2)\mu_1'(a_1)\mu_2(b_2)\mu_2'(b_1)$$

for all $a_1 a_2 = b_1 b_2$, thus

$$\mu_1^2 = \mu_1'^{\,2} = \mu_2^{-2} = \mu_2'^{\,2},$$

which replaces the requirement $\mu_1^2 = \mu_2^2 = 1$ in the case of $\mathbf{C}_0$.

As for a representation $(\mu_1,\ \mu_1') \times \pi_2$ of the Levi subgroup $\mathbf{M}_\beta$, thus $\mu_1 \mu_2' \omega_\pi = 1$, if it is $s_\alpha$-fixed then its value at $\operatorname{diag}(a, b) \times g$, $ab = \det g$, which is $\mu_1(a)\mu_1'(b)\pi_2(g)$, is equal to its value at $\operatorname{diag}(b, a) \times g$, which is $\mu_1(b)\mu_1'(a)\pi_2(g)$. Here $ab = \det g$, so we conclude that $\frac{\mu_1'}{\mu_1}(\frac{\det g}{a^2}) = 1$ for all $a, g$, so $\mu_1' = \mu_1$.

Since all of the representations which contribute to the spectral sides of the trace formulae of $\mathbf{H}$, $\mathbf{C}$, $\mathbf{C}_0$ associated to proper parabolic subgroups are induced and are irreducible, except in the cases of $1 \rtimes \pi_2$, the intertwining operator $M(s, 0)$ in each case where the representation is irreducible is a scalar which comes outside the trace. Hence our assumption on the components of the test function $f$, hence also on the matching functions $f_H$, $f_C$, $f_{C_0}$, implies the vanishing of the contributions from the properly induced representations to the spectral sides of the trace formulae of $\mathbf{H}$, $\mathbf{C}$, $\mathbf{C}_0$.

## 4. Trace Formula Identity

We now review the trace formula identity for a test function $f = \otimes f_v$ on $\mathbf{G}(\mathbb{A}) = \mathrm{PGL}(4, \mathbb{A})$, and matching functions $f_H = \otimes f_{Hv}$ on $\mathbf{H}(\mathbb{A}) = \mathrm{PGSp}(2, \mathbb{A})$, $f_{C_0}$ on $\mathbf{C}_0(\mathbb{A}) = \mathrm{PGL}(2, \mathbb{A}) \times \mathrm{PGL}(2, \mathbb{A})$, $f_C$ on

$$\mathbf{C}(\mathbb{A}) = [\mathrm{GL}(2, \mathbb{A}) \times \mathrm{GL}(2, \mathbb{A})]'/\mathbb{A}^\times$$

where the prime indicates $(g_1, g_2) = ((g_{1v}), (g_{2v}))$ with $\det g_{1v} = \det g_{2v}$ in $F_v^\times$ for all $v$, and $f_{C_E} = \otimes f_{C_E, v}$ on $\mathbf{C}_E(\mathbb{A}) = \mathbb{A}_E^\times \times \mathbb{A}_E^\times$. The $\theta$-elliptic

$\theta$-semisimple strongly $\theta$-regular part of the geometric side of the $\theta$-twisted trace formula, $\mathrm{T}_e(f, \mathbf{G}, \theta)$, is

$$\mathrm{T}_e(f_H, \mathbf{H}) - \frac{1}{4}\mathrm{T}_e(f_{C_0}, \mathbf{C}_0) + \frac{1}{2}[\mathrm{T}_e(f_C, \mathbf{C}) - \frac{1}{4}\sum_{[E:F]=2} \mathrm{T}_E(f_{C_E})].$$

The assumption that at the place $v_0$ of $F$ the components $f_{v_0}$, $f_{Hv_0}, \dots$ vanish on the $(\theta)-$ singular set of $\mathbf{G}(F_{v_0})$, $\mathbf{H}(F_{v_0}), \dots$, and the assumptions that the components of $f$, $f_H$, ... at $v = v_1$, $v_2$, $v_3$ have ($\theta$-) orbital integrals which vanish on the strongly-($\theta$-) regular non-($\theta$-) elliptic sets, imply that the geometric sides of the ($\theta$-twisted) trace formulae are equal to the ($\theta$-) elliptic parts. The geometric sides are equal to the spectral sides – these are the trace formulae for each of the groups under consideration.

The spectral side $\mathrm{T}_{\mathrm{sp}}(f, \mathbf{G}, \theta)$ of the $\theta$-trace formula for $G$ and $f$ will be equal to the (weighted) sum of the spectral sides of the trace formulae:

$$\mathrm{T}_{\mathrm{sp}}(f_H, \mathbf{H}) - \frac{1}{4}\mathrm{T}_{\mathrm{sp}}(f_{C_0}, \mathbf{C}_0) + \frac{1}{2}[\mathrm{T}_{\mathrm{sp}}(f_C, \mathbf{C}) - \frac{1}{4}\sum_{[E:F]=2} \mathrm{T}_E(f_{C_E})].$$

Here is a summarized expression of the form of the spectral side of the $\theta$-twisted trace formula for $f$ on $\mathbf{G}(\mathbb{A}) = \mathrm{PGL}(4, \mathbb{A})$:

$$\mathrm{T}_{\mathrm{sp}}(f, \mathbf{G}, \theta) = I + \frac{1}{2}I_{(3,1)} + \frac{1}{2}I_{(2,2)} + \frac{1}{4}I_{(2,1,1)} + I_1,$$

where

$$I = \sum_{\pi} \operatorname{tr} \pi(f \times \theta),$$

$\pi$ ranges over the (equivalence classes of) discrete spectrum representations $\pi$ of $\mathbf{G}(\mathbb{A})$ which are $\theta$-invariant. Note that each of these $\pi$ occurs with multiplicity 1 in the discrete spectrum of $L^2(\mathbf{G}(F)\backslash\mathbf{G}(\mathbb{A}))$.

Further,

$$I_{(3,1)} = \sum_{\chi^2=1} \sum_{\tau \simeq {}^\theta\tau} \operatorname{tr} M(\alpha_3\alpha_2\alpha_1, \tau)(\chi I_{\mathbf{P}_{(3,1)}}(\tau, 1))(f \times \theta),$$

where $\tau$ ranges over the discrete spectrum representations of $\mathrm{PGL}(3, \mathbb{A})$ which satisfy $\tau \simeq \tau^\theta$; here $\theta(g) = J^{-1t}g^{-1}J$ and $J$ is $(\delta_{i,3-j})$, and $\chi$ is any quadratic character of $\mathbb{A}^\times/F^\times$ or 1.

Furthermore, $I_{(2,2)}$ is the sum of $I'_{(2,2)}$,

$$\frac{1}{2}\sum_{[E:F]=2}\sum_{\mu\in(\mathbb{A}_E^\times/\mathbb{A}^\times E^\times)^\wedge} \operatorname{tr} M(J,\pi(\tilde{\mu}),\pi(\tilde{\mu}))I_{\mathbf{P}_{(2,2)}}(\pi(\tilde{\mu}),\pi(\tilde{\mu}); f\times\theta),$$

$$\sum_{[E:F]=2}\sum_{\tilde{\mu}_1\neq\tilde{\mu}_2\in(\mathbb{A}_E^\times/\mathbb{A}^\times E^\times)^\wedge} \operatorname{tr} M(J,\pi(\tilde{\mu}_1),\pi(\tilde{\mu}_2))I_{\mathbf{P}_{(2,2)}}(\pi(\tilde{\mu}_1),\pi(\tilde{\mu}_2); f\times\theta),$$

where

$$I'_{(2,2)} = \frac{1}{2}\sum_{\tau} \operatorname{tr} M(J,\tau,\tau)I_{\mathbf{P}_{(2,2)}}(\tau,\tau; f\times\theta)$$
$$+\sum_{\tau_1\neq\tau_2} \operatorname{tr} M(J,\tau_1,\tau_2)I_{\mathbf{P}_{(2,2)}}(\tau_1,\tau_2; f\times\theta).$$

Here we put $\tilde{\mu}(z)=\mu(z/\overline{z})$ for a character $\mu$ of $\mathbb{A}_E^1/E^1$; $\tilde{\mu}$ is a character of $\mathbb{A}_E^\times/\mathbb{A}^\times E^\times$; $\tau_1$ and $\tau_2$ (and $\tau$) are discrete spectrum representations of $\mathrm{PGL}(2,\mathbb{A})$ and the sum over $\tau_1\neq\tau_2$ is over the unordered pairs; the sum over $\tilde{\mu}_1\neq\tilde{\mu}_2$ is over the unordered pairs too. Note that for representations of PGL(2) we have $\check{\tau}\simeq\tau$.

Next $I_{(2,1,1)}$ is the sum of

$$\frac{1}{2}\sum_{\chi^2=1}\sum_{\tau_1} \operatorname{tr}[M(J,(\tau_1,1,1))(\chi I_{\mathbf{P}_{(2,1,1)}}(\tau_1,1,1))](f\times\theta)$$

and

$$\sum_{[E:F]=2}\sum_{\chi^2=1}\sum_{\mu_0\in(\mathbb{A}_E^1/E^1)^\wedge} X$$

where

$$X=\operatorname{tr}\left[M(J,(\pi(\tilde{\mu}_0),\chi_E,1))I_{P_{(2,1,1)}}(\chi(\pi(\tilde{\mu}_0),\chi_1,1))\right](f\times\theta).$$

Here $\chi$ is a quadratic character of $\mathbb{A}^\times/F^\times$, $\tau_1$ is a discrete spectrum representation of $\mathrm{PGL}(2,\mathbb{A})$, and $\chi_E$ signifies the character $\neq 1$ on $\mathbb{A}^\times/F^\times$ which is trivial on $N_{E/F}\mathbb{A}_E^\times$.

Finally,

$$I_1 = \frac{1}{8} \sum_{\substack{x_i^2=1=\prod \chi_i \\ \chi_i \neq \chi_j}} \operatorname{tr} M(J,0) I(\boldsymbol{\chi}; f \times \theta)$$

$$+ \frac{1}{4 \cdot 8} \sum_{\substack{\chi_1 \neq \chi_2 \\ \chi_i^2=1}} \operatorname{tr} M(J,0) I((\chi_1, \chi_1, \chi_2, \chi_2); f \times \theta)$$

$$+ \frac{1}{4!8} \sum_{\chi^2=1} \operatorname{tr} M(J,0) I((\chi, \chi, \chi, \chi); f \times \theta)$$

$$+ \frac{1}{6} \sum_{\chi^2=1} \operatorname{tr} M((321)J,0) I((\chi, \chi, \chi, \chi); f \times \theta).$$

The twisted trace formula for $f$ on $\mathbf{G}(\mathbb{A})$ is equal to a sum of trace formulae listed below. First we have $\mathrm{T_{sp}}(f_H, \mathbf{H})$, which is

$$\sum_{\pi_H} \operatorname{tr} \pi_H(f_H) + \frac{1}{4} \sum_{\pi_2 \text{ of } \mathrm{PGL}(2,\mathbb{A})} \operatorname{tr} \otimes_v R_v \cdot (1 \rtimes \pi_{2v})(f_{Hv})$$

$$+ \frac{1}{4} \sum_{\mu \neq 1 = \mu^2} \sum_{\{\pi_2; \mu\pi_2 = \pi_2\}} \operatorname{tr} M I_{\mathbf{P}_\beta}(\mu, \pi_2; f_H)$$

$$+ \frac{1}{4} \sum_{\mu^2=1} \sum_{\pi_2 \text{ of } \mathrm{PGL}(2,\mathbb{A})} \operatorname{tr} M I_{\mathbf{P}_\alpha}(\pi_2, \mu; f_H) + \dots$$

The fourth contribution here involves a properly induced representation $I_{\mathbf{P}_\alpha}(\pi_2, \mu)$, which is irreducible. Consequently the intertwining operator $M(s_{\alpha+\beta}, 0)$ is a scalar which can be taken outside the trace. Then $\operatorname{tr} I_{\mathbf{P}_\alpha}(\pi_2, \mu; f_H)$ is a product of local terms, and those local terms at $v = v_1$, $v_2$, $v_3$ are zero by the assumption that we made, that the orbital integrals of $f_{Hv_i}$ vanish at the regular nonelliptic orbits. Similar observation applies to the third term in the spectral side of the trace formula of $\mathbf{H}$ and $f_H$ ($I_{\mathbf{P}_\beta}(\mu, \pi_2)$, $\mu \neq 1 = \mu^2$), as well as to the contributions from $\mathbf{P}_0$ that we did not write out here: they vanish for our test function $f_H$. Only the first two terms remain under our local assumption.

From this we subtract $\frac{1}{4}$ of the spectral side $\mathrm{T_{sp}}(f_{C_0}, \mathbf{C}_0)$ of the trace formula of $\mathbf{C}_0$ and $f_{C_0}$:

$$-\frac{1}{4}\Bigg[\sum_{\pi_1,\pi_2} \operatorname{tr}(\pi_1\times\pi_2)(f_{C_0}) + \frac{1}{4}\sum_{\mu_1^2=1,\pi_2} \operatorname{tr} M(s_\alpha,0) I_{\mathbf{P}_\beta}(\mu_1,\pi_2;f_{C_0})$$

$$+\frac{1}{4}\sum_{\mu_1^2=1,\pi_2} \operatorname{tr} M(s_\beta,0) I_{\mathbf{P}_\alpha}(\pi_2,\mu_1;f_{C_0})$$

$$+\frac{1}{16}\sum_{\mu_1^2=1=\mu_2^2} \operatorname{tr} M(s_\alpha s_\beta,0) I_{\mathbf{P}_0}(\mu_1,\mu_2;f_{C_0})\Bigg].$$

The representations $\quad I_{\mathbf{P}_\beta}(\mu_1,\pi_2) = I(\mu_1)\times\pi_2,$

$$I_{P_\alpha}(\pi_2,\mu_1) = \pi_2\times I(\mu_1), \quad I_{\mathbf{P}_0}(\mu_1,\mu_2) = I(\mu_1)\times I(\mu_2)$$

are properly induced, where $I(\mu)$ denotes the representation of $\mathrm{PGL}(2,\mathbb{A})$ induced from the character $\begin{pmatrix} a & b \\ 0 & a^{-1}\end{pmatrix} \mapsto \mu(a)$ of the proper parabolic subgroup. Since the group in question is PGL(2) they are irreducible, hence the operator $M(s,0)$ is a scalar, can be taken in front of the trace (in fact it is equal to 1), and the $\operatorname{tr} I_{\mathbf{P}}(\tau;f_{C_0})$ are products of local factors, those indexed by $v = v_1$, $v_2$, $v_3$ are zero by our assumption on the vanishing of the orbital integrals of the components $f_{C_0,v_i}$ on the regular elliptic set, hence the only contribution is the first:

$$-\frac{1}{4}\sum \operatorname{tr}(\pi_1\times\pi_2)(f_{C_0}).$$

To this we add $\frac{1}{2}$ of the spectral side of the stabilized trace formula for $\mathbf{C}$ and $f_C$; it is stabilized by subtracting $\frac{1}{4}\sum_E T_E(f_{C_E})$. To explain this trace formula, recall that a representation of $\mathbf{C}(\mathbb{A})$ is an equivalence class of representations $\pi_1\times\pi_2$ of $\mathrm{GL}(2,\mathbb{A})\times\mathrm{GL}(2,\mathbb{A})$ with $\omega_{\pi_1}\omega_{\pi_2}=1$ under the equivalence relation $\pi_1\times\pi_2 \simeq \chi\pi_1\times\chi^{-1}\pi_2$ for any character $\chi$ of $\mathbb{A}^\times/F^\times$. Thus the terms

$$\frac{1}{4}\operatorname{tr} M(s_\alpha,0) I_{\mathbf{P}_\beta}(\mu_1,\mu_1,\pi_2;f),$$

sum over the discrete spectrum representations $\pi_2$ of $\mathrm{GL}(2,\mathbb{A})$ and characters $\mu_1$ of $\mathbb{A}^\times/F^\times$, which appear in the trace formula for $\mathrm{GL}(2)\times\mathrm{GL}(2)$, would contribute to the trace formula of $C$ precisely one term: $I(\mu_1, \mu_1, \pi_2)$ would make a representation of $\mathbf{C}(\mathbb{A})$ precisely when $\mu_1^2 \omega_{\pi_2} = 1$, and $I(\mu_1, \mu_2, \pi_2) \simeq I(1, 1, \mu_1^{-1}\pi_2)$ for any $\mu_1$ as a representation of $\mathbf{C}(\mathbb{A})$. For any representation $\pi_2$ of $\mathrm{PGL}(2, \mathbb{A})$ (thus the central character $\omega_{\pi_2}$ is trivial), $I(1, 1, \pi_2)$ is irreducible. Hence the intertwining operator $M$ is a scalar, which can be evaluated to be equal to one by standard arguments.

Similarly, the terms

$$\frac{1}{4} \operatorname{tr} M(s_\beta, 0) I_{\mathbf{P}_\alpha}(\pi_2, \mu_1, \mu_1; f) \qquad (\mu_1^2 \omega_{\pi_2} = 1)$$

contribute just

$$\frac{1}{4} \operatorname{tr} I_{\mathbf{P}_\alpha}(\pi_2, 1, 1; f_C),$$

which is in fact equal to $\frac{1}{4} \operatorname{tr} I_{\mathbf{P}_\beta}(1, 1, \pi_2; f_C)$. Thus we have

$$+\frac{1}{2}\Bigg[\sum_{\pi_1 \times \pi_2} \operatorname{tr}(\pi_1 \times \pi_2)(f_C) + 2 \cdot \frac{1}{4} \sum_{\pi_2} \operatorname{tr} I_{\mathbf{P}_\beta}(1, 1, \pi_2; f_C)$$
$$+ \frac{1}{16} \sum_{\mu_1^2 = \mu_2^2 = 1} \operatorname{tr} M(s_\alpha s_\beta, 0) I_{\mathbf{P}_0}(1, \mu_1, \mu_2, \mu_1\mu_2; f_C) - \frac{1}{4} \sum_{[E:F]=2} \mathbf{T}_E(f_{C_E})\Bigg].$$

The first sum ranges over all discrete spectrum representations

$$\pi_1 \times \pi_2 \quad (\simeq \chi\pi_1 \times \chi^{-1}\pi_2, \qquad \omega_{\pi_1}\omega_{\pi_2} = 1)$$

of $\mathbf{C}(\mathbb{A})$. The second sum ranges over all discrete spectrum representations $\pi_2$ of $\mathrm{PGL}(2, \mathbb{A})$.

As the group $\mathbf{C} = [\mathrm{GL}(2) \times \mathrm{GL}(2)]'/\mathbb{G}_m$ is a proper subgroup of $[\mathrm{GL}(2) \times \mathrm{GL}(2)]/\mathbb{G}_m$, the induced representations $I_{\mathbf{P}_0}$ might be reducible (I have not checked this). Recall that a representation $I(\mu)$ of $\mathrm{SL}(2, F)$, induced from a character $\begin{pmatrix} a & b \\ 0 & a^{-1} \end{pmatrix} \mapsto \mu(a)$ is reducible precisely when $\mu$ has order 2 (or $\mu(x) = |x|^{\pm 1}$), in which case $I(\mu)$, $\mu^2 = 1 \neq \mu$, is the direct sum of two tempered constituents.

To derive lifting consequences from this identity of trace formulae – or rather their spectral sides – we use a usual argument of "generalized linear independence of characters", which is based on the "fundamental lemma". Almost all components of a representation $\pi = \otimes\pi_v$ of $\mathbf{G}(\mathbb{A})$ are unramified and for any spherical function $f_v$ on $\mathbf{G}(F_v)$ we have $\operatorname{tr}\pi_v(f_v \times \theta) = f_v^\vee(t(\pi_v) \times \theta)$ where $f_v^\vee$ is the Satake transform of $f_v$ and $t(\pi_v) \times \theta$ is the semisimple conjugacy class in $\widehat{G} \times \theta$ parametrizing the unramified representation $\pi_v$.

A standard argument (see, e.g., [FK2]) shows that the spherical functions provide a sufficiently large family to separate the classes $t(\pi_v) \times \theta$. The trace identity takes then the form where we fix a finite set $V$ of places of $F$ including the archimedean places, and an irreducible unramified representation $\pi_v$ of $\mathbf{G}(F_v)$ at each place $v$ outside $V$, and then all sums range only over the $\pi$ (or $I(\tau)$) whose component at $v$ is $\pi_v$, while the sums of representations of the groups $\mathbf{H}(\mathbb{A})$, $\mathbf{C}(\mathbb{A})$, $\mathbf{C}_0(\mathbb{A}), \dots$ range over the representations $\pi_H = \otimes\pi_{Hv}$, $\pi_C = \otimes\pi_{Cv}$, $\pi_{C_0} = \otimes\pi_{C_0v}$, etc., whose components at $v$ outside $V$ are unramified and satisfy

$$\lambda(t(\pi_{Hv})) = t(\pi_v), \quad \lambda_1(t(\pi_{Cv})) = t(\pi_v), \quad \lambda(\lambda_0(t(\pi_{C_0,v}))) = t(\pi_v).$$

Note that by multiplicity one and rigidity theorem for discrete spectrum representations for $\mathrm{PGL}(4,\mathbb{A})$, there exists at most one nonzero term in all the sums in $\mathrm{T_{sp}}(f,\mathbf{G},\theta)$. However, fixing $t(\pi_v)$ at all $v \notin V$ does not fix the $t(\pi_{Cv})$ and at this stage it is not even clear that the number of $\pi_H$, $\pi_C$, $\pi_{C_0}$ which appear in the trace formulae is finite.

The terms themselves in the sum are replaced by a finite product of local terms over the places $v$ at $V$, taking the forms

$$\prod_{v\in V} \operatorname{tr}\pi_v(f_v \times \theta), \qquad m(s,\tau)\prod_{v\in V} \operatorname{tr} R(\tau_v) I(\tau_v; f_v \times \theta).$$

The intertwining operator $M(s,\tau)$ is of the form $m(s,\tau)\prod_v R(s,\tau_v)$, where $m(s,\tau)$ is a normalizing global scalar valued function of the inducing representation $\tau$ on the Levi subgroup, and the $R(s,\tau_v)$ are local normalized intertwining operators, normalized by the property that they map the normalized (nonzero) $K_v$-fixed vector in the unramified representation to such a vector.

We shall view the identity of spectral sides of trace formulae stated above for matching test functions as stated for a choice of a finite set $V$, unramified representations $\pi_v$ at each $v$ outside $V$, and matching test functions $f_v$, $f_{Hv}, \dots$ at the places $v$ in $V$, where the terms in the sum are such finite products over $v$ in $V$.

For the statement of the fundamental lemma in our context and its proof we refer to [F5]. The existence of matching functions follows by a general argument of Waldspurger [W3] from the fundamental lemma. The statement ("generalized fundamental lemma") that corresponding (via the dual groups homomorphisms) spherical functions have matching orbital integrals, follows from the fundamental lemma (which deals only with unit elements in the Hecke algebras of spherical functions, namely with those functions which are supported and are constant on the standard maximal compact subgroups) by a well-known local-global argument, which uses the trace formula. We do not elaborate on this here, but simply use the ("generalized") fundamental lemma and the existence of matching functions.

# IV. LIFTING FROM SO(4) TO PGL(4)

## 1. From SO(4) to PGL(4)

We begin with the study of the lifting $\lambda_1$, and employ the trace identity

$$\mathrm{T}_{\mathrm{sp}}(f,\mathbf{G},\theta) = \mathrm{T}_{\mathrm{sp}}(f_H,\mathbf{H}) - \frac{1}{4}\mathrm{T}_{\mathrm{sp}}(f_{C_0},\mathbf{C}_0)$$

$$+\frac{1}{2}[\mathrm{T}_{\mathrm{sp}}(f_C,\mathbf{C}) - \frac{1}{4}\sum_{[E:F]=2} T_E(f_E)]$$

with data of a term in $\mathrm{T}_{\mathrm{sp}}(f_C,\mathbf{C})$. We choose the components at $v$ outside $V$ to be those of the trivial representation $\mathbf{1}_C = \mathbf{1}_2 \times \mathbf{1}_2$ of $\mathbf{C}(\mathbb{A})$. The parameters

$$t_C(\mathbf{1}_{2v} \times \mathbf{1}_{2v}) = [\mathrm{diag}(q_v^{1/2}, q_v^{-1/2}) \times \mathrm{diag}(q_v^{1/2}, q_v^{-1/2})]/\{\pm I\}$$

of its local components are mapped by $\lambda_1$ to $t = \mathrm{diag}(q_v, 1, 1, q_v^{-1})$, thus to the class of $I_{3,1}(\mathbf{1}_3, 1)$, the unramified irreducible representation of $\mathrm{PGL}(4, F_v)$ normalizedly induced from the trivial representation of the standard parabolic subgroup of type (3,1). Consequently the only nonzero contribution to $\mathrm{T}_{\mathrm{sp}}(f,\mathbf{G},\theta)$ is to $\frac{1}{2}I_{(3,1)}$. Had there been a nonzero contribution to $\mathrm{T}_{\mathrm{sp}}(f_H,\mathbf{H})$, almost all of its local components would have the parameters $t$, associated to $\pi_{\mathrm{PGSp}(2)}(\nu, 1) = L(\nu \times \nu \rtimes \nu^{-1})$, which is not unitarizable by [ST], Theorem 4.4. However, all components of an automorphic representation of $\mathrm{PGSp}(2,\mathbb{A})$ are unitarizable, hence there is no contribution to $\mathrm{T}_{\mathrm{sp}}(f_H,\mathbf{H})$.

Similarly there is no contribution to $\mathrm{T}_{\mathrm{sp}}(f_{C_0},\mathbf{C}_0)$, since had there been a contribution its local components would have to be $\pi_2(\nu,\nu^{-1}) \times \pi_2(1,1)$ at almost all places, but the irreducible $\pi_2(\nu,\nu^{-1})$ is not unitarizable. There is no contribution to any of the $T_E(f_E)$, since a contribution from a pair $\mu_1, \mu_2$ of characters of $\mathbf{C}_E(\mathbb{A}) = \mathbb{A}_E^\times \times \mathbb{A}_E^\times$ corresponds to a (cuspidal) representation $\pi_2(\mu_1) \times \pi_2(\mu_2)$ of $\mathbf{C}(\mathbb{A})$. But we fixed the parameters of the trivial representation $\mathbf{1}_C = \mathbf{1}_2 \times \mathbf{1}_2$ of $\mathbf{C}(\mathbb{A}) = [\mathrm{PGL}(2,\mathbb{A}) \times \mathrm{PGL}(2,\mathbb{A})]'/\mathbb{A}^\times$.

Moreover, since a discrete spectrum representation of $\mathrm{PGL}(2,\mathbb{A})$ with a trivial component is necessarily trivial, the only contribution to $\mathrm{T}_{\mathrm{sp}}(f_C,\mathbf{C})$ is from $\mathbf{1}_C$. We conclude that the trace formula reduces in our case to

$$(1)\qquad \prod_{v\in V}\operatorname{tr} I_{4v}(\mathbf{1}_3,1;f_v\times\theta)=\prod_{v\in V}\operatorname{tr}\mathbf{1}_{C,v}(f_{C,v}).$$

Since each of the representations $\mathbf{1}_{C,v}$ is elliptic – its character is not zero on the regular elliptic set in $C_v=\mathbf{C}(F_v)$ – we can choose three of the functions $f_v$ so that $f_{C,v}$ not only has orbital integrals which vanish on the regular nonelliptic set but moreover be supported on the regular elliptic set of $C_v$, and $\operatorname{tr}\mathbf{1}_{C,v}(f_{C,v})\neq 0$. The equality (1) then implies that $\operatorname{tr} I_{4v}(\mathbf{1}_3,1;f_v\times\theta)$ is a nonzero multiple of $\operatorname{tr}\mathbf{1}_{C,v}(f_{C,v})$ for all matching $f_v$, $f_{C,v}$.

1.1 PROPOSITION. *For every place $v$ of $F$, and for all matching functions $f_v$ and $f_{C,v}$ we have*

$$\operatorname{tr} I_{(3,1),v}(\mathbf{1}_3,1;f_v\times\theta)=\operatorname{tr}\mathbf{1}_{C,v}(f_{C,v}).$$

PROOF. Name the place $v$ of the proposition $v_0$. We apply the displayed identity with a set $V$ containing at least 3 places, but not the place $v_0$, and use $f_v$ such that $f_{Cv}$ is supported on the regular elliptic set for 3 places $v$ in $V$. We then apply the displayed identity with the set $V\cup\{v_0\}$, and with the same functions $f_v, f_{Cv}$ for $v\in V$. Of course we use $f_v, f_{Cv}$ with $\operatorname{tr}\mathbf{1}_{C,v}(f_{C,v})\neq 0$ for all $v$ in $V$. Taking the quotient, the proposition follows. □

Let us derive a character relation for the $\theta$-elliptic $\theta$-regular elements $t$ from the equality of the proposition, using the Weyl integration formula.

1.2 PROPOSITION. *We have the character identity*

$$\Delta(t\theta)\chi_\pi(t^{\mathbf{r}}\theta)=\iota\kappa(\mathbf{r})\Delta_C(Nt)\chi_{\pi_C}(Nt)\quad(\pi=I(\mathbf{1},1),\quad \pi_C=\mathbf{1}_C),$$

*where $t^{\mathbf{r}}$ denotes the element stably $\theta$-conjugate but not $\theta$-conjugate to $t$, and $\mathbf{r}$ ranges over $F^\times/N_{E/F}E^\times$ in case* I, *$E_3^\times/N_{E/E_3}E^\times$ in case* III, *and*

*$\kappa$ denotes the nontrivial character of this group. Here $\iota$ is* 2 *in case* I, 1 *in case* III, *and* 0 *in cases* II *and* IV.

PROOF. In local notations,

$$\operatorname{tr} I(\mathbf{1},1;f\times\theta)=\sum_T \frac{1}{[W^\theta(T)]}\int_{T/T^{1-\theta}}\Delta(t\theta)\chi(t\theta)\cdot F_f(t)dt$$

is equal to

$$\operatorname{tr}\mathbf{1}_C(f_C)=\sum_{T_C}\frac{1}{[W(T_C)]}\int_{T_C}\Delta_C(Nt)\chi_{\pi_C}(Nt)\cdot F_{f_C}(Nt)d(Nt)$$

for test functions $f$ and $f_C$ with matching orbital integrals. Matching means that for $\theta$-elliptic $\theta$-regular $t$ of type I or III, the $(\kappa,\theta)$-orbital integral of $f$ on $G$, denoted

$$F_f^\kappa(t)=F_f(t)-F_f(t')$$

(here $t'$ is an element stably $\theta$-conjugate but not $\theta$-conjugate to the $\theta$-regular $t$; $\kappa$ indicates the nontrivial character on the group of $\theta$-conjugacy classes within the stable $\theta$-conjugacy class) is equal to the stable orbital integral of $f_C$ on $C$ at the norm $Nt$ of $t$, denoted

$$F_{f_C}^{\mathrm{st}}(Nt)=F_{f_C}(Nt)+F_{f_C}((Nt)'),$$

where $(Nt)'$ denotes an element stably conjugate but not conjugate to $Nt$. Implicitly we use the fact that the norm map $N$ is onto. It is defined for elliptic elements only in types I and III, as recalled in chapter II, section 5. The notation $F_f(t)$ and $F_{f_C}(t)$ for the ($\theta$-) orbital integral multiplied by the $\Delta$-factor was introduced in Definition II.2.4.

To determine the group of conjugacy classes within the stable class of $T_C=NT$ in $C$ where $T$ is of type I or III, we compute $H^{-1}(F,X_*(T_C))$. It is the quotient of the lattice $\{X\in X_*(T_C);NX=0\}$

$$=\{(x_1,y_1;x_2,y_2)\bmod(x,x;y,y);x,x_i,y,y_i\in\mathbb{Z},x_1+y_1\equiv x_2+y_2(\mathrm{mod}\,2)\}$$

by

$$\langle X-\tau X;\tau\in\mathrm{Gal}(\overline{F}/F)\rangle=\{(x,-x;y,-y);x,y\in\mathbb{Z}\},$$

namely $\mathbb{Z}/2$. Indeed, in case I the Galois group is $\mathrm{Gal}(E/F) = \langle\sigma\rangle$, with

$$\sigma(x_1, y_1; x_2, y_2) = (y_1, x_1; y_2, x_2).$$

In case III the Galois group is $\mathrm{Gal}(E/F) = \langle\sigma, \tau\rangle$ with

$$\sigma(x_1, y_1; x_2, y_2) = (y_1, x_1; y_2, x_2) \quad \text{and} \quad \tau(x_1, y_1; x_2, y_2) = (x_1, y_1; y_2, x_2).$$

We choose the function $f_C$ to be supported only on the regular elliptic set, and stable, namely such that $F_{f_C}(Nt)$ and $F_{f_C}((Nt)')$ be equal. We choose the function $f$ on $G$ to be supported on the $\theta$-regular $\theta$-elliptic set and unstable, thus $F_f(t) = -F_f(t')$ if $t$, $t'$ are stably $\theta$-conjugate but not $\theta$-conjugate. Thus we choose $f, f_C$ related on elements of type I by

$$F_f\begin{pmatrix} a_1\mathbf{r} & 0 & 0 & a_2D\mathbf{r} \\ 0 & \mathbf{s}b_1 & \mathbf{s}Db_2 & 0 \\ 0 & \mathbf{s}b_2 & \mathbf{s}b_1 & 0 \\ a_2\mathbf{r} & 0 & 0 & a_1\mathbf{r} \end{pmatrix} = \kappa_E(\mathbf{rs})F_{f_C}(\boldsymbol{\alpha}, \boldsymbol{\beta}),$$

$\boldsymbol{\alpha} = \begin{pmatrix} \alpha_1 & \alpha_2 D \\ \alpha_2 & \alpha_1 \end{pmatrix}$ if $\alpha = \alpha_1 + \alpha_2\sqrt{D}$, where

$$\alpha = a_1b_1 + Da_2b_2 + (b_1a_2 + a_1b_2)\sqrt{D} (= (a_1 + a_2\sqrt{D})(b_1 + b_2\sqrt{D}))$$

and

$$\beta = a_1b_1 - Da_2b_2 + (b_1a_2 - a_1b_2)\sqrt{D} (= (a_1 + a_2\sqrt{D})(b_1 - b_2\sqrt{D})),$$

and similarly in case III. Taking $f$, $f_C$ with $F_f(t)$ supported on a small neighborhood of a $\theta$-regular $t_0$, the proposition and the Weyl integration formulae imply – since the characters are locally constant functions on the $\theta$-regular set – the character identity

$$\Delta(t\theta)\chi_\pi(t^{\mathbf{r}}\theta) = \frac{[W^\theta(T)]}{[W(T_C)]}\kappa(\mathbf{r})\Delta_C(Nt)\chi_{\pi_C}(Nt) \quad (\pi = I(\mathbf{1}, 1), \quad \pi_C = \mathbf{1}_C),$$

where $t^{\mathbf{r}}$ denotes the element stably $\theta$-conjugate but not $\theta$-conjugate to $t$, and $\mathbf{r}$ ranges over $F^\times/N_{E/F}E^\times$ in case I, $E_3^\times/N_{E/E_3}E^\times$ in case III, and $\kappa$ denotes the nontrivial character of this group.

Since the $\theta$-conjugacy classes of type II and IV are not related by the norm map to conjugacy classes in $C$, whatever the choice of $f$ is on these classes, the integral

$$\int_{T/T^{1-\theta}} \Delta(t\theta)\chi_\pi(t\theta)F_f(t)dt$$

is zero, hence $\chi_\pi(t\theta)$ vanishes on the $\theta$-regular $\theta$-conjugacy classes of type II or IV.

It remains to compute the numbers $[W^\theta(T)]$ and $[W(T_C)]$. The torus $T_C$ consists of elements

$$\left(\begin{pmatrix} c_1 & c_2 D \\ c_2 & c_1 \end{pmatrix}, \begin{pmatrix} d_1 & d_2 D \\ d_2 & d_1 \end{pmatrix}\right), \quad c_1^2 - Dc_2^2 = d_1^2 - Dd_2^2 \bmod F^{\times 2}.$$

Its normalizer (modulo centralizer) in $\mathbf{C}(\overline{F})$ is generated by

$$(\mathrm{diag}(i,-i), I), \qquad (I, \mathrm{diag}(i,-i)),$$

where $i \in \overline{F}^\times$ with $i^2 = -1$. Hence $[W(T_C)]$ is 4.

The $\theta$-normalizer modulo the $\theta$-centralizer of the torus $T$ is generated by $\begin{pmatrix} w & 0 \\ 0 & w \end{pmatrix}$, $w = \begin{pmatrix} 0 & 1 \\ 1 & 0 \end{pmatrix}$, and $\mathrm{diag}(i,1,1,-i)$ in case I. Hence $[W^\theta(T)]$ is 8, and $[W^\theta(T)]/[W(T_C)] = 2$. In case III the $\theta$-normalizer modulo the $\theta$-centralizer is $\mathbb{Z}/2 \times \mathbb{Z}/2$, generated by the matrices $\mathrm{diag}(-1,-1,1,1)$ and $\mathrm{diag}(-1,1,-1,1)$, hence $[W^\theta(T)]/[W(T_C)] = 1$. □

REMARKS. (1) The computation of the twisted character $\chi_{I(\mathbf{1},1)}(t^{\mathbf{r}}\theta)$ is reached by purely local means in the paper [FZ] with D. Zinoviev.
(2) The propositions remain true when the local field $F_{v_1}$ is archimedean. Indeed, we choose the global field $F$ to be $\mathbb{Q}$ or an imaginary quadratic extension thereof, and apply the global identity (1) once with a set $V$ consisting of 3 nonarchimedean places (and $f_v$, $f_{Cv}$ supported on the ($\theta$-) elliptic set for $v \in V$), and once with $V \cup \{v_0\}$. In the real case, where $F_{v_0} = \mathbb{R}$, the only $\theta$-elliptic elements are of type I, and we obtain the character relation

$$\Delta(t\theta)\chi_{I(\mathbf{1},1)}(t^{\mathbf{r}}\theta) = 2\kappa(\mathbf{r})\Delta_C(Nt)\chi_{\mathbf{1}_C}(Nt), \quad \kappa : \mathbb{R}^\times/\mathbb{R}_+^\times \tilde{\to} \{\pm 1\}.$$

In the complex case there are no $\theta$-elliptic elements, and all $\theta$-regular elements are $\theta$-conjugate to elements in the diagonal torus $T$. For $t =$

diag$(a,b,c,d)$ and $Nt = (\mathrm{diag}(ab,cd), \mathrm{diag}(ac,bd))$, noting that $W^\theta(T) = D_4$ has cardinality 8, and $W(T_C)$ is $\mathbb{Z}/2 \times \mathbb{Z}/2$, generated by $(w, I)$ and $(I, w)$, $w = \begin{pmatrix} 0 & -1 \\ 1 & 0 \end{pmatrix}$, of cardinality 4, we have

$$\Delta(t\theta)\chi_\pi(t\theta) = 2\Delta_C(Nt)\chi_{\pi_C}(Nt),$$

when $F_{v_0} = \mathbb{C}$, $\pi = I(\mathbf{1}_3, 1)$ and $\pi_C = \mathbf{1}_C$, or $F$ is any local field, $\pi_C = I(\mu_1, \mu_1^{-1}) \times I(\mu_2, \mu_2^{-1})$ and $\pi = \lambda_1(\pi_C) = I(\mu_1\mu_2, \mu_1/\mu_2, \mu_2/\mu_1, 1/\mu_1\mu_2)$ are induced.

1.3 Definition. The admissible representation $\pi_C = \pi_1 \times \pi_2$ of

$$C = [\mathrm{GL}(2,F) \times \mathrm{GL}(2,F)]'/F^\times$$

*lifts* to the admissible representation $\pi$ of $G = \mathrm{PGL}(4,F)$, $F$ local, and we write $\pi = \lambda_1(\pi_C)$, if for all matching functions $f$, $f_C$ we have

$$\mathrm{tr}\,\pi(f \times \theta) = \mathrm{tr}\,\pi_C(f_C).$$

Equivalently we have the character relations $\chi_\pi(t\theta) = 0$ for $\theta$-regular elements without norm in $C$ (type II and IV for $\theta$-elliptic elements, as well as non-$\theta$-elliptic elements of type (2), (3) of [F5], p. 15 and p. 9, where $T^* = \{\mathrm{diag}(a,b,\sigma a,\sigma b); a,b \in E^\times\}$), and

$$\Delta(t\theta)\chi_\pi(t^{\mathbf{r}}\theta) = ([W^\theta(T)]/[W_C(NT)])\kappa(\mathbf{r})\Delta_C(Nt)\chi_{\pi_C}(Nt)$$

for $\theta$-regular $t$ in $G$ with norm in $C$, thus of type I and III for $\theta$-elliptic $t$, for split $t$ and for $t$ of type (1) and (1′) of [F5], p. 15 (and p. 9).

Type (1) has $T^* = \{\mathrm{diag}(a,\sigma a,b,\sigma b); a,b \in E^\times\}$, type (1′) has $T^* = \{\mathrm{diag}(a,b,\sigma b,\sigma a); a,b \in E^\times\}$, $[E:F] = 2$. The norms are

$$(\mathrm{diag}(a\sigma a, b\sigma b), \mathrm{diag}(ab, \sigma a\sigma b)) \quad \text{and} \quad (\mathrm{diag}(ab, \sigma a\sigma b), \mathrm{diag}(a\sigma a, b\sigma b)),$$

in cases (1) and (1′), and stable $\theta$-conjugacy coincides with $\theta$-conjugacy in cases (1), (1′) and the split elements. Thus $\kappa = 1$ and $\mathbf{r} = 1$ in these cases. In the case of split $t$, $W^\theta(T) = D_4$ has 8 elements while $W_C(NT) = \mathbb{Z}/2 \times \mathbb{Z}/2$ has 4. For $t$ of type (1), $W^\theta(T^*)$ consists of 1, (12)(34), (13)(24), (14)(23) ($W^\theta(T)$ is generated by diag$(1,-1,1,-1)$ and antidiag (I,I)), and

$W_C(NT) = \mathbb{Z}/2 \times \mathbb{Z}/2$ too, generated by $\left(\left(\begin{smallmatrix} 0 & -1 \\ 1 & 0 \end{smallmatrix}\right), I\right)$ and (diag$(-1,1)$, diag$(-1,1)$). In type (1′) $W^\theta(T)$ is generated by diag$(-1,-1,1,1)$ and diag$(w,w)$, and $W_C(NT)$ by (diag$(-1,1)$, diag$(-1,1)$) and $\left(I, \left(\begin{smallmatrix} 0 & -1 \\ 1 & 0 \end{smallmatrix}\right)\right)$.

Then in cases (1) and (1′) we have $[W^\theta(T)]/[W_C(NT)] = 1$, and $= 2$ for split $T$ or $T$ of type I. In type III, $W^\theta(T)$ is generated by diag$(1,-1,1,-1)$ and diag$(-I,I)$ (which act on diag$(\alpha,\tau\alpha,\sigma\tau\alpha,\sigma\alpha)$ in $T^*$ as (43)(21) and (32)(41)), and $W_C(NT)$ by (diag$(i,-i), I$) and ($I$, diag$(i,-i)$), hence the quotient $[W^\theta(T)]/[W_C(NT)]$ is 1 in type III.

Given a representation $\pi_C = \pi_1 \times \pi_2$ of $C = [\mathrm{GL}(2,F) \times \mathrm{GL}(2,F)]'/F^\times$ – thus the central characters $\omega_1, \omega_2$ of $\pi_1, \pi_2$ satisfy $\omega_1\omega_2 = 1$ – and characters $\chi_1, \chi_2$ of $F^\times$ with $\chi_1^2\chi_2^2 = 1$, $F$ local, we write $\chi_1\pi_1 \times \chi_2\pi_2$ for the representation $(g_1, g_2) \mapsto (\pi_1(g_1) \otimes \pi_2(g_2))\chi_1(g_1)\chi_2(g_2)$; note that $\chi_1(g_1)\chi_2(g_2) = \chi_1\chi_2(g_1) = \chi_1\chi_2(g_2)$ since $\det g_1 = \det g_2$. The character relation implies

1.4 PROPOSITION. *If $\pi_1 \times \pi_2$ lifts to a representation $\pi$ of the group $G = \mathrm{PGL}(4,F)$, then $\chi_1\pi_1 \times \chi_2\pi_2$ lifts to $\chi_1\chi_2\pi$.*

PROOF. The characters $\chi_1, \chi_2$ depend only on the determinant. As the norm map is

$$N(\mathrm{diag}(a,b,c,d)) = (\mathrm{diag}(ab,cd), \mathrm{diag}(ac,bd)),$$

we have

$$(\chi_1\chi_2)(abcd) = \chi_1(ab \cdot cd)\chi_2(ac \cdot bd). \qquad \Box$$

Denote by $\mathrm{sp}_2$ or $\mathrm{St}_2$ the special (= Steinberg) square integrable subrepresentation of the induced representation $I(\nu^{1/2}, \nu^{-1/2})$ of $\mathrm{PGL}(2,F)$, and by $\mathrm{St}_3$ the Steinberg square integrable subrepresentation of the induced representation $I(\nu, 1, \nu^{-1})$ of $\mathrm{PGL}(3,F)$. Put also $\mathrm{sp}_2(\chi) = \chi \otimes \mathrm{sp}_2$ for a character $\chi$ of $F^\times/F^{\times 2}$.

Since $\chi_{I(\nu^{1/2},\nu^{-1/2})} = \chi_{\mathrm{sp}_2} + \chi_{\mathbf{1}_2}$ vanishes on the regular elliptic set of $\mathrm{PGL}(2,F)$, for a function $h$ on the regular elliptic set of $\mathrm{PGL}(2,F)$ we have $\mathrm{tr}\,\mathrm{sp}_2(h) = -\mathrm{tr}\,\mathbf{1}_2(h)$. Hence for a function $f_C$ on $C$, supported on the regular elliptic set of $C$, we have

$$\mathrm{tr}(\mathbf{1}_2 \times \mathbf{1}_2)(f_C) = -\mathrm{tr}(\mathrm{sp}_2 \times \mathbf{1}_2)(f_C)$$

$$= -\operatorname{tr}(\mathbf{1}_2 \times \mathrm{sp}_2)(f_C) = \operatorname{tr}(\mathrm{sp}_2 \times \mathrm{sp}_2)(f_C).$$

Let $\pi_2$ denote an irreducible unitarizable representation of $\mathrm{PGL}(2, F)$. Let $J(\nu^{1/2}\pi_2, \nu^{-1/2}\pi_2)$ be the unique ("Langlands") quotient of the induced representation $I = I(\nu^{1/2}\pi_2, \nu^{-1/2}\pi_2)$ of $\mathrm{PGL}(4, F)$. It is unramified if $\pi_2$ is, in fact it is the unique unramified constituent of $I$ if $I$ is unramified.

1.5 PROPOSITION. *The representations $\pi_{C,v_0} = \pi_{v_0} \times \mathbf{1}_{v_0}$ and $\mathbf{1}_{v_0} \times \pi_{v_0}$ of $C_{v_0}$ lift via $\lambda_1$ to $J(\nu_{v_0}^{1/2}\pi_{v_0}, \nu_{v_0}^{-1/2}\pi_{v_0})$ for every square integrable representation $\pi_{v_0}$ of* $\mathrm{PGL}(2, F_{v_0})$.

PROOF. We choose a number field $F$ whose completion at a place $v_0$ is our nonarchimedean field $F_{v_0}$, and 3 other nonarchimedean places: $v_1$, $v_2$, $v_3$. Fix a cuspidal representation $\pi_{v_1}$ of $\mathrm{PGL}(2, F_{v_1})$, and let $\pi_{v_2}$ and $\pi_{v_3}$ be the special representations $\mathrm{sp}_2$ at $v_2$ and $v_3$. Using the trace formula for $\mathrm{PGL}(2, \mathbb{A})$ one constructs a cuspidal representation $\pi$ whose components at $v_1, v_2, v_3$ are our $\pi_{v_i}$, which is unramified outside $V = \{v_1, v_2, v_3, \infty\}$, and a cuspidal representation $\pi'$ whose components at $v$ in $V' = \{v_0, v_1, v_2, v_3, \infty\}$ are our $\pi_v$, which is unramified outside $V'$.

We use the trace identity with the sets $V$ (resp. $V'$), such that $\pi \times \mathbf{1}_2$ and $\mathbf{1}_2 \times \pi$ (resp. $\pi' \times \mathbf{1}_2$ and $\mathbf{1}_2 \times \pi'$) are the only contributions to the trace formula of $\mathbf{C}(\mathbb{A})$. We choose test functions $f$ (resp. $f'$) such that their components at $v_2, v_3$ are supported on the $\theta$-regular elliptic set, and such that the stable $\theta$-orbital integral of $f_{v_2}$ and $f_{v_3}$ are zero. This guarantees that in the trace identity there are no contributions from $H$.

Now in the trace identity, for $v$ outside $V$ we fix the class

$$t_{C,v}(\mathbf{1}_2 \times \pi_{2v}) = [\operatorname{diag}(q_v^{1/2}, q_v^{-1/2}) \times \operatorname{diag}(\mu_{2v}^\bullet, \mu_{2v}^{\bullet-1})]/\{\mathbb{C}^\times\},$$

where $\pi_{2v}$ is the unramified component of $I(\mu_{2v}, \mu_{2v}^{-1})$, and $\mu_{2v}^\bullet = \mu_{2v}(\boldsymbol{\pi}_v)$. This class is mapped by $\lambda_1$ to

$$t_v = \operatorname{diag}(q_v^{1/2}\mu_{2v}^\bullet, q_v^{1/2}\mu_{2v}^{\bullet-1}, q_v^{-1/2}\mu_{2v}^\bullet, q_v^{-1/2}\mu_{2v}^{\bullet-1}),$$

which is the parameter of $J(\nu_v^{1/2}\pi_{2v}, \nu_v^{-1/2}\pi_{2v})$.

Thus the unique contribution to the trace formula of $G = \mathrm{PGL}(4, F)$ is the discrete spectrum noncuspidal representation

$$J(\nu^{1/2}\pi, \nu^{-1/2}\pi).$$

Since the trace formula for $C$ appears in our identity with coefficient $\frac{1}{2}$, and both $\mathbf{1} \times \pi$ and $\pi \times \mathbf{1}$ make equal contribution, we conclude the equalities

$$\prod_v \operatorname{tr} J(\nu_v^{1/2}\pi_v, \nu_v^{-1/2}\pi_v; f_v \times \theta) = \prod_v \operatorname{tr}(\mathbf{1}_v \times \pi_v)(f_{C,v}),$$

where the products range over $V$, and over $V'$. Since we can choose $f_{C,v}$ for which the terms on the right are nonzero, dividing the equality for $V'$ by that for $V$, and noting that $f_{v_0}$ is arbitrary, the proposition follows. □

## 2. Symmetric Square

The diagonal case of the lifting $\lambda_1 : \pi_C = \pi_1 \times \pi_2 \mapsto \pi$, that is when $\check{\pi}_1 = \pi_2$, coincides with the symmetric square lifting from SL(2) to PGL(3) established in [F3]. More precisely we shall use here the results of [F3] to relate terms in our identity of trace formulae, and in particular obtain the (new) character relation $\lambda_1(\pi_1 \times \check{\pi}_1) = I_{(3,1)}(\operatorname{Sym}^2 \pi_1, 1)$ for admissible representations $\pi_1$. The global and local results of [F3] are considerably stronger than what we need here. Not only that we work in [F3] with arbitrary cuspidal representations $\pi_1$, and put no local restrictions (that 3 local components $\pi_{1v}$ of $\pi_{1v}$ be elliptic) as here, but more significantly, [F3] lifts representations of SL(2) – rather than of GL(2). Consequently [F3] proves in particular multiplicity one theorem for discrete spectrum representations of $\mathrm{SL}(2,\mathbb{A})$ as well as the rigidity theorem for packets of such representations, as well as it characterizes all representations of $\mathrm{PGL}(3,\mathbb{A})$ which are invariant under transpose-inverse as lifts from $\mathrm{SL}(2,\mathbb{A})$ (or $\mathrm{PGL}(2,\mathbb{A})$).

For our purposes here we simply observe that the restriction of a representation of $\mathrm{GL}(2,\mathbb{A})$ (resp. of $\mathrm{GL}(2,F)$, $F$ local) to $\mathrm{SL}(2,\mathbb{A})$ (resp. $\mathrm{SL}(2,F)$) defines a packet of representations on $\mathrm{SL}(2,\mathbb{A})$ (resp. $\mathrm{SL}(2,F)$). At almost all places of a number field, the unramified components of $\pi = \otimes \pi_v$ satisfy $\lambda_1(\pi_v \times \check{\pi}_v) = I_{(3,1)}(\operatorname{Sym}^2 \pi_v, 1)$, where if $\pi_v = I(a_v, b_v)$ then $\operatorname{Sym}^2(a_v, b_v) = (a_v/b_v, 1, b_v/a_v)$.

Here is a summary of the symmetric square case in our context of PGL(4).

2.1 PROPOSITION. (1) *For each cuspidal representation* $\pi_2$ *of* $\mathrm{GL}(2,\mathbb{A})$ *there exists an automorphic representation* $\pi = \mathrm{Sym}^2(\pi_2)$ *of* $\mathrm{PGL}(3,\mathbb{A})$ *which is invariant under the transpose-inverse involution* $\theta_3$ *such that* $\lambda_1(\pi_2 \times \check{\pi}_2) = I_{(3,1)}(\mathrm{Sym}^2(\pi_2),1)$.
(2) *If* $\pi_2$ *is of the form* $\pi_2(\mu)$, *related to a character* $\mu : \mathbb{A}_E^\times/E^\times \to \mathbb{C}^\times$ *where* $E/F$ *is a quadratic extension of number fields, then* $\mathrm{Sym}^2(\pi_2)$ *is* $I_{(2,1)}(\pi_2(\mu/\overline{\mu}),\chi_E)$, *where* $\overline{\mu}(x) = \mu(\overline{x})$, $x \mapsto \overline{x}$ *denotes the action of the nontrivial automorphism of* $E/F$ *and* $\chi_E$ *is the quadratic character of* $\mathbb{A}^\times/F^\times$ *trivial on the norm subgroup* $N_{E/F}\mathbb{A}_E^\times$.
(3) *If* $\pi_2$ *is cuspidal but not of the form* $\pi_2(\mu)$, *then* $\mathrm{Sym}^2(\pi_2)$ *is cuspidal.*
(4) *If* $\mathrm{Sym}^2(\pi_2) = \mathrm{Sym}^2(\pi_2')$ *then* $\pi_2' = \chi\pi_2$ *for some character* $\chi$ *of* $\mathbb{A}^\times/F^\times$.
(5) *Each* $\theta_3$*-invariant cuspidal* $\pi_3$ *is of the form* $\mathrm{Sym}^2(\pi_2)$.

*The analogous results hold locally. For each admissible irreducible representation* $\pi_{2v}$ *of* $\mathrm{GL}(2,F_v)$ *there is an irreducible representation* $\pi_{3v} = \mathrm{Sym}^2(\pi_{2v})$ *of* $\mathrm{PGL}(3,F_v)$, *invariant under the transpose-inverse involution* $\theta_3$, *such that the character relation* $\lambda_1(\pi_{2v} \times \check{\pi}_{2v}) = I_{(3,1)}(\mathrm{Sym}^2(\pi_{2v}),1)$ *holds. If* $\mathrm{Sym}^2(\pi_{2v}') = \mathrm{Sym}^2(\pi_{2v})$ *then* $\pi_{2v}' = \chi_v\pi_{2v}$ *for some character* $\chi_v$ *of* $F_v^\times$. *Each* $\theta_3$*-invariant cuspidal* $\pi_{3v}$ *is of the form* $\mathrm{Sym}^2(\pi_{2v})$. *As* $\mathrm{Sym}^2(\mathrm{sp}_2) = \mathrm{St}_3$, *we have* $\lambda_1(\mathrm{sp}_2 \times \mathrm{sp}_2) = I_{(3,1)}(\mathrm{St}_3,1)$.

PROOF. The global claims (1)-(5) are consequences of the results of [F3]. The new claim here is the character relation. Note that the character relation has already been proven by direct computation for $\pi_{2v}$ which is an induced representation, as well as for the trivial representation $\pi_{2v} = \mathbf{1}_{2v}$. Thus we need to prove the character relation for square integrable $\pi_{2v}$.

We fix a global field $F$ which is $\mathbb{Q}$ if $F_v = \mathbb{R}$ or totally imaginary if $F_v$ is nonarchimedean , whose completion at a place $v_0$ is our $F_v$, cuspidal representations $\pi_{2v_1}$, $\pi_{2v_2}$, and the special representation $\pi_{2v_3}$ of $\mathrm{GL}(2,F_v)$ at the nonarchimedean places $v = v_1, v_2, v_3$ of $F$, and construct a cuspidal representation $\pi_2$ whose components at $v_i$ $(0 \le i \le 3)$ are those specified, while those outside the set $V$ consisting of the archimedean places and $v_i$ $(0 \le i \le 3)$, are unramified.

We apply the trace formula identity with the set $V$ and a contribution $\pi_C = \pi_2 \times \check{\pi}_2$ to the trace formula identity. We take the test function $f_{v_3}$ to be supported on the $\theta$-elliptic regular set, such that $\mathrm{tr}\,\pi_{C,v_3}(f_{C,v_3}) \neq 0$ and with $f_{H,v_3} = 0$ (thus the stable $\theta$-orbital integrals of $f_{v_3}$ are zero).

This choice is possible by the character identity $\mathrm{tr}(\mathbf{1}_{2v} \times \mathbf{1}_{2v})(f_{Cv}) = \mathrm{tr}\, I_{(3,1)}(\mathbf{1}_{3v}, 1_v)(f_v \times \theta)$. Consequently we get a trace identity with no contributions from the trace formula of $H$, while the contribution to the $\theta$-twisted trace formula of $G$ is only $I_{(3,1)}(\mathrm{Sym}^2 \pi_2, 1)$, by the rigidity theorem for GL(4). Note that the coefficient of $\mathrm{T_{sp}}(f_C, \mathbf{C})$ is $\frac{1}{2}$, and so is the coefficient of the term $I_{(3,1)}$ in $\mathrm{T_{sp}}(f, \mathbf{G}, \theta)$.

Denoting by $V_f$ the set $\{v_0, v_1, v_2, v_3\}$, we conclude, for all matching functions $f_{v_0}$ and $f_{C,v_0}$, the identity

$$m(\alpha_3\alpha_2\alpha_1, \tau) \prod_{v \in V_f} \mathrm{tr}\, I_{(3,1)}(\tau_v, 1)(f_v \times \theta) = \prod_{v \in V_f} \mathrm{tr}\, \pi_{C,v}(f_{C,v}).$$

Here we wrote the intertwining operator $M(\alpha_3\alpha_2\alpha_1, \tau)$, $\tau = \mathrm{Sym}^2(\pi_2)$ where $\pi_C = \pi_2 \times \check{\pi}_2$, as a product of local factors $R(\alpha_3\alpha_2\alpha_1, \tau_v)$ over all places $v$ and a global normalizing factor $m(\pi_2) = m(\alpha_3\alpha_2\alpha_1, \tau)$, and incorporated the local factor in the definition of the operator $\theta$, thus $\mathrm{tr}\, I_{(3,1)}(\tau_v, 1)(f_v \times \theta)$ stands for

$$\mathrm{tr}\, R(\alpha_3\alpha_2\alpha_1, \tau_v) I_{(3,1)}(\tau_v, 1)(f_v \times \theta).$$

Note that $R(\tau_v) = R(\alpha_3\alpha_2\alpha_1, \tau_v)$ is normalized by the property that $R(\tau_v)\pi_v(\theta)$ fixes the $K_v$-fixed vector when $\pi_v = I_{(3,1)}(\tau_v, 1)$ is unramified.

We now repeat our argument with the set $V_f' = V_f - \{v_0\}$ and construct a cuspidal $\pi_2'$ unramified outside $V_f'$ whose components at $v_1, v_2, v_3$ are as above (we are assuming that $v_0$ is nonarchimedean). Dividing the identity for $V_f$ by the new identity for $V_f'$ we get

$$\frac{m(\alpha_3\alpha_2\alpha_1, \tau)}{m(\alpha_3\alpha_2\alpha_1, \tau')} \mathrm{tr}\, I_{(3,1)}(\tau_{v_0}, 1)(f_{v_0} \times \theta) = \mathrm{tr}\, \pi_{C,v_0}(f_{C,v_0}).$$

The constant $m(\alpha_3\alpha_2\alpha_1, \tau)/m(\alpha_3\alpha_2\alpha_1, \tau')$ is independent of the global representations $\tau, \tau'$; it depends only on the local representation $\pi_{2v_0}$, and will be denoted $m(\pi_{2v_0})$. It is equal to 1 for $\pi_{2v} = \mathbf{1}_{2v}$, the trivial representation, by Proposition 1.1, hence also for the special representation $\mathrm{sp}_{2v}$.

Hence $m(\pi_2) = \prod m(\pi_{2v})$, product only over the cuspidal components $\pi_{2v}$ of $\pi_2$, and we replace $R(\tau_v)$ by $m(\pi_{2v})R(\tau_v)$ when $\pi_{2v}$ is cuspidal to obtain the character relation as claimed. □

# 3. Induced Case

We then turn to the study of the $\lambda_1$-lifting of $\pi_1 \times \pi_2$, $\omega_1\omega_2 = 1$ ($\omega_i$ is the central character of $\pi_i$), when $\pi_2$ is not the contragredient $\check{\pi}_1$ of $\pi_1$. Note that $\check{\pi}_1(A) = \pi_1(\check{A})$ where $\check{A} = w^t A^{-1} w$. This $\check{\pi}_1$ is equivalent to $\omega_1^{-1}\pi_1$.

3.1 Proposition. *Let $\pi_2$ be an admissible representation of* GL(2, $F$) *($F$ is a local field) with central character $\omega$. Let $\pi_1 = I(\mu_1, \mu_1')$ be an induced representation of* GL(2, $F$) *with $\mu_1\mu_1'\omega = 1$. Then $\lambda_1(I_2(\mu_1, \mu_1') \times \pi_2) = \pi$, where $\pi = I_4(\mu_1\pi_2, \mu_1'\pi_2)$ is the representation of* PGL(4, $F$) *induced from the parabolic subgroup of type* (2, 2) *as indicated.*

Proof. Since $\lambda_1(\mu\pi_1 \times \mu^{-1}\pi_2) = \lambda_1(\pi_1 \times \pi_2)$, it suffices to show that $\lambda_1(I_2(1, \omega^{-1}) \times \pi_2) = I_4(\pi_2, \check{\pi}_2)$, as the contragredient $\check{\pi}_2$ of $\pi_2$ is $\omega^{-1}\pi_2$. We then compute the $\theta$-twisted character of the induced representation $\pi = \pi_4 = I_4(\pi_2, \check{\pi}_2)$. Put $\rho = \pi_2 \otimes \check{\pi}_2$. Write

$$\rho(\mathrm{diag}(A, C)) \quad \text{for} \quad \rho(A, C) = \pi_2(A) \otimes \check{\pi}_2(C).$$

Its space consists of $\phi : G \to \rho$ with

$$\phi(nmk) = \delta^{1/2}(m)(\pi_2 \times \check{\pi}_2)(m)\phi(k),$$

$m = \mathrm{diag}(A, C)$ with $A, C$ in GL(2, $F$) and $n$ is a unipotent matrix (upper triangular, type (2,2)),

$$\delta(m) = |\det(\mathrm{Ad}(m)|\,\mathrm{Lie}\,N)| = |\det(AC^{-1})|^2,$$

and $\pi$ acts by right translation.

Note that $\pi = \pi_4$ is $\theta$-invariant. Namely there exists an intertwining operator $s' : \pi \to \pi$ with $s'\pi(g) = \pi(\theta g)s'$. Fix $s'$ to be $(s\phi)(g) = \rho(\theta)(\phi(\theta g))$. Here $\rho(\theta)$ intertwines $\pi_2 \otimes \check{\pi}_2$ with $\check{\pi}_2 \otimes \pi_2$ by $\rho(\theta)(\xi \otimes \check{\xi}) = \check{\xi} \otimes \xi$ and

$$\rho(\theta)(\pi_2(A) \otimes \check{\pi}_2(C)) = (\check{\pi}_2(C) \otimes \pi_2(A))\rho(\theta).$$

Note that $s$ is well defined. When $\pi$ is irreducible (this is the case unless $\pi_2 = \nu^{1/2}\pi_2'$, $\pi_2' \simeq \check{\pi}_2'$), $s'^2$ is a scalar by Schur's lemma, so we can multiply $s'$ by a scalar to assume $s'^2 = I$, and so $s'$ is unique up to a sign. It is easy to see that our choice of $s' = s$ here is the same as our usual choice of $\pi(\theta)$, preserving Whittaker models or a $K$-fixed vector if $\pi_2$ is unramified.

Extend $\rho$, by $\rho(\theta) = s$, to a representation of $[\mathrm{GL}(2,F)\times\mathrm{GL}(2,F)]\rtimes\langle\theta\rangle$. Note that $\quad \operatorname{tr}\rho(\theta)(\pi_2(A)\otimes\check{\pi}_2(C))$

$$= \operatorname{tr}[\pi_2(A)\otimes 1\cdot\rho(\theta)\cdot(1\otimes\check{\pi}_2(C))] = \operatorname{tr}(\pi_2(A\check{C})\otimes 1)\rho(\theta).$$

To compute $\operatorname{tr}\rho(\theta)(\pi_2(A)\otimes 1)$ choose an orthogonal basis $v_i$ for $\pi_2$, and a dual basis $\check{v}_i$ for $\check{\pi}_2$. (It is standard to "smooth" our argument on using test functions). Then $\pi_2(A)\otimes 1$ takes $v_i\otimes\check{v}_j$ to $\pi_2(A)v_i\otimes\check{v}_j$, and $\rho(\theta)$ takes $\pi_2(A)v_i\otimes\check{v}_j$ to $\check{v}_j\otimes\pi_2(A)v_i$. The trace $\operatorname{tr}\rho(\theta)(\pi_2(A)\otimes 1)$ is then

$$\sum_{ij}\langle\check{v}_j\otimes\pi_2(A)v_i, v_i\otimes\check{v}_j\rangle = \sum_{ij}\langle\check{v}_j, v_i\rangle\langle\pi_2(A)v_i,\check{v}_j\rangle$$

$$= \sum_i\langle\pi_2(A)v_i,\check{v}_i\rangle = \operatorname{tr}\pi_2(A).$$

As usual, $\quad (\pi(\theta f\,dg)\phi)(h) \quad$ is

$$= \int_G f(g)\rho(\theta)(\phi(\theta(h)g))dg = \int_G f(\theta(h)^{-1}g)\rho(\theta)(\phi(g))dg$$
$$= \iiint f(\theta(h)^{-1}nmk)\delta^{1/2}(m)(\pi_2\times\check{\pi}_2)(\theta m)\phi(k)\delta^{-1}(m)dn\,dm\,dk.$$

Write $m = \operatorname{diag}(A,C)$ as $\theta(m_1^{-1})m_0m_1$ with $m_1 = \operatorname{diag}(I,C)$ and $m_0 = \operatorname{diag}(A',I)$, where $A' = w^tC^{-1}wA$. We have $\operatorname{tr}(\pi_2\times\check{\pi}_2)(\theta\operatorname{diag}(A,C)) = \operatorname{tr}\pi_2(Aw^tC^{-1}w)$. Put

$$M_0 = \{\operatorname{diag}(X,I); X\in\mathrm{GL}(2,F)\},\qquad M_1 = \{\operatorname{diag}(I,X); X\in\mathrm{GL}(2,F)\}$$

as well as for the images of these groups in $\mathrm{PGL}(4,F)$. Note that $\delta(m) = \delta(m_0)$. Then, putting $m = \theta(m_1^{-1})m_0m_1$, $m_0 = \operatorname{diag}(A,I)$, we have

$$\operatorname{tr}\pi(\theta f\,dg) = \iiint f(\theta(k^{-1})n_1mk)\delta^{-1/2}(m)\operatorname{tr}\pi_2(A)dn_1dm\,dk.$$

Change variables $n_1\mapsto n$, where $n_1 = nm\theta(n^{-1})m^{-1}$. This has the Jacobian $|\det(1-\operatorname{Ad}(m\theta))|\operatorname{Lie}N|$. Replace $n$ by $\theta(n)^{-1}$ and note that

$$\operatorname{Ad}(\theta(m_1^{-1})m_0m_1\cdot\theta) = \operatorname{Ad}(\theta(m_1^{-1}))\operatorname{Ad}(m_0\theta)\operatorname{Ad}(\theta(m_1)).$$

We obtain $\operatorname{tr}\pi(\theta f\ dg) = \int_{M_0} \Delta_M(m_0\theta)\operatorname{tr}\pi_2(A)X dm_0$, where

$$X = \int_K \int_N \int_{M_1} f(\theta(k^{-1}n^{-1}m_1^{-1})m_0m_1nk)dndk,$$

and

$$\Delta_M(m_0\theta) = \delta^{-1/2}(m_0)|\det(1 - \operatorname{Ad}(m_0\theta))|\operatorname{Lie}(N)|.$$

Note that if $m_0 = \operatorname{diag}(A, I)$ and $A$ has eigenvalues $a$, $b$, then $\delta(m_0) = |ab|^2$, and

$$\Delta_M(m_0\theta) = |(1-a)(1-b)(1-ab)/ab|$$

has the same value at $A$ and at $w{}^tA^{-1}w$ (or $A^{-1}$). Writing the trace again as

$$\operatorname{tr}\pi(\theta f dg) = \int_{M_0} \Delta_M(m_0\theta)\chi_{\pi_2}(A)\int_{M_0\backslash G} f(\theta(g^{-1})m_0g)d\dot{g}\ dm_0,$$

$m_0 = \operatorname{diag}(A, I)$, we use the fact that the $\theta$-normalizer of $M$ in $G$ is generated by $M$ and $J$. Since

$$\theta(J^{-1})\begin{pmatrix} A & 0 \\ 0 & I \end{pmatrix} J = \begin{pmatrix} I & 0 \\ 0 & wAw \end{pmatrix} = \theta(m_1^{-1})m_0m_1,$$

$$m_1 = \operatorname{diag}(I, wAw), \qquad m_0 = \operatorname{diag}({}^tA^{-1}, I),$$

and since $\chi_{\pi_2}({}^tA^{-1}) = \chi_{\pi_2}(\det A^{-1}\cdot A) = \omega^{-1}(\det A)\chi_{\pi_2}(A)$, we finally conclude that $\operatorname{tr}\pi(\theta f\ dg)$ is

$$= \int_{M_0} \frac{1}{2}\Delta_M(m_0\theta)(1 + \omega^{-1}(\det A))\chi_{\pi_2}(A)\int_{M_0\backslash G} f(\theta(g^{-1})m_0g)d\dot{g}\ dm_0.$$

On the other hand, using the 2-fold submersion

$$M_{\theta\text{-reg}} \times M\backslash G \to G_{\theta\text{-reg}}, \qquad (m, g) \mapsto \theta(g^{-1})mg,$$

whose Jacobian is

$$|\det(1 - \operatorname{Ad}(m\theta))|\operatorname{Lie}(G/M)| = \delta^{-1}(m)|\det(1 - \operatorname{Ad}(m\theta))|\operatorname{Lie} N|^2,$$

and noting that the $\theta$-Weyl group $W_\theta(M) = \{g \in G; \theta(g)^{-1}Mg = M\}/M$ is represented by $I$ and $J$, if $g \mapsto \chi_\pi(g\theta)$ denotes the $\theta$-character of $\pi$ then we have

$$\operatorname{tr}\pi(\theta f\ dg) = \int_G f(g)\chi_\pi(g\theta)dg = \int_M \frac{1}{2}\delta^{-1}(m)|\det(1 - \operatorname{Ad}(m\theta))|\operatorname{Lie} N|^2$$

$$\cdot \chi_\pi(m\theta) \int_{M\backslash G} f(\theta(g)^{-1}mg)d\dot{g}\, dm.$$

On writing $m = \theta(m_1)^{-1}m_0m_1$ this becomes

$$= \int_{M_0} \frac{1}{2}\Delta_M(m_0\theta)^2\chi_\pi(m_0\theta) \int_{M_0\backslash G} f(\theta(g)^{-1}m_0g)d\dot{g}\, dm_0.$$

We conclude that

$$\Delta_M(m_0\theta)\chi_\pi(m_0\theta) = (1+\omega^{-1}(\det A))\chi_{\pi_2}(A), \qquad m_0 = \mathrm{diag}(A, I),$$

and that $\chi_\pi(g\theta)$ is supported on the set $\theta(g^{-1})m_0g, m_0 \in M_0, g \in G$. In particular it vanishes on the $\theta$-elliptic stable conjugacy classes of types I, II, III, IV.

Now

$$\Delta_M(m_0\theta) = \frac{\Delta(m_0\theta)}{\Delta_C(Nm_0)}\Delta_2(\mathrm{diag}(\det A, 1)),$$

as

$$\Delta(m_0\theta)/\Delta_C(Nm_0) = |\frac{(1-a)^2(1-b)^2}{ab}|^{1/2}$$

if $A$ has eigenvalues $a$ and $b$, and $\Delta_2$ is the usual Jacobian of GL(2) : $\Delta_2(A) = |\frac{(a-b)^2}{ab}|^{1/2}$. We rewrite our conclusion as

$$\Delta(m_0\theta)\chi_\pi(m_0\theta) = \Delta_C(Nm_0)\chi_{I(1,\omega^{-1})}(\mathrm{diag}(\det A, 1))\chi_{\pi_2}(A)$$
$$= \Delta_C(Nm_0)\chi(I(1,\omega^{-1}) \times \pi_2)(Nm_0)$$

where $N(\mathrm{diag}(A, I)) = \left(\begin{pmatrix} \det A & 0 \\ 0 & 1 \end{pmatrix}, A\right)$, since

$$\chi_{I(\mu_1,\mu_2)}(\mathrm{diag}(a,b)) = (\mu_1(\mu)\mu_2(b) + \mu_1(b)\mu_2(a))/\Delta_2(\mathrm{diag}(a,b))$$

(and it is zero on the elliptic element in GL(2)). But this is precisely the statement that $I(1,\omega^{-1}) \times \pi_2$ $\lambda_1$-lifts to $\pi = I(\pi_2, \check{\pi}_2)$. □

REMARK. The character relation implies that $\chi_\pi$ vanishes on the $\theta$-elliptic conjugacy classes.

3.2 COROLLARY. *Let $F$ be a local field. For every cuspidal representation $\pi_2$ of* $\mathrm{PGL}(2,F)$ *the representations $\pi_C = \pi_2 \times \mathrm{sp}_2$ and $\mathrm{sp}_2 \times \pi_2$ of $C$ $\lambda_1$-lift to the subrepresentation*

$$S(\nu^{1/2}\pi_2, \nu^{-1/2}\pi_2)(= \ker[I(\nu^{1/2}\pi_2, \nu^{-1/2}\pi_2) \to J(\nu^{1/2}\pi_2, \nu^{-1/2}\pi_2)])$$

*of the fully induced $I(\nu^{1/2}\pi_2, \nu^{-1/2}\pi_2)$.*

PROOF. To simplify the notations we write simply $0 \to S \to I \to J \to 0$ omitting the $(\nu^{1/2}\pi_2, \nu^{-1/2}\pi_2)$. Since $\lambda_1(I(\nu^{1/2}, \nu^{-1/2}) \times \pi_2) = I$ and $\lambda_1(\mathbf{1}_2 \times \pi_2) = J$, and since the composition series of $I(\nu^{1/2}, \nu^{-1/2})$ consists of $\mathbf{1}_2$ and $\mathrm{sp}_2$, while the composition series of $I$ consists of $J$ and $S$, the claim of the corollary follows from the additivity of the character of a representation: $\chi_{\pi_1+\pi_2} = \chi_{\pi_1} + \chi_{\pi_2}$. □

# 4. Cuspidal Case

It remains to $\lambda_1$-lift cuspidal representations of $C$.

4.1 PROPOSITION. *Let $F$ be a local field. Let $\pi_2'$ and $\pi_2''$ be (irreducible) cuspidal representations of* $\mathrm{GL}(2,F)$ *with central characters $\omega'$, $\omega''$ with $\omega'\omega'' = 1$ so that $\pi_C = \pi_2' \times \pi_2''$ is a cuspidal representation of $C$. Then $\lambda_1(\pi_2' \times \pi_2'')$ exists as an irreducible $\theta$-invariant representation $\pi$ of $G$.*

*This $\pi$ is cuspidal unless (1) $\pi_2'' = \check{\pi}_2'\chi$, $\chi^2 = 1$, where*

$$\lambda_1(\pi_2' \times \check{\pi}_2'\chi) = \chi I_{(3,1)}(\mathrm{Sym}^2 \pi_2', 1),$$

*or (2) there is a quadratic extension $E$ of $F$ and characters $\mu_1$ and $\mu_2$ of $E^\times$ with $\mu_1\mu_2|F'^\times = 1$ such that $\pi_2' = \pi_E(\mu_1)$, $\pi_2'' = \pi_E(\mu_2)$, in which case*

$$\lambda_1(\pi_E(\mu_1) \times \pi_E(\mu_2)) = I_{(2,2)}(\pi_E(\mu_1\overline{\mu}_2), \pi_E(\mu_1\mu_2)).$$

*In particular, if $\pi_2'$ and $\pi_2''$ are monomial but not associated to the same quadratic extension, then $\lambda_1(\pi_2' \times \pi_2'')$ is cuspidal.*

*When $F$ is a global field and $\pi_2'$, $\pi_2''$ are automorphic cuspidal representations of* $\mathrm{GL}(2,\mathbb{A})$ *(with $\omega'\omega'' = 1$) the analogous global results hold. In particular $\lambda_1(\pi_2' \times \pi_2'')$ exists as an irreducible automorphic $\theta$-invariant* *representation* $\pi$ *of* $\mathbf{G}(\mathbb{A})$, *which is cuspidal except at the indicated cases. In this global case we require that at least at* 3 *places the components of* $\pi_2'$, $\pi_2''$ *be square integrable.*

PROOF. We denote the local fields of the proposition by $F'$, $E'$. Suppose $F'$ is nonarchimedean. Choose a totally imaginary global field $F$ whose completion at a place $v_0$ is our $F'$. Fix four nonarchimedean places $v_1$, $v_2$, $v_3$, $v_4$ ($\neq v_0$) of $F$, and cuspidal representations $\pi'_{v_i}$ ($i = 1, 2, 3, 4$) of $\mathrm{PGL}(2, F_{v_i})$. Let $V$ be the set of places of $F$ consisting of $v_i$ ($0 \leq i \leq 4$) and the archimedean places. Construct cuspidal representations $\pi_1, \pi_2$ of $\mathrm{GL}(2, \mathbb{A})$ (with $\omega_1\omega_2 = 1$) which are unramified outside $V$ whose components at $v_0$ are $\pi_2', \pi_2''$ of the proposition, at $v_1$ are $\mathrm{sp}_{2,v_1}$ (resp. $\pi'_{v_1}$), and at $v_2$, $v_3$, $v_4$ are $\pi'_{v_2}$, $\pi'_{v_3}$, $\pi'_{v_4}$ (resp. $\mathrm{sp}_{2,v_2}$, $\mathrm{sp}_{2,v_3}$, $\mathrm{sp}_{2,v_4}$).

Set up the trace formula identity with the set $V$ such that $\pi_1 \times \pi_2$ (and $\pi_2 \times \pi_1$) contribute to the side of $C$. Take the components $f_{C,v_i}$ ($i = 1, 2, 3$) to have orbital integrals equal zero outside the elliptic set. Consequently we may and do choose the matching $f_{v_i}$ ($i = 1, 2, 3$) to have zero stable $\theta$-orbital integrals. Hence $f_{H,v_i}$ ($i = 1, 2, 3$) are zero, and there is no contribution to the trace formulae of $H$ and $C_0$.

We need to show that there is a contribution $\pi$ to the $\theta$-twisted trace formula of $G$. If there is then it is unique, by rigidity theorem for automorphic representations on $\mathrm{GL}(n)$, and it is cuspidal, since $\lambda_1(\mathrm{sp}_{2,v_4} \times \pi'_{v_4}) = S(\nu_{v_4}^{1/2}\pi'_{v_4}, \nu_{v_4}^{-1/2}\pi'_{v_4})$ is not induced from any proper parabolic subgroup.

We may apply "generalized linear independence" of characters at the archimedean places of $F$. There the completion is the complex numbers. Hence the local components are induced and the local lifting known. All matching functions $f_v$, $f_{Cv}$ are at our disposal. There remain in our trace identity only products over the set $V_f = \{v_i; 0 \leq i \leq 4\}$ of finite places in $V$.

Note that both $\pi_1 \times \pi_2$ and $\pi_2 \times \pi_1$ contribute to the side of $C$, which has coefficient $\frac{1}{2}$, while the coefficient of the cuspidal contribution to the $\theta$-twisted trace formula of $G$ is 1. Choosing the $f_{C,v_i}(0 \leq i \leq 4)$ to be pseudo coefficients of the $\pi_{C,v_i}$ we obtain on the side of $C$ a sum of 1's. We conclude that the side of $G$ is also nonzero, hence $\pi$ exists.

Next we make this choice only at the places $v_i$ ($i = 1, 2, 3, 4$). Observe that $\mathrm{tr}\,\pi_{C,v_i}(f_{C,v_i})$ is 1, and $\mathrm{tr}\,\pi_{v_i}(f_{v_i}) = 1$, since for $i = 1, 2, 3, 4$ the component $\pi_{v_i}$ of $\pi$ is $S(\nu_{v_i}^{1/2}\pi'_{v_i}, \nu_{v_i}^{-1/2}\pi'_{v_i})$, which is the $\lambda_1$-lift of $\pi_{C,v_i} =$

$\pi'_{v_i} \times \mathrm{sp}_{2,v_i}$. In fact the character relation $\lambda_1(\pi'_{v_i} \times \mathrm{sp}_{2,v_i}) = S_{v_i}$ on the $\theta$-elliptic set alone, and the orthogonality relations for twisted characters (used below), would imply that the component $\pi_{v_i}$ is $S_{v_i}$ or $J_{v_i}$. Had the component $\pi_{v_i}$ been $J_{v_i}$ for an odd number of places $v_i$ $(1 \leq i \leq 4)$ we would get a coefficient $-1$, and a contradiction. We obtain, for all matching functions $f_{v_0}$ and $f_{C,v_0}$, the identity $(v = v_0)$

$$(1) \qquad \operatorname{tr} \pi_v(f_v \times \theta) = n_v \operatorname{tr} \pi_{Cv}(f_{Cv}) + \sum_{n(\pi'_{Cv})>0} n(\pi'_{Cv}) \operatorname{tr} \pi'_{C,v}(f_{Cv}).$$

Here $\pi'_{C,v}$ are representations of $C_v$ not equivalent to $\pi_{C,v} = \pi'_2 \times \pi''_2$ of the proposition, under the equivalence relation generated by $\pi' \times \pi'' \simeq \pi'' \times \pi'$ and $\pi'\chi \times \pi''\chi^{-1} \simeq \pi' \times \pi''$. The coefficient $n_v$ counts the number of equivalence classes of global automorphic cuspidal representations whose components at each $w$ are in the equivalence class of $\pi_{1w} \times \pi_{2w}$ (where $\pi_1$, $\pi_2$ are our global representations).

*We claim that all the $\pi'_{Cv}$ which appear on the right are cuspidal.* For this we use the central exponents of the representations which appear in our identity. Since all of the $n(\pi'_{Cv})$ are nonnegative real numbers, a familiar ([FK1], [F4;II]) argument of linear independence of central exponents, based on a suitable choice of the functions $f_{C,v}$, implies that if one of the $\pi'_{C,v}$ which appears with $n(\pi'_{Cv}) > 0$ has nonzero central exponents – namely it is not cuspidal – then $\pi_v$ must have matching $\theta$-twisted central exponents. This means that $\pi_v$ is the $\lambda_1$-lift of some $\pi'_2 \times \pi''_2$ where $\pi'_2$ of $\pi''_2$ are not cuspidal, since we already know to $\lambda_1$-lift $\pi'_2 \times \pi''_2$ where $\pi'_2$ is fully induced or special. Linear independence of characters (after replacing $\operatorname{tr} \pi_v(f_v \times \theta)$ on the left by $\operatorname{tr}(\pi'_2 \times \pi''_2)(f_{Cv})$) gives a contradiction which implies that all the $\pi'_{Cv}$ which appear on the right are cuspidal, as claimed.

Note that the identity exists for each local cuspidal $\pi_{C,v}$.

*We claim that $\pi_v$ is uniquely determined by $\pi_{C,v}$, and that the identity defines a partition of the cuspidal representations of $C_v$.* For this we use the orthogonality relations for characters of elliptic representations of Kazhdan [K2] in its twisted form [F1;II]. These assert the existence of *pseudo coefficients*: if $f_v$ is a pseudo coefficient of a $\theta$-elliptic $\pi'_v$ inequivalent to $\pi_v$ then $\operatorname{tr} \pi_v(f_v \times \theta) = 0$; this is $\neq 0$ if $\pi'_v = \pi_v$, and $= 1$ if $\pi'_v = \pi_v$ is cuspidal. Now let $f_v$ be a pseudo coefficient of $\pi'_v$ inequivalent to $\pi_v$ for which we have the identity (sum over the $\pi''_{Cv}$ with $n(\pi''_{Cv}) > 0$)

$$\operatorname{tr}\pi'_v(f'_v) = \sum n(\pi''_{C,v})\operatorname{tr}\pi''_{C,v}(f'_{C,v}),$$

where $f'_{Cv}$ is the matching function on $C_v$. We then have that the stable orbital integrals of $f'_{C,v}$ are equal to $\sum n(\pi''_{Cv})\chi_{\pi''_{C,v}}$ on the elliptic set of $C_v$. Evaluating our identity (1) at $f'_v$ and $f'_{C,v}$, we note that only finite number of terms in (1) can be nonzero. Indeed, $\pi'_{C,v}$ of (1) is a component of an automorphic representation of $\mathbf{C}(\mathbb{A})$ which is unramified outside $V$, its archimedean components (hence their infinitesimal characters) lie in a finite set, and the ramification of the remaining components is bounded (fixed at $v_1$, $v_2$, $v_3$, $v_4$; bounded by $f_{v_0}$ – which is biinvariant under some small compact open subgroup – at $v_0$). Hence $0 = \operatorname{tr}\pi_v(f_v \times \theta)$ is

$$= n_v\langle\pi_{Cv}, \sum n(\pi''_{Cv})\pi''_{Cv}\rangle + \sum_{n(\pi'_{Cv})>0} n(\pi'_{Cv})\langle\pi'_{Cv}, \sum n(\pi''_{Cv})\pi''_{Cv}\rangle.$$

The inner products $\langle\pi_{Cv}, \pi''_{Cv}\rangle$, or $\langle\chi_{\pi_{Cv}}, \chi_{\pi''_{Cv}}\rangle$, are nonnegative integers, hence no $\pi''_{Cv}$ can equal $\pi_{Cv}$ or $\pi'_{Cv}$, as claimed.

Given a cuspidal $\pi_{Cv}$ we now denote the $\pi_v$ specified by (1) by $\lambda''_1(\pi_{Cv})$. *We claim that*

$$\lambda''_1(\pi_E(\mu_1) \times \pi_E(\mu_2)) = I_{(2,2)}(\pi_E(\mu_1\overline{\mu}_2), \pi_E(\mu_1\mu_2)),$$

where $\mu_1, \mu_2$ are characters of the local quadratic extension $E'/F'$ of the proposition, with $\mu_1 \neq \overline{\mu}_1, \mu_2 \neq \overline{\mu}_2$ and $\mu_1\mu_2|F'^{\times} = 1$. As in the beginning of this proof, we choose a totally imaginary global quadratic extension $E/F$ such that at the place $v_0$ of $F$ the completion $E_{v_0}/F_{v_0}$ is our $E'/F'$.

Then we choose global characters $\mu_1$, $\mu_2$ of $\mathbb{A}_E^{\times}/E^{\times}$ with our local components at $v_0$, with $\mu_1\mu_2|\mathbb{A}^{\times} = 1$, which are unramified outside a set $V$ consisting of the archimedean places of $F$, $v_0$ and three finite places $v_1$, $v_2$, $v_3$ ($\neq v_0$) which do not split in $E$, where the components are taken to satisfy $\mu_{iv_j} \neq \overline{\mu}_{iv_j}$ (bar indicates the action of the nontrivial automorphism of $E/F$). The existence of $\mu_1, \mu_2$ is shown on using the summation formula for $\mathbb{A}_E^{\times}/E^{\times}$, which is the trace formula for GL(1). First we construct $\mu_1$, and then $\mu_2$ – which is known to be $\mu_1^{-1}$ on $\mathbb{A}^{\times}/F^{\times}$ – has to be constructed on $\mathbb{A}_E^{\times}/E^{\times}\mathbb{A}^{\times}$.

Set up the trace formula identity with the set $V_f = \{v_i; 0 \le i \le 3\}$ such that $\pi_E(\mu_1) \times \pi_E(\mu_2)$ contributes to the trace formula of $C$. We choose the components $f_{Cv_i}$ of the test function to have stable orbital integrals which vanish on the regular nonelliptic set. Hence we may take $f_{v_i}$ $(i = 1, 2, 3)$ to have zero stable orbital integrals, so that we can choose $f_{Hv_i}$ to be zero, hence there is no contribution to the trace formulae of $H$ and $C_0$.

The trace formula of $G$ will have the (unique) contribution

$$I_{(2,2)}(\pi_E(\mu_1\overline{\mu}_2), \pi_E(\mu_1\mu_2)).$$

Indeed, we follow the homomorphisms of the Weil group

$$W_{E/F} = \langle z, \sigma; z \in C_E, \sigma z \sigma^{-1} = \overline{z}, \sigma^2 \in C_F - N_{E/F}C_E \rangle$$

which define $\pi_E(\mu_1)$, $\pi_E(\mu_2)$, and their composition with $\lambda_1$ (put $\mu_i = \mu_i(z)$, $\overline{\mu}_i = \mu_i(\overline{z})$):

$$\begin{aligned}
z &\mapsto \begin{pmatrix} \mu_1 & 0 \\ 0 & \overline{\mu}_1 \end{pmatrix} \times \begin{pmatrix} \mu_2 & 0 \\ 0 & \overline{\mu}_2 \end{pmatrix} \\
&\overset{\lambda_1}{\mapsto} \begin{pmatrix} \mu_1\mu_2 & & & \\ & \mu_1\overline{\mu}_2 & & \\ & & \mu_2\overline{\mu}_1 & \\ & & & \overline{\mu}_1\overline{\mu}_2 \end{pmatrix} \overset{(13)}{\longmapsto} \begin{pmatrix} \mu_2\overline{\mu}_1 & & & \\ & \mu_1\overline{\mu}_2 & & \\ & & \mu_1\mu_2 & \\ & & & \overline{\mu}_1\overline{\mu}_2 \end{pmatrix}, \\
\sigma &\mapsto \begin{pmatrix} 0 & 1 \\ \mu_1(\sigma^2) & 0 \end{pmatrix} \times \begin{pmatrix} 0 & 1 \\ \mu_2(\sigma^2) & 0 \end{pmatrix} \\
&\overset{\lambda_1}{\mapsto} \begin{pmatrix} & & & 1 \\ & & \mu_2(\sigma^2) & \\ & \mu_1(\sigma^2) & & \\ 1 & & & \end{pmatrix} \overset{(13)}{\longmapsto} \begin{pmatrix} 0 & \mu_1(\sigma^2) & & \\ \mu_2(\sigma^2) & 0 & & \\ & & 0 & 1 \\ & & 1 & 0 \end{pmatrix}.
\end{aligned}$$

This homomorphism implies that

$$\lambda_1(\pi_E(\mu_1) \times \pi_E(\mu_2)) = I_{(2,2)}(\pi_E(\overline{\mu}_1\mu_2), \pi_E(\mu_1\mu_2))$$

at all places where (E/F and) $\mu_1, \mu_2$ are unramified. Following the arguments leading to (1), we obtain (1) for our $\pi_{C,v} = \pi_E(\mu_1) \times \pi_E(\mu_2)$, where $\pi_v$ is $I_{(2,2)}(\pi_E(\overline{\mu}_1\mu_2), \pi_E(\mu_1\mu_2))$, as claimed.

Note that when $\mu_1 = \mu_2^{-1}$ we have $\pi_E(\mu_1\overline{\mu}_2) = \pi_E(\mu_1/\overline{\mu}_1)$ and $\pi_E(\mu_1\mu_2)$ $= I(\chi_{E/F}, 1)$, thus $\quad \lambda_1(\pi_E(\mu_1) \times \pi_E(\mu_1)^\vee) \quad$ is

$$= I_{(2,1,1)}(\pi_E(\mu_1/\overline{\mu}_1), \chi_{E/F}, 1) = I_{(3,1)}(\mathrm{Sym}^2(\pi(\mu_1)), 1),$$

as is known already.

At this stage we note that we dealt with all square integrable representations $\pi_{Cv}$ of $C_v$, except pairs $\pi_1 \times \pi_2 = \pi_{E_1}(\mu_1) \times \pi_{E_2}(\mu_2)$, where $E_1, E_2$ are two distinct quadratic extensions of the local field $F'$, and $\mu_i$ are characters of $E_i^\times$ with $\mu_i \neq \overline{\mu}_i$ and $\mu_1\mu_2|F'^\times = 1$, and $\pi_{E_i}(\mu_i)$ is not monomial from $E_j$ ($\{i,j\} = \{1,2\}$). In fact, in residual characteristic 2 there are also "extraordinary" representations, which are not monomial; we shall deal with these later.

*We claim that* $\pi_v$ *in* (1) *for such a* $\pi_{E_1}(\mu_1) \times \pi_{E_2}(\mu_2)$ *is cuspidal.*

Let us review the homomorphisms of the Weil group which define the product $\pi_{E_1}(\mu_1) \times \pi_{E_2}(\mu_2)$. Denote by $E$ the compositum of $E_1$ and $E_2$ and by $\langle\sigma,\tau\rangle$ the Galois group of $E/F$, so that $E_1 = E^\tau$, $E_2 = E^\sigma$ are the fixed points fields of $\tau$, $\sigma$ respectively (thus $\mathrm{Gal}(E_1/F) = \langle\sigma\rangle$ and $\mathrm{Gal}(E_2/F) = \langle\tau\rangle$). To multiply $\pi_{E_1}(\mu_1)$ and $\pi_{E_2}(\mu_2)$ we view their parameters as homomorphisms of $W_{E/F}$, an extension of $\mathrm{Gal}(E/F)$ by $E^\times$, which factorize through $W_{E_1/F}$ and $W_{E_2/F}$. For $\pi_{E_1}(\mu_1)$ we have:

$$z\ (\in E^\times) \overset{N_{E/E_1}}{\longmapsto} z\tau z \quad (\in N_{E/E_1}E^\times) \mapsto \begin{pmatrix} \mu_1(z\cdot\tau z) & 0 \\ 0 & \mu_1(\sigma z\cdot\sigma\tau z)\end{pmatrix},$$

$$\tau\ (\in \mathrm{Gal}(E/F)) \mapsto \tau^2 \quad (\in E_1^\times - N_{E/E_1}E^\times) \mapsto \begin{pmatrix} \mu_1(\tau^2) & 0 \\ 0 & \mu_1(\sigma\tau^2)\end{pmatrix},$$

$$\sigma\ (\in \mathrm{Gal}(E/F),\ \text{viewed in } W_{E_1/F}) \mapsto \begin{pmatrix} 0 & 1 \\ \mu_1(\sigma^2) & 0\end{pmatrix}.$$

We simply pull $\mathrm{Ind}(\mu_1; W_{E_1/F}, W_{E_1/E_1})$ from $W_{E_1/F}$ to $W_{E/F}$ using the diagram

$$\begin{array}{ccccc} W_{E/E_1} = W_{E_1}/W_E^c = \langle C_E, \tau\rangle & \hookrightarrow & W_{E/F} = W_F/W_E^c & \twoheadrightarrow & \mathrm{Gal}(E_1/F) = \langle\sigma\rangle \\ \downarrow & & \downarrow & & \| \\ W_{E_1/E_1} = W_{E_1}/W_{E_1}^c = C_{E_1} & \hookrightarrow & W_{E_1/F} = W_F/W_{E_1}^c & \twoheadrightarrow & \mathrm{Gal}(E_1/F). \end{array}$$

The middle vertical (surjective) arrow is the quotient by

$$(W_{E_1}/W_E^c)^c = \{z\tau z^{-1}\tau^{-1}; z \in C_E\} = C^1_{E/F_1}.$$

The arrow on the left is also surjective. Its restriction to $C_E \subset W_{E/E_1}$ is

$$z \mapsto N_{E/E_1}z \quad (\in N_{E/E_1}C_E \subset C_{E_1}),$$

and $\tau \in W_{E/E_1}$ maps to $\tau^2 \in C_{E_1} - N_{E/E_1}C_E$.

For $\pi_{E_2}(\mu_2)$ we have:

$$z(\in E^\times) \overset{N_{E/E_2}}{\longmapsto} z\sigma z \quad (\in N_{E/E_2}E^\times) \mapsto \begin{pmatrix} \mu_2(z\cdot\sigma z) & 0 \\ 0 & \mu_2(\tau z\cdot\tau\sigma z) \end{pmatrix},$$

$$\sigma(\in \mathrm{Gal}(E/F)) \mapsto \sigma^2 \quad (\in E_2^\times - N_{E/E_2}E^\times) \mapsto \begin{pmatrix} \mu_2(\sigma^2) & 0 \\ 0 & \mu_2(\tau\sigma^2) \end{pmatrix},$$

$$\tau(\in \mathrm{Gal}(E/F),\ \text{viewed in } W_{E_2/F}) \mapsto \begin{pmatrix} 0 & 1 \\ \mu_2(\tau^2) & 0 \end{pmatrix}.$$

Composing these two representations by $\lambda_1$, we obtain the 4-dimensional representation $\rho$ of $W_{E/F}$:

$$z(\in E^\times) \mapsto \mathrm{diag}(\mu_1{}^\tau\mu_1\mu_2{}^\sigma\mu_2, \mu_1{}^\tau\mu_1{}^\tau\mu_2{}^{\tau\sigma}\mu_2, {}^\sigma\mu_1{}^{\sigma\tau}\mu_1\mu_2{}^\sigma\mu_2, {}^\sigma\mu_1{}^{\sigma\tau}\mu_1{}^\tau\mu_2{}^{\tau\sigma}\mu_2)$$

where ${}^\alpha\mu$ means $\mu(\alpha(z))$, or $z \mapsto \mathrm{diag}(\mu, {}^\tau\mu, {}^\sigma\mu, {}^{\sigma\tau}\mu)$ where

$$\mu = \mu_1(z)\mu_1(\tau z)\mu_2(z)\mu_2(\sigma z),$$

$$\tau \mapsto \begin{pmatrix} 0 & \mu_1(\tau^2) & 0 & 0 \\ \mu_1\mu_2(\tau^2) & 0 & 0 & 0 \\ 0 & 0 & 0 & \mu_1(\sigma\tau^2) \\ 0 & 0 & \mu_1(\sigma\tau^2)\mu_2(\tau^2) & 0 \end{pmatrix},$$

$$\sigma \mapsto \begin{pmatrix} 0 & 0 & \mu_2(\sigma^2) & 0 \\ 0 & 0 & 0 & \mu_2(\tau\sigma^2) \\ \mu_1\mu_2(\sigma^2) & 0 & 0 & 0 \\ 0 & \mu_1(\sigma^2)\mu_2(\tau\sigma^2) & 0 & 0 \end{pmatrix}.$$

When $\mu_1 \neq {}^\sigma\mu_1$ and $\mu_2 \neq {}^\tau\mu_2$ this 4-dimensional representation $\rho$ is irreducible, hence – repeating the global construction employed twice already in this proof – we conclude that (1) is obtained with $\pi_v$ cuspidal, as had $\pi_v$ been induced from a proper parabolic subgroup of $G_v$, the representation $\rho$ would have had to be reducible. This establishes the claim.

After completing the study of the lifting from PGSp(2) to PGL(4) we shall conclude the same result – that $\lambda_1(\pi_1 \times \pi_2)$ has to be cuspidal if $\pi_1$ is not $\chi\pi_2$, $\chi^2 = 1$, and $\pi_1$, $\pi_2$ are cuspidal but not $\pi_E(\mu_1)$, $\pi_E(\mu_2)$, by showing that there are no induced $G$-modules that such $\pi_1 \times \pi_2$ can $\lambda_1$-lift to.

• Since $\lambda_1''(\pi_{E_1}(\mu_1) \times \pi_{E_2}(\mu_2)) = \pi_v$ is cuspidal, the orthonormality relations for twisted characters of cuspidal representations on $G_v$, and the orthonormality relations for characters on $C_v$, imply that the identity (1) reduces to only one contribution on the right side, namely our $\pi_{Cv} = \pi_{E_1}(\mu_1) \times \pi_{E_2}(\mu_2)$, with coefficient $n_v = 1$, thus $\lambda_1(\pi_{E_1}(\mu_1) \times \pi_{E_2}(\mu_2)) = \pi_v$, a cuspidal representation of $G_v$.

• The orthogonality relations now imply (in odd residual characteristic) that for $\pi_{Cv} = \pi_E(\mu_1) \times \pi_E(\mu_2)$, the trace identity (1) has only the term $\pi_{Cv}$ on the right, and it becomes

$$\operatorname{tr} I_{(2,2)}(\pi_E(\mu_1\overline{\mu}_2), \pi_E(\mu_1\mu_2))(f_v \times \theta) = n_v \operatorname{tr}(\pi_E(\mu_1) \times \pi_E(\mu_2))(f_{Cv}).$$

Clearly when one of $\pi_E(\mu_i)$ is induced (thus $\mu_i = \overline{\mu}_i$), we have this identity with $n_v = 1$.

*We claim that* $n_v$ *is* 1 *for all* $\pi_E(\mu_1) \times \pi_E(\mu_2)$.

But first let us explain the meaning of the $n_v$. A global (cuspidal, automorphic) representation $\pi_C = \pi_1 \times \pi_2$ (with at least 3 square integrable components) defines an automorphic representation $\pi$ of $\mathbf{G}(\mathbb{A})$ on using the trace formula identity, by the arguments used repeatedly above (we choose test functions such that the function $f_H$ is zero at one of the places where $\pi_C$ is square integrable, and such that $\operatorname{tr} \pi_{Cv}(f_{Cv})$ is 1 for square integrable $\pi_{Cv}$). Note that both $\pi_C = \pi_1 \times \pi_2$ and $\tilde{\pi}_C = \pi_2 \times \pi_1$ contribute to the trace formula of $C$ when $\pi_1 \not\simeq \chi\check{\pi}_2$, $\chi^2 = 1$, as we now assume. The trace formula identity (for a suitable finite set $V$) then takes the form

$$\prod_{v \in V} \operatorname{tr} \pi_v(f_v \times \theta) = \sum_{\{\pi'_C, \tilde{\pi}'_C\}} \prod_{v \in V} \operatorname{tr} \pi'_{Cv}(f_{Cv}).$$

On the other hand we have the local character relations

$$\operatorname{tr} \pi_v(f_v \times \theta) = n_v \operatorname{tr} \pi_{Cv}(f_{Cv})$$

for each $\pi_v$ on the left, where $n_v = 1$ unless $\pi_{Cv} = \pi_E(\mu_1) \times \pi_E(\mu_2)$. Replacing then the left side by $\prod_{v \in V} n_v \operatorname{tr} \pi_{Cv}(f_{Cv})$ we conclude (applying linear independence of characters on $C_v$) that there are $\prod_{v \in V} n_v$ pairs $\pi'_C$, $\tilde{\pi}'_C$ of cuspidal representations of $\mathbf{C}(\mathbb{A})$ whose local components belong to the pair $\{\pi_{Cv}, \tilde{\pi}_{Cv}\}$. In other words, the representations $\pi'_C = \pi_1 \times \pi_2$ which contribute to the right side are obtained from each other on interchanging the local components of $\pi_1$ and $\pi_2$ at a set $S$ of places of $F$ which is infinite and whose complement is infinite (if $\pi_1$, $\pi'_1$ differ by only finitely many components and both are cuspidal then they are equal by rigidity theorem for GL(2)). If all $n_v$ are 1, we would have on the right side only the contributions $\prod \operatorname{tr} \pi_{Cv}(f_{Cv})$ and $\prod \operatorname{tr} \tilde{\pi}_{Cv}(f_{Cv})$.

Let $E'/F'$ be a quadratic extension of local fields, and $\pi''_C = \pi_{E'}(\mu_1) \times \pi_{E'}(\mu_2)$ a cuspidal representation of $\mathbf{C}(F')$, where $\mu_i \neq \overline{\mu}_i$, $\mu_1\mu_2|F' = 1$, $\mu_1\mu_2 \neq \chi \circ N_{E'/F'}$ for any quadratic character $\chi$ of $F'$. *Our claim is that the associated* $n_v$ *is* 1.

For this we construct first a totally imaginary number field $F$ whose completion at a place $v_0$ is $F'$, and a cuspidal representation $\pi_C$ of $\mathbf{C}(\mathbb{A})$ which is unramified outside the set $V$ consisting of the archimedean places and the finite places $v_0, \dots, v_3$. The component of $\pi_C$ at $v_0$ is our $\pi''_C$ and at $v_1$, $v_2$, $v_3$ it is $\mathrm{sp}_{2v_i} \times \mathrm{sp}_{2v_i}$. Then there are $n_{v_0}$ (a positive integer depending on $\pi''_C$) pairs $\pi'_C$, $\tilde{\pi}'_C$ of cuspidal representations of $\mathbf{C}(\mathbb{A})$ whose components at $v_1$, $v_2$, $v_3$ are $\mathrm{sp}_2 \times \mathrm{sp}_2$ and at each $v$ the components belong to the pair $\{\pi_{Cv}, \tilde{\pi}_{Cv}\}$. Now we apply the theory of basechange for GL(2) for a quadratic extension $E$ of $F$ whose completion $E \otimes_F F_{v_1} = E_{v_1}$ is the local quadratic extension $E'$ of $F' = F_{v_1}$. Then $\pi_{C,v_1} = \pi''_C = \pi_{E'}(\mu_1) \times \pi_{E'}(\mu_2)$ lifts to a fully induced representation $\pi^E_{C,v_1} = I(\mu_1, \overline{\mu}_1) \times I(\mu_2, \overline{\mu}_2)$ of $\mathbf{C}(E_{v_1})$, and the global $\pi_C$ lifts to the cuspidal representation $\pi^E_C$ whose components at the places of $E$ above $v_1$, $v_2$, $v_3$ are $\mathrm{sp}_2 \times \mathrm{sp}_2$, at $v_0$ it is $\pi^E_{C,v_0}$ specified above, and it is unramified at the places outside $V$.

Since $\pi^E_C$ has no components of the form $\pi_M(\mu'_1) \times \pi_M(\mu'_2)$, where $\mu'_1$, $\mu'_2$ are characters of $M^\times$, $M$ a quadratic extension of $E$, and it has at least three square integrable components, it and its companion $\tilde{\pi}^E_C$ are the only cuspidal representations (with the indicated components at the places of $E$ above $v_1$, $v_2$, $v_3$) which $\lambda_1$-lift to $\pi^E = \lambda_1(\pi^E_C)$. Consequently, each of the $n_{v_0}$ pairs $\pi'_C$, $\tilde{\pi}'_C$ of cuspidal representations of $\mathbf{C}(\mathbb{A})$ which $\lambda_1$-lift to $\lambda_1(\pi'_C)$, basechange from $F$ to $E$ to the pair $\pi^E_C$, $\tilde{\pi}^E_C$ of cuspidal representations of $\mathbf{C}(\mathbb{A}_E)$, which $\lambda_1$-lift to $\pi^E = \lambda_1(\pi^E_C)$. But the fiber of the base change map $BC_{E/F}$, which takes $\pi_C$ to $\pi^E_C$, consists only of $\pi_C$ and $\chi_{E/F}\pi_C$, where $\chi_{E/F}$ is the quadratic character of $\mathbb{A}^\times/F^\times N_{E/F}\mathbb{A}^\times_E$. Consequently each pair $\{\pi'_C, \tilde{\pi}'_C\}$ is equal to the pair $\{\pi_C, \tilde{\pi}_C\}$, up to multiplication by $\chi_{E/F}$. But this implies that $n_{v_1} = 1$, as asserted.

In residual characteristic two there are also the "extraordinary" cuspidal representations, which are not associated with a character of a quadratic extension. But since the relation (1) defines a partition of the set of representations of $C$, and we already handled the monomial representations, the orthogonality relations imply the lifting and character relation, and the proof of the proposition is complete. □

We obtain the following rigidity theorem for representations of SO(4), that is of $\mathbf{C}(\mathbb{A})$.

Note that $\lambda_1(\chi\pi_1 \times \chi^{-1}\pi_2) = \lambda_1(\pi_1 \times \pi_2) = \lambda_1(\pi_2 \times \pi_1)$.

4.2 COROLLARY. *Let $\pi_1$, $\pi_2$, $\pi'_1$, $\pi'_2$, be cuspidal representations of* $\mathrm{GL}(2,\mathbb{A})$ *with central characters $\omega_1$, $\omega_2$, $\omega'_1$, $\omega'_2$, satisfying $\omega_1\omega_2 = 1$, $\omega'_1\omega'_2 = 1$. Suppose that there is a set $S$ of places of $F$ such that $(\pi'_{1v}, \pi'_{2v}) = (\pi_{1v}\chi_v, \pi_{2v}\chi_v^{-1})$ for all $v$ in $S$ and $(\pi'_{1v}, \pi'_{2v}) = (\pi_{2v}\chi_v, \pi_{1v}\chi_v^{-1})$ for all $v$ outside $S$, for some character $\chi_v$ of $F_v^\times$ (for each $v$). Then the pair $(\pi'_1, \pi'_2)$ is $(\pi_1\chi, \pi_2\chi^{-1})$ or $(\pi_2\chi, \pi_1\chi^{-1})$ for some character $\chi$ of $\mathbb{A}^\times/F^\times$.*

A considerably weaker result, where the notion of equivalence is generated only by $\pi_{1v} \times \check{\pi}_{2v} \simeq \check{\pi}_{2v} \times \pi_{1v}$ but not by $\pi_{1v} \times \check{\pi}_{2v} \simeq \chi_v\pi_{1v} \times \chi_v^{-1}\check{\pi}_{2v}$, follows also on using the Jacquet-Shalika [JS] theory of $L$-functions, comparing the poles at $s = 1$ of the partial, product $L$-functions

$$L^V(s, \pi'_1 \times \check{\pi}_1)L^V(s, \pi'_2 \times \check{\pi}_1) = L^V(s, \pi_1 \times \check{\pi}_1)L^V(s, \pi_2 \times \check{\pi}_1).$$

Moreover, such a proof assumes the theory of $L$-functions.

This has a consequence purely for characters.

4.3 COROLLARY. *Let $E/F$ be a quadratic extension of number fields, and $\mu_1$, $\mu_2$, $\mu'_1$, $\mu'_2$ characters of $\mathbb{A}_E^\times/E^\times$ such that the restriction to $\mathbb{A}^\times/F^\times$ of the products $\mu_1\mu_2$ and $\mu'_1\mu'_2$ is trivial. Suppose that at 3 places $v$ of $F$ which do not split in $E$ we have that $\overline{\mu}_{iv} \neq \mu_{iv}$ $(i = 1, 2)$. Suppose that there is a set $S$ of places of $F$, and characters $\chi_v$ of $F_v^\times$ for each place $v$ of $F$, such that if $\mu_{iv}$ are the local components of $\mu_i$ on $E_v^\times = (E \otimes_F F_v)^\times$, then*

$$(\mu'_{1v}, \mu'_{2v}) = (\mu_{1v} \cdot \chi_v \circ N, \mu_{2v} \cdot (\chi_v \circ N)^{-1})$$

*for all $v$ in $S$, and*

$$(\mu'_{1v}, \mu'_{2v}) = (\mu_{2v} \cdot \chi_v \cdot N, \mu_{1v} \cdot (\chi_v \circ N)^{-1})$$

*for all $v$ outside $S$ (where $N$ is the norm map from $E_v$ to $F_v$). Then there is a character $\chi$ of $\mathbb{A}^\times/F^\times$ such that*

$$(\mu'_1, \mu'_2) = (\mu_1{\cdot}\chi{\circ}N, \mu_2{\cdot}(\chi{\circ}N)^{-1}) \quad \textit{or} \quad (\mu'_1, \mu'_2) = (\mu_2{\cdot}\chi{\circ}N, \mu_1{\cdot}(\chi{\cdot}N)^{-1}).$$

PROOF. Consider the cuspidal representations $\pi_E(\mu_1)$, $\pi_E(\mu_2)$. Note that they are cuspidal at least at three places, and that $\chi\pi_E(\mu) = \pi_E(\mu \cdot \chi \circ N)$. Apply the previous corollary. □

4.4 PROPOSITION. *Let $\pi_{v_0}$ be a square integrable $\theta$-invariant representation of the group* PGL$(4, F_{v_0})$. *Its $\theta$-character is not identically zero on the $\theta$-elliptic regular set (by the orthonormality relations). Suppose it is not a $\theta$-stable function on the $\theta$-elliptic regular set. Then it is a $\lambda_1$-lift of a square integrable representation $\pi_{2v_0} \times \pi'_{2v_0}$ of $\mathbf{C}(F_{v_0})$, and its $\theta$-character is a $\theta$-unstable function.*

PROOF. Let $F$ be a totally imaginary global field such that $F_{v_i} = F_{v_0}$ $(i = 0, 1, 2, 3)$. We use a test function $f = \otimes f_v$ such that $f_{v_i}$ $(i = 1,$ $2, 3)$ is a pseudo-coefficient of a $\theta$-invariant cuspidal representation $\pi_{v_i} = \lambda_1(\pi_{E'_1}(\mu_1) \times \pi_{E'_2}(\mu_2))$, $E'_1$, $E'_2$ are quadratic extensions of $F_{v_i}$ and $\mu_1\mu_2|F_v^\times = 1$, and $f_{v_0}$ is a pseudo coefficient of $\pi_{v_0}$. At all finite $v \neq v_i$ $(0 \leq i \leq 3)$ we take $f_v$ to be spherical, such that for $\kappa \neq 1$ which corresponds to the endoscopic group $\mathbf{C}$ and with $f^\infty = \otimes f_v$, $v$ finite, the $\kappa$-$\theta$-orbital integral $\Phi_\gamma^\kappa(f^\infty)$ is not zero at some $\theta$-regular elliptic $\gamma$ in $\mathbf{G}(F)$; this simply requires taking the support of the $f_v \geq 0$ for $v \neq v_i$ $(0 \leq i \leq 3)$ to be large enough. Since the $\theta$-stable orbital integrals of $f_{v_i}$ $(1 \leq i \leq 3)$ are 0, the $\theta$-elliptic regular part of the $\theta$-trace formula consists entirely of $\kappa$-$\theta$-orbital integrals, by a standard stabilization argument.

As $\mathbf{G}(F)$ is discrete in $\mathbf{G}(\mathbb{A})$, for every $f_\infty = \otimes_v f_v$, $v$ archimedean, $f = f_\infty f^\infty$ is compactly supported, we can choose $f_\infty$ to have small enough support around $\gamma \in \mathbf{G}(F)$ with $\Phi_\gamma^\kappa(f^\infty) \neq 0$ in the $\theta$-regular set of $\mathbf{G}(F_\infty)$, to guarantee that $\Phi_\gamma^\kappa(f) \neq 0$ for a single $\theta$-stable $\theta$-regular conjugacy class $\gamma$ in $\mathbf{G}(F)$, which is necessarily $\theta$-elliptic. Hence the geometric part of the $\theta$-trace formula reduces to the single term $\Phi_\gamma^\kappa(f)$, which is nonzero, hence the geometric part is nonzero, and so is the spectral side.

The choice of the pseudo coefficients $f_{v_i}$ implies that in the spectral side we have a $\theta$-invariant cuspidal representation $\pi$ of $\mathbf{G}(\mathbb{A})$ with the cuspidal components $\pi_{v_i}$ $(i = 1, 2, 3)$ and the square integrable component $\pi_{v_0}$ of the proposition (note that $\pi$ is cuspidal since it has cuspidal components at $v_i$ $(i = 1, 2, 3)$, hence it is generic). The components of $\pi$ at any other finite place are spherical. Since the $\theta$-stable orbital integrals of $f_{v_i}$ $(i = 1, 2, 3)$ are zero, we may take $f_{Hv_i}$ and so $f_H$ to be zero. Hence there is no contribution to the spectral form of the trace formula identity from the trace formulae of $H$ and $C_0$.

Using generalized linear independence of characters we get the form of the trace formula identity with only our $\pi$ as the single term on the spectral side of $G$, while the only contributions to the other side – $\pi_C$ – depend only on $f_C$. Any unramified component of $\pi_C$ $\lambda_1$-lifts to the corresponding component of $\pi$, and similar statement holds for the archimedean places.

Using the pseudo-coefficients $f_{v_i}$ at the places $v_i$ ($i = 1, 2, 3$) we see that $\pi_{C,v_i} = \pi_{E_1'}(\mu_1) \times \pi_{E_2'}(\mu_2)$. We are left with an identity of $\operatorname{tr} \pi_{v_0}(f_{v_0} \times \theta)$ with a sum $\sum m(\pi_C) \operatorname{tr} \pi_{C,v_0}(f_{C,v_0})$ for all matching $f_{v_0}$, $f_{C,v_0}$, from which we conclude as usual using the character relations that the $\pi_{C,v_0}$ are square integrable, finite in number, and in fact consist of a single square-integrable $\pi_{2v_0} \times \pi_{2v_0}'$ which $\lambda_1$-lifts to $\pi_{v_0}$. This has already been treated by our complete description of the $\lambda_1$-lifting. $\square$

Remark. The central character of a monomial $\pi_E(\mu)$ is

$$\chi \cdot \mu | F^\times \qquad (\chi : F^\times / N_{E/F} E^\times \tilde{\to} \{\pm 1\}).$$

If $\pi_E(\mu_1) \times \pi_E(\mu_2)$ defines a representation of $C$ then the product of the central characters is 1, thus $\mu_1 \mu_2 | F^\times = 1$. Hence $\pi_E(\mu_1 \overline{\mu}_2)$, $\pi_E(\mu_1 \mu_2)$ have central characters $\chi \cdot \mu_1 \overline{\mu}_2 | F^\times = \chi = \chi \cdot \mu_1 \mu_2 | F^\times$. Thus

$$I(\pi_E(\mu_1 \overline{\mu}_2), \pi_E(\mu_1 \mu_2))$$

will not be in the image of $\lambda$ – see Proposition V.5 below: it is not $I(\pi_1, \pi_2)$, $\pi_1$, $\pi_2$ on PGL(2).

# V. LIFTING FROM PGSp(2) TO PGL(4)

## 1. Characters on the Symplectic Group

Next we proceed with preliminaries on the lifting of representations of $\mathbf{H} = \mathrm{PGSp}(2)$ to those on $\mathbf{G} = \mathrm{PGL}(4)$. Recall that the norm map $N : \mathbf{G} \to \mathbf{H}$ is defined on the diagonal tori $N : \mathbf{T}^* \to \mathbf{T}^*_{\mathbf{H}}$ by

$$N(\mathrm{diag}(a,b,c,d)) = \mathrm{diag}(ab, ac, db, dc),$$

and on the Levi factors on the other two proper parabolic subgroups by

$$N(\mathrm{diag}(A,B)) = \mathrm{diag}(\det A, \varepsilon B \varepsilon A, \det B) \quad \text{where} \quad \varepsilon = \begin{pmatrix} 1 & 0 \\ 0 & -1 \end{pmatrix},$$

and $N(\mathrm{diag}(a,A,d)) = \mathrm{diag}(aA, d\varepsilon A\varepsilon)$. The dual, lifting, map of representations takes the induced-from-the-Borel representation

$$I_H(\mu_1,\mu_2) = \mu_1\mu_2 \times \mu_1/\mu_2 \rtimes \mu_1^{-1} \qquad \text{to} \qquad I_G(\mu_1,\mu_2,\mu_2^{-1},\mu_1^{-1}),$$

where $H = \mathrm{PGSp}(2,F)$, $G = \mathrm{PGL}(4,F)$, $F$ a local field. Lifting is defined by means of the character relation.

Before continuing, let us verify

1.1 Lemma. *The Jacobians satisfy* $\Delta_G(t\theta) = \Delta_H(Nt)$.

Proof. We take $t = \mathrm{diag}(\alpha,\beta,\gamma,\delta)$, $\alpha\delta = \beta\gamma$, and compute

$$\Delta_H(t) = |\det(\mathrm{Ad}(t)|\,\mathrm{Lie}\,N)|^{-1/2}|\det(1-\mathrm{Ad}(t))|\,\mathrm{Lie}\,N|,$$

where $N$ denotes the upper triangular unipotent subgroup in $H$. The Lie algebra $\mathrm{Lie}\,N$ consists of $X \in \mathrm{Lie}\,H = \{X = -J^{-1}XJ\}$ of the form

$$\begin{pmatrix} 0 & x & y & z \\ 0 & 0 & u & y \\ 0 & 0 & 0 & -x \\ 0 & 0 & 0 & 0 \end{pmatrix},$$

and the effect of $1-\mathrm{Ad}(t)$ is

$$x \mapsto (1-\alpha/\beta)x, \quad y \mapsto (1-\alpha/\gamma)y, \quad z \mapsto (1-\alpha/\delta)z, \quad u \mapsto (1-\beta/\gamma)u.$$

Thus

$$\Delta_H(t) = \left|\frac{\beta\gamma\delta\gamma}{\alpha\alpha\alpha\beta}\right|^{1/2} \left|\left(1-\frac{\alpha}{\beta}\right)\cdot\left(1-\frac{\alpha}{\gamma}\right)\cdot\left(1-\frac{\alpha}{\delta}\right)\cdot\left(1-\frac{\beta}{\gamma}\right)\right|$$

gives $\Delta_G(h\theta)$ when $\Delta_H(t)$ is evaluated at $t=Nh$. □

We are now ready to extend the basic lifting result from the minimal parabolic to the other two proper parabolic subgroups of $H$.

1.2 PROPOSITION. *We have that $\omega_\pi \rtimes \check{\pi} - \omega_\pi^{-1} \rtimes \pi$ ($\lambda$-) lifts to $\pi_4 = I_G(\pi, \check{\pi})$, where $\omega_\pi$ is the central character of the representation $\pi = \pi_2$ of* $\mathrm{GL}(2,F) = \mathrm{GSp}(1,F)$, *and $\check{\pi} = \omega_\pi^{-1}\pi$ is the contragredient of $\pi$.*

*For the representation $\pi_2$ of* $\mathrm{PGL}(2,F)$ *we have that $\mu_1\pi_2 \rtimes \mu_1^{-1}$ ($\lambda$-) lifts to $\pi_4 = I_G(\mu_1, \pi_2, \mu_1^{-1})$, and $I_2(\mu_1, \mu_1^{-1}) \times \pi_2$ $\lambda_0$-lifts from $C_0$ to $\mu_1\pi_2 \rtimes \mu_1^{-1}$ on $H$.*

PROOF. Recall that at $m_0 = \mathrm{diag}(A, I)$, $A \in \mathrm{GL}(2,F)$, the value of the character $\chi_{\pi_4}(m_0\theta)$, where $\pi_4 = I_4(\pi, \check{\pi})$, has been computed to be

$$(1+\omega_\pi^{-1}(\det A))\chi_\pi(A)/\left|\frac{(1-a)(1-b)(1-ab)}{ab}\right|.$$

Since $N(\mathrm{diag}(A,I))$ is $\mathrm{diag}(\boldsymbol{\lambda}, A, 1)$, $\boldsymbol{\lambda} = \det A$, we have to compute the character $\chi_{\omega^{-1}\rtimes\pi}$ at $\mathrm{diag}(\boldsymbol{\lambda}, A, 1)$. A general element $m$ of the Levi $M_H$ of type (1,2,1) in $H$ has the form $m = \mathrm{diag}(a, A, \boldsymbol{\lambda}/a)$, $a \in F^\times$. If $N = N_H$ is the corresponding upper triangular unipotent subgroup then

$$\delta_N(m) = |\det(\mathrm{Ad}(m)|\,\mathrm{Lie}\,N)| = |a^2/\det A|^2$$

(using the $X$ of the proof of Lemma 1.1 with $u=0$). The usual argument, using the measure decomposition $dg = \delta_N^{-1}(m)dndmdk$, shows that $(\pi_H(fdg)\phi)(h)$ is

$$= \int_N \int_M \int_K f(h^{-1}n_1mk)\delta_N^{1/2}(m)(\omega_\pi^{-1} \rtimes \pi)(m)\phi(k)\delta_N^{-1}(m)dndmdk.$$

Hence

$$\operatorname{tr}\pi_H(fdg) = \iiint f(k^{-1}n_1mk)\delta_N^{-1/2}(m)\chi_\pi(a^{-1}A)dn_1dmdk.$$

The change $n_1 = nmn^{-1}m^{-1}$ of variables has $|\det(1-\mathrm{Ad}(m))|\operatorname{Lie}N|$ as Jacobian. Hence with

$$\Delta_{M_H}(m) = \delta_N^{-1/2}(m)|\det(1-\mathrm{Ad}(m))|\operatorname{Lie}N| = \left|\frac{bc}{a^2}\right| \cdot \left|\frac{a-b}{b}\;\frac{a-c}{c}\;\frac{a-d}{d}\right|$$

(we denoted the eigenvalues of $A$ by $b$ and $c$), the trace is

$$\begin{aligned} &= \int_{M_H} \Delta_{M_H}(m)\chi_\pi(a^{-1}A) \int_K \int_N f(k^{-1}n^{-1}mnk)dndkdm \\ &= \int_{M_H} \Delta_M(m)\chi_\pi(a^{-1}A) \int_{M_H\backslash H} f(g^{-1}mg)d\dot{g}dm. \end{aligned}$$

Now the Weyl group of $M_H$ in $H$ (normalizer$/M_H$) is represented by 1 and $J$. Changing variables $g \mapsto Jg$ has the effect of mapping $m = \operatorname{diag}(a, A, \boldsymbol{\lambda}/a)$ to $m' = \operatorname{diag}(\boldsymbol{\lambda}/a, A, a)$. We have $\chi_{\omega_\pi^{-1}\rtimes\pi}(m) = \chi_\pi(a^{-1}A)$ and

$$\chi_{\omega_\pi^{-1}\rtimes\pi}(m') = \chi_\pi(\frac{a}{\boldsymbol{\lambda}}A) = \omega_\pi(\det(a^{-1}A))^{-1}\chi_\pi(a^{-1}A).$$

The trace becomes

$$\int_{M_H} \frac{1}{2}\Delta_{M_H}(m)(1+\omega_\pi(\det(a^{-1}A))^{-1})\chi_\pi(a^{-1}A) \int_{M_H\backslash H} f(g^{-1}mg)d\dot{g}dm.$$

Hence

$$\chi_{\omega_\pi^{-1}\rtimes\pi}(\operatorname{diag}(\boldsymbol{\lambda}, A, 1)) = (1+\omega_\pi^{-1}(\det A))\chi_\pi(A)/\Delta_M(\operatorname{diag}(\boldsymbol{\lambda}, A, 1)),$$

where

$$\Delta_{M_H}(\operatorname{diag}(ab, A, 1)) = |\frac{a-1}{a} \cdot \frac{b-1}{b} \cdot (ab-1)|$$

(where $a, b$ are the eigenvalues of $A$), and we recover $\chi_{\pi_4}(m_0\theta)$. We are done by Lemma 1.1: $\Delta_G(t\theta) = \Delta_H(Nt)$.

To show that $\lambda(\mu_1\pi_2 \rtimes \mu_1^{-1}) = I_G(\mu_1, \pi_2, \mu_1^{-1})$, we first compute the $\theta$-character of $\pi_4 = I_G(\mu_1, \pi_2, \mu_1^{-1})$. Note that $\phi \in \pi_4$ takes $nmk$ to

$\delta_M^{1/2}(m)\mu_1(a/d)\pi_2(A)\phi(k)$, where $m = \mathrm{diag}(a, A, d)$ is in the standard Levi $M$ of $G$ of type (1,2,1). The measure decomposition contributes a factor $\delta_M(m)^{-1}$, so we have

$$\mathrm{tr}\,\pi(\theta f\ dg) = \iiint f(\theta(k)^{-1} n_1 mk)\delta_M^{-1/2}(m)\mu_1(a/d)\chi_{\pi_2}(A)dn_1 dmdk.$$

The Jacobian of $n_1 \mapsto n$, $n_1 = nm\theta(n^{-1})m^{-1}$ is $|\det(1 - \mathrm{Ad}(m\theta))|\,\mathrm{Lie}\,N|$. Putting

$$\Delta_M(m\theta) = \delta_M^{-1/2}(m)|\det(1 - \mathrm{Ad}(m\theta))|\,\mathrm{Lie}\,N|$$

we get

$$= \int_M \Delta_M(m0)\mu_1(a/d)\chi_{\pi_2}(A)\int_{M\backslash G} f(\theta(g)^{-1}mg)d\dot{g}\ dm.$$

The $\theta$-Weyl group $W^\theta(M)$ of $M$ in $G$ ($\theta$-normalizer$/M$) is represented by $\{I, J\}$. Hence the trace is

$$= \int_M \frac{1}{2}\Delta_M(m\theta)[\mu_1(a/d) + \mu_1(d/a)]\chi_{\pi_2}(A)\int_{M\backslash G} f(\theta(g)^{-1}mg)d\dot{g}\ dm,$$

and the character is $\frac{1}{2}[\mu_1(a/d) + \mu_1(d/a)]\chi_{\pi_2}(A)/\Delta_M(m\theta)$.

This we compare with the character of $\pi_H = \mu_1\pi_2 \rtimes \mu_1^{-1}$, the representation of $H$ normalizedly induced from the representation

$$\begin{pmatrix} A & * \\ 0 & \frac{\boldsymbol{\lambda}}{\det A}\varepsilon A\varepsilon \end{pmatrix} = \begin{pmatrix} A & * \\ 0 & \boldsymbol{\lambda} w {}^t A^{-1} w \end{pmatrix} \mapsto \mu_1(\boldsymbol{\lambda}^{-1}\det A)\pi_2(A)$$

of the standard parabolic subgroup of $H$ whose Levi $M_H$ is of type (2,2). As usual we have that $\mathrm{tr}\,\pi_H(f\ dg)$

$$= \int_{M_H} \Delta_{M_H}(m)\mu_1(\boldsymbol{\lambda}^{-1}\det A)\chi_{\pi_2}(A)\int_{M_H\backslash H} f(h^{-1}mh)d\dot{h}\ dm.$$

The Weyl group of $M_H$ in $H$ (normalizer $/M_H$) is represented by $\{I, J\}$. Writing $|A|$ for $\det A$, and $X$ for $\int_{M_H\backslash H} f(h^{-1}mh)dh$,

$$\begin{pmatrix} 0 & w \\ -w & 0 \end{pmatrix}\begin{pmatrix} A & 0 \\ 0 & \frac{\boldsymbol{\lambda}}{\det A}\varepsilon A\varepsilon \end{pmatrix}\begin{pmatrix} 0 & -w \\ w & 0 \end{pmatrix} = \begin{pmatrix} \boldsymbol{\lambda}|A|^{-1}\omega A\omega & 0 \\ 0 & wAw \end{pmatrix}$$

$$\mapsto \mu_1(\boldsymbol{\lambda}/|A|)\pi_2(\boldsymbol{\lambda}|A|^{-1}\omega A\omega),$$

and we obtain

$$= \int_{M_H} \frac{1}{2}\Delta_{M_H}(m)[\mu_1(\boldsymbol{\lambda}^{-1}\det A) + \mu_1(\boldsymbol{\lambda}/\det A)]\chi_{\pi_2}(A)X dm.$$

The character of $\pi_H = \mu_1\pi_2 \rtimes \mu_1^{-1}$ is then

$$\frac{1}{2}[\mu_1(\boldsymbol{\lambda}^{-1}\det A) + \mu_1(\boldsymbol{\lambda}/\det A)]\chi_{\pi_2}(A)/\Delta_{M_H}(m).$$

We need to compare the characters at an element $g = \operatorname{diag}(a, A, d)$ whose norm is $h = Ng = \operatorname{diag}(aA, d\varepsilon A\varepsilon)$. Note that on this $h$, the character from which $\pi_H$ is induced takes the value

$$\begin{pmatrix} aA & * \\ 0 & \frac{d}{a}a\varepsilon A\varepsilon \end{pmatrix} \mapsto \mu_1(a/d)\pi_2(aA) = \mu_1(a/d)\pi_2(A),$$

the last equality since $\pi_2$ has trivial central character. As $\Delta_G(g\theta) = \Delta_H(Ng)$ our character identity be complete once we show

1.3 LEMMA. *We have* $\Delta_M(g\theta) = \Delta_{M_H}(Ng)$.

PROOF. As these factors depend only on the eigenvalues of $A$, we may take $t = \operatorname{diag}(a, b, c, d)$, and $Nt = \operatorname{diag}(\alpha, \beta, \gamma, \delta) = \operatorname{diag}(ab, ac, db, dc)$. Then $\Delta_{(1,2,1)}(t\theta)$ is the product of $\delta_M(t\theta)^{-1}$, where

$$\delta_M(t\theta) = \left|\frac{a}{b}\,\frac{c}{d}\cdot\frac{a}{c}\,\frac{b}{d}\cdot\frac{a}{d}\right|,$$

with

$$\left|\det(1 - \operatorname{Ad}(t\theta))\middle| \operatorname{Lie} N_{(1,2,1)}\right| = \left|\left(1 - \frac{a}{b}\,\frac{c}{d}\right)\left(1 - \frac{a}{c}\,\frac{b}{d}\right)\left(1 - \frac{a}{d}\right)\right|,$$

namely

$$\Delta_{(1,2,1)}(t\theta) = \left|\frac{(ab-cd)^2(ac-bd)^2(a-d)^2}{a^3b^2c^2d^3}\right|^{1/2}.$$

Similarly

$$\Delta^H_{(2,2)}(Nt) = \left|\frac{\gamma^2\delta}{\alpha^2\beta}\right|^{1/2}\left|\left(1 - \frac{\alpha}{\gamma}\right)\cdot\left(1 - \frac{\alpha}{\delta}\right)\cdot\left(1 - \frac{\beta}{\gamma}\right)\right|$$

$$= \left| \frac{(\alpha-\gamma)^2(\alpha-\delta)^2(\beta-\gamma)^2}{\alpha^2\beta\gamma^2\delta} \right|^{1/2} = \left| \frac{(ab-db)^2(ab-cd)^2(ac-bd)^2}{a^2b^2acd^2b^2cd} \right|^{1/2}$$

as $(\alpha,\beta,\gamma,\delta)=(ab,ac,db,dc)$. We conclude that $\Delta_{(1,2,1)}(t\theta)=\Delta^H_{(2,2)}(Nt)$. This completes the proof of the lemma, hence also of the proposition. □

Let $\chi$ denote a character (multiplicative function) of $F^\times/F^{\times 2}$. It defines one dimensional representation $\chi_H$ of $H$ by $h \mapsto \chi(\boldsymbol{\lambda}(h))$. If $h = Ng$ (on diagonal matrices, if $g=\operatorname{diag}(a,b,c,d)$ then $h=\operatorname{diag}(ab,ac,db,dc)$) then $\boldsymbol{\lambda}(h)=\det\ g$. Hence

1.4 LEMMA. *The one dimensional representation $\chi_H$, or $\chi\cdot 1_H$, of $H$, $\lambda$-lifts to the one dimensional representation $\chi : g \mapsto \chi(\det\ g)$ of $G$. The trivial representation of $H$ lifts to the trivial representation of $G$.* □

We conclude

1.5 COROLLARY. *The Steinberg representation of $H$ $\lambda$-lifts to the Steinberg representation of $G$.*

PROOF. We use Lemma 3.5 of [ST], which asserts the following decomposition result: $\nu^2 \times \nu \rtimes \nu^{-3/2}\sigma$ is equal to

$$= \nu^{3/2}\operatorname{sp}_2 \rtimes \nu^{-3/2}\sigma + \nu^{3/2}\mathbf{1}_2 \rtimes \nu^{-3/2}\sigma = \nu^2 \rtimes \nu^{-1}\sigma\operatorname{sp}_2 + \nu^2 \rtimes \nu^{-1}\sigma\mathbf{1}_2$$

in the Grothendieck group ($\mathbb{Z}$-module generated by the irreducible representations) of $H$. Here $\sigma^2=1$ to have trivial central character, and as usual $\nu(x)=|x|$. The terms on the right decompose into irreducibles (on the right of the following four equations, which in fact define the square integrable Steinberg representation of $H=\operatorname{PGSp}(2,F)$):

a) $\nu^{3/2}\mathbf{1}_2 \rtimes \nu^{-3/2}\sigma = \sigma\cdot\mathbf{1}_{\operatorname{GSp}(2)} + L(\nu^2,\nu^{-1}\sigma\cdot\operatorname{sp}_2)$,
b) $\nu^2 \rtimes \nu^{-1}\sigma\cdot\operatorname{sp}_2 = \sigma\cdot\operatorname{St}_{\operatorname{GSp}(2)} + L(\nu^2,\nu^{-1}\sigma\cdot\operatorname{sp}_2)$,
c) $\nu^2 \rtimes \nu^{-1}\sigma\cdot\mathbf{1}_2 = \sigma\cdot\mathbf{1}_{\operatorname{GSp}(2)} + L(\nu^{3/2}\operatorname{sp}_2,\nu^{-3/2}\sigma)$,
d) $\nu^{3/2}\operatorname{sp}_2 \rtimes \nu^{-3/2}\sigma = \sigma\cdot\operatorname{St}_{\operatorname{GSp}(2)} + L(\nu^{3/2}\operatorname{sp}_2,\nu^{-3/2}\sigma)$.

We can apply the $\lambda$-lifting to a), as the lifts of two of its term is known:

$$\sigma I(\nu^{-3/2},\mathbf{1}_2,\nu^{3/2}) = \sigma\cdot\mathbf{1}_4 + \lambda(L(\nu^2,\nu^{-1}\sigma\cdot\operatorname{sp}_2)).$$

Next we apply the $\lambda$-lifting to b) and note that $\sigma I(\nu^{-1}\,\mathrm{sp}_2 \times \nu\,\mathrm{sp}_2)$, the left side, is known to be of length two, consisting of the Steinberg representation $\sigma\cdot\mathrm{St}_4$, and an irreducible which lies in the composition series of $\sigma I(\nu^{-3/2}, \mathbf{1}_2, \nu^{3/2})$ (which is also of length two, the other irreducible being $\sigma \cdot \mathbf{1}_4$). Hence the common irreducible is $\lambda(L(\nu^2, \nu^{-1}\sigma \cdot \mathrm{sp}_2))$, and the $\lambda$-lift of $\sigma\,\mathrm{St}_{\mathrm{GSp}(2)}$ is $\sigma\,\mathrm{St}_4$.

An alternative proof is obtained on $\lambda$-lifting c) to get

$$\sigma I(\nu^{-1}\mathbf{1}_2 \times \nu\mathbf{1}_2) = \sigma\mathbf{1}_4 + \lambda(L(\nu^{3/2}\,\mathrm{sp}_2, \nu^{-3/2}\sigma)),$$

the last irreducible is the one common with $\sigma I(\nu^{-3/2}, \mathrm{sp}_2, \nu^{3/2})$, which is the $\lambda$-lift of the left side of d); the latter has $\sigma\,\mathrm{St}_4$ as the other irreducible in its composition series, hence the $\lambda$-lift of the $\sigma\,\mathrm{St}_{\mathrm{GSp}(2)}$ on the right of d) has to be $\sigma\,\mathrm{St}_4$. □

# 2. Reducibility

It will be useful to record the results of [ST], Lemmas 3.3, 3.7, 3.4, 3.9, 3.6, 3.8, on reducibility of induced representations of $H$. This we do next. Note that the case of $\nu^2 \times \nu \rtimes \nu^{-3/2}\sigma$ is discussed in the proof of Corollary 1.5 above.

2.1 PROPOSITION. (*a*) *The representation $\chi_1 \times \chi_2 \rtimes \sigma$ of $H$, where $\chi_1, \chi_2, \sigma$ are characters of $F^\times$, is reducible precisely when $\chi_1$, $\chi_2$, $\chi_1\chi_2$ or $\chi_1/\chi_2$ equals $\nu$ or $\nu^{-1}$ (its central character is $\chi_1\chi_2\sigma^2$).*
(*b*) *If $\chi \notin \{\xi\nu^{\pm 1/2}, \nu^{\pm 3/2}\}$ for any character $\xi$ with $\xi^2 = 1$, then $\chi \cdot \mathrm{sp}_2 \rtimes \sigma$ and $\chi \cdot \mathbf{1}_2 \rtimes \sigma$ are irreducible and*

$$\nu^{1/2}\chi \times \nu^{-1/2}\chi \rtimes \sigma = \chi \cdot \mathbf{1}_2 \rtimes \sigma + \chi \cdot \mathrm{sp}_2 \rtimes \sigma.$$

*If $\chi \neq 1$, $\nu^{\pm 1}$, $\nu^{\pm 2}$ then $\chi \rtimes \sigma \cdot \mathrm{sp}_2$ and $\chi \rtimes \sigma \cdot \mathbf{1}_2$ are irreducible and*

$$\chi \times \nu \rtimes \nu^{-1/2}\sigma = \chi \rtimes \sigma \cdot \mathrm{sp}_2 + \chi \rtimes \sigma \cdot \mathbf{1}_2.$$

*For any character $\sigma$ we have that $\nu \rtimes \nu^{-1/2}\sigma \cdot \mathrm{sp}_2$ and $\nu \rtimes \nu^{-1/2}\sigma \cdot \mathbf{1}_2$ are irreducible and*

$$\nu \times \nu \rtimes \nu^{-1}\sigma = \nu \rtimes \nu^{-1/2}\sigma \cdot \mathrm{sp}_2 + \nu \rtimes \nu^{-1/2}\sigma \cdot \mathbf{1}_2.$$

(c) *If* $\xi \neq 1 = \xi^2$, $\nu\xi \times \xi \rtimes \nu^{-1/2}\sigma$ *contains a unique essentially square integrable subrepresentation denoted* $\delta(\xi\nu^{1/2}\,\mathrm{sp}_2, \nu^{-1/2}\sigma)$. *Since*

$$\xi \rtimes \nu^{-1/2}\sigma = I(\xi\sigma\nu^{-1/2}, \sigma\nu^{-1/2}) = \xi \rtimes \nu^{-1/2}\sigma\xi,$$

*we have*

$$\nu\xi \times \xi \rtimes \nu^{-1/2}\sigma = \nu^{1/2}\xi\,\mathrm{sp}_2 \rtimes \nu^{-1/2}\sigma + \nu^{1/2}\xi\mathbf{1}_2 \rtimes \nu^{-1/2}\sigma$$
$$= \nu\xi \times \xi \rtimes \nu^{-1/2}\sigma\xi = \nu^{1/2}\xi\,\mathrm{sp}_2 \rtimes \nu^{-1/2}\sigma\xi + \nu^{1/2}\xi\mathbf{1}_2 \rtimes \nu^{-1/2}\sigma\xi,$$

$$\delta(\xi\nu^{1/2}\,\mathrm{sp}_2, \nu^{-1/2}\sigma) = \delta(\xi\nu^{1/2}\,\mathrm{sp}_2, \nu^{-1/2}\sigma\xi),$$

*as well as*

$$\nu^{1/2}\xi\,\mathrm{sp}_2 \rtimes \nu^{-1/2}\sigma = \delta(\xi\nu^{1/2}\,\mathrm{sp}_2, \nu^{-1/2}\sigma) + L(\nu^{1/2}\xi\,\mathrm{sp}_2, \nu^{-1/2}\sigma),$$
$$\nu^{1/2}\xi\mathbf{1}_2 \rtimes \nu^{-1/2}\sigma = L(\nu^{1/2}\xi\,\mathrm{sp}_2, \nu^{-1/2}\sigma\xi) + L(\nu\xi, \xi \rtimes \nu^{-1/2}\sigma);$$

*the* 4 *representations on the right of the last two lines are irreducible.*
(d) *The representations* $1 \rtimes \sigma \cdot \mathrm{sp}_2$ *and* $\nu^{1/2}\,\mathrm{sp}_2 \rtimes \nu^{-1/2}\sigma$ (*resp.* $\nu^{1/2}\mathbf{1}_2 \rtimes \nu^{-1/2}\sigma$) *have a unique irreducible subquotient in common; it is essentially tempered, denoted by*
$\tau(\nu^{1/2}\,\mathrm{sp}_2, \nu^{-1/2}\sigma)$ (*resp.* $\tau(\nu^{1/2}\mathbf{1}_2, \nu^{-1/2}\sigma)$). *These two* $\tau$*'s are inequivalent, and we have*

$$\nu \times 1 \rtimes \nu^{-1/2}\sigma = \nu^{1/2}\,\mathrm{sp}_2 \rtimes \nu^{-1/2}\sigma + \nu^{1/2}\mathbf{1}_2 \rtimes \nu^{-1/2}\sigma$$

$$= 1 \times \nu \rtimes \nu^{-1/2}\sigma = 1 \rtimes \sigma \cdot \mathrm{sp}_2 + 1 \rtimes \sigma \cdot \mathbf{1}_2,$$

*as well as the following decomposition into irreducibles:*

$$\nu^{1/2}\,\mathrm{sp}_2 \rtimes \nu^{-1/2}\sigma = \tau(\nu^{1/2}\,\mathrm{sp}_2, \nu^{-1/2}\sigma) + L(\nu^{1/2}\,\mathrm{sp}_2, \nu^{-1/2}\sigma),$$
$$\nu^{1/2}\mathbf{1}_2 \rtimes \nu^{-1/2}\sigma = \tau(\nu^{1/2}\mathbf{1}_2, \nu^{-1/2}\sigma) + L(\nu, 1 \rtimes \nu^{-1/2}\sigma),$$
$$1 \rtimes \sigma\,\mathrm{sp}_2 = \tau(\nu^{1/2}\,\mathrm{sp}_2, \nu^{-1/2}\sigma) + \tau(\nu^{1/2}\mathbf{1}_2, \nu^{-1/2}\sigma),$$
$$1 \rtimes \sigma\mathbf{1}_2 = L(\nu^{1/2}\,\mathrm{sp}_2, \nu^{-1/2}\sigma) + L(\nu, 1 \rtimes \nu^{-1/2}\sigma). \qquad \square$$

Note that the $4 \times 4$ matrix representing the last four equations is not invertible, hence the irreducibles on the right cannot be expressed as linear combinations of the representations on the left.

Note that the $\lambda$-lift of $\nu\xi \times \xi \rtimes \nu^{-1/2}\sigma$ is $\sigma I(\nu^{1/2}, \nu^{1/2}\xi, \nu^{-1/2}\xi, \nu^{-1/2})$

$$\begin{aligned} &= \sigma I(\nu^{1/2}, \xi\,\mathrm{sp}_2, \nu^{-1/2}) + \sigma I(\nu^{1/2}, \xi\mathbf{1}_2, \nu^{-1/2}) \\ &= \sigma I(\mathrm{sp}_2 \times \xi\,\mathrm{sp}_2) + \sigma I(\mathbf{1}_2 \times \xi\,\mathrm{sp}_2) + \sigma I(\mathrm{sp}_2 \times \xi\mathbf{1}_2) + \sigma I(\mathbf{1}_2 \times \xi\mathbf{1}_2). \end{aligned}$$

It is invariant under multiplication by $\xi$ ($\xi^2 = 1$). To determine the liftings of the constituents of $\nu\xi \times \xi \rtimes \nu^{-1/2}\sigma$ we shall use the trace formula identity.

We shall also state the results of [Sh2], Proposition 8.4, [Sh3], Theorem 6.1, as recorded in [ST], Propositions 4.6-4.9, on reducibility of representations of $H$ supported on the proper maximal parabolics $P_{(2)}$ of type (2,2) and $P_{(1)}$ of type (1,2,1).

2.2 Proposition. (a) *Let $\pi_2$ be a cuspidal representation of* $\mathrm{PGL}(2,F)$ *and $\sigma$ a character of $F^\times$. Then $\nu^{1/2}\pi_2 \rtimes \nu^{-1/2}\sigma$ has a unique irreducible subrepresentation, which is square integrable. Inequivalent $(\pi_2, \sigma)$ define inequivalent square integrables, and each square integrable representation of $H$ supported in $P_{(2)}$ is so obtained (with $\omega_{\pi_2}\sigma^2 = 1$).*
(b) *All irreducible tempered non square integrable representations of $H$ supported in $P_{(2)}$ are of the form $\pi_2 \rtimes \sigma$ where $\pi_2$ is cuspidal unitarizable and $\sigma$ is a unitary character (with $\omega_{\pi_2}\sigma^2 = 1$). The only relation is $\pi_2 \rtimes \sigma = \check{\pi}_2 \rtimes \omega_{\pi_2}\sigma$.*
(c) *The unitarizable nontempered irreducible representations of $H$ supported in $P_{(2)}$ are $L(\nu^\beta\pi_2, \sigma)$, $0 < \beta \le \frac{1}{2}$, $\sigma$ a unitary character of $F^\times$, $\pi_2$ a cuspidal representation of* $\mathrm{PGL}(2,F)$. □

2.3 Proposition. (a) *Let $\pi_2$ be a cuspidal unitarizable representation of* $\mathrm{GL}(2,F)$ *such that $\pi_2\xi = \pi_2$ for a character $\xi \ne 1 = \xi^2$ of $F^\times$. Then $\nu\xi \rtimes \nu^{-1/2}\pi_2$ has a unique subrepresentation, which is square integrable. Inequivalent $(\pi_2, \xi)$ define inequivalent square integrables. All irreducible square integrable representations of $H$ supported in $P_{(1)}$ are so obtained, with $\xi\omega_{\pi_2} = 1$.*
(b) *All tempered irreducible non square integrable representations of $H$ supported in $P_{(1)}$ are either of the form $\chi \rtimes \pi_2$, $\pi_2$ cuspidal unitarizable representation of* $\mathrm{GL}(2,F)$ *and $\chi \ne 1$, as well as $\chi\omega_{\pi_2} = 1$ (the only equivalence relation on this set is $\chi \rtimes \pi_2 \simeq \chi^{-1} \rtimes \chi\pi_2$), or one of the two inequivalent constituents of $1 \rtimes \pi_2$.*

(c) *The irreducible unitarizable representations of* $H$ *supported on* $P_{(1)}$ *which are not tempered are* $L(\nu^\beta \xi, \nu^{-\beta/2}\pi_2)$, $0 < \beta \leq 1$, $\xi \neq 1 = \xi^2$, *and* $\pi_2$ *a cuspidal unitarizable representation of* $\mathrm{GL}(2, F)$ *with* $\pi_2 \xi \simeq \pi_2$ *and* $\xi\omega_{\pi_2} = 1$. □

# 3. Transfer of Distributions

In relating characters on the group $C_0 = \mathrm{PGL}(2, F) \times \mathrm{PGL}(2, F)$ with those on $H = \mathrm{PGSp}(2, F)$, $F$ local, we need a transfer $D_0 \to D_H$ of distributions which is dual to the transfer of orbital integrals $f_H \to f_0$ for functions on $H$ and on $C_0$. This transfer is crucial to the orthogonality relations of characters, a main tool in our work.

Let us recall (from chapter II, section 5) some basic definitions. Two regular elements $h$, $h'$ of $H$, and two tori $T_H$, $T'_H$ of $\mathbf{H}$, are called *stably conjugate* if they are conjugate in $\mathbf{H}(\overline{F})$; $\overline{F}$ is a separable algebraic closure of $F$.

Let $A(T_H/F)$ be the set of $x$ in $\mathbf{H}(\overline{F})$ such that $T'_H = T_H^x = x^{-1}T_Hx$ is defined over $F$. The set $B(T_H/F) = \mathbf{T}_H(\overline{F})\backslash A(T_H/F)/H$ parametrizes the morphisms of $\mathbf{T}_H$ into $\mathbf{H}$ over $F$, up to inner automorphisms by elements of $H$. If $\mathbf{T}_H$ is the centralizer of $h$ in $\mathbf{H}$ then $B(T_H/F)$ parametrizes the set of conjugacy classes within the stable conjugacy class of $h$ in $H$. The map

$$x \mapsto \{\tau \mapsto x_\tau = \tau(x)x^{-1}; \quad \tau \text{ in } \mathrm{Gal}(\overline{F}/F)\}$$

defines a bijection

$$B(T_H/F) \simeq \ker[H^1(F, T_H) \to H^1(F, H)].$$

Since $F$ is nonarchimedean, $H^1(F, H_{\mathrm{sc}}) = \{0\}$. Hence

$$\ker[H^1(F, T_H) \to H^1(F, H)] = \mathrm{Im}[H^1(F, T_{H,\mathrm{sc}}) \to H^1(F, H)].$$

Consequently it is a group, which is isomorphic – by the Tate-Nakayama theory – to

$$C(T_H/F) = \mathrm{Im}[H^{-1}(X_*(T_{H,\mathrm{sc}})) \to H^{-1}(X_*(T_H))].$$

Stable conjugacy for regular elliptic elements of $H = \mathrm{PGSp}(2,F)$ differs from conjugacy only for elements in tori of types I and II, where the stable conjugacy class consists of two conjugacy classes.

Denote by $W(T_H)$ the Weyl group of $T_H$ in $H$, and by $W'(T_H)$ the Weyl group of $T_H$ in $A(T_H/F)$.

Let $d_H$ be a locally integrable conjugacy invariant complex valued function on $H$. The Weyl integration formula asserts that

$$\int_H f(h)d_H(h)dh = \sum_{\{T_H\}} \frac{1}{[W(T_H)]} \int_{T_H} \Delta_H(t)^2 \Phi(t, f_H) d_H(t)dt.$$

The sum ranges over a set of representatives $T_H$ for the conjugacy classes of tori in $H$; $[X]$ denotes the cardinality of a set $X$.

Suppose $t$ is a regular element of $H$ which lies in $T_H$. Then the number of $\delta$ in $B(T_H/F)$ such that $t^\delta$ is conjugate to an element of $T_H$ is $[W'(T_H)]/[W(T_H)]$. Hence when the function $d_H$ is invariant under stable conjugacy, we have

$$\int_H f(h)d_H(h)dh = \sum_{\{T_H\}_s} \frac{1}{[W'(T_H)]} \int_{T_H} \Delta_H(t)^2 \Phi^{\mathrm{st}}(t, f_H) d_H(t)dt.$$

Here $\{T_H\}_s$ is a set of representatives for the stable conjugacy classes of tori in $H$.

3.1 Definition. Given a distribution $D_0$ on $C_0$, let $D_H = D_H(D_0)$ be the distribution on $H$ defined by $D_H(f_H) = D_0(f_0)$, where $f_0$ is the function on $C_0$ matching $f_H$ on $H$.

Our next aim is to compute $D_H$ if $D_0$ is represented by a locally integrable function. We first state the result, and explain the notations at the beginning of the proof.

3.2 Proposition. *Suppose that $D_0$ is a distribution on $C_0$ represented by the locally integrable function $d_0$. Then the corresponding distribution $D_H = D_H(D_0)$ on $H$ is given by a locally integrable function $d_H$ defined on the regular elliptic set of $H$ by $d_H(t) = 0$ if $t$ lies in a torus of type III or IV, and by*

$$\Delta_H(t)d_H(t^r) = \chi(r)\kappa(t)\Delta_0(t_0)[d_0(t_0) + d_0(t_0^w)]$$

*if $t$ is of type I or II, where $r$ ranges over $F^\times/N_{E/F}E^\times$ or $E_3^\times/N_{E/E_3}E^\times$, and $\chi$ is the nontrivial character of this group; if $r \neq 1$ then $t^r$ indicates the element stably conjugate but not conjugate to $t$. If $t_0 = t_0' \times t_0'' \in C_0 =$ $\mathrm{PGL}(2,F) \times \mathrm{PGL}(2,F)$, then $t_0^w$ indicates $t_0'' \times t_0'$.*

REMARK. If $D_0$ is represented by $d_0$ and $D_H$ by $d_H$, we shall also write $d_H = d_H(d_0)$ for $D_H = D_H(D_0)$.

PROOF. We need to recall the description of elements of types I (and later II) and their properties. A torus $T_H$ of type I splits over a quadratic extension $E = F(\sqrt{D})$ of $F$, and we choose explicit representatives for the two tori in the stable conjugacy class:

$$T_H^r = \{t^r = \begin{pmatrix} \alpha_1 & 0 & 0 & \beta_1 D \\ 0 & \alpha_2 & \beta_2 Dr & 0 \\ 0 & r^{-1}\beta_2 & \alpha_2 & 0 \\ \beta_1 & 0 & 0 & \alpha_1 \end{pmatrix} = h_r^{-1} t^* h_r; t = \mathrm{diag}(x_1, x_2, \alpha x_2, \alpha x_1)\}.$$

Here $r$ ranges over a set of representatives for $F^\times/N_{E/F}E^\times$, $\sigma$ is the nontrivial automorphism of $E$ over $F$, $x_i = \alpha_i + \beta_i\sqrt{D} \in E^\times$ are the eigenvalues of $t^r$, and $h_r$ are suitable matrices in $\mathrm{Sp}(2,\overline{F})$, described in [F5], p. 11. The norm map relates the elliptic torus $T_0$ of $C_0$ which splits over $E$ to $T_H^r$, on the level of eigenvalues it is given by

$$t_0^* = (\mathrm{diag}(t_1, \sigma t_1), \mathrm{diag}(t_2, \sigma t_2)) \overset{N}{\mapsto}$$

$$t^* = \mathrm{diag}(x_1 = t_1 t_2, x_2 = t_1 \sigma t_2, \sigma x_2 = \sigma t_1 \cdot t_2, \sigma x_1 = \sigma t_1 \cdot \sigma t_2).$$

Now the Weyl group of an elliptic torus in $\mathrm{PGL}(2,F)$ is $\mathbb{Z}/2$, hence the Weyl group $W(T_0)$ of $T_0$ in $C_0$ is $\mathbb{Z}/2 \times \mathbb{Z}/2$. The Weyl group $W(T_H)$ of $T_H$ in $H$ (of type I) contains $\mathbb{Z}/2 \times \mathbb{Z}/2$: it contains $s_1 = (12)(34)$ (acting on the diagonal matrix $t^*$), which is represented by $\mathrm{diag}\left(\begin{pmatrix} 0 & 1 \\ 1 & 0 \end{pmatrix}, \begin{pmatrix} 0 & 1 \\ 1 & 0 \end{pmatrix}\right)$ in $H$ (acting on $t \in T_H$), and $(14)(23)$, which is represented by $\mathrm{diag}(1,-1,1,-1)$ (and hence $W(T_H)$ contains also $(13)(24)$).

To use the Weyl integration formula we need to compute $W'(T_H)$. There are two cases for $T_H$ of type I. In case $\mathrm{I}_1$, $-1 \notin N_{E/F}E^\times$ (this happens when $E/F$ is ramified and $-1 \notin F^{\times 2}$). Then we can take $r \neq 1$ to be $-1$. Choose $i \in \overline{F}$ with $i^2 = -1$, and put $w = \mathrm{diag}(1, i, -i, 1)$. It lies in $\mathrm{Sp}(2,\overline{F})$, and $w^{-1} t w = t^r$. Then $w$ represents $\delta \neq 1$ in $B(T_H/F)$, and

$W'(T_H) = D_4$ ($w$ acts as (23) on $t^*$, and $s_2 = (23)$, $s_1 = (12)(34)$ generate $D_4$) contains $W(T_H) = \mathbb{Z}/2 \times \mathbb{Z}/2 = W(T_0)$ as a subgroup of index 2.

In case $\mathrm{I}_2$ we have $-1 \in N_{E/F}E^\times$, hence we can write $\mathrm{diag}(-1,1)$ as $cs$, $s \in \mathrm{SL}(2,F)$, $c \in C_{\mathrm{GL}(2,F)}(T_E)$ (= centralizer in $\mathrm{GL}(2,F)$ of an elliptic torus $T_E$ which splits over $E$), and $w = \mathrm{diag}(1,-1,1,1)$ as $ch$ with $h$ in $\mathrm{Sp}(2,F)$ and $c$ in $C_{\mathrm{GL}(4,F)}(T_H)$. Then $(ch)^{-1}tch = t^r$, $t' = h^{-1}t'^*h$, $h'^* = \mathrm{diag}(x_1, \sigma x_2, x_2, \sigma x_1)$. Hence in case $\mathrm{I}_2$ we have $W'(T_H) = W(T_H) = D_4$, as $w \in W'(T_H)$ is represented by $h \in \mathrm{Sp}(2,F)$, and it acts as (23) on $t^*$.

Note that the action of $w$ in both cases $\mathrm{I}_1$ and $\mathrm{I}_2$ is to interchange $x_2$ and $\sigma x_2$, namely $t_0 = (\mathrm{diag}(t_1, \sigma t_1), \mathrm{diag}(t_2, \sigma t_2))$ with $(t^w)_0 = (\mathrm{diag}(t_2, \sigma t_2), \mathrm{diag}(t_1, \sigma t_1))$. Then

$$\kappa(t) = \chi_{E/F}((x_1 - \sigma x_1)(x_2 - \sigma x_2)/D)$$

and

$$\kappa(t^w) = \chi_{E/F}((x_1 - \sigma x_1)(\sigma x_2 - x_2)/D) = \chi_{E/F}(-1)\kappa(t).$$

Let now $f_H$ be a function on $H$ such that the orbital integral $\Phi(t, f_H)$ is supported on the conjugacy class of a single torus of type I. Then

$$\begin{aligned} D_H(f_H) = D_0(f_0) &= \frac{1}{[W(T_0)]} \int_{T_0} \Delta_0(t_0)^2 \Phi(t_0, f_0) d_0(t_0) dt_0 \\ &= \frac{1}{[W(T_0)]} \int_{T_H} \Delta_0(t_0)\kappa(t)\Delta_H(t)[\Phi(t, f_H) - \Phi(t^\delta, f_H)] d_0(t_0) dt. \end{aligned}$$

Note that the norm map $N : T_0 \to T_H$ is an isomorphism. Now in case $\mathrm{I}_1$, $w$ represents $\delta \neq 1$, and $\kappa(t^w) = \chi_{E/F}(-1)\kappa(t) = -\kappa(t)$, and $W(T_H) = W(T_0)$, so we get

$$\begin{aligned} &\frac{1}{[W(T_H)]} \int_{T_H} \Delta_0(t_0)\Delta_H(t)[\kappa(t)\Phi(t, f_H) + \kappa(t^w)\Phi(t^w, f_H)] d_0(t_0) dt \\ &= \frac{1}{[W(T_H)]} \int_{T_H} \Delta_0(t_0)\Delta_H(t)\kappa(t)\Phi(t, f_H)[d_0(t_0) + d_0((t^w)_0)] dt, \end{aligned}$$

and since $\Phi(t, f_H)$ is any function (locally constant) on the regular set of $T_H$, we conclude – by the Weyl integration formula – that

$$\Delta_H(t) d_H(t^r) = \chi_{E/F}(r)\Delta_0(t_0)\kappa(t)[d_0(t_0) + d_0((t^w)_0)].$$

Again, $t^r$ is $t^w$ when $r \neq 1$ in $F^\times/N_{E/F}E^\times$, and $d_H(t^w) = -d_H(t)$, $t \in T_H$.

In case $\mathrm{I}_2$ we have that $t^w$ is conjugate to $t$ in $H$, and $[W(T_H)] = 2[W(T_0)]$, and $\Phi(t, f_H)$ is supported on $T_H^r$ for a single $r$. Hence $D_H(f_H)$ is

$$\frac{1}{[W(T_H)]}\int_{T_H} 2\Delta_0(t_0)\Delta_H(t)\kappa(t)\chi_{E/F}(r)\Phi(t^r, f_H)d_0(t_0)dt.$$

Since $\Phi(t^r, f_H)$ is any locally constant function on the regular set of $T_H$, we obtain

$$\begin{aligned}\Delta_H(t)d_H(t^r) &= \chi_{E/F}(r)\Delta_0(t_0)\kappa(t)2d_0(t_0)\\ &= \chi_{E/F}(r)\kappa(t)\Delta_0(t_0)[d_0(t_0) + d_0((t^w)_0)].\end{aligned}$$

Tori of type II split over a biquadratic extension $E = E_1E_2E_3$ of $F$, where $E_1 = E^\tau = F(\sqrt{D})$, $E_2 = E^{\sigma\tau} = F(\sqrt{AD})$, $E_3 = E^\sigma = F(\sqrt{A})$; $A$, $D$, $AD$ are in $F - F^2$, and we write $t_1 = \alpha_1 + \beta_1\sqrt{D}$ for elements of $E_1$, $t_2 = \alpha_2 + \beta_2\sqrt{AD}$ for elements of $E_2$. The norm map takes

$$t_0^* = (\mathrm{diag}(t_1, \sigma t_1), \mathrm{diag}(t_2, \tau t_2))$$

to

$$t^* = \mathrm{diag}(x_1 = t_1t_2, \tau x_1 = t_1\tau t_2, \sigma\tau x_1 = \sigma t_1 \cdot t_2, \sigma x_1 = \sigma t_1 \cdot \tau t_2).$$

Thus $T_0 \simeq E_1^\times/F^\times \times E_2^\times/F^\times$ (in contrast to case I where $E_1 = E_2 = E$ is quadratic over $F$) has Weyl group $W(T_0) = \mathbb{Z}/2 \times \mathbb{Z}/2$. The tori $T_H^r$ consist of

$$T_H^r = \left\{t^r = h_r^{-1}t^*h_r = \begin{pmatrix} \mathbf{a} & \mathbf{b}D\mathbf{r} \\ \mathbf{r}^{-1}\mathbf{b} & \mathbf{a}\end{pmatrix}; \mathbf{a} = \begin{pmatrix} a_1 & a_2A \\ a_2 & a_1\end{pmatrix}, \mathbf{b} = \begin{pmatrix} b_1 & b_2A \\ b_2 & b_1\end{pmatrix}\right\},$$

where $x_1 = a + b\sqrt{D}$; $a = a_1 + a_2\sqrt{A}$, $b = b_1 + b_2\sqrt{A} \in E_3^\times$; $\sigma x_1 = a - b\sqrt{D}$, $\tau x_1 = \tau a + \tau b \cdot \sqrt{D}$, thus

$$\begin{aligned}x_1 &= a_1 + a_2\sqrt{A} + b_1\sqrt{D} + b_2\sqrt{AD},\\ \tau x_1 &= a_1 - a_2\sqrt{A} + b_1\sqrt{D} - b_2\sqrt{AD},\\ \sigma\tau x_1 &= a_1 - a_2\sqrt{A} - b_1\sqrt{D} + b_2\sqrt{AD},\\ \sigma x_1 &= a_1 + a_2\sqrt{A} - b_1\sqrt{D} - b_2\sqrt{AD}.\end{aligned}$$

Further, $r$ ranges over $E_3^\times/N_{E/E_3}E^\times$, and if $r = r_1 + r_2\sqrt{A}$ we put $\mathbf{r} = \begin{pmatrix} r_1 & r_2A \\ r_2 & r_1 \end{pmatrix}$. Then $h_r = h\begin{pmatrix} I & 0 \\ 0 & \mathbf{r} \end{pmatrix}$, and $h = \mathbf{h}_D = \begin{pmatrix} h_A & 0 \\ 0 & \varepsilon h_A \varepsilon \end{pmatrix}\begin{pmatrix} I & \sqrt{D} \\ I & -\sqrt{D} \end{pmatrix}$, $h_A = \begin{pmatrix} 1 & \sqrt{A} \\ 1 & -\sqrt{A} \end{pmatrix}$, $\varepsilon = \begin{pmatrix} 1 & 0 \\ 0 & -1 \end{pmatrix}$, see [F5], p. 12.

The Weyl group $W'(T_H^r)$ of $T_H^r$ in $A(T_H^r/F)$ is $\mathbb{Z}/2 \times \mathbb{Z}/2$, generated by $\tilde{\sigma} = (14)(23)$, which maps the eigenvalue $x_1$ of $t^*$ to $\sigma x_1$, and $\tilde{\tau} = (12)(34)$, which maps $x_1$ to $\tau x_1$. It is equal to the Weyl group $W(T_H^r)$ of $T_H^r$ in $H$, since $(14)(23)$ is represented in $H$ by $\operatorname{diag}(I, -I)$, and $(12)(34)$ is represented by $\operatorname{diag}(1, -1, 1, -1)$.

We shall compare the norm map $T_0 \xrightarrow{\sim} T_H$ with that for $\tilde{T}_0 \xrightarrow{\sim} \tilde{T}_H$, where the tilde indicates that the roles of $E_1$ and $E_2$ are interchanged. Thus

$$\tilde{t}_0^* = ((\operatorname{diag}(t_2, \tau t_2), \operatorname{diag}(t_1, \sigma t_1))$$

$$\mapsto \tilde{t}^* = \operatorname{diag}(x_1 = t_2 t_1, \sigma\tau x_1 = t_2 \cdot \sigma t_1, \tau x_1 = \tau t_2 \cdot t_1, \sigma x_1 = \tau t_2 \cdot \sigma t_1)$$

and with $\quad \mathbf{b}' = \begin{pmatrix} b_2 & b_1 \\ b_1/A & b_2 \end{pmatrix}$,

$$\tilde{t}^r = \begin{pmatrix} I & 0 \\ 0 & \mathbf{r} \end{pmatrix}^{-1} \mathbf{h}_{AD}^{-1}\tilde{t}^*\mathbf{h}_{AD} \cdot \tilde{t}^* \cdot \mathbf{h}_{AD}\begin{pmatrix} I & 0 \\ 0 & \mathbf{r} \end{pmatrix} = \begin{pmatrix} \mathbf{a} & \mathbf{b}'AD\mathbf{r} \\ \mathbf{r}^{-1}\mathbf{b}' & \mathbf{a} \end{pmatrix}.$$

Note that $\tilde{t}^*$ is obtained from $t^*$ by the transposition (23).

We claim that $t$ and $\tilde{t}$ are stably conjugate. For this choose $\alpha$ in $\overline{F}$ with $\alpha^4 = -A/4$, thus $2\alpha^2 + A/2\alpha^2 = 0$. Put

$$Y = \begin{pmatrix} \frac{1}{\det y}\varepsilon y \varepsilon & 0 \\ 0 & y \end{pmatrix}, \quad y = \begin{pmatrix} \alpha & A/2\alpha \\ 1/2\alpha & \alpha \end{pmatrix}, \quad \varepsilon = \begin{pmatrix} 1 & 0 \\ 0 & -1 \end{pmatrix}.$$

Then $Y \in \operatorname{Sp}(2, \overline{F})$ satisfies

$$Y^{-1}\begin{pmatrix} \mathbf{a} & \mathbf{b}D \\ \mathbf{b} & \mathbf{a} \end{pmatrix} Y = \begin{pmatrix} \mathbf{a} & \mathbf{b}'AD \\ \mathbf{b}' & \mathbf{a} \end{pmatrix}.$$

These $t$ and $\tilde{t}$ are conjugate if $-1 \notin F^{\times 2}$ and $|A| = 1$ (we normalize $A$ and $D$ to lie in $R^\times$ or $\pi R^\times$). Indeed in this case we may choose $A = -1$. Then either $2 \in F^{\times 2}$ or $-2 \in F^{\times 2}$, and there is $\alpha \in F^\times$ with $\alpha^2 = 1/2$ or $= -1/2$ (respectively), hence $\alpha^4 = 1/4 = -A/4$, and $t$, $\tilde{t}$ are conjugate in $\operatorname{GSp}(2, F)$.

If $-1 \in F^{\times 2}$, say $-1 = i^2$, $i \in F^\times$, then $(2i\alpha^2)^2 = A$ has no solution with $\alpha$ in $F^\times$. If $|A| = q^{-1}$ then $(2\alpha^2)^2 = -A$ has no solutions with $\alpha$ in $F^\times$, so $\tilde{t}$ is conjugate to $t^r$, $r \neq 1$.

The transfer factor $\kappa(t)$ is

$$\kappa(t) = \chi_{E_1/F}((x_1-\sigma x_1)(\tau x_1-\tau\sigma x_1)/D) = \chi_{E_1/F}(b\tau b) = \chi_{E_1/F}(b_1^2-b_2^2A).$$

The transfer factor $\kappa(\tilde{t})$ is

$$\chi_{E_2/F}((x_1-\sigma x_1)(\sigma\tau x_1-\tau x_1)/AD) = \chi_{E_2/F}(b'\tau b') = \chi_{E_2/F}(b_2^2-b_1^2/A)$$
$$= \chi_{E_2/F}(-A)\chi_{E_2/F}(b_1^2-b_2^2A) = \chi_{E_2/F}(-A)\chi_{E_1/F}(b_1^2-b_2^2A).$$

For the last equality note that $b_1^2-b_2^2A \in N_{E_3/F}E_3^\times$ lies in $N_{E_2/F}E_2^\times$ iff it lies in $N_{E_1/F}E_1^\times$ iff it lies in $F^{\times 2}$.

Note that $\chi_{E_2/F}(-A)=1$ if $A=-1\notin F^{\times 2}$. If $-1\in F^{\times 2}$ and $|A|=1$ then $\chi_{E_2/F}(-A)=-1$, since then $E_2/F$ is ramified, $R^\times \cap N_{E_2/F}E_2^\times = R^{\times 2}$ and $A\notin R^{\times 2}$ so $-A\notin N_{E_2/F}E_2^\times$. Moreover $\chi_{E_2/F}(-A)=-1$ if $|A|=q^{-1}$: if $E_2/F$ is unramified then $-A\notin N_{E_2/F}E_2^\times$; and if $E_2/F$ is ramified we may assume that $D=u\in R^\times - R^{\times 2}$, and then $N_{E_2/F}E_2^\times = \{x^2-y^2u\boldsymbol{\pi}\}$ where we write $\boldsymbol{\pi}$ for $A$. But if $-A=-\boldsymbol{\pi}=x^2-y^2u\boldsymbol{\pi}$ is solvable then $x\in\boldsymbol{\pi}R$ and $y^2u\equiv 1 \bmod \boldsymbol{\pi}$ and $u$ be a square in $R^\times$.

We conclude that $\kappa(\tilde{t})=\kappa(t)$ if $t$, $\tilde{t}$ are conjugate, and $\kappa(\tilde{t})=-\kappa(t)$ if $t$, $\tilde{t}$ are not conjugate. Consequently

$$\Delta_0(t_0)\Phi(t_0,f_0) = \Delta_H(t)\kappa(t)\sum_{r\in E_3^\times/N_{E/E_3}E^\times}\chi_{E/E_3}(r)\Phi(t^r,f_H)$$

is equal to the expression obtained on replacing $t$ by $\tilde{t}$, which we denote by $t^w$ from now on to be consistent with the case of type I. It follows that each stable conjugacy class of $t\in T_H$ of type II is obtained twice, from $t_0$ and $\tilde{t}_0$ (or $t_0^w$). As $W(T_H)=W(T_0)$, we conclude that $D_H(f_H)$ is equal to

$$\frac{1}{[W(T_H)]}\int_{T_H} 2\Delta_0(t_0)\kappa(t)\chi_{E/E_3}(r)d_0(t_0)\cdot\Delta_H(t)\Phi(t^r,f_H)dt$$

if $\Phi(t,f_H)$ is supported on the conjugacy class of $T_H^r$, and hence

$$\Delta_H(t)d_H(t^r) = \chi_{E/E_3}(r)\kappa(t)\Delta_0(t_0)[d_0(t_0)+d_0((t^w)_0)].$$

It is clear that tori of types III and IV do not contribute to $D_H(f_H)$, which is equal to $D_0(f_0)$. □

# 4. Orthogonality Relations

We are interested in relating the distributions $D_H$ and $D_0$ since we need to relate orthogonality relations on $H$ and on $C_0$.

4.1 DEFINITION. (1) Let $d_H$, $d'_H$ be conjugacy invariant functions on the elliptic set of $H$. Put

$$\langle d_H, d'_H\rangle_H = \sum_{\{T_H\}_e} \frac{1}{[W(T_H)]} \int_{T_H} \Delta_H(t)^2 d_H(t)\overline{d}'_H(t)dt$$
$$= \sum_{\{T_H\}_{e,s}} \frac{1}{[W'(T_H)]} \sum_{\delta\in B(T_H/F)} \int_{T_H} \Delta_H(t)^2 d_H(t^\delta)\overline{d}'_H(t^\delta)dt.$$

Here $\{T_H\}_e$ (resp. $\{T_H\}_{e,s}$) is a set of representatives for the (resp. stable) conjugacy classes of elliptic tori $T_H$ in $H$.
(2) Let $d_0$, $d'_0$ be conjugacy invariant functions on the elliptic set of $C_0$. Put

$$\langle d_0, d'_0\rangle_0 = \sum_{\{T_0\}_e} \frac{1}{[W(T_0)]} \int_{T_0} \Delta_0(t)^2 d_0(t)\overline{d}'_0(t)dt,$$

where $\{T_0\}_e$ is a set of representatives for the conjugacy classes of elliptic tori in $C_0$.
(3) Write $d_0^w(t)$ for $d_0(t^w)$, where if $t = t'\times t'' \in C_0$ then $t^w$ or $\tilde{t}$ is $t''\times t'$.

4.2 PROPOSITION. *Let $d_0$, $d'_0$ be locally integrable class functions on the elliptic set of $C_0$, and $d_H = d_H(d_0)$, $d'_H = d_H(d'_0)$ the associated class function on the elliptic regular set of $H$. Then*

$$\langle d_H, d'_H\rangle_H = 2\langle d_0, d'_0\rangle_0 + 2\langle d_0, d'^w{}_0\rangle_0.$$

PROOF. By definition $\langle d_H, d'_H\rangle_H$ is a sum over $\{T_H\}_{e,s}$. For tori of type I we have $[W'(T_H)] = 2[W(T_0)]$, so the contribution is

$$\sum_{\{T_0\}_{e,I}} \frac{[F^\times : N_{E/F}E^\times]}{2[W(T_0)]} \int_{T_0} \chi_{E/F}(r)^2\kappa(t)^2\Delta_0(t_0)^2$$

$$\cdot[d_0(t_0) + d_0(t_0^w)][\overline{d}'_0(t_0) + \overline{d}'_0(t_0^w)]dt_0$$

where $\chi^2_{E/F} = 1$ and the set of $r$, $F^\times/N_{E/F}E^\times$, has cardinality two. For tori of type II we have $W'(T_H) = W(T_0)$, but each $t$ in $T_H$ is obtained – up to stable conjugacy – twice: once from $T_0$ and once from $T_0^w$. Hence the integral over $T_H$ has to be expressed as the sum of integrals over $T_0$ and $T_0^w$, divided by 2. Then the contribution to $\langle d_H, d'_H\rangle_H$ will be the sum over $\{T_0\}_{e,II}$ of

$$\frac{[E_3^\times : N_{E/E_3}E^\times]}{2[W(T_0)]}\int_{T_0} \chi_{E/E_3}(r)^2\kappa(t)^2\Delta_0(t_0)^2$$

$$\cdot[d_0(t_0)+d_0(t_0^2)][\overline{d}'_0(t_0)+\overline{d}'(t_0^w)]dt_0,$$

where $\chi^2_{E/E_3} = 1$ and $\kappa^2 = 1$. The cardinality of the $r$ is $|E_3^\times/N_{E/E_3}E^\times| = 2$. We then obtain a sum over all $T_0$, of types I (splitting over a quadratic extension $E$ of $F$) and II (splitting over a biquadratic extension):

$$= \sum_{\{T_0\}_e} \frac{1}{[W(T_0)]}\int_{T_0} \Delta_0(t_0)^2(d_0+d_0^w)(t_0)(\overline{d}'_0+\overline{d}'^{\,w}_0)(t_0)dt_0$$

$$= \langle d_0 + d_0^w, d'_0 + d'^{\,w}_0\rangle_0 = 2\langle d_0, d'_0\rangle_0 + 2\langle d_0^w, d'_0\rangle_0,$$

since $\langle d_0^w, d'^{\,w}_0\rangle_0 = \langle d_0, d'_0\rangle_0$. □

4.3 Corollary. *Let $\pi_i$, $\pi'_i$ $(i = 1,2)$ denote square integrable representations of* PGL$(2,F)$. *Put*

$$d_0(t_1,t_2) = \chi_{\pi_1}(t_1)\chi_{\pi_2}(t_2), \qquad d'_0(t_1,t_2) = \chi_{\pi'_1}(t_1)\chi_{\pi'_2}(t_2),$$

*where $\chi_\pi$ denotes the character of $\pi$. Then*

$$d_0^w(t_1,t_2) = d_0(t_2,t_1), \qquad \langle d_0, d'_0\rangle_0 = \delta(\pi_1,\pi'_1)\delta(\pi_2,\pi'_2)$$

*and*

$$\langle d_0^w, d'_0\rangle_0 = \delta(\pi_2,\pi'_1)\delta(\pi_1,\pi'_2),$$

*where $\delta(\pi,\pi')$ is 1 if $\pi$ and $\pi'$ are equivalent and 0 otherwise, so that*

$$\langle d_H, d'_H\rangle_H = \begin{cases} 0, & \pi_i \not\simeq \pi'_i \text{ and } \pi_i \not\simeq \pi'_j; \\ 2, & \pi_i \simeq \pi'_i \text{ and } \pi_i \not\simeq \pi'_j, \text{ or } \pi_i \not\simeq \pi'_i \text{ and } \pi_i \simeq \pi'_j; \\ 4, & \pi_i \simeq \pi'_i \simeq \pi_j \end{cases}$$

*where $\{i,j\} = \{1,2\}$.* □

4.4 PROPOSITION. *Let $d_H$ be a locally integrable class function on the elliptic set of $H$. Then $d_H$ is stable if and only if $\langle d_H, d_H(d_0)\rangle_H$ is 0 for every class function $d_0 = \chi_{\pi_0}$, where $\pi_0$ ranges over the square integrable irreducible representations of $C_0$.*

PROOF. We have that $\langle d_H, d_H(d_0)\rangle_H$ is equal to

$$\sum_{\{T_H\}_{s,I,II}} \frac{1}{[W'(T_H)]} \sum_r \int_{T_H} \Delta_H(t) d_H(t^r) \cdot \chi(r)\kappa(t)\Delta_0(t_0)[\bar{d}_0(t_0)+\bar{d}_0(t_0^w)]dt,$$

where the first sum ranges over a set of representatives for the stable conjugacy classes of tori in $H$ of types I and II, and $r$ ranges over a set of representatives for the conjugacy classes within the stable classes ($F^\times/NE^\times$ or $E_3^\times/N_{E/E_3}E^\times$), is 0 if $d_H(t^r) = d_H(t)$ for all $r$ and $t$. If $d_H$ is not stable, note that

$$\Delta_H(t)[\sum_r \chi(r) d_H(t^r)]\kappa(t)$$

is a nonzero class function on the elliptic set of $H$ which is invariant under $t \mapsto t^w$, and introduce a function $d_{H,0}$ on the elliptic set of $C_0$ by

$$\Delta_0(t_0) d_{H,0}(t_0) = \Delta_H(t)\kappa(t) \sum_r \chi(r) d_H(t^r).$$

Then $\langle d_H, d_H(d_0)\rangle_H$ becomes

$$\sum_{\{T_0\}_e} \frac{1}{2[W(T_0)]} \int_{T_0} \Delta_0(t_0)^2 d_{H,0}(t_0)[\bar{d}_0(t_0) + \bar{d}_0(t_0^w)]dt_0$$

$$= \sum_{\{T_0\}_e} \frac{1}{[W(T_0)]} \int_{T_0} \Delta_0(t_0)^2 d_{H,0}(t_0)\bar{d}_0(t_0)dt_0 \quad (\text{as } d_{H,0}(t_0^w) = d_{H,0}(t_0)).$$

But since $d_{H,0}$ is a nonzero conjugacy class function on the elliptic set of $C_0$ there is a square integrable irreducible representation $\pi_0$ of $C_0$ such that $\langle d_{H,0}, \chi_{\pi_0}\rangle_{C_0} \neq 0$, and the proposition follows. □

Let us review several $\lambda$-lifting facts, used in the study of the character relations below.

(1) The representation $1 \rtimes \pi_1$ of $H$, where $\pi_1$ is a PGL$(2,F)$-module, $\lambda$-lifts to the $\theta$-invariant $G$-module $I_G(\pi_1, \pi_1)$ (Proposition V.1.2). If $\pi_1$

is cuspidal then $1 \rtimes \pi_1$ is the direct sum of two irreducible inequivalent tempered representations $\pi_H^+ = \pi_H^+(\pi_1)$ and $\pi_H^- = \pi_H^-(\pi_1)$ (Proposition V.2.3(b)). Then

$$\operatorname{tr} I_G(\pi_1, \pi_1; f \times \theta) = \operatorname{tr} \pi_H^+(f_H) + \operatorname{tr} \pi_H^-(f_H)$$

for all matching $f$, $f_H$. The same assertion holds when $\pi_1$ is $\xi\,\mathrm{sp}_2$, $\xi^2 = 1$, see (3) below.

(2) For any PGL$(2, F)$-module $\pi_2$, the $H$-module $\xi\pi_2\nu^{1/2} \rtimes \xi\nu^{-1/2}$ $\lambda$-lifts to the $G$-module $I_G(\xi\nu^{1/2}, \pi_2, \xi\nu^{-1/2})$ (Proposition V.1.2), which has composition series consisting of $I_G(\xi\,\mathrm{sp}_2, \pi_2)$ and $I_G(\xi_2\mathbf{1}_2, \pi_2)$. The $H$-module $\xi\pi_2\nu^{1/2} \rtimes \xi\nu^{-1/2}$ has a unique irreducible subrepresentation, which we denote by $\delta(\xi\pi_2\nu^{1/2}, \xi\nu^{-1/2})$. It is square integrable, and a unique quotient $L(\pi_2\xi\nu^{1/2}, \xi\nu^{-1/2})$ which is nontempered, when $\pi_2$ is cuspidal (or is $\mathrm{sp}_2$, see (3) below); see Proposition V.2.2. Thus

$$\operatorname{tr} I_G(\xi\,\mathrm{sp}_2, \pi_2; f \times \theta) + \operatorname{tr} I_G(\xi\mathbf{1}_2, \pi_2; f \times \theta)$$

$$= \operatorname{tr} \delta(\xi\pi_2\nu^{1/2}, \xi\nu^{-1/2})(f_H) + \operatorname{tr} L(\pi_2\xi\nu^{1/2}, \xi\nu^{-1/2})(f_H)$$

for all matching $f$ and $f_H$.

(3) We have the following decomposition into irreducibles, where the $\tau$ are tempered and $\delta$ is square integrable:

$$\begin{aligned}
&1 \rtimes \sigma\,\mathrm{sp}_2 = \tau(\nu^{1/2}\,\mathrm{sp}_2, \nu^{-1/2}\sigma) + \tau(\nu^{1/2}\mathbf{1}_2, \nu^{-1/2}\sigma),\\
&\nu^{1/2}\,\mathrm{sp}_2 \rtimes \nu^{-1/2}\sigma = \tau(\nu^{1/2}\,\mathrm{sp}_2, \nu^{-1/2}\sigma) + L(\nu^{1/2}\,\mathrm{sp}_2, \nu^{-1/2}\sigma),\\
&\nu^{1/2}\mathbf{1}_2 \rtimes \nu^{-1/2}\sigma = \tau(\nu^{1/2}\mathbf{1}_2, \nu^{-1/2}\sigma) + L(\nu, 1 \rtimes \nu^{-1/2}\sigma),\\
&1 \rtimes \sigma\mathbf{1}_2 = L(\nu^{1/2}\,\mathrm{sp}_2, \nu^{-1/2}\sigma) + L(\nu, 1 \rtimes \nu^{-1/2}\sigma),\\
&\nu^{1/2}\xi\,\mathrm{sp}_2 \rtimes \nu^{-1/2}\sigma = \delta(\xi\nu^{1/2}\,\mathrm{sp}_2, \nu^{-1/2}\sigma) + L(\nu^{1/2}\xi\,\mathrm{sp}_2, \nu^{-1/2}\sigma),\\
&\nu^{1/2}\xi\mathbf{1}_2 \rtimes \nu^{-1/2}\sigma = L(\nu^{1/2}\xi\,\mathrm{sp}_2, \nu^{-1/2}\sigma\xi) + L(\nu\xi, \xi \rtimes \nu^{1/2}\sigma),
\end{aligned}$$

(Proposition V.2.1). Here $\sigma$ and $\xi$ are quadratic characters of $F^\times$.

# 5. Character Relations

Our main local results of character relations are derived from the trace formula identity.

5. PROPOSITION. *Let $\pi_1$, $\pi_2$ be two inequivalent cuspidal (resp. cuspidal or special) representations of* PGL$(2,F)$, *$F$ a local p-adic field. Then there are two cuspidal (resp. square integrable) representations of $H =$* PGSp$(2,F)$, *$\pi_H^+$ and $\pi_H^-$, such that for all matching functions $f$, $f_H$, $f_0$ on $G$, $H$, $C_0$, we have*

$$\operatorname{tr}(\pi_1 \times \pi_2)(f_{C_0}) = \operatorname{tr} \pi_H^+(f_H) - \operatorname{tr} \pi_H^-(f_H)$$

*and*

$$\operatorname{tr} I(\pi_1, \pi_2; f \times \theta) = \operatorname{tr} \pi_H^+(f_H) + \operatorname{tr} \pi_H^-(f_H).$$

*The same identities hold when $\pi_1 = \pi_2$ is square-integrable, but then $\pi_H^+$ and $\pi_H^-$ are the two irreducible constituents of $1 \rtimes \pi_1$. They are tempered and $\pi_H^+ + \pi_H^- = 1 \rtimes \pi_1$.*

PROOF. This is our main local assertion in this work. The long proof will be cut into a sequence of assertions, most of which we name "Lemmas". The case of $\pi_1 = \pi_2$ is in 6.5. Elsewhere we assume that $\pi_1 \neq \pi_2$, unless otherwise specified.

Let $(\pi, V)$ be an admissible representation of a p-adic reductive group $G$. As in [BZ1], we introduce the

5.1 DEFINITION. (1) Let $N$ denote the unipotent radical of a parabolic subgroup of $G$ with Levi subgroup $M$. Then the quotient $V_N$ of $V$ by the span of the vectors $\pi(n)v - v$ as $n$ ranges over $N$ and $v$ over $V$, is an admissible $M$-module $\tilde{\pi}_N$. Its tensor product with $\delta_N^{1/2}$ ($\delta_N(m) = |\det(\mathrm{Ad}(m)|\,\mathrm{Lie}\,N)|$) is called the *normalized $M$-module $(\pi_N, V_N)$ of $N$-coinvariants* of $\pi$.
(2) The representation $\pi$ is called cuspidal when $\pi_N = \{0\}$ for all $N \neq \{1\}$, that is when $\operatorname{tr} \pi_N(\phi) = 0$ for every test measure $\phi$ on $M$. Analogously,
(3) If $\theta$ is an automorphism of $G$ and $\pi$ is $\theta$-invariant ($\pi \simeq {}^\theta\pi$), we say that $\pi$ is *$\theta$-cuspidal* if for every $\theta$-invariant proper parabolic subgroup of $G$ we have $\operatorname{tr} \pi_N(\phi \times \theta) = 0$ for every test measure $\phi$.

5.2 LEMMA. *The representation* $\pi = I(\pi_1, \pi_2)$, $\pi_1 \neq \pi_2$, *is* $\theta$*-invariant and* $\theta$*-cuspidal.*

PROOF. If $N$ is the unipotent radical of a proper parabolic subgroup of $\mathrm{PGL}(4,F)$ then $\pi_N$ is zero unless the parabolic is of type (2,2), in which case $\pi_N = \pi_1 \times \pi_2 + \pi_2 \times \pi_1$.

However, the irreducible constituents $\pi_1 \times \pi_2$ and $\pi_2 \times \pi_1$ of this $\pi_N$ are interchanged by $\theta$ so that the trace $\mathrm{tr}\,\pi_N(\phi \times \theta)$ vanishes for any test measure $\phi$ on $M$. □

When $\pi_1 \neq \pi_2$ but one or two of them is square integrable noncuspidal (special) representation of $\mathrm{PGL}(2,F)$, the representation $I(\pi_1,\pi_2)$ is ($\theta$-invariant and) tempered subquotient of the induced $I(\nu^{1/2}, \pi_2, \nu^{-1/2})$ (if $\pi_1 = \mathrm{sp}_2$). This is the $\lambda$-lift of $\pi_2\nu^{1/2} \rtimes \nu^{-1/2}$, whose composition series consists of the square integrable $\delta(\pi_2\nu^{1/2}, \nu^{-1/2})$ and the nontempered $L(\pi_2\nu^{1/2}, \nu^{-1/2})$ (if $\pi_2$ is cuspidal), or square integrable $\delta(\xi\nu^{1/2}\,\mathrm{sp}_2, \nu^{-1/2})$ and nontempered $L(\nu^{1/2}\xi\,\mathrm{sp}_2, \nu^{-1/2})$ (if $\pi_2$ is the special $\xi\,\mathrm{sp}_2$, where $\xi^2 = 1 \neq \xi$). Since the functor of $N$-coinvariants is exact ([BZ1]), the central exponents (central characters of constituents) of $I(\mathrm{sp}_2, \pi_2)_N$ correspond to those of $\delta(\pi_2\nu^{1/2}, \nu^{-1/2})$ ($\pi_2$ cuspidal) or $\delta(\xi\nu^{1/2}\,\mathrm{sp}_2, \nu^{-1/2})$ (if $\pi_2 = \xi\,\mathrm{sp}_2$), which are decaying.

The twisted analogue of the orthogonality relations of Kazhdan [K2], Theorem K, implies in our case where $\pi_1 \neq \pi_2$ are square integrable $\mathrm{PGL}(2,F)$-modules:

5.3 LEMMA. *There exists a* $\theta$*-pseudo-coefficient* $f^1$ *of* $\pi = I(\pi_1, \pi_2)$.

A $\theta$-pseudo-coefficient $f^1$ is a test measure with the property that $\mathrm{tr}\,I(\pi_1,\pi_2; f^1 \times \theta) = 1$ but $\mathrm{tr}\,\pi'(f^1 \times \theta) = 0$ for every irreducible $G$-module $\pi'$ inequivalent to (*i*) $I(\pi_1,\pi_2)$ if $\pi_1$, $\pi_2$ are cuspidal, or to (*ii*) any constituent of $I(a,b)$ if $\pi_1$ (or $\pi_2$) is $\xi\,\mathrm{sp}_2$, $\xi^2 = 1$, in which case $a$ is $\xi I(\nu^{1/2}, \nu^{-1/2})$ (or $b$ is such).

Note that the $\theta$-orbital integral $\Phi(g, f^1 \times \theta)$ of $f^1$ is supported on the $\theta$-elliptic set, and is equal to the complex conjugate of the $\theta$-twisted character $\chi_I(g \times \theta)$ of $I = I(\pi_1, \pi_2)$.

We shall show below that $\chi_I(g \times \theta)$ depends only on the stable $\theta$-conjugacy class of $g$, hence $\Phi(g, f^1 \times \theta)$ depends only on the stable $\theta$-conjugacy class of $g$.

We now pass to global notations. Thus we fix a totally imaginary number field $F$ whose completion at the places $v_i$ $(0 \leq i \leq 3)$ is our local field, denoted now $F_{v_0}$. Denote our local representations by $\pi_{jv_0}$, $j = 1, 2$. Fix $\pi_{jv_i} \simeq \pi_{jv_0}$ $(j = 1, 2;\ i = 1, 2, 3)$ under the isomorphism $F_{v_i} \simeq F_{v_0}$.

5.4 LEMMA. *There exist cuspidal representation $\pi_1$ and $\pi_2$ which are unramified outside the places $v_i$ $(1 \leq i \leq 3)$ of $F$ whose components at $v_i$ are our $\pi_{1v_i}$ and $\pi_{2v_i}$ (respectively).*

PROOF. This is done using the nontwisted trace formula for PGL(2), and a test measure $f$ whose components $f_{v_i}$ are pseudo coefficients of $\pi_{jv_i}$, and whose components $f_v$ at all other finite places are spherical. At one of these $v \neq v_i$ take $f_v$ with $\operatorname{tr} \pi_v(f_v) = 0$ for all one-dimensional representations $\pi_v$ of $\mathrm{PGL}(2, F_v)$ (the trivial representation and its twist by a quadratic character). Most of the $f_v$ are the unit element of the Hecke algebra, but the remaining finite set can be taken to have the property that the orbital integral of $f^\infty = \otimes_{v<\infty} f_v$ is nonzero at a rational (in $\mathrm{PGL}(2, F)$) elliptic regular element $\gamma$. The coefficients of the characteristic polynomial of the conjugacy classes of rational conjugacy classes are discrete and lie in a compact, once $f_\infty$ is chosen. We can choose $f_\infty$ so that the orbital integral of $f = f_\infty \otimes f^\infty$ is nonzero at $\gamma$, but zero at any other rational conjugacy class (in particular, choose $f_\infty$ to vanish on the singular set). For such $f$ the geometric side of the trace formula reduces to a single nonzero term (the weighted orbital integrals vanish as two components $f_{v_i}$ are elliptic, the singular orbital integrals vanish by choice of $f_\infty$).

On the spectral side the logarithmic derivatives of the intertwining operators and the contributions from the continuous spectrum vanish as two components $f_{v_i}$ are elliptic. If $\pi$ occurs with $\operatorname{tr} \pi(f) \neq 0$, its components at finite $v \neq v_i$ are unramified, and its components at $v_i$ are our chosen $\pi_{jv_i}$, since $f_{v_i}$ are their pseudo coefficients. In the case that the $\pi_{jv_i}$ are special we chose some $f_v$ to be spherical with trace zero at each one dimensional representation $\pi_v$. In this last case the global $\pi$ will not be one dimensional, so it has to be cuspidal. □

Once we have the cuspidal representations $\pi_1$ and $\pi_2$, we use our usual trace formula identity where the contribution to the trace formula of $C_0$ is the cuspidal representation $\pi_0 = \pi_1 \times \pi_2$. There is no contribution to the trace formula of the $\theta$-twisted endoscopic group $C$, and by the rigidity theorem for PGL(4) the only contribution to the $\theta$-twisted trace formula is $I(\pi_1, \pi_2)$. The contributions to the trace formula of $H$ are some discrete spectrum representations $\pi_H$.

Applying generalized linear independence of characters at all places $v \neq v_i$ $(0 \leq i \leq 3)$ where $\pi_{1v} \times \pi_{2v}$ is unramified or $F_v$ is $\mathbb{C}$, we obtain the identity

$$\prod_v \operatorname{tr} I(\pi_{1v}, \pi_{2v}; f_v \times \theta) + \prod_v \operatorname{tr}(\pi_{1v} \times \pi_{2v})(f_{C_0 v}) \tag{1}$$
$$= 2 \sum_{\pi_H} m(\pi_H) \prod_v \operatorname{tr} \pi_{Hv}(f_{Hv}).$$

The products range over $v = v_i$ $(0 \leq i \leq 3)$; $m(\pi_H)$ are the multiplicities of the discrete spectrum representations $\pi_H$. The identity holds for all triples $(f_v, f_{C_0 v}, f_{Hv})$ of matching measures such that at 3 out of the 4 places the orbital integrals vanish on the nonelliptic set.

It is clear that

5.5 LEMMA. *The distribution $f_v \mapsto \operatorname{tr} I(\pi_{1v}, \pi_{2v}; f_v \times \theta)$ depends only on $f_{Hv}$, namely only on the stable $\theta$-orbital integrals of $f_v$.*

Consequently the $\theta$-twisted character of $I(\pi_{1v}, \pi_{2v})$ is a $\theta$-stable function. In particular, the $\theta$-twisted orbital integral of a $\theta$-pseudo-coefficient of $I(\pi_{1v}, \pi_{2v})$ is not identically zero on the $\theta$-elliptic set.

This establishes a fact which is used in the derivation of the identity (2) below.

5.6 LEMMA. *Fix $v \in \{v_i\}$. The right side of* (1) *is not identically zero as $f_{Hv}$ ranges over the functions whose orbital integrals vanish outside the elliptic set of $H$.*

PROOF. Had it been zero we could choose $f_{Hv}$ whose stable orbital integrals are zero (and so $f_v = 0$) but with unstable orbital integrals, that is $f_{C_0,v}$, with $\operatorname{tr}(\pi_{1v} \times \pi_{2v})(f_{C_0 v}) \neq 0$. Hence for each $v$ there are $\pi_{Hv}$ on the right whose character is nonzero on the elliptic set. □

Using the $\theta$-twisted trace formula and a totally imaginary field $F$ whose completion at $v_0$ is the local field of the proposition, we construct a representation $\pi$ as follows.

5.7 LEMMA. *There exists a cuspidal $\theta$-invariant automorphic representation $\pi$ with the following properties. Its component at $v_0$ is our $\pi_{v_0} = I(\pi_{1v_0}, \pi_{2v_0})$, where $\pi_{1v_0} \neq \pi_{2v_0}$ are square integrable representations of* $\mathrm{PGL}(2, F_{v_0})$. *At three nonarchimedean places $v_1$, $v_2$, $v_3$ the component is the Steinberg (square integrable) representation $\mathrm{St}_{v_i}$. At all other nonarchimedean places $v$ the component is unramified.*

PROOF. We construct $\pi$ on using the stable $\theta$-twisted trace formula and a test function $f = \otimes f_v$ whose component $f_v$ is unramified at $v \neq v_i$ ($i =$ 0, 1, 2, 3), our pseudo coefficient at $v_0$, and the pseudo coefficient of $\mathrm{St}_{v_i}$ ($i = 1, 2, 3$) at $v_i$.

Since the $\theta$-character of $\mathrm{St}_{v_i}$ is $\theta$-stable (being the $\lambda$-lift of $\mathrm{St}_{H,v_i}$), the $\theta$-twisted trace formula for $f$ with such a component is $\theta$-stable, namely its geometric part depends only on the $\theta$-stable orbital integrals. As we showed in 5.5 above, the $\theta$-stable orbital integral of the pseudo-coefficient $f_{v_0}$ of $\pi_{v_0}$ does not vanish identically on the $\theta$-elliptic set. This determines the nonarchimedean components of $f$.

The geometric side of the stable $\theta$-trace formula consists of orbital integrals. We choose the archimedean components to be supported on a small enough neighborhood of a single $\theta$-regular stable elliptic $\theta$-conjugacy class, such that there will be only one rational stable $\theta$-conjugacy class $\gamma$, which is in the support of the global $f$ and there $\Phi^{\mathrm{st}}(\gamma, f \times \theta) \neq 0$.

Then the geometric side of the stable $\theta$-trace formula reduces to a single nonzero term, namely $\Phi^{\mathrm{st}}(\gamma, f \times \theta)$, and so the spectral side of the $\theta$-trace formula is nonzero.

By the choice of $f$ there is a representation $\pi$ of $\mathbf{G}(\mathbb{A})$ which is $\theta$-invariant, whose components outside $v_i$ ($i = 0, 1, 2, 3$) are unramified and at $v_i$ are $I(\pi_{1v_0}, \pi_{2v_0})$ and $\mathrm{St}_{v_i}$ ($i = 1, 2, 3$).

In fact, $\pi$ cannot have at $v_i$ ($i = 1, 2, 3$) components other than $\mathrm{St}_{v_i}$ because the choice of $f_{v_i}$ and the fact that $\mathrm{tr}\, \pi_{v_i}(f_{v_i} \times \theta) \neq 0$ imply that $\pi_{v_i}$ is a constituent in the composition series of the induced representation $I_{v_i}$ (from the Borel subgroup) containing $\mathrm{St}_{v_i}$. Since $\pi$ has the component $I(\pi_{1v_0}, \pi_{2v_0})$, had it not been a discrete series, it could only be induced $I(\pi_1, \pi_2)$ from a cuspidal representation $\pi_1 \times \pi_2$ of the Levi subgroup of type (2,2), and its components at $v_i$ ($i = 1, 2, 3$) would have to be $I(\nu\, \mathrm{sp}_2, \nu^{-1} \mathrm{sp}_2)$ or $I(\nu \mathbf{1}_2, \nu^{-1} \mathbf{1}_2)$, which are not unitarizable.

Of course $\pi_{v_i}$ ($i = 1, 2, 3$) cannot be the trivial representation, since then $\pi$ would be trivial, but is has the component $I(\pi_{1v_0}, \pi_{2v_0})$.

Now since $\pi$ has components $\mathrm{St}_{v_i}$ it has to be cuspidal. Indeed, having the component $I(\pi_{1v_0}, \pi_{2v_0})$ prevents $\pi$ from being a noncuspidal discrete spectrum representation, a complete list of which is given in [MW1]. In particular $\pi$ is generic. □

Having the representation $\pi$ we can use the trace formulae identity, and a standard argument of generalized linear independence of characters, applied at all places where $\pi_v$ is unramified, to obtain an identity

$$\prod \operatorname{tr} \pi_v(f_v \times \theta) = \sum_{\pi_H} m(\pi_H) \prod \operatorname{tr} \pi_{Hv}(f_{Hv}).$$

The products range over $v = v_i$ $(0 \leq i \leq 3)$ and the archimedean places. By rigidity and multiplicity one theorem for $\mathbf{G} = \mathrm{PGL}(4)$ the only contribution on the left side is our cuspidal $\pi$. Since it has Steinberg components, the only contribution on the right is of discrete spectrum representations $\pi_H$ of $\mathbf{H}(\mathbb{A})$; there can be no contributions from the endoscopic group $\mathbf{C}_0$ of $\mathbf{H}$, and contributions from the $\theta$-endoscopic group $\mathbf{C}$ of $\mathbf{G}$ have been dealt with already.

We now apply generalized linear independence of characters at the archimedean places $v$ of $F$, where $F_v = \mathbb{C}$ and the representation $\pi_v$ is fully induced (from the Borel subgroup). At the places $v_i$ $(i = 1, 2, 3)$ we use $f_{Hv_i}$ which is a matrix coefficient of $\mathrm{St}_{Hv_i}$ and $f_{v_i}$ which is a $\theta$-matrix coefficient of $\mathrm{St}_{v_i}$. These functions are matching since $\mathrm{St}_{Hv_i}$ $\lambda$-lifts to $\mathrm{St}_{v_i}$ and $\Phi(f_{Hv_i}) = \overline{\chi}_{\mathrm{St}_{Hv_i}}$, $\Phi(f_{v_i}) = \overline{\chi}_{St_{v_i}}$. Their orbital integrals vanish on the non ($\theta$-) elliptic set.

On the side of $H$ we have that $\operatorname{tr} \pi_{Hv_i}(f_{Hv_i})$ is 0 unless $\pi_{Hv_i}$ is a subquotient of $\nu^2 \times \nu \rtimes \nu^{-3/2}$ (see Corollary V.1.5). Since a component of an automorphic representation $\pi_H$ has to be unitarizable, the subquotients of $\nu^2 \times \nu \rtimes \nu^{-3/2}$ which may occur are $\mathrm{St}_{Hv_i}$ and the trivial representation $\mathbf{1}_{Hv_i}$. But a discrete spectrum $\pi_H$ which has a trivial component is trivial (by the weak approximation theorem), and writing $v$ for $v_0$ we conclude that

5.8 LEMMA. *There are $H_v$-modules $\pi_{Hv}$ and positive integers $m'(\pi_{Hv})$ such that for any matching functions $f_v$ and $f_{Hv}$ we have*

$$\operatorname{tr} I(\pi_{1v}, \pi_{2v}; f_v \times \theta) = \sum_{\pi_{Hv}} m'(\pi_{Hv}) \operatorname{tr} \pi_{Hv}(f_{Hv}). \tag{2}$$

*The $\pi_H$ which occur are cuspidal or square integrable.*

PROOF. Let $N$ denote the unipotent radical of any proper parabolic subgroup of $H$. Let $\pi_{H,N}$ be the module of $N$-coinvariants of $\pi_H$. Since (*i*) the representation $I(\pi_1, \pi_2)$ is $\theta$-cuspidal if $\pi_1 \neq \pi_2$ are cuspidal, and (*ii*) its $\theta$-central exponents decay if $\pi_1 \neq \pi_2$ are square integrable, and (*iii*) each $N$ corresponds to the unipotent radical of a proper $\theta$-invariant parabolic subgroup of $G$, the character relation (2) implies that $\sum m'(\pi_H)\chi_{\pi_{HN}}$ is zero if $\pi_1 \neq \pi_2$ are cuspidal, and it decays if $\pi_1 \neq \pi_2$ are square integrable. But the $m'(\pi_H)$ are positive. Hence all $\pi_{H,N}$ are zero if $\pi_1 \neq \pi_2$ are cuspidal, and decay if $\pi_1 \neq \pi_2$ are square integrable, namely the $\pi_H$ which occur are cuspidal or square integrable, respectively. □

5.9 LEMMA. *The sum* (2) *with coefficients* $m'$ *is finite.*

PROOF. To see this, write it in the form

$$\operatorname{tr} I(f \times \theta) = \sum_{i=1}^{b} m_i \operatorname{tr} \pi_{Hi}(f_H), \qquad I = I(\pi_1, \pi_2).$$

where $1 \leq b \leq \infty$. Let $f_i$ be a pseudo coefficient of the square integrable $\pi_{Hi}$, and for any finite $a \leq b$ put $f^a = \sum^a f_i$, where $\sum^a$ indicates the sum over $i$ $(1 \leq i \leq a)$. Then

$$a^2 \leq (\sum^a m_i)^2 = \big[\sum^a m_i \operatorname{tr} \pi_{Hi}(f^a)\big]^2 = [\operatorname{tr} I(f(f^a) \times \theta)]^2$$
$$\leq \langle \chi_I, \sum^a \chi_{Hi}\rangle^2 \leq \langle \chi_I, \chi_I\rangle_H \cdot \langle \sum^a \chi_{Hi}, \sum^a \chi_{Hi}\rangle = a\langle \chi_I, \chi_I\rangle_H.$$

Here $\chi_I(Ng) = \chi_I(g \times \theta)$ is a function on the space of stable conjugacy classes in $H$, since $\chi_I(g \times \theta)$ depends only on the stable $\theta$-conjugacy class of $g$ in $G$. The orthogonality relations for twisted characters, which are locally integrable, imply that $\langle \chi_I, \chi_I\rangle_H$ is finite, hence $a$ is bounded, and the sum is finite. □

## 6. Fine Character Relations

In this section we conclude the proof of Proposition 5. Using (2), we can rewrite our identity in the form

$$\prod_v \operatorname{tr}(\pi_{1v} \times \pi_{2v})(f_{C_0 v}) = \sum m(\otimes_v \pi_{Hv}) \prod_v \operatorname{tr} \pi_{Hv}(f_{Hv}),$$

where the $m$ here are integers, possibly negative. As the left side is nonzero, there are nonzero contributions on the right whose character is nonzero on the elliptic regular set, for each $v$.

Once again using (2) we write our identity – but on choosing $f_{Hv}$ to be a matrix coefficient of $\pi_{Hv}$ which occurs on the right side, at 3 out of our 4 places, so that $\operatorname{tr} \pi_{Hv}(f_{Hv})$ is 1 or 0 (or $-1$) at this places, we get an identity of the form

$$c \operatorname{tr}(\pi_{1,v_0} \times \pi_{2,v_0})(f_{C_0,v_0})$$
$$= \sum m(\pi_{H,v_0}) \operatorname{tr} \pi_{H,v_0}(f_{H,v_0}) - \sum m'(\pi_{H,v_0}) \operatorname{tr} \pi_{H,v_0}(f_{H,v_0}).$$

The term following the negative sign is $\operatorname{tr} I(\pi_{1v_0}, \pi_{2v_0}; f_{v_0} \times \theta)$. Here $f_{Hv_0}$, $f_{C_0 v_0}$ are arbitrary matching functions, and $c \neq 0$

$$(c = \prod_{v \neq v_0} \operatorname{tr}(\pi_{1v} \times \pi_{2v})(f_{C_0 v})).$$

Write $\tilde{m}$ for $m/c$, and note that the 4 places are the same: $F_{v_0} = F_{v_i}$ $(i = 1, 2, 3)$, and so are the components $\pi_{1v} \times \pi_{2v}$, by our construction. Multiplying the last identity over the 4 places (not only $v_0$, but also $v_1$, $v_2$, $v_3$) we obtain $\prod_v \operatorname{tr}(\pi_{1v} \times \pi_{2v})(f_{C_0 v})$

$$= \prod_v \left[\sum \tilde{m}(\pi_{Hv}) \operatorname{tr} \pi_{Hv}(f_{Hv}) - \sum \tilde{m}'(\pi_{Hv}) \operatorname{tr} \pi_{Hv}(f_{Hv})\right].$$

Comparing this with the original identity for the left side we deduce that the complex number $\prod_v \tilde{m}(\pi_{Hv})$ is an integer, namely $(m(\pi_{Hv})/c)^4$ is an integer, hence $c$ divides each of the $m$. We finally get – for all matching functions $f_{Hv}$ and $f_{C_0 v}$ – the identity

$$\operatorname{tr}(\pi_{1v} \times \pi_{2v})(f_{C_0 v}) = \sum m(\pi_{Hv}) \operatorname{tr} \pi_{Hv}(f_{Hv}) - \sum m'(\pi_{Hv}) \operatorname{tr} \pi_{Hv}(f_{Hv})$$

$$= \sum m''(\pi_{Hv}) \operatorname{tr} \pi_{Hv}(f_{Hv}),$$

where the $m''(\pi_{Hv})$ are integers, positive or negative.

Of course had we used the $\theta$-trace formula with no restrictions at 3 places the derivation of the last identity from (2) would be easier, but we do not use the unrestricted trace formula identity.

6.1 LEMMA. *The sum with coefficients $m''$ is finite. It consists of $\leq 2$ summands if $\pi_1 \neq \pi_2$.*

PROOF. To see this, write it in the form

$$\operatorname{tr}(\pi_1 \times \pi_2)(f_{C_0}) = \sum_{i=1}^{b} m_i'' \operatorname{tr} \pi_{Hi}(f_H)$$

where $1 \leq b \leq \infty$. Let $f_i$ be a pseudo coefficient of the square integrable $\pi_{Hi}$, and for a finite $a \leq b$ put $f^a = \sum^a \frac{m_i''}{|m_i''|} f_i$, where $\sum^a$ indicates sum over $i$ $(1 \leq i \leq a)$. Then

$$\begin{aligned}
a^2 &\leq (\sum^a |m_i''|)^2 = \big[\sum^a m_i'' \operatorname{tr} \pi_{Hi}(f^a)\big]^2 = \big[\operatorname{tr}(\pi_1 \times \pi_2)(f_0(f^a))\big]^2 \\
&= \langle d_H(\chi_{\pi_1 \times \pi_2}), \sum^a \frac{m_i''}{|m_i''|} \chi_{\pi Hi} \rangle_H^2 \\
&\leq \langle d_H(\chi_{\pi_1 \times \pi_2}), d_H(\chi_{\pi_1 \times \pi_2}) \rangle_H \cdot \langle \sum^a \chi_{\pi_{Hi}}, \sum^a \chi_{\pi_{Hi}} \rangle_H \\
&= 2(1 + \delta(\pi_1, \pi_2))a.
\end{aligned}$$

The last equality follows from Corollary 4.3. Hence $a \leq 2$ if $\pi_1$, $\pi_2$ are inequivalent. □

6.2 LEMMA. *The $m''$ take both positive and negative values.*

PROOF. To see this, we write $\chi = d_H(\chi_{\pi_1 \times \pi_2})$. Then $\chi$ is an unstable conjugacy function on $H$, thus zero on the elliptic tori of types III and IV, and its value at one conjugacy class of type I or II is negative its value at the other conjugacy class within the stable class.

The $\pi_{Hi}$ $(1 \leq i \leq a)$ which occur in the identity for $\operatorname{tr}(\pi_1 \times \pi_2)(f_{C_0})$ with $m_i \neq 0$ are cuspidal or square integrable or constituents of $1 \rtimes \pi_2$, $\pi_2$ square integrable (see Propositions 2.1(d) and 2.3(b)), by Casselman

[C] (compare the central exponents), since $\pi_1 \times \pi_2$ is cuspidal or square integrable. Hence, choosing $F$ to be totally imaginary, and using pseudo-coefficients and the trace formula as usual, we can construct a global discrete spectrum representation $\pi_H$ with (1) a component $\pi^0_{Hv_0}$ which occurs in the trace identity of our local $\pi_1 \times \pi_2$ at $v_0$, (2) a Steinberg component $\mathrm{St}_{Hv_i}$ at $v_1$, $v_2$, $v_3$, (3) the nonarchimedean components of $\pi_H$ away from $v_i$ $(0 \leq i \leq 3)$ are unramified. This $\pi_H$ contributes to the trace formula identity, where the contribution $\pi$ to the twisted formula of $G$ is necessarily cuspidal, as it has a Steinberg component.

We apply as usual generalized linear independence of characters at the unramified components and the archimedean ones, and use coefficients of $\mathrm{St}_{Hv_i}$ at $v_1$, $v_2$, $v_3$. We deduce that there is a $\theta$-invariant generic representation $\pi$ of $\mathbf{G}(F_{v_0})$ with an identity

$$\operatorname{tr}\pi(f \times \theta) = \sum \tilde{m}(\pi_H) \operatorname{tr}\pi_H(f_H),$$

where $\tilde{m}(\pi_H) \geq 0$ for all $\pi_H$ and $> 0$ for the $\pi^0_H$ with which we started, which occurs in the trace identity for $\operatorname{tr}\pi_1 \times \pi_2$, and which is square integrable or elliptic tempered constituent of $1 \rtimes \pi_2$, square integrable $\pi_2$.

Clearly $\chi_\pi(Nt) = \chi_\pi(t \times \theta)$ is a stable class function on $H$, hence perpendicular to the unstable function $\chi$, that is

$$0 = \langle \chi, \chi_\pi \rangle_H = \langle \sum_{\pi_H} m''(\pi_H)\chi_{\pi_H}, \sum_{\pi'_H} \tilde{m}(\pi'_H)\chi_{\pi'_H} \rangle = \sum_{\pi_H} m''(\pi_H)\tilde{m}(\pi_H).$$

Now the $\tilde{m}$ are nonnegative and $\tilde{m}(\pi^0_H) > 0$ for the $\pi^0_H$ for which $m''(\pi^0_H) \neq 0$. Hence $m''$ takes both positive and negative values. □

In particular we see that $a$, the number of irreducible $\pi_H$ with $m''(\pi_H) \neq 0$, is at least two, hence $a = 2$ when $\pi_1$ and $\pi_2$ are inequivalent.

6.3 Lemma. *Suppose that $\pi_1$ and $\pi_2$ are (irreducible) cuspidal (resp. square integrable) inequivalent representations of* PGL$(2, F)$. *Then there are (irreducible) cuspidal (resp. square integrable) representations $\pi^+_H = \pi^+_H(\pi_1 \times \pi_2)$ and $\pi^-_H = \pi^-_H(\pi_1 \times \pi_2)$ such that for all matching functions $f_H$, $f_{C_0}$ we have*

$$\operatorname{tr}(\pi_1 \times \pi_2)(f_{C_0}) = \operatorname{tr}\pi^+_H(f_H) - \operatorname{tr}\pi^-_H(f_H). \tag{3}$$

PROOF. We have $a = 2$ and $(\sum^{a} |m''|)^2 \leq 4$. As the $m''$ are integers we see that $|m''| = 1$. □

We now return to the identity (1), and evaluate it at $f_{Hv_i}$ $(i = 1, 2, 3)$ which are matrix coefficients of $\pi^{-}_{Hv_i}(\pi_{1v_i} \times \pi_{2v_i})$. This choice determines $f_{C_0 v_i}$ and $f_{v_i}$ (or rather their orbital integrals), and our identity becomes

$$c \operatorname{tr} I(\pi_1, \pi_2; f \times \theta) = (2m(\pi_H^+) + 1) \operatorname{tr} \pi_H^+(f_H)$$

$$+(2m(\pi_H^-) + 1) \operatorname{tr} \pi_H^-(f_H) + 2 \sum_{\pi_H} m(\pi_H) \operatorname{tr} \pi_H(f_H).$$

Here we deleted the subscript $v_0$ to simplify the notations, as usual. The $\pi_H$ are inequivalent (pairwise and to $\pi_H^{\pm}$), hence $c \neq 0$. Since

$$\operatorname{tr} I(\pi_1, \pi_2; f \times \theta)$$

is a linear combination with positive coefficients of some $\operatorname{tr} \pi_H(f_H)$, we conclude that $c = 1$ (in fact there is so far a possibility that $c = \frac{1}{2}$, but this will be ruled out later). Once again, the $\pi_H$ which occur with $m(\pi_H) \neq 0$ are cuspidal and finite in number.

6.4 LEMMA. *Suppose that $\pi_1$ and $\pi_2$ are (irreducible) cuspidal (resp. square integrable) inequivalent representations of* PGL$(2, F)$. *Then $m(\pi_H^+) = m(\pi_H^-)$. We write $m(\pi_1 \times \pi_2)$ for the joint value.*

PROOF. The twisted character $\chi(Nt) = \chi_{I(\pi_1, \pi_2)}(t \times \theta)$ of $I(\pi_1, \pi_2)$ is a stable function, while $\chi^+ - \chi^-$, $\chi^{\pm} = \chi_{\pi_H^{\pm}}$, is unstable. Hence their inner product is zero:

$$0 = \langle \chi, \chi^+ - \chi^- \rangle_H = (2m(\pi_H^+) + 1) - (2m(\pi_H^-) + 1). \qquad \square$$

Let us discuss the case of a square integrable $\pi_1 = \pi_2$ on PGL$(2, F)$.

6.5 LEMMA. *If $\pi_1 = \pi_2$ are square integrable, they satisfy the conclusion of Proposition* 5.

PROOF. The representation $1 \rtimes \pi_2$ is reducible (see V.2.1(d), V.2.3(b)). It is the direct sum of its two irreducible constituents, $\pi_H^+$ and $\pi_H^-$, which are tempered. The induced representation $1 \rtimes \pi_2$ of $H$ $\lambda$-lifts to the induced representation $I(\pi_2,\pi_2)$ of $G$ by V.1.2, namely we have, for matching $f$, $f_H$,

$$\operatorname{tr} I(\pi_2,\pi_2; f\times\theta) = \operatorname{tr}(1\rtimes\pi_2)(f_H) = \operatorname{tr}\pi_H^+(f_H) + \operatorname{tr}\pi_H^-(f_H).$$

For the other identity of Proposition 5 we denote our representation by $\pi_{2v_0}$, choose a totally imaginary number field $F$ whose completion at $v_0$ is our local field, $F_{v_0}$, and construct two cuspidal representations, $\pi_1$ and $\pi_2$, of PGL$(2,\mathbb{A})$, which have the same cuspidal components at three places $v_1$, $v_2$, $v_3$ ($\neq v_3$), which are unramified outside the set $V=\{v_0,v_1,v_2,v_3\}$, such that $\pi_{1v_0}$ is unramified while the component at $v_0$ of $\pi_2$ is our square integrable $\pi_{2v_0}$.

We use the trace formulae identity, and the set $V$, such that the only contributions are those associated with $I(\pi_2,\pi_2)$. These contributions are precisely those associated with $I(\pi_2,\pi_2)$, $1\rtimes\pi_2$ on $\mathbf{H}(\mathbb{A})$ and $\pi_2\times\pi_2$ on $\mathbf{C}_0(\mathbb{A})$. Note that at the three places $v_1$, $v_2$, $v_3$ we work with $f_{v_i}$ whose twisted orbital integrals vanish outside the $\theta$-elliptic set, while $I(\pi_{2v_i},\pi_{2v_i})$ is not $\theta$-elliptic. Hence the contribution from $I(\pi_2,\pi_2)$ to the trace formula identity vanishes for our test functions.

Now $1\rtimes\pi_2$ enters the trace formula of $\mathbf{H}(\mathbb{A})$ as

$$\frac{1}{4}\prod_{0\le i\le 3}\operatorname{tr} R(\pi_{2v_i})(1\rtimes\pi_{2v_i})(f_{Hv_i}),$$

where $R(\pi_{2v_i})$ is the normalized intertwining operator on $1\rtimes\pi_{2v_i}$, while $\pi_2\times\pi_2$ enters the trace formula of $\mathbf{C}_0(\mathbb{A})$ as $\prod_{0\le i\le 3}\operatorname{tr}(\pi_{2v_i}\times\pi_{2v_i})(f_{C_0v_i})$.

In the identity of trace formulae, the trace formula of $\mathbf{C}_0(\mathbb{A})$ enters with coefficient $-\frac{1}{4}$ (see e.g. first formula in Chapter IV). We conclude that

$$\prod_{0\le i\le 3}\operatorname{tr} R(\pi_{2v_i})(1\rtimes\pi_{2v_i})(f_{Hv_i}) = \prod_{0\le i\le 3}\operatorname{tr}(\pi_{2v_i}\times\pi_{2v_i})(f_{C_0v_i}).$$

Repeating the same argument with $\pi_1$ instead of $\pi_2$ we get the same identity but where the product ranges over $1\le i\le 3$ instead. In both cases $f_{C_0v_i}$ $(1\le i\le 3)$ can be any functions supported on the elliptic set of $C_{v_i}$. Taking the quotient we conclude that

$$\operatorname{tr} R(\pi_{2v_0})(1\rtimes\pi_{2v_0})(f_{Hv_0}) = \operatorname{tr}(\pi_{2v_0}\times\pi_{2v_0})(f_{C_0v_0})$$

for all matching functions $f_{C_0 v_0}$, $f_{Hv_0}$. The normalized intertwining operator $R(\pi_{2v_0})$ has order 2 but it is not a scalar on the reducible $1 \rtimes \pi_{2v_0}$. It is 1 on one of the two constituents, which we now name $\pi^+_{Hv_0}$, and $-1$ on the other, which we name $\pi^-_{Hv_0}$, as required. □

We can now continue the discussion of the case of square integrable $\pi_1 \neq \pi_2$. We claim that

6.6 LEMMA. *The (finite) sum over* $\pi_H (\neq \pi^\pm_H)$ *in our identity (for all matching* $f$, $f_H$, *where the* $m$ *are nonnegative)*

$$\operatorname{tr} I(\pi_1, \pi_2; f \times \theta) = (2m(\pi^+_H) + 1) \operatorname{tr} \pi^+_H(f_H) + (2m(\pi^-_H) + 1) \operatorname{tr} \pi^-_H(f_H)$$

$$+2 \sum_{\pi_H} m(\pi_H) \operatorname{tr} \pi_H(f_H)$$

*is empty.*

PROOF. To show this we introduce the class functions on the elliptic set of $H$

$$\chi^1 = (2m(\pi^+_H) + 1)\chi_{\pi^+_H} + (2m(\pi^-_H) + 1)\chi_{\pi^-_H}$$

and

$$\chi^0 = 2 \sum_{\pi_H} m(\pi_H)\chi_{\pi_H}.$$

Also write $\chi^\theta_{I(\pi_1,\pi_2)}$ for the class function on the regular set of $H$ whose value at the stable conjugacy class $Ng$ is $\chi_{I(\pi_1,\pi_2)}(g \times \theta)$.

Our first claim is that $\chi^1$ (and $\chi^0$) is stable. It suffices to show that $\langle \chi^1, d_H(\pi'_1 \times \pi'_2)\rangle_H$ is 0 for all square integrable $\pi'_1 \times \pi'_2$ on $C_0$. By (3) and since $m^+ = m^-$ this holds when $\pi'_1 \times \pi'_2$ is equivalent to $\pi_1 \times \pi_2$ (or $\pi_2 \times \pi_1$). When $\pi'_1 \times \pi'_2$ is inequivalent to $\pi_1 \times \pi_2$ or $\pi_2 \times \pi_1$, the twisted orthogonality relations for twisted characters imply that $\langle \chi^\theta_{I(\pi_1,\pi_2)}, \chi^\theta_{I(\pi'_1,\pi'_2)}\rangle_H$ is zero. Since the coefficients $m$ are nonnegative, if $\pi_H \in \{\pi^+_H(\pi_1 \times \pi_2)\} \cup \{\pi^-_H(\pi_1 \times \pi_2)\}$ then it is perpendicular to $d_H(\pi'_1 \times \pi'_2)$, and the claim follows.

Next we claim that $\chi^0$ is zero. If not, $\chi = \langle \chi^1 + \chi^0, \chi^1\rangle_H \cdot \chi^0 - \langle \chi^1 + \chi^0, \chi^0\rangle_H \cdot \chi^1$ is a nonzero stable function on the elliptic set of $H$. (Note that $\langle \chi^0, \chi^1\rangle_H = 0$). Choose $f'_{v_0}$ on $G_{v_0}$ such that $\Phi(t, f'_{v_0} \times \theta) = \chi(Nt)$ on the $\theta$-elliptic set of $G_{v_0}$ and it is zero outside the $\theta$-elliptic set. As usual fix a totally imaginary field $F$ and create a cuspidal $\theta$-invariant representation

$\pi$ which is unramified outside $v_0$, $v_1$, $v_2$, $v_3$, has the component $\mathrm{St}_{v_i}$ at $v_i$ $(i = 1, 2, 3)$, and $\operatorname{tr}\pi_{v_0}(f'_{v_0} \times \theta) \neq 0$. Since $\pi$ is cuspidal as usual by generalized linear independence of characters we get the local identity

$$\operatorname{tr}\pi_{v_0}(f_{v_0} \times \theta) = \sum_{\pi_{H,v_0}} m^1(\pi_{H,v_0}) \operatorname{tr}\pi_{H,v_0}(f_{H,v_0})$$

for all matching $f_{v_0}$, $f_{H,v_0}$. The local representation $\pi = \pi_{v_0}$ is perpendicular to $I(\pi_1, \pi_2)$ since $\langle \chi, \chi^0 + \chi^1 \rangle_H = 0$, and $\chi^0 + \chi^1 = \chi^\theta_{I(\pi_1,\pi_2)}$.

Since $\chi^1 + \chi^0$ is perpendicular to the $\theta$-twisted character $\chi^\theta_\Pi$ of any $\theta$-invariant representation $\Pi$ inequivalent to $I(\pi_1, \pi_2)$, $\chi$ is also perpendicular to all $\chi^\theta_\Pi$, hence $\operatorname{tr}\Pi(f'_{v_0} \times \theta) = 0$ for all $\theta$-invariant representations $\Pi$, contradicting the construction of $\pi_{v_0}$ with $\operatorname{tr}\pi_{v_0}(f'_{v_0} \times \theta) \neq 0$. Hence $\chi = 0$, which implies that $\chi^0 = 0$, namely that for $\pi_1 \neq \pi_2$ we have

$$\begin{aligned} &\operatorname{tr} I(\pi_1, \pi_2; f \times \theta) \\ (4) \qquad &= (2m(\pi_H^+) + 1) \operatorname{tr}\pi_H^+(f_H) + (2m(\pi_H^-) + 1) \operatorname{tr}\pi_H^-(f_H). \end{aligned}$$

Since the character on the left is stable, it is perpendicular to the unstable character on the left of (3). So the right sides of (3) and (4) are orthogonal, hence $m(\pi_H^+) = m(\pi_H^-)$. □

6.7 LEMMA. *The integer* $m = m(\pi_H^+) = m(\pi_H^-)$ *is* 0.

We show at the end of section 10 that precisely one out of $\pi_H^+$, $\pi_H^-$ is generic.

Our proof of the vanishing of $m(\pi_H^+) = m(\pi_H^-)$ is global. It is based on the theory of generic representations. This latter theory implies that given automorphic cuspidal (generic) representations $\pi_1$ and $\pi_2$ of $\mathrm{PGL}(2, \mathbb{A})$ there exists a generic cuspidal representation $\pi_H$ of $\mathrm{PGSp}(2, \mathbb{A})$ which is a $\lambda_0$-lift of $\pi_1 \times \pi_2$, namely $\lambda_0(\pi_{1v} \times \pi_{2v}) = \pi_{Hv}$ at almost all places $v$ of $F$, where $\pi_1$, $\pi_2$ and $\pi_H$ are unramified and the local lifting $\lambda_0$ is defined formally by the dual group homomorphism $\lambda_0 : \hat{C}_0 \to \hat{H}$.

Moreover, in Corollary 7.2 below we prove that $\pi_H$ occurs in the discrete spectrum of $\mathrm{PGSp}(2, \mathbb{A})$ with multiplicity one.

To use this, beginning with our local square integrable representations $\pi'_{1v_0}$ and $\pi'_{2v_0}$, we construct a totally imaginary field $F$ with $F_{v_i} = F_{v_0}$ at three places $v_1$, $v_2$, $v_3$ and global cuspidal representations $\pi_1$ and $\pi_2$

of PGL(2, $\mathbb{A}$), which are unramified outside $v_i$ $(0 \le i \le 3)$, with cuspidal components $\pi_{1v_3}$ and $\pi_{2v_3}$, and $\pi_{jv_i} \simeq \pi'_{jv_0}$ $(i = 0, 1, 2; j = 1, 2)$.

We set up the identity (2), which in view of (3) and (4) takes the form

$$\prod_v (2m_v+1)[\operatorname{tr} \pi^+_{H,v}(f_{H,v})+\operatorname{tr} \pi^-_{H,v}(f_{H,v})]+\prod_v[\operatorname{tr} \pi^+_{H,v}(f_{H,v})-\operatorname{tr} \pi^-_{H,v}(f_{H,v})]$$

$$= 2\sum_{\pi_H} m(\pi_H)\prod_v \operatorname{tr} \pi_{H,v}(f_{H,v}),$$

where $v$ ranges over the finite set $\{v_i;\ 0 \le i \le 3\}$. Corollary 7.2 below asserts that $m(\pi_H)$ is 1 for at least one $\pi_H = \otimes\pi_{Hv}$ (product over all places $v$ of $F$). Hence the corresponding number $\prod_i(2m_{v_i}+1) \pm 1$ is $2m(\pi_H) = 2$. Since $m_{v_0} = m_{v_1} = m_{v_2}$, and $3^3 \pm 1 > 2$, $m_{v_0}$ is zero. The proposition follows. □

REMARK. Our proof is global. It resembles (but is strictly different from) the second attempt at a proof of multiplicity one theorem for the discrete spectrum of U(3) in [F4;II], Proposition 3.5, p. 48, which is also based on the theory of generic representations.

However, the proof of [F4;II], p. 48, is not complete. Indeed, the claim in Proposition 2.4(i) in reference [GP] to [F4;II], that "$L^2_{0,1}$ has multiplicity 1", is interpreted in [F4;II] as asserting that generic representations of U(3) occur in the discrete spectrum with multiplicity one. But it should be interpreted as asserting that irreducible $\pi$ in $L^2_{0,1}$ have multiplicity one *only in the subspace* $L^2_{0,1}$ of the discrete spectrum. This claim does not exclude the possibility of having a cuspidal $\pi'$ perpendicular and equivalent to $\pi \subset L^2_{0,1}$.

Multiplicity one for the generic spectrum would follow via this global argument from the statement that a (locally generic) representation equivalent to a globally generic one is globally generic (multiplicity one implies this statement too). In our case of PGSp(2) this follows from [KRS], [GRS], [Sh1]. A proof for U(3) still needs to be written down.

The usage of the theory of generic representations in the proof above is not natural. A purely local proof of multiplicity one theorem for the discrete spectrum of U(3) based only on character relations is proposed in [F4;II], Proof of Proposition 3.5, p. 47. It is based on Rodier's result [Ro1] that the number of Whittaker models is encoded in the character of the representation near the origin. Details of this proof are given in [F4;IV] in odd residual characteristic for the basechange lifting from $\mathrm{U}(3, E/F)$ to $\mathrm{GL}(3, E)$. It implies that in a tempered packet of representations of $\mathrm{U}(3, E/F)$ there is precisely one generic representation. We carried out this proof in the case of the symmetric square lifting from SL(2) to PGL(3) ([F3]) but not yet for our lifting from PGSp(2) to PGL(4).

## 7. Generic Representations of PGSp(2)

We proceed to explain the result quoted at the end of the global proof of Proposition 5 above (after Lemma 6.7) and attributed to the theory of generic representations.

We start with a result of [GRS] which asserts: the weak (in terms of almost all places) lifting establishes a bijection from the set of equivalence classes of (irreducible automorphic) cuspidal *generic* representations $\pi_H$ of the split group $\mathrm{SO}(2n+1, \mathbb{A})$, to the set of representations of $\mathrm{PGL}(2n, \mathbb{A})$ of the form $\pi = I(\pi_1, \ldots, \pi_r)$, normalized induction from the standard parabolic subgroup of the type $(2n_1, \ldots, 2n_r)$, $n = n_1 + \cdots + n_r$, where $\pi_i$ are cuspidal representations of $\mathrm{GL}(2n_i, \mathbb{A})$ such that $L(S, \pi_i, \Lambda^2, s)$ has a pole at $s = 1$ and $\pi_i \neq \pi_j$ for all $i \neq j$. The partial $L$-function is defined as a product outside a finite set $S$ where all $\pi_i$ are unramified.

Moreover, if $\pi_H$ is a cuspidal generic representation (in the space of cusp forms) of $\mathrm{SO}(2n+1, \mathbb{A})$ which weakly lifts to $\pi$ as above, and $\pi'_H$ is a cuspidal representation of $\mathrm{SO}(2n+1, \mathbb{A})$ which weakly lifts to $\pi$ and is orthogonal to $\pi_H$, then $\pi'_H$ is not generic (has zero Whittaker coefficients with respect to any nondegenerate character).

Note that this result does not rule out the possibility that there exists a cuspidal representation $\pi'_H$ of $\mathrm{SO}(2n+1, \mathbb{A})$ which is both orthogonal and equivalent to the generic cuspidal $\pi_H$, and consequently is locally generic everywhere, but is not (globally) generic. Hence $\pi_H$ may occur in the discrete, in fact cuspidal, spectrum of $\mathrm{SO}(2n+1, \mathbb{A})$ with multiplicity $m(\pi_H)$ greater than one.

Of course we are interested in the case $n = 2$, where

7.0 LEMMA. $\mathrm{PGSp}(2) \xrightarrow{\sim} \mathrm{SO}(5)$.

PROOF. This well-known isomorphism can be constructed as follows. Let $U$ be the 5-dimensional space of $4 \times 4$ matrices $u$ such that $\mathrm{tr}(u) = 0$ and ${}^t J {}^t u J = u$. Then PGSp(2) acts on $U$ by conjugation: $g : u \mapsto gug^{-1}$, and the action preserves the nondegenerate form $(u_1, u_2) \mapsto \mathrm{tr}(u_1 u_2)$ on $U$. The action embeds PGSp(2) as the connected component SO(5) of the identity of the orthogonal group O(5) preserving this form. □

A related result is Theorem 8.1 of [KRS]. It asserts that if $\pi_0$ is a cuspidal representation of $\mathrm{Sp}(2, \mathbb{A})$ which is locally generic everywhere, and the partial $L$-function $L(S, \pi_0, \mathrm{id}_5, s)$ is nonzero at $s = 1$ then $\pi_0$ is (globally) generic. Here $L$ is the degree 5 $L$-function associated with the 5-dimensional representation $\mathrm{id}_5 : \mathrm{SO}(5, \mathbb{C}) \hookrightarrow \mathrm{GL}(5, \mathbb{C})$ of the dual group $\mathrm{SO}(5, \mathbb{C})$ of Sp(2). When $\pi_0$ is generic this $L$-function is nonzero at $s = 1$ by Shahidi [Sh1], Theorem 5.1, since $\mathrm{id}_5$ can also be obtained by the adjoint action of the $\mathrm{SO}(5, \mathbb{C})$-factor in the Levi subgroup $\mathrm{GL}(1, \mathbb{C}) \times \mathrm{SO}(5, \mathbb{C})$ of $\mathrm{SO}(7, \mathbb{C})$ on the 5-dimensional Lie algebra of the unipotent radical. This is case (xx) of Langlands [L2]. Together, [KRS], Theorem 8.1, and [Sh1], Theorem 5.1, although do not yet imply that a locally generic cuspidal representation of $\mathrm{Sp}(2, \mathbb{A})$ is generic, do assert that:

7.1 PROPOSITION. *Let $\pi_0$, $\pi_0'$ be cuspidal representations of* $\mathrm{Sp}(2, \mathbb{A})$. *Suppose that $\pi_0'$ is generic, $\pi_{0v}$ is generic for all $v$, and $\pi_{0v} \simeq \pi_{0v}'$ for almost all $v$. Then $\pi_0$ is generic.*

I wish to thank S. Rallis for pointing out to me [KRS], [GRS] and [Sh1] in the context used above, and F. Shahidi for the reference to [L2], (xx).

We need this result for $\mathrm{PGSp}(2, \mathbb{A})$:

7.2 COROLLARY. *Any generic cuspidal representation $\pi$ occurs in the discrete spectrum of the group* $\mathrm{PGSp}(2, \mathbb{A}) = \mathrm{SO}(5, \mathbb{A})$ *with multiplicity one.*

In view of the results of [GRS] quoted above it suffices to show that:

7.3 LEMMA. *Let $\pi_H$, $\pi_H'$ be cuspidal representations of* $\mathrm{PGSp}(2, \mathbb{A})$. *Suppose that $\pi_H'$ is generic, $\pi_{Hv}$ is generic for all $v$, and $\pi_{Hv} \simeq \pi_{Hv}'$ for almost all $v$. Then $\pi_H$ is generic.*

To see this, let us explain the difference between the group $\mathrm{PGSp}(2, F)$ (which is equal to $\mathrm{GSp}(2, F)/Z(F)$) and the group $\mathrm{Sp}(2, F)/\{\pm I\}$.

Note that PGSp(2) = PSp(2) as algebraic groups (over an algebraic closure $\overline{F}$ of the base field $F$). We have the exact sequences

$$1 \to \mathbb{G}_m \to \mathrm{GSp}(2) \to \mathrm{PGSp}(2) \to 1,$$

$$1 \to \{\pm I\} \to \mathrm{Sp}(2) \to \mathrm{PSp}(2) \to 1,$$

since the center $Z$ of GSp(2) is $\mathbb{G}_m$ while that $Z_S$ of Sp(2) is $\{\pm I\}$. Since $H^1(F, \mathbb{G}_m) = \{0\}$ and $H^1(F, \mathbb{Z}/2) = F^\times/F^{\times 2}$, the associate exact sequences of Galois cohomology give

$$1 \to F^\times \to \mathrm{GSp}(2,F) \to \mathrm{PGSp}(2,F) \to 1,$$

thus $\mathrm{PGSp}(2,F) = \mathrm{GSp}(2,F)/F^\times$, and

$$1 \to \{\pm I\} \to \mathrm{Sp}(2,F) \to \mathrm{PSp}(2,F) \to F^\times/F^{\times 2}.$$

Hence $\mathrm{Sp}(2,F)/\{\pm I\} = \ker[\mathrm{PGSp}(2,F) \to F^\times/F^{\times 2}]$ (as $\mathrm{PGSp}(2,F) = \mathrm{PSp}(2,F)$). The kernel is induced from the map $\boldsymbol{\lambda} : \mathrm{GSp}(2) \to \mathbb{G}_m$, associating to $g$ its factor of similitudes. Globally we have

$$\mathrm{Sp}(2,\mathbb{A})/Z_S(\mathbb{A}) = \ker[\mathrm{GSp}(2,\mathbb{A})/Z(\mathbb{A}) \to \mathbb{A}^\times/\mathbb{A}^{\times 2}],$$

where $Z_S(\mathbb{A})$ is the group of idèles $(z_v) \in \mathbb{A}^\times$ with $z_v \in \{\pm I\}$ for all $v$. It will be simpler to work with the group $\mathrm{Zp}(2,\mathbb{A}) = Z(\mathbb{A})\,\mathrm{Sp}(2,\mathbb{A})$, with center $Z(\mathbb{A})$, and $\mathrm{Zp}(2,F) = Z(F)\,\mathrm{Sp}(2,F)$. Note that $\mathrm{Zp}(2,\mathbb{A})/Z(\mathbb{A}) = \mathrm{Sp}(2,\mathbb{A})/Z_S(\mathbb{A})$ and

$$\mathrm{Zp}(2,F)/F^\times = \mathrm{Sp}(2,F)/\{\pm I\}.$$

An automorphic representation of $\mathrm{Sp}(2,\mathbb{A})$ with trivial central character is the same as an automorphic representation of $\mathrm{Zp}(2,\mathbb{A})$ with trivial central character.

Let us also explain the passage from representations of $\mathrm{GSp}(2,F)$ to those on $F^\times\,\mathrm{Sp}(2,F)$.

7.4 LEMMA. *Put* $H = \mathrm{GSp}(2,F)$ *and* $S = \mathrm{Sp}(2,F)Z(F)$.
(i) *Let* $\pi$ *be an irreducible admissible representation of* $H$. *Then the restriction* $\mathrm{Res}_S^H\,\pi$ *of* $\pi$ *to* $S$ *is the direct sum of finitely many irreducible representations.*

(ii) *Let* $\pi^S$ *be an irreducible admissible representation of* $S$. *Then there is an irreducible admissible representation* $\pi$ *of* $H$ *whose restriction to* $S$ *contains* $\pi^S$.

PROOF. The map $\mathrm{GSp}(2,F) \to F^\times$ associating to $h$ its factor $\boldsymbol{\lambda}(h)$ of similitudes defines the isomorphism $H/S \xrightarrow{\sim} F^\times/F^{\times 2} = (\mathbb{Z}/2)^r$, $r$ finite ($r = 2$ if $F$ has odd residual characteristic). By induction, it suffices to show (i), (ii) with $H$, $S$ replaced by $H'$, $S'$ with $S \subset S' \subset H' \subset H$, $H'/S' = \mathbb{Z}/2$.

(i): Let $(\pi, V)$ be an admissible irreducible representation of $H'$. Then $\mathrm{Res}_{S'}^{H'} \pi$ is admissible. If it is irreducible, (i) follows for $\pi$. If not, $V$ contains a nontrivial subspace $W$ invariant and irreducible under $S'$. For $h \in H' - S'$ we have $V = W + \pi(h)W$. Since $W \cap \pi(h)W$ is invariant under $H'$, it is zero, and so $V = W \oplus \pi(h)W$ where $W$ and $\pi(h)W$ are irreducible $S'$-modules. (i) follows.

(ii): Given an irreducible admissible representation $(\pi^{S'}, W)$ of $S'$, put $\pi_I = \mathrm{Ind}_{S'}^{H'}(\pi^{S'})$. For $h \in H' - S'$, if $s \mapsto \pi^{S'}(h^{-1}sh)$ ($s \in S'$) is not equivalent to $\pi^{S'}$ then $\pi_I$ is irreducible and $\mathrm{Res}_{S'}^{H'}(\pi_I)$ contains $\pi^{S'}$. Otherwise there exists an intertwining operator $A : (\pi^{S'}, W) \to (\pi^{S'}, W)$ with $\pi^{S'}(h^{-1}sh) = A^{-1}\pi^{S'}(s)A$ ($s \in S'$) and $A^2 = \pi^{S'}(h^2)$ (by Schur's lemma). We can then extend $\pi^{S'}$ to a representation $\pi$ on the space $W$ of $\pi^{S'}$ by $\pi(h) = A$. We have $(\pi, W) \hookrightarrow \pi_I$ by $w \mapsto f_w(g) = \pi(g)w$ ($g \in H'$), and $\pi_I \simeq \pi \oplus \pi\omega$, where $\omega$ is the nontrivial character of $H'/S' = \mathbb{Z}/2$. □

REMARK. The restriction of a generic admissible irreducible $\pi$ of $H$ to $S$ contains no irreducible representation $\pi^S$ with multiplicity $> 1$.

Indeed, $\pi$ is generic if $\pi \hookrightarrow \mathrm{Ind}_N^H \psi$ for some generic character $\psi$ of the unipotent radical $N = N(F)$ of $H$. Note that $N \subset S$. Since $H = \cup\, \mathrm{diag}(I, \boldsymbol{\lambda} I)S$, $\boldsymbol{\lambda} \in F^\times/F^{\times 2}$, and $\mathrm{diag}(I, \boldsymbol{\lambda} I)$ normalizes $N$, each $\pi^S \subset \mathrm{Res}_S^H \pi$ is a constituent of $\mathrm{Ind}_N^S \psi^\lambda$ for some generic character $\psi^\lambda$ of $N$. Now $\pi_I = \mathrm{Ind}_S^H(\pi^S) \subset \mathrm{Ind}_S^H \mathrm{Ind}_N^S \psi^\lambda = \mathrm{Ind}_N^H \psi^\lambda$. The uniqueness of the embedding ("Whittaker model") of $\pi$ in $\mathrm{Ind}_N^H \psi$ implies the uniqueness of the embedding of $\pi$ in $\pi_I$, hence of $\pi^S$ in $\pi$, since by Frobenius reciprocity: $\mathrm{Hom}_S(\pi^S, \mathrm{Res}_S^H \pi) = \mathrm{Hom}_H(\pi_I, \pi)$, and the complete reducibility (i) above, $\pi^S$ is contained in $\pi$ with the same multiplicity that $\pi$ is contained in $\pi_I$. □

PROOF OF LEMMA 7.3. Let us then take a cuspidal representation $\pi =$

$\otimes\pi_v$ of $\mathrm{PGSp}(2,\mathbb{A})$ which is locally generic. Thus for each $v$ there is a nondegenerate character $\psi_v$ of the unipotent radical $N_v$ of the Borel subgroup of $H_v = \mathrm{PGSp}(2,F_v)$ (and of $S_v = \mathrm{Zp}(2,F_v)/F_v^\times$) such that $\pi_v \hookrightarrow \mathrm{Ind}_{N_v}^{H_v}(\psi_v)$. Applying the exact functor $\mathrm{Res}_{S_v}^{H_v}$ of restriction from $H_v$ to $S_v$ we see that $\mathrm{Res}_{S_v}^{H_v}\pi_v \hookrightarrow \oplus_\gamma \mathrm{Ind}_{N_v}^{S_v}(\psi_v^\gamma)$, where $\psi_v^\gamma$ are the translates of $\psi_v$ under $H_v/S_v \simeq F_v^\times/F_v^{\times 2}$. Thus each irreducible constituent $\pi_v^S$ of $\mathrm{Res}_{S_v}^{H_v}\pi_v$ is generic.

Since $\pi$ is a submodule of the space $L_0^2(\mathrm{PGSp}(2,F)\backslash \mathrm{PGSp}(2,\mathbb{A}))$, the restriction map $\phi \mapsto \phi|(\mathrm{Zp}(2,\mathbb{A})/Z(\mathbb{A}))$ defines a subspace $\pi_0^S$ of

$$L_0^2(\mathrm{Zp}(2,F)Z(\mathbb{A})\backslash \mathrm{Zp}(2,\mathbb{A})).$$

Choose an irreducible (under the right action of $\mathrm{Zp}(2,\mathbb{A})$) subspace $\pi^S$ of $\pi_0^S$. Then $\pi^S = \otimes\pi_v^S$ is a cuspidal representation of $\mathrm{Zp}(2,\mathbb{A})$ whose components are all generic. The same construction, applied to the cuspidal generic $\pi'$, gives a cuspidal generic $\pi'^S$, locally equivalent to $\pi^S$ at almost all places. By Proposition 7.1 (namely the results of [KRS] and [Sh1] for $\mathrm{Zp}(2,\mathbb{A}) = Z(\mathbb{A})\,\mathrm{Sp}(2,\mathbb{A})$), $\pi^S$ is generic. This means that for some nondegenerate character $\psi$ of $N(F)\backslash N(\mathbb{A})$, we have $\pi^S \hookrightarrow \mathrm{Ind}_{N(\mathbb{A})}^{\mathrm{Zp}(2,\mathbb{A})/\mathbb{A}^\times}\psi$. But $\pi \subset \mathrm{Ind}_{\mathrm{Zp}(2,\mathbb{A})/\mathbb{A}^\times}^{\mathrm{PGSp}(2,\mathbb{A})}(\pi^S)$, and induction is transitive: $\mathrm{Ind}_B^C\,\mathrm{Ind}_A^B = \mathrm{Ind}_A^C$, and exact, hence $\pi \hookrightarrow \mathrm{Ind}_{N(\mathbb{A})}^{\mathrm{PGSp}(2,\mathbb{A})}\psi$. In other words, $\pi$ is generic. □

Once we complete our global results on the lifting $\lambda$ from the group $\mathrm{PGSp}(2,\mathbb{A})$ to the group $\mathrm{PGL}(4,\mathbb{A})$ in section 10, we deduce from [GRS] that each local tempered packet contains precisely one generic member, and each packet which lifts to a cuspidal representation of $\mathrm{PGL}(4,\mathbb{A})$, or to an induced $I(\pi_1,\pi_2)$ where $\pi_1$, $\pi_2$ are cuspidal on $\mathrm{PGL}(2,\mathbb{A})$, contains precisely one representation which is everywhere locally generic. The latter is generic if it lifts to $I(\pi_1,\pi_2)$.

# 8. Local Lifting from PGSp(2)

One more case remains to be dealt with.

8.1 PROPOSITION. *Let $\pi_{v_0}$ be a $\theta$-invariant irreducible square integrable representation of $G_{v_0}$ over a local field $F_{v_0}$ which is not a $\lambda_1$-lift. Then* *there exists a square integrable irreducible representation $\pi_{H,v_0}$ of $H_{v_0}$ which $\lambda$-lifts to $\pi_{v_0}$, thus* $\operatorname{tr}\pi_{v_0}(f_{v_0}\times\theta)=\operatorname{tr}\pi_{H,v_0}(f_{H,v_0})$ *for all matching $f_{v_0}$ and $f_{H,v_0}$.*

*In particular the $\theta$-character of $\pi_{v_0}$ is $\theta$-stable, and the character of $\pi_{H,v_0}$ is a stable function on $H_{v_0}$.*

PROOF. Since $\pi_{v_0}$ is $\theta$-invariant, square-integrable and not a $\lambda_1$-lift, its character is $\theta$-stable by Proposition IV.4.4. We choose a totally imaginary global field $F$ and a function $f=\otimes f_v$ whose components $f_v$ at 4 places $v_i$ $(i=0,1,2,3)$ with $F_{v_i}=F_{v_0}$ are pseudo matrix coefficients of $\pi_v$, where at $v=v_0$ this $\pi_v$ is the $\pi_{v_0}$ of the proposition, at $v_1$ it is $\pi_{v_1}=I_G(\pi_{1v_1},\pi_{2v_1})$, where $\pi_{iv_1}$ are distinct cuspidal PGL$(2,F_{v_1})$-modules, while $\pi_{v_2}$ and $\pi_{v_3}$ are the Steinberg PGL$(4,F_v)$-modules. The other components $f_v$ at finite $v$ are taken to be spherical and $\geqslant 0$. Since the $\theta$-orbital integrals of $f_{v_0}$ (in fact also $f_{v_i}, i=1,2,3$) are $\theta$-stable functions (supported on the $\theta$-elliptic set), the geometric part of the $\theta$-trace formula is $\theta$-stable: it is the sum of $\theta$-stable orbital integrals, $\Phi_\gamma^{\rm st}(f)$. We choose $f_\infty=\otimes_v f_v$, $v$ archimedean, to vanish on the non $\theta$-regular set.

Since $\mathbf{G}(F)$ is discrete in $\mathbf{G}(\mathbb{A})$ and $f=\otimes f_v$ is compactly supported, $\Phi_\gamma^{\rm st}(f)\neq 0$ for only finitely many $\theta$-stable conjugacy classes ($\theta$-elliptic and regular) $\gamma$ in $\mathbf{G}(F)$. Restricting the support of $f_\infty$ we can arrange that $\Phi_\gamma^{\rm st}(f)\neq 0$ for a single $\theta$-stable class $\gamma$. Hence the geometric side of the $\theta$-trace formula is nonzero. Consequently the spectral side is nonzero.

The choice of $f_{v_i}$ $(i=0,1,2,3)$ as a pseudo-coefficient can be used now to show the existence of a $\theta$-invariant $\pi$ whose component at $v_1$ is $\pi_{v_1}=I_G(\pi_{1v_1},\pi_{2v_1})$, and consequently $\pi_{v_2}$ and $\pi_{v_3}$ are Steinberg (note that $\operatorname{tr}\pi_{v_2}(f_{v_2}\times\theta)\neq 0$ does not assure us that $\pi_{v_2}$ is Steinberg, but given that $\pi_{v_1}$ is $I_G(\pi_{1v_1},\pi_{2v_1})$, the global $\pi$ must be generic, hence $\pi_{v_2}$ is the square integrable generic constituent in the fully induced

$$I_G(\nu_v^{3/2},\nu_v^{1/2},\nu_v^{-1/2},\nu_v^{-3/2})).$$

It follows that $\pi$ is generic and cuspidal (it contributes to the sum $I$ in the spectral side of the $\theta$-trace formula, not to $I_{(2,2)}$, etc.). Since the components $\pi_{v_i}$ $(i=1,2,3)$ and by the same argument also $\pi_{v_0}$, are our $\theta$-stable ones, $\pi$ is not a $\lambda_1$-lift, nor it is a $\lambda$-lift of the form $I(\pi_1,\pi_2)$. Thus when we write the trace formula identity fixing all finite components to be those of $\pi$ at all $v \neq v_i$ $(i = 0, 1, 2, 3)$, the only contribution other than $\pi$ would be from $H$, namely

$$\prod_v \operatorname{tr} \pi_v(f_v \times \theta) = \sum_{\pi_H} m(\pi_H) \prod_v \operatorname{tr} \pi_{Hv}(f_{Hv}).$$

Thus $\pi_H$ are discrete spectrum representations of $\mathbf{H}(\mathbb{A})$ whose components at each finite $v \neq v_i$ $(i = 0, 1, 2, 3)$ are unramified and $\lambda$-lift to $\pi_v$. The products range over $v_i$ $(i = 0, 1, 2, 3)$ and the archimedean places.

Next we apply the generalized linear independence argument at the archimedean places. Consequently we can and do omit the archimedean $v$ from the product, and restrict the sum to $\pi_H$ with $\lambda(\pi_{H\infty}) = \pi_\infty$.

Evaluating at $v_1$, $v_2$, $v_3$ with the pseudo coefficient $f_{v_i}$, which is $\theta$-elliptic, we can delete these $v$ from the product, but now the sum ranges over the $\pi_H$ which in addition have the Steinberg component at $v_2$ and $v_3$, and $\pi^+_{Hv_1}$ or $\pi^-_{Hv_1}$ at $v_1$.

Omitting the index $v_0$, we finally get for our $\pi = \pi_{v_0}$ the equality

$$\operatorname{tr} \pi(f \times \theta) = \sum_{\pi_H} m(\pi_H) \operatorname{tr} \pi_H(f_H)$$

for all matching $f$ and $f_H$.

Since $\pi$ is square integrable and the $m(\pi_H)$ are nonnegative, the theorem of [C] on modules of coinvariants implies that all $\pi_H$ on the right are cuspidal, except for one square integrable noncuspidal $\pi_H$ if $\pi$ is the square integrable constituent of $I_G(\nu^{1/2}\pi_2, \nu^{-1/2}\pi_2)$, for a cuspidal $\pi_2 = \pi_2(\mu)$, where $\mu$ is a character of $E^\times/F^\times$, $E/F$ being a local quadratic extension.

Evaluating at $f_H = f_H^a = \sum_{i=1}^a f(\pi_{Hi})$, where we list the $\pi_H$ and $f(\pi_{Hi})$ denotes a pseudo coefficient of $\pi_{Hi}$, we conclude from the orthonormality relations for twisted characters that the sum over $\pi_H$ is finite.

The resulting character relation

$$\chi_\pi(g \times \theta) = \sum m(\pi_H) \chi_{\pi_H}(Ng)$$

and the orthonormality relations for $\theta$-characters of square integrable representations, i.e.: $\langle \chi_\pi^\theta, \chi_\pi^\theta \rangle = 1$, imply that

$$\langle \sum m(\pi_H) \chi_{\pi_H}, \sum m(\pi_H) \chi_{\pi_H} \rangle$$

is 1, thus $\sum m(\pi_H)^2$ is 1. Hence there is only one term on the right with coefficient $m = 1$. □

8.2 COROLLARY. *Let $\pi_2$ be a cuspidal (irreducible) representation of $\mathrm{GL}(2,F)$, $F$ local, with $\xi\pi_2 = \pi_2$ and central character $\xi \neq 1 = \xi^2$. The square integrable subrepresentation $\delta(\nu\xi, \nu^{-1/2}\pi_2)$ of the $H$-module $\nu\xi \rtimes \nu^{-1/2}\pi_2$, $\lambda$-lifts to the square integrable submodule $S(\nu^{1/2}\pi_2, \nu^{-1/2}\pi_2)$ of the $G$-module $I_G(\nu^{1/2}\pi_2, \nu^{-1/2}\pi_2)$.*

PROOF. This follows from the proof of the proposition. Note that the only noncuspidal non Steinberg selfcontragredient square integrable representation of $\mathbf{G}(F)$ is of the form $S(\nu^{1/2}\pi_2, \nu^{-1/2}\pi_2)$, where $\pi_2$ is a cuspidal representation of $\mathrm{GL}(2,F)$ with central character $\xi$, $\xi^2 = 1$, and $\xi\pi_2 = \pi_2$.

The square integrable $S(\nu^{1/2}\pi_2, \nu^{-1/2}\pi_2)$, where $\pi_2$ is a cuspidal representation of $\mathrm{PGL}(2,F)$, is the $\lambda_1$-lift of $\mathrm{sp}_2 \times \pi_2$. If the central character of $\pi_2$ is $\xi \neq 1 = \xi^2$, it is associated with a quadratic extension $E$ of $F$, and since $\xi\pi_2 = \pi_2$ there is a character $\mu$ of $E^\times$, trivial on $F^\times$, such that $\pi_2 = \pi_2(\mu)$. The only square integrable representations of $\mathrm{PGSp}(2,F)$ not accounted for so far are $\delta(\nu\xi, \nu^{-1/2}\pi_2)$, $\omega_{\pi_2} = \xi \neq 1 = \xi^2$, $\xi\pi_2 = \pi_2$.

Since $\nu\xi \rtimes \nu^{-1/2}\pi_2$ $\lambda$-lifts to $I_G(\nu^{1/2}\check{\pi}_2, \nu^{-1/2}\pi_2)$, and $\check{\pi}_2 = \xi\pi_2$, the decaying central exponents in these fully induced representations correspond, hence $\delta(\nu\xi, \nu^{-1/2}\pi_2)$ $\lambda$-lifts to $S(\nu^{1/2}\pi_2, \nu^{-1/2}\pi_2)$ from the proof of the proposition. □

8.3 COROLLARY. *The nontempered quotient $L(\nu\xi, \nu^{-1/2}\pi_2)$ in the composition series of the $\mathbf{H}(F)$-module $\nu\xi \rtimes \nu^{-1/2}\pi_2$, where $\pi_2$ is a cuspidal $\mathrm{GL}(2,F)$-module with central character $\xi \neq 1 = \xi^2$ and $\xi\pi_2 = \pi_2$, $\lambda$-lifts to the nontempered quotient $J(\nu^{1/2}\pi_2, \nu^{-1/2}\pi_2)$ in the composition series of the induced $I_G(\nu^{1/2}\pi_2, \nu^{-1/2}\pi_2)$.*

PROOF. This follows from

$$\operatorname{tr} L(\nu\xi, \nu^{-1/2}\pi_2)(f_H) = \operatorname{tr}(\nu\xi \rtimes \nu^{-1/2}\pi_2)(f_H) - \operatorname{tr} \delta(\nu\xi, \nu^{-1/2}\pi_2)(f_H)$$

and $\quad \operatorname{tr} J(\nu^{1/2}\pi_2, \nu^{-1/2}\pi_2; f \times \theta)$

$$= \operatorname{tr} I_G(\nu^{1/2}\pi_2, \nu^{-1/2}\pi_2; f \times \theta) - \operatorname{tr} S(\nu^{1/2}\pi_2, \nu^{-1/2}\pi_2; f \times \theta). \qquad \square$$

For any irreducible square integrable $\mathrm{PGL}(2,F)$-modules $\pi_1$ and $\pi_2$ we have

$$\operatorname{tr}(\pi_1 \times \pi_2)(f_{C_0}) = \operatorname{tr} \pi_H^+(f_H) - \operatorname{tr} \pi_H^-(f_H),$$
$$\operatorname{tr} I_G(\pi_1, \pi_2; f \times \theta) = \operatorname{tr} \pi_H^+(f_H) + \operatorname{tr} \pi_H^-(f_H),$$

for all matching functions $f$, $f_H$, $f_{C_0}$, where $\pi_H^+$, $\pi_H^-$ are tempered irreducible (square integrable if $\pi_1 \neq \pi_2$) representations of $H$ determined by the unordered pair $\pi_1$, $\pi_2$.

If $\pi_1 = \pi_2$ is cuspidal, $\pi_H^+$ and $\pi_H^-$ are the two inequivalent constituents of $1 \rtimes \pi_1$.

If $\pi_1 = \pi_2$ is $\xi\,\mathrm{sp}_2$, where $\xi$ is a character of $F^\times$ with $\xi^2 = 1$, $\pi_H^+$ and $\pi_H^-$ are the two tempered inequivalent constituents $\tau(\nu^{1/2}\,\mathrm{sp}_2, \xi\nu^{-1/2})$ and $\tau(\nu^{1/2}\mathbf{1}_2, \xi\nu^{-1/2})$ of $1 \rtimes \xi\,\mathrm{sp}_2$.

If $\pi_1 = \xi\,\mathrm{sp}_2$, $\xi^2 = 1$, and $\pi_2$ is cuspidal, then $\pi_H^+$ is the square integrable constituent $\delta(\pi_2\xi\nu^{1/2}, \xi\nu^{-1/2})$ of the induced $\pi_2\xi\nu^{1/2} \rtimes \xi\nu^{-1/2}$, while $\pi_H^-$ is cuspidal, which we denote by $\delta^-(\xi\nu^{1/2}\pi_2, \xi\nu^{-1/2})$.

If $\pi_1 = \sigma\,\mathrm{sp}_2$ and $\pi_2 = \xi\sigma\,\mathrm{sp}_2$, $\xi(\neq 1 = \xi^2)$ and $\sigma$ are characters of $F^\times$, then $\pi_H^+$ is the square integrable constituent $\delta(\xi\nu^{1/2}\,\mathrm{sp}_2, \nu^{-1/2}\sigma)$ of the induced $\mathrm{sp}_2\,\xi\nu^{1/2} \rtimes \sigma\nu^{-1/2}$, while $\pi_H^-$ is cuspidal, which we denote by $\delta^-(\xi\nu^{1/2}\,\mathrm{sp}_2, \nu^{-1/2}\sigma)$.

We made this explicit list in order to describe the character relations where in the last three paragraphs $\mathrm{sp}_2$ is replaced by the nontempered trivial representation $\mathbf{1}_2$ of $\mathrm{PGL}(2, F)$.

8.4 PROPOSITION. *For any cuspidal representation $\pi_2$ of* $\mathrm{PGL}(2, F)$ *and character $\xi$ of $F^\times$ with $\xi^2 = 1$, we have*

$$\begin{aligned}
&\mathrm{tr}(\xi\mathbf{1}_2 \times \pi_2)(f_{C_0}) \\
&\qquad = \mathrm{tr}\, L(\pi_2\xi\nu^{1/2}, \xi\nu^{-1/2})(f_H) + \mathrm{tr}\,\delta^-(\pi_2\xi\nu^{1/2}, \xi\nu^{-1/2})(f_H), \\
&\mathrm{tr}\, I_G(\xi\mathbf{1}_2, \pi_2; f \times \theta) \\
&\qquad = \mathrm{tr}\, L(\pi_2\xi\nu^{1/2}, \xi\nu^{-1/2})(f_H) - \mathrm{tr}\,\delta^-(\pi_2\xi\nu^{1/2}, \xi\nu^{-1/2})(f_H),
\end{aligned}$$

*for all matching $f$, $f_H$, $f_{C_0}$.*

PROOF. This follows from

$$\mathrm{tr}\, I(\xi\,\mathrm{sp}_2 \times \pi_2; f \times \theta) = \mathrm{tr}\,\delta(f_H) + \mathrm{tr}\,\delta^-(f_H),$$

$$\mathrm{tr}(\xi\,\mathrm{sp}_2 \times \pi_2)(f_{C_0}) = \mathrm{tr}\,\delta(f_H) - \mathrm{tr}\,\delta^-(f_H),$$

and

$$\begin{aligned}
&\mathrm{tr}\, I_G(\xi\,\mathrm{sp}_2, \pi_2; f \times \theta) + \mathrm{tr}\, I_G(\xi\mathbf{1}_2, \pi_2; f \times \theta) \\
&= \mathrm{tr}\, I_G(\xi\nu^{1/2}, \pi_2, \xi\nu^{-1/2}; f \times \theta) \\
&= \mathrm{tr}(\pi_2\xi\nu^{1/2} \rtimes \xi\nu^{-1/2})(f_H) = \mathrm{tr}\,\delta(f_H) + \mathrm{tr}\, L(f_H) \\
&= \mathrm{tr}(\xi I_2 \times \pi_2)(f_{C_0}) = \mathrm{tr}(\xi\mathbf{1}_2 \times \pi_2)(f_{C_0}) + \mathrm{tr}(\xi\,\mathrm{sp}_2 \times \pi_2)(f_{C_0}),
\end{aligned}$$

where $I_2 = I(\nu^{1/2}, \nu^{-1/2})$. □

8.5 PROPOSITION. *For any characters* $\xi \neq 1 = \xi^2$ *and* $\sigma$ $(\sigma^2 = 1)$ *of* $F^\times$, *for all matching* $f$, $f_H$, $f_{C_0}$ *we have*

$$\begin{aligned}
&\operatorname{tr}(\sigma\mathbf{1}_2 \times \sigma\xi\,\mathrm{sp}_2)(f_{C_0})\\
&\qquad = \operatorname{tr} L(\nu^{1/2}\xi\,\mathrm{sp}_2, \sigma\nu^{-1/2})(f_H) + \operatorname{tr}\delta^-(\xi\nu^{1/2}\,\mathrm{sp}_2, \sigma\nu^{-1/2})(f_H),\\
&\operatorname{tr} I_G(\sigma\mathbf{1}_2, \sigma\xi\,\mathrm{sp}_2; f \times \theta)\\
&\qquad = \operatorname{tr} L(\nu^{1/2}\xi\,\mathrm{sp}_2, \sigma\nu^{-1/2})(f_H) - \operatorname{tr}\delta^-(\xi\nu^{1/2}\,\mathrm{sp}_2, \sigma\nu^{-1/2})(f_H),\\
&\operatorname{tr}(\sigma\xi\mathbf{1}_2 \times \sigma\mathbf{1}_2)(f_{C_0})\\
&\qquad = \operatorname{tr} L(\nu\xi, \xi \rtimes \nu^{-1/2}\sigma)(f_H) - \operatorname{tr}\delta^-(\xi\nu^{1/2}\,\mathrm{sp}_2, \xi\sigma\nu^{-1/2})(f_H),\\
&\operatorname{tr} I_G(\sigma\xi\mathbf{1}_2, \sigma\mathbf{1}_2; f \times \theta)\\
&\qquad = \operatorname{tr} L(\nu\xi, \xi \rtimes \nu^{-1/2}\sigma)(f_H) + \operatorname{tr}\delta^-(\xi\nu^{1/2}\,\mathrm{sp}_2, \xi\sigma\nu^{-1/2})(f_H).
\end{aligned}$$

PROOF. We use the identities displayed above for

$$\operatorname{tr} I(\sigma\,\mathrm{sp}_2, \sigma\xi\,\mathrm{sp}_2; f \times \theta) \quad \text{and} \quad \operatorname{tr}(\sigma\,\mathrm{sp}_2 \times \sigma\xi\,\mathrm{sp}_2)(f_{C_0}),$$

and

$$\begin{aligned}
&\operatorname{tr} I_G(\sigma\mathbf{1}_2, \sigma\xi\,\mathrm{sp}_2; f \times \theta) + \operatorname{tr} I_G(\sigma\,\mathrm{sp}_2, \sigma\xi\,\mathrm{sp}_2; f \times \theta)\\
&= \operatorname{tr} I_G(\sigma\nu^{1/2}, \sigma\xi\,\mathrm{sp}_2, \sigma\nu^{-1/2}; f \times \theta)\\
&= \operatorname{tr}(\nu^{1/2}\xi\,\mathrm{sp}_2 \rtimes \sigma\nu^{-1/2})(f_H)\\
&= \operatorname{tr}\delta(\xi\nu^{1/2}\,\mathrm{sp}_2, \sigma\nu^{-1/2})(f_H) + \operatorname{tr} L(\nu^{1/2}\xi\,\mathrm{sp}_2, \sigma\nu^{-1/2})(f_H)\\
&= \operatorname{tr}(\sigma I_2 \times \sigma\xi\,\mathrm{sp}_2)(f_{C_0})\\
&= \operatorname{tr}(\sigma\,\mathrm{sp}_2 \times \sigma\xi\,\mathrm{sp}_2)(f_{C_0}) + \operatorname{tr}(\sigma\mathbf{1}_2 \times \sigma\xi\,\mathrm{sp}_2)(f_{C_0}).
\end{aligned}$$

For the last two identities we use the first two, and

$$\begin{aligned}
&\operatorname{tr} I_G(\sigma\mathbf{1}_2, \sigma\xi\mathbf{1}_2; f \times \theta) + \operatorname{tr} I_G(\sigma\,\mathrm{sp}_2, \sigma\xi\mathbf{1}_2; f \times \theta)\\
&= \operatorname{tr} I_G(\sigma\nu^{1/2}, \sigma\xi\mathbf{1}_2, \sigma\nu^{-1/2}; f \times \theta) = \operatorname{tr}(\xi\nu^{1/2}\mathbf{1}_2 \rtimes \sigma\nu^{-1/2})(f_H)\\
&= \operatorname{tr} L(\xi\nu^{1/2}\,\mathrm{sp}_2, \sigma\xi\nu^{-1/2})(f_H) + \operatorname{tr} L(\nu\xi, \xi \rtimes \sigma\nu^{-1/2})(f_H)\\
&= \operatorname{tr}(\sigma I_2 \times \sigma\xi\mathbf{1}_2)(f_{C_0})\\
&= \operatorname{tr}(\xi\sigma\mathbf{1}_2 \times \sigma\mathbf{1}_2)(f_{C_0}) + \operatorname{tr}(\xi\sigma\mathbf{1}_2 \times \sigma\,\mathrm{sp}_2)(f_{C_0}).
\end{aligned}$$

□

8.6 PROPOSITION. *For all matching $f_H$, $f_{C_0}$, and characters $\xi$ of $F^\times$ with $\xi^2 = 1$ we have*

$$\begin{aligned}
&\operatorname{tr}(\xi \mathbf{1}_2 \times \xi \operatorname{sp}_2)(f_{C_0}) \\
&\qquad = \operatorname{tr} L(\nu^{1/2} \operatorname{sp}_2, \xi\nu^{-1/2})(f_H) + \operatorname{tr} \tau(\nu^{1/2}\mathbf{1}_2, \xi\nu^{-1/2})(f_H), \\
&\operatorname{tr}(\xi\ \mathbf{1}_2 \times \xi\mathbf{1}_2)(f_{C_0}) \\
&\qquad = \operatorname{tr} L(\nu, 1 \rtimes \xi\nu^{-1/2})(f_H) - \operatorname{tr} L(\nu^{1/2} \operatorname{sp}_2, \xi\nu^{-1/2})(f_H).
\end{aligned}$$

PROOF. The first equality follows from

$$\begin{aligned}
&\operatorname{tr}(\xi \operatorname{sp}_2 \times \xi \operatorname{sp}_2)(f_{C_0}) \\
&\qquad = \operatorname{tr} \tau(\nu^{1/2} \operatorname{sp}_2, \xi\nu^{-1/2})(f_H) - \operatorname{tr} \tau(\nu^{1/2}\mathbf{1}_2, \xi\nu^{-1/2})(f_H)
\end{aligned}$$

and

$$\begin{aligned}
&\operatorname{tr}(\xi\mathbf{1}_2 \times \xi \operatorname{sp}_2)(f_{C_0}) + \operatorname{tr}(\xi \operatorname{sp}_2 \times \xi \operatorname{sp}_2)(f_{C_0}) = \operatorname{tr}(\xi I_2 \times \xi \operatorname{sp}_2)(f_{C_0}) \\
&= \operatorname{tr}(\nu^{1/2} \operatorname{sp}_2 \rtimes \xi\nu^{-1/2})(f_H) \\
&= \operatorname{tr} \tau(\nu^{1/2} \operatorname{sp}_2, \xi\nu^{-1/2})(f_H) + \operatorname{tr} L(\nu^{1/2} \operatorname{sp}_2, \xi\nu^{-1/2})(f_H).
\end{aligned}$$

The second equality follows from this as well as from

$$\begin{aligned}
&\operatorname{tr}(\xi\mathbf{1}_2 \times \xi\mathbf{1}_2)(f_{C_0}) + \operatorname{tr}(\xi \operatorname{sp}_2 \times \xi\mathbf{1}_2)(f_{C_0}) = \operatorname{tr}(\xi I_2 \times \xi\mathbf{1}_2)(f_{C_0}) \\
&= \operatorname{tr}(\nu^{1/2}\mathbf{1}_2 \rtimes \xi\nu^{-1/2})(f_H) \\
&= \operatorname{tr} \tau(\nu^{1/2}\mathbf{1}_2, \xi\nu^{-1/2})(f_H) + \operatorname{tr} L(\nu, 1 \rtimes \xi\nu^{-1/2})(f_H). \qquad \square
\end{aligned}$$

Recall (Proposition V.1.2) that for any admissible representation $\pi$ of PGL$(2, F)$ we have that $1 \rtimes \pi$ $\lambda$-lifts to $I_G(\pi, \pi)$, thus

$$\operatorname{tr} I_G(\pi, \pi; f \times \theta) = \operatorname{tr}(1 \rtimes \pi)(f_H)$$

for all matching $f$ and $f_H$. When $\pi = \pi_1 + \pi_2$ we get

$$\begin{aligned}
&\operatorname{tr} I_G(\pi_1, \pi_1; f \times \theta) + \operatorname{tr} I_G(\pi_1, \pi_2; f \times \theta) \\
&+ \operatorname{tr} I_G(\pi_2, \pi_1; f \times \theta) + \operatorname{tr} I_G(\pi_2, \pi_2; f \times \theta) \\
&= \operatorname{tr} I_G(\pi, \pi; f \times \theta) = \operatorname{tr}(1 \rtimes \pi)(f_H) = \operatorname{tr}(1 \rtimes \pi_1)(f_H) + \operatorname{tr}(1 \rtimes \pi_2)(f_H) \\
&= \operatorname{tr} I_G(\pi_1, \pi_1; f \times \theta) + \operatorname{tr} I_G(\pi_2, \pi_2; f \times \theta).
\end{aligned}$$

It follows that the normalization of $\Pi(\theta)$ on $\Pi = I_G(\pi, \pi)$, which is unique only up to a sign on any irreducible $\theta$-invariant representation of $G$, as $\theta^2 = 1$, induces a normalization of $\Pi_{2,1}(\theta)$, $\Pi_{2,1} = I_G(\pi_2, \pi_1)$, which is different in sign than the normalization of $\Pi_{1,2}(\theta)$, $\Pi_{1,2} = I_G(\pi_1, \pi_2)$ ($\simeq I_G(\pi_2, \pi_1)$ when $\pi_1$, $\pi_2$ are irreducible), with the consequence of

$$\operatorname{tr} I_G(\pi_1, \pi_2; f \times \theta) + \operatorname{tr} I_G(\pi_2, \pi_1; f \times \theta) = 0$$

for all $f$. A similar phenomenon is encountered in the following.

8.7 PROPOSITION. *For all matching $f$ and $f_H$ we have*

$$\begin{aligned}
&\operatorname{tr} I_G(\mathbf{1}_2, \mathbf{1}_2; f \times \theta) \\
&\qquad = \operatorname{tr} L(\nu, 1 \rtimes \nu^{-1/2})(f_H) + \operatorname{tr} L(\nu^{1/2} \operatorname{sp}_2, \nu^{-1/2})(f_H), \\
&\operatorname{tr} I_G(\operatorname{sp}_2, \mathbf{1}_2; f \times \theta) = \\
&\qquad \operatorname{tr} \tau(\nu^{1/2}\mathbf{1}_2, \nu^{-1/2})(f_H) - \operatorname{tr} L(\nu^{1/2} \operatorname{sp}_2, \nu^{-1/2})(f_H).
\end{aligned}$$

PROOF. The first identity follows from $\lambda(1 \rtimes \pi_1) = I_G(\pi_1, \pi_1)$ and the fact that the composition series of $1 \rtimes \pi_1$ for $\pi_1 = \mathbf{1}_2$ consists of the two irreducible representations $L$. The second identity is a consequence of the first, as well as

$$\begin{aligned}
&\operatorname{tr} \tau(\nu^{1/2}\mathbf{1}_2, \nu^{-1/2})(f_H) + \operatorname{tr} L(\nu, 1 \rtimes \nu^{-1/2})(f_H) = \operatorname{tr}(\nu^{1/2}\mathbf{1}_2 \rtimes \nu^{-1/2})(f_H) \\
&= \operatorname{tr} I_G(I_2, \mathbf{1}_2; f \times \theta) = \operatorname{tr} I_G(\mathbf{1}_2, \mathbf{1}_2; f \times \theta) + \operatorname{tr} I_G(\operatorname{sp}_2, \mathbf{1}_2; f \times \theta). \qquad \square
\end{aligned}$$

REMARK. On $\Pi = I_G(I_2, \mathbf{1}_2)$ we normalize the intertwining operator $\Pi(\theta)$, whose square is the identity, by the property that it maps the unramified ($K$-fixed) vector to itself. This coincides with the normalization of $\theta$ on the quotient $I(\mathbf{1}_2 \times \mathbf{1}_2)$ of $I_G(I_2, \mathbf{1}_2)$, and induces a normalization of $\theta$ on the subrepresentation $I_G(\operatorname{sp}_2, \mathbf{1}_2)$.

On the other hand, we could normalize $\Pi'(\theta)$ on $\Pi' = I_G(I_2, \operatorname{sp}_2)$ by mapping the Whittaker vector to itself ($W \mapsto {}^\theta W$). This coincides with the normalization of $\theta$ on the subrepresentation $I_G(\operatorname{sp}_2, \operatorname{sp}_2)$ of $\Pi'$, and induces a normalization of $\theta$ on the quotient $I_G(\mathbf{1}_2, \operatorname{sp}_2)$ of $\Pi'$ which is the *negative* of the normalization of $\theta$ on

$$I_G(\operatorname{sp}_2, \mathbf{1}_2) \quad (\simeq I_G(\mathbf{1}_2, \operatorname{sp}_2))$$

viewed as a subrepresentation of $\Pi$. Indeed, using

$$\operatorname{tr} I_G(\mathrm{sp}_2, \mathrm{sp}_2; f \times \theta) = \operatorname{tr} \tau(\nu^{1/2} \mathrm{sp}_2, \nu^{-1/2})(f_H) + \operatorname{tr} \tau(\nu^{1/2} \mathbf{1}_2, \nu^{-1/2})(f_H)$$

and

$$\begin{aligned} &\operatorname{tr} \tau(\nu^{1/2} \mathrm{sp}_2, \nu^{-1/2})(f_H) + \operatorname{tr} L(\nu^{1/2} \mathrm{sp}_2, \nu^{-1/2})(f_H) \\ &= \operatorname{tr}(\nu^{1/2} \mathrm{sp}_2 \rtimes \nu^{-1/2})(f_H) = \operatorname{tr} I_G(I_2, \mathrm{sp}_2; f \times \theta) \\ &= \operatorname{tr} I_G(\mathrm{sp}_2, \mathrm{sp}_2; f \times \theta) + \operatorname{tr} I_G(\mathbf{1}_2, \mathrm{sp}_2; f \times \theta) \end{aligned}$$

we conclude that

$$\operatorname{tr} I_G(\mathbf{1}_2, \mathrm{sp}_2; f \times \theta) = \operatorname{tr} L(\nu^{1/2} \mathrm{sp}_2, \nu^{-1/2})(f_H) - \operatorname{tr} \tau(\nu^{1/2} \mathbf{1}_2, \nu^{-1/2})(f_H).$$

This does not contradict the Proposition, but reinforces it, yet with a different normalization of the intertwining operator $\theta$ on $I_G(\mathbf{1}_2, \mathrm{sp}_2)$.

## 9. Local Packets

These character relations permit us to define the notion of a packet of tempered representations, and that of a quasi-packet, locally. The packet of a nontempered representation $\pi_H$ is defined to consist of $\pi_H$ alone.

9.1 DEFINITION. Let $F$ be a local field. The *packet* of (an irreducible) tempered $H$-module $\pi_H$ consists of $\pi_H$ alone unless $\pi_H$ is $\pi_H^+$ or $\pi_H^-$ for some pair $\pi_1$, $\pi_2$ of (irreducible) square integrable $\mathrm{PGL}(2,F)$-modules, in which case the packet consists of $\pi_H^+$ and $\pi_H^-$.

For example, if $\pi_2$ is a cuspidal representation of $\mathrm{GL}(2,F)$ with central character $\xi \neq 1 = \xi^2$, the packet of $\delta(\xi\nu, \nu^{-1/2}\pi_2)$ consists of a single element.

We write $\operatorname{tr}\{\pi_H\}$ for the sum of $\operatorname{tr} \pi'_H$ as $\pi'_H$ ranges over the packet $\{\pi_H\}$ of $\pi_H$.

9.2 DEFINITION. The *quasi-packet* of a nontempered (irreducible) $H$-module $\pi_H$ is defined only for such an $H$-module which occurs in the character relation for $\sigma\mathbf{1}_2 \times \pi_2$, where $\pi_2$ is a square integrable or one dimensional $\mathrm{PGL}(2,F)$-module and $\sigma$ is a character of $F^\times/F^{\times 2}$. It is defined to be the pair $\pi_H^\times$, $\pi_H^-$ which occurs in this character relation (which also defines $\pi_H^\times$).

Thus when $\pi_2$ is square integrable the quasi-packets are defined to be

$$\{L(\pi_2\sigma\nu^{1/2}, \sigma\nu^{-1/2}), \delta^-(\pi_2\sigma\nu^{1/2}, \sigma\nu^{-1/2})\},$$

$$\{L(\nu^{1/2}\xi\,\mathrm{sp}_2, \sigma\nu^{-1/2}), \delta^-(\xi\nu^{1/2}\,\mathrm{sp}_2, \sigma\nu^{-1/2})\}$$

and

$$\{L(\nu^{1/2}\,\mathrm{sp}_2, \sigma\nu^{-1/2}), \tau(\nu^{1/2}\mathbf{1}_2, \sigma\nu^{-1/2})\},$$

for any characters $\xi \neq 1$, $\sigma$ of $F^\times/F^{\times 2}$ and cuspidal $\pi_2$. Note that the $\pi_H^-$ in the last packet is tempered, but not square integrable.

Correspondingly we write $\lambda_0(\pi_1 \times \pi_2) = \{\pi_H^+, \pi_H^-\}$ and $\lambda(\{\pi_H^+, \pi_H^-\}) = I_G(\pi_1, \pi_2)$ when $\pi_1$, $\pi_2$ are square integrable, $\lambda_0(\sigma\mathbf{1}_2 \times \pi_2) = \{\pi_H^\times, \pi_H^-\}$ and $\lambda(\{\pi_H^\times, \pi_H^-\}) = I_G(\sigma\mathbf{1}_2, \pi_2)$ when $\pi_2$ is square integrable and $\sigma^2 = 1$. This notation applies also when $\pi_2$ is $\mathrm{sp}_2$, or $\xi\mathbf{1}_2$, in the following sense.

The quasi-packet $\lambda_0(\sigma\xi\mathbf{1}_2 \times \sigma\mathbf{1}_2)$, $\xi \neq 1 = \xi^2$, $\sigma^2 = 1$, is defined to consist of

$$\{\pi_H^\times = L(\nu\xi, \xi \rtimes \nu^{-1/2}\sigma), \qquad \pi_H^- = \delta^-(\xi\nu^{1/2}\,\mathrm{sp}_2, \xi\sigma\nu^{-1/2})\}.$$

We observe that $\pi_H^-$ of $\lambda_0(\sigma\xi\mathbf{1}_2 \times \sigma\mathbf{1}_2)$ and of $\lambda_0(\sigma\xi\mathbf{1}_2 \times \sigma\,\mathrm{sp}_2)$ are the same, although the corresponding $\pi_H^\times$ are not. Thus *it is the* $\pi_H^\times$ *which determines the quasi-packet, and not the* $\pi_H^-$.

The quasi-packet $\lambda_0(\sigma\mathbf{1}_2 \times \sigma\mathbf{1}_2)$, $\sigma^2 = 1$, consists of

$$\{\pi_H^\times = L(\nu, 1 \rtimes \sigma\nu^{-1/2}), \qquad \pi_H^- = L(\nu^{1/2}\,\mathrm{sp}_2, \sigma\nu^{-1/2})\}.$$

Here we observe our $\pi_H^-$ is not tempered, and is in fact $\pi_H^\times$ in the quasi-packet $\lambda_0(\sigma\mathbf{1}_2 \times \sigma\,\mathrm{sp}_2)$.

# 10. Global Packets

This description of local packets of representations of $H$ will now be used together with the trace formula identity to describe the automorphic representations of $\mathbf{H}(\mathbb{A})$. Taking into account the complete results on the lifting $\lambda_1$ from $\mathbf{C}(\mathbb{A})$ to $\mathbf{G}(\mathbb{A})$, the trace formula identity can be phrased as follows:

$$I' + \frac{1}{2}I'_{(2,2)} = \mathrm{T_{sp}}(f_H, \mathbf{H}) - \frac{1}{4}\mathrm{T_{sp}}(f_{C_0}, \mathbf{C}_0)$$

$$+\frac{1}{4}\sum_{\pi_2 \text{ of } \mathrm{PGL}(2)} \mathrm{tr}(I_2(1,1)\times\pi_2)(f_C).$$

Here $I'$ is the subsum of $I$, namely $\sum_\pi \mathrm{tr}\,\pi(f\times\theta)$, over those discrete spectrum representations $\pi \simeq {}^\theta\pi$ of $\mathrm{PGL}(4,\mathbb{A})$ which are not $\lambda_1$-lifts (from $\mathbf{C}(\mathbb{A})$).

Similarly, $I'_{(2,2)}$ is the subsum of $I_{(2,2)}$ which consists of those induced representations $I_{(2,2)}(\pi_1,\pi_2)$ which are not lifts via $\lambda_1$ from $\mathbf{C}(\mathbb{A})$.

10.1 Lemma. *If $I_{(2,2)}(\pi_1,\pi_2)$ appears in $I'_{(2,2)}$ then the $\pi_i$ have trivial central characters, namely are representations of* $\mathrm{PGL}(2,\mathbb{A})$.

Proof. The $\pi_1$ and $\pi_2$ are representations of $\mathrm{GL}(2,\mathbb{A})$ with $\pi_i \simeq \check{\pi}_i$. If $\omega$ denotes the central character of $\pi_1$ (hence also of $\pi_2$, since $I_{(2,2)}(\pi_1,\pi_2)$ has trivial central character), then $\check{\pi}_i \simeq \omega\pi_i$, and so $\omega^2 = 1$. If $\omega \neq 1$, thus $\omega = \chi_{E/F}$ for some quadratic extension $E$ of $F$, then $\pi_1 = \pi_E(\mu'_1)$ and $\pi_2 = \pi_E(\mu'_2)$, where $\mu'_i$ are characters of $\mathbf{A}_E^\times/E^\times$.

The central character of such an $\pi_E(\mu)$ is $\chi_{E/F}\cdot\mu|\mathbb{A}^\times$. So for our $\pi_E(\mu'_i)$ of central character $\chi_{E/F}$, we have $\mu'_i|\mathbb{A}^\times = 1$, which means (since the kernel of $z\mapsto z/\overline{z}$ in $\mathbb{A}_E^\times$ is $\mathbb{A}^\times$) that there are $\mu_1$, $\mu_2$, characters of $\mathbb{A}_E^\times$, with $\mu'_i(z) = \mu_i(z/\overline{z})$. Now

$$\lambda_1(\pi_E(\mu)\times\pi_E(\mu')) = I_{(2,2)}(\pi_E(\mu\overline{\mu}'),\pi_E(\mu\mu')),$$

so

$$\lambda_1(\pi_E(\mu_1\mu_2)\times\pi_E(1/\mu_1\overline{\mu}_2)) = I_{(2,2)}(\pi_E(\mu_1/\overline{\mu}_1),\pi_E(\mu_2/\overline{\mu}_2)).$$

Hence the $I_{(2,2)}(\pi_1,\pi_2)$ with $\omega_{\pi_i}\neq 1$ are lifts from $\mathbf{C}(\mathbb{A})$ via $\lambda_1$. The lemma follows. □

We shall use – as usual – the form of the trace formula identity where the local component is fixed to be a fixed unramified representation at all places outside a finite set.

By the rigidity theorem for PGL(4) at most one of $I'$ and $I'_{(2,2)}$ would have a (single nonzero) contribution.

Let $\pi_1\times\pi_2$ be a discrete spectrum representation of the group $\mathbf{C}_0(\mathbb{A}) = \mathrm{PGL}(2,\mathbb{A})\times\mathrm{PGL}(2,\mathbb{A})$. It makes a contribution in $\mathrm{T_{sp}}(f_{C_0},\mathbf{C}_0)$ as well as in $I'_{(2,2)}$.

**10.2** *Suppose first that* $\pi_2 = \pi_1$. Then the contribution to $\frac{1}{2}I'_{(2,2)}$ is

$$\frac{1}{4}\operatorname{tr} I_G(\pi_2, \pi_2; f \times \theta).$$

This is equal to the contribution

$$\frac{1}{4}\operatorname{tr}(I_2(1,1) \times \pi_2)(f_C)$$

to the trace formula of $C$ (see Proposition IV 3.1). Thus these two cancel each other (of course for matching $f$, $f_C$, and $f_H$, $f_{C_0}$ below).

The corresponding contribution (determined by fixing all unramified components) to the trace formula of $H$ is

$$\frac{1}{4}\prod_v \operatorname{tr} R_v \circ (1 \rtimes \pi_{2v})(f_{Hv}).$$

The corresponding contribution to the trace formula of $C_0$ is

$$\frac{1}{4}\prod_v \operatorname{tr}(\pi_{2v} \times \pi_{2v})(f_{C_0,v}).$$

At all places $v$ where $\pi_{2v}$ is properly induced (and irreducible), $R_v$ is the scalar 1, and $\operatorname{tr}(1 \rtimes \pi_{2v})(f_{Hv}) = \operatorname{tr}(\pi_{2v} \times \pi_{2v})(f_{C_0,v})$, as $\pi_{2v}$ is a representation of $\mathrm{PGL}(2, F_v)$ (see Proposition V 1.2).

If $\pi_{2v}$ is square integrable (or one dimensional), our local results (Propositions V 5 and 2.3(b) for square integrable $\pi_{2v}$, Propositions V 8.6 and 2.1(d) for one dimensional $\pi_{2v}$) assert that the two constituents of the composition series of $1 \rtimes \pi_{2v}$ can be labeled $\pi_{Hv}^+$ and $\pi_{Hv}^-$ ($= L(\nu, 1 \rtimes \sigma\nu^{-1/2})$ and $L(\nu^{1/2}\mathrm{sp}_2, \sigma\nu^{-1/2})$ when $\pi_{2v}$ is one dimensional $\sigma\mathbf{1}_2$), such that for matching functions

$$\operatorname{tr}(\pi_{2v} \times \pi_{2v})(f_{C_0,v}) \quad \text{is} \quad \operatorname{tr}\pi_{Hv}^+(f_{Hv}) - \operatorname{tr}\pi_{Hv}^-(f_{Hv}).$$

Moreover, $R_v$ acts on $\pi_{Hv}^+$ as 1 and on $\pi_{Hv}^-$ as $-1$ (this follows for example from the global comparison). Hence these contributions to the formula of $H$ and of $C_0$ cancel each other.

**10.3** We can then *assume that* $\pi_1 \neq \pi_2$. *Suppose that* $\pi_1$, $\pi_2$ *are discrete spectrum representations of* $\mathrm{PGL}(2, \mathbb{A})$. Note that the pairs $(\pi_1, \pi_2)$ and

$(\pi_2, \pi_1)$ make the same contribution to the formulae of $C_0$ and of $G$ (in $I'_{(2,2)}$), hence the coefficient $\frac{1}{4}$ is replaced by $\frac{1}{2}$.

When $\pi_1$ and $\pi_2$ are cuspidal the corresponding part of the trace formulae identity asserts

$$\sum m(\pi_H) \prod_v \operatorname{tr} \pi_{Hv}(f_{Hv})$$

$$(1) \qquad = \frac{1}{2} \prod_v \operatorname{tr}(\pi_{1v} \times \pi_{2v})(f_{C_0,v}) + \frac{1}{2} \prod_v \operatorname{tr} I_G(\pi_{1v}, \pi_{2v}; f_v \times \theta).$$

The products are over the finite set $V$ of places where both $\pi_{1v}$ and $\pi_{2v}$ are square integrable. The sum ranges over all equivalence classes of irreducible discrete spectrum representations $\pi_H$ of $\mathbf{H}(\mathbb{A})$ ($\pi_H$ occurs with multiplicity $m(\pi_H) \geq 1$ in the discrete spectrum) whose component at each $v$ outside $V$ is $\lambda_0(\pi_{1v} \times \pi_{2v})$. Recall that

$$\lambda_0(I(\mu_1, \mu_1^{-1}) \times \pi_2) = \mu_1 \pi_2 \rtimes \mu_1^{-1} \quad \text{and} \quad \lambda(\mu_1 \pi_2 \rtimes \mu_1^{-1}) = I_G(\mu_1, \pi_2, \mu_1^{-1}).$$

Now at the places $v$ in $V$ the representations $\pi_{1v}$, $\pi_{2v}$ are square integrable and the character relations permit us to rewrite the right side of the formula as

$$= \frac{1}{2} \prod_{v \in V} [\operatorname{tr} \pi^+_{Hv}(f_{Hv}) - \operatorname{tr} \pi^-_{Hv}(f_{Hv})] + \frac{1}{2} \prod_{v \in V} [\operatorname{tr} \pi^+_{Hv}(f_{Hv}) + \operatorname{tr} \pi^-_{Hv}(f_{Hv})],$$

where $\pi^{\pm}_{Hv} = \pi^{\pm}_{Hv}(\pi_{1v} \times \pi_{2v})$ are the tempered representations of $H_v$ determined by $\pi_{1v}$ and $\pi_{2v}$. It follows that the discrete spectrum representations $\pi_H$ of $\mathbf{H}(\mathbb{A})$ with components $\lambda_0(\pi_{1v} \times \pi_{2v})$ at all $v \notin V$ have components

$$\pi^+_{Hv}(\pi_{1v} \times \pi_{2v}) \quad \text{or} \quad \pi^-_{Hv}(\pi_{1v} \times \pi_{2v})$$

at all places $v \in V$, and the multiplicity $m(\pi_H)$ of such $\pi_H = \otimes \pi_{Hv}$ in the discrete spectrum of $\mathbf{H}(\mathbb{A})$ is

$$m(\pi_H) = \frac{1}{2}(1 + (-1)^{n(\pi_H)}),$$

where $n(\pi_H)$ is the number of components of $\pi_H$ of the form $\pi^-_{Hv}$. The $\pi_H$ with $m(\pi_H) = 1$ are all cuspidal as there are no residual representations with components $\lambda_0(\pi_{1v} \times \pi_{2v})$ for almost all $v$.

In fact since we work with test functions $f = \otimes f_v$ with 3 elliptic components we can deduce only a weaker statement, which applies only when the set $V$ has at least 3 members. Namely we cannot exclude the possibility that there exist discrete spectrum $\pi_H$ with properly induced components at all $v \in V$ (and components $\lambda_0(\pi_{1v} \times \pi_{2v})$ at all $v \notin V$). So our global results be complete only after removal of the 3-places constraint on the test functions of $f$, $f_H$.

**10.4** Next we deal with the case where $\pi_2$ *is cuspidal but* $\pi_1$ *is one dimensional,* $\xi\mathbf{1}_2$, $\xi$ is a character of $\mathbb{A}^\times/F^\times\mathbb{A}^{\times 2}$. The trace formula identity reduces to

$$\sum m(\pi_H)\prod_v \operatorname{tr}\pi_{Hv}(f_{Hv})$$

$$=\frac{1}{2}\prod_v \operatorname{tr}(\xi_v\mathbf{1}_2\times\pi_{2v})(f_{C_0,v})+\frac{1}{2}\varepsilon(\xi\mathbf{1}_2\times\pi_2)\prod_v \operatorname{tr} I_G(\xi_v\mathbf{1}_2,\pi_{2v};f_v\times\theta),$$

where the product ranges over the set $V$ of places where $\pi_{2v}$ is square integrable, and the sum ranges over the discrete spectrum $\pi_H$ whose component $\pi_{Hv}$ at $v \notin V$ is

$$\pi_{Hv}^\times = \lambda_0(I(\mu_{1v},\mu_{1v}^{-1})\times\xi_v\mathbf{1}_2) = \mu_{1v}\xi_v\mathbf{1}_2 \rtimes \mu_{1v}^{-1} \;\text{ if }\; \pi_{2v} = I(\mu_{1v},\mu_{1v}^{-1}).$$

Note that the involution $\theta$ defined by $\theta(g) = J^{-1t}g^{-1}J$ on $\mathbf{G}(\mathbb{A})$ and its automorphic forms, induces an involution $\pi(\theta)$ on each automorphic representation. However, abstractly there are two choices of an intertwining operator $\pi \xrightarrow{\sim} {}^\theta\pi$ whose square $(\pi \xrightarrow{\sim} \pi)$ is 1, and they differ by a sign.

We observe that on a generic representation $\pi$, the global involution equals the product of the local involutions $\pi_v(\theta)$ which act on the Whittaker functions of $\pi_v$ by $\theta$. This coincides with the choice of the intertwining operator $\pi_v \xrightarrow{\sim} {}^\theta\pi_v$, when $\pi_v$ is unramified, which maps the $K_v$-fixed vector to itself. Our representation $\pi = I(\xi\mathbf{1}_2, \pi_2)$ is not generic, nor it is everywhere unramified (unless so is $\pi_2$).

Hence the global involution $\pi(\theta)$ is the product of the local involutions $\pi_v(\theta)$, and a sign, which we denote by $\varepsilon(\xi\mathbf{1}_2 \times \pi_2)$. The presence of this sign was first noticed in a different context by G. Harder ([Ha], p. 173).

Our local character relations express $\operatorname{tr}(\xi_v\mathbf{1}_2 \times \pi_{2v})(f_{C_0,v})$ as the sum of traces at $f_H$ of the nontempered constituent

$$\pi_{Hv}^\times = L(\xi_v\nu_v^{1/2}\pi_{2v}, \xi_v\nu_v^{-1/2})$$

of the indicated induced $H_v$-module, and of a cuspidal (if $\pi_{2v}$ is), square integrable (if $\pi_{2v} = \xi'_v \operatorname{sp}_{2v}, \xi'_v \neq \xi_v$) or tempered (if $\pi_{2v} = \xi_v \operatorname{sp}_{2v}$) representation $\pi^-_{Hv}$. The trace

$$\operatorname{tr} I_G(\xi_v \mathbf{1}_2, \pi_{2v}; f_v \times \theta)$$

is the difference of these two traces. Thus

$$= \frac{1}{2} \prod_v [\operatorname{tr} \pi^\times_{Hv}(f_{Hv}) + \operatorname{tr} \pi^-_{Hv}(f_{Hv})]$$

$$+ \frac{1}{2} \varepsilon(\xi \mathbf{1}_2 \times \pi_2) \prod_v [\operatorname{tr} \pi^\times_{Hv}(f_{Hv}) - \operatorname{tr} \pi^-_{Hv}(f_{Hv})].$$

We conclude that if there is a discrete spectrum $\pi_H$ with components $\lambda_0(\xi_v \mathbf{1}_2 \times \pi_{2v})$ at all places where $\pi_{2v}$ is fully induced, then its component at each $v$ in the remaining finite set $V$ lies in the quasi-packet $\{\pi^\times_{Hv}, \pi^-_{Hv}\}$. Its multiplicity is

$$m(\pi_H) = \frac{1}{2}[1 + \varepsilon(\xi \mathbf{1}_2 \times \pi_2)(-1)^{n(\pi_H)}],$$

where $n(\pi_H)$ is the number of components $\pi^-_{Hv}$ in $\pi_H$.

10.5 Lemma. *For any cuspidal $\pi_2$ and quadratic character $\xi$ we have $\varepsilon(\xi \mathbf{1}_2 \times \pi_2) = \varepsilon(\xi\pi_2, \frac{1}{2})$.*

Proof. Here $\varepsilon(\pi_2, s)$ is the epsilon factor in the functional equation of the $L$-function of $\pi_2$. Note that $\varepsilon(\xi \mathbf{1}_2 \times \pi_2)$ is 1 iff $\pi^\times_{Hv} = \otimes_v \pi^\times_{Hv}$ is discrete spectrum. It is known from the theory of Eisenstein series ([A2], p. 32; [Kim], Theorem 7.1) that this representation is residual, namely discrete spectrum and generated by residues of Eisenstein series, precisely when the $L$-function $L(\xi\pi_2, s)$ of $\xi\pi_2$ is nonzero at $s = \frac{1}{2}$.

The case of $\xi \neq 1$ reduces to that of $\xi = 1$ as

$$\xi\nu^{1/2}\pi_2 \rtimes \xi\nu^{-1/2} = \xi(\nu^{1/2}\xi\pi_2 \rtimes \nu^{-1/2}).$$

To repeat: in this case where $L(\xi\pi_2, \frac{1}{2}) \neq 0$, $\varepsilon(\xi \mathbf{1}_2 \times \pi_2)$ is 1, as is $\varepsilon(\xi\pi_2, \frac{1}{2})$. To determine, when $L(\xi\pi_2, \frac{1}{2}) = 0$, whether the quotient $\pi^\times_H$ of $\xi\nu^{1/2}\pi_2 \rtimes \xi\nu^{-1/2}$ is cuspidal or not, we appeal to the theory of the theta correspondence. As noted above, it suffices to assume $\xi = 1$. Then $\varepsilon(\xi\pi_2, \frac{1}{2}) = 1$ implies that $\pi_H^\times$ is cuspidal by [W2], Proposition 24, p. 305. The converse follows from [PS1], Theorem 2.2 ($1 \Rightarrow 4$).

To explain this, recall that the theta $\theta = \theta_\psi$ and the Waldspurger's Wald $=$ Wald$_\psi$ correspondences depend on a nontrivial additive character $\psi : \mathbb{A} \bmod F \to \mathbb{C}^1$, which we now fix. These correspondences fit in the chart:

$$\begin{array}{ccccc} \mathrm{PGSp}(2,\mathbb{A}) = \mathrm{SO}(5,\mathbb{A}) & \overset{\theta}{\leftarrow} & \widetilde{\mathrm{SL}}(2,\mathbb{A}) & \overset{\mathrm{Wald}}{\longrightarrow} & \mathrm{PGL}(2,\mathbb{A}) = \mathrm{SO}(3,\mathbb{A}) \\ & & \| & & \uparrow \mathrm{JL} \\ & & \widetilde{\mathrm{SL}}(2,\mathbb{A}) & \overset{\theta}{\leftarrow} & D_{\mathbb{A}}^\times \end{array}$$

Here $\widetilde{\mathrm{SL}}(2,\mathbb{A})$ is the metaplectic two fold covering group of $\mathrm{SL}(2,\mathbb{A})$, and JL denotes the Jacquet-Langlands correspondence from the multiplicative group $D_{\mathbb{A}}^\times$ of the quaternion algebra $D_{\mathbb{A}}$. Given $\pi_1 = \otimes_v \pi_{1v}$ on $\mathrm{PGL}(2,\mathbb{A})$, $\mathrm{Wald}^{-1}(\pi_{1v})$ is $\tilde{\pi}_{v,\mathrm{gen}}$ if $\pi_{1v}$ is principal series, $\{\tilde{\pi}_{v,\mathrm{gen}}, \tilde{\pi}_{v,\mathrm{ng}}\}$ if $\pi_{1v}$ is discrete series. Here $\tilde{\pi}_{v,\mathrm{gen}}$ is the $\theta$-image of $\pi_{1v}$ while $\tilde{\pi}_{v,\mathrm{ng}}$ is the $\theta$-image of $\pi_{1v}^D = \mathrm{JL}^{-1}(\pi_{1v})$ at the places $v$ where $D$ ramifies. The product $\otimes_v \tilde{\pi}_{v,\mathrm{gen}}$ defines a representation of $\widetilde{\mathrm{SL}}(2,\mathbb{A})$ when $\varepsilon(\pi_1, \frac{1}{2}) = 1$, and the $\theta$-lifting $\widetilde{\mathrm{SL}}(2,\mathbb{A}) \to \mathrm{PGSp}(2,\mathbb{A})$ maps $\otimes_v \tilde{\pi}_{v,\mathrm{gen}}$ to

$$\pi_H^\times = \otimes_v \pi_{Hv}^\times = L(\nu^{1/2}\pi_1, \nu^{-1/2}).$$

This $\pi_H^\times$ is cuspidal when $L(\pi_1, \frac{1}{2}) = 0$ and $\varepsilon(\pi_1, \frac{1}{2}) = 1$ by [W2], Proposition 24, p. 305.

Now suppose that $\pi_H^\times$ is cuspidal. Then $L(\pi_1, \frac{1}{2}) = 0$. We claim that $\varepsilon(\pi_1, \frac{1}{2}) = 1$. By definition, $\pi_H^\times$ is in $\Omega_P$ of [PS1], p. 315. Hence there is a cuspidal irreducible representation $\sigma$ of $\widetilde{\mathrm{SL}}(2,\mathbb{A})$ which $\theta$-lifts to $\pi_H^\times$ by [PS1], Theorem 2.2 ($1 \Rightarrow 4$). Moreover $\mathrm{Wald}(\sigma) = \pi_1$ by the rigidity theorem for $\mathrm{GL}(2,\mathbb{A})$. If $\varepsilon(\pi_1, \frac{1}{2}) = -1$, the representation $\sigma$ in $\mathrm{Wald}^{-1}(\pi_1)$ which $\theta$-lifts to $\mathrm{PGSp}(2,\mathbb{A})$ must have a component $\tilde{\pi}_{v,\mathrm{ng}}$: it cannot have the component $\tilde{\pi}_{v,\mathrm{gen}}$ at all places. But the local $\theta$-lift takes $\tilde{\pi}_{v,\mathrm{ng}}$ to a tempered representation of $\mathrm{PGSp}(2, F_v)$, contradicting the assumption that $\pi_H^\times$, with which we started, has no tempered components.

As already noted, the case of $\xi \neq 1$ follows from this and the equality of $\xi\nu^{1/2}\pi_2 \rtimes \xi\nu^{-1/2}$ and $\xi(\xi\nu^{1/2}\pi_2 \rtimes \nu^{-1/2})$. Thus $\pi_H^\times$ is cuspidal iff $\varepsilon(\xi\pi_2, \frac{1}{2}) = 1$ and $L(\xi\pi_1, \frac{1}{2}) = 0$. It is non discrete series iff $\varepsilon(\xi\pi_2, \frac{1}{2}) = -1$.

In summary: $\varepsilon(\xi\mathbf{1}_2\times\pi_2)=\varepsilon(\xi\pi_2,\frac{1}{2})$. Further details on Waldspurger's correspondence can be found in Schmidt [Sch]. □

**10.6** Similarly, for characters $\xi\neq 1$, $\sigma$ of $\mathbb{A}^\times/F^\times\mathbb{A}^{\times 2}$ we have the following part of the traces identity

$$\sum m(\pi_H)\prod_v \operatorname{tr}\pi_{Hv}(f_{Hv})=\frac{1}{2}\prod_v \operatorname{tr}(\sigma_v\xi_v\mathbf{1}_2\times\sigma_v\mathbf{1}_2)(f_{C_0,v})$$

$$+\frac{1}{2}\varepsilon(\sigma\xi\mathbf{1}_2\times\sigma\mathbf{1}_2)\prod_v \operatorname{tr} I_G(\sigma_v\xi_v\mathbf{1}_2,\sigma_v\mathbf{1}_2;f_v\times\theta).$$

The product ranges over a set $V$ such that $\sigma_v$, $\xi_v$ are unramified for $v\notin V$. The sum ranges over the discrete spectrum $\pi_H$ of $\mathbf{H}(\mathbb{A})$ whose component at $v\notin V$ is $\pi_{Hv}^\times=L(\xi_v\nu_v,\xi_v\rtimes\sigma_v\nu_v^{-1/2})$. We also let $\pi_{Hv}^-$ be the cuspidal

$$\delta^-(\xi_v\nu_v^{1/2}\operatorname{sp}_{2v},\sigma_v\xi_v\nu_v^{-1/2})\quad\text{if}\quad\xi_v\neq 1$$

and

$$L(\nu_v^{1/2}\operatorname{sp}_{2v},\sigma_v\nu_v^{-1/2})\quad\text{if}\quad\xi_v=1.$$

We conclude that for each $v$ the component of $\pi_H$ is $\pi_{Hv}^\times$ or $\pi_{Hv}^-$. The multiplicity is again determined by the formula

$$m(\pi_H)=\frac{1}{2}(1+\varepsilon(\sigma\xi\mathbf{1}_2\times\sigma\mathbf{1}_2)(-1)^{n(\pi_H)}),$$

where $n(\pi_H)$ is the number of components $\pi_{Hv}^-$ of $\pi_H$. The sign $\varepsilon$ here is in fact 1 since

$$\pi_H^\times=\otimes_v\pi_{Hv}^\times=\otimes_v L(\xi_v\nu_v,\xi_v\rtimes\sigma_v\nu_v^{-1/2}),$$

which we denote also by $L(\xi\nu,\xi\rtimes\sigma\nu^{-1/2})$, thus $n(\pi_H)=0$, is discrete spectrum, in fact a residual representation, by [Kim], 7.3(2).

The representations whose components are almost all $\pi_{Hv}^\times$ and which have a cuspidal component $\pi_{Hv}^-$ are cuspidal (if they are automorphic). They make counterexamples to the generalized Ramanujan conjecture, as almost all of their components are the nontempered $\pi_{Hv}^\times$.

With the complete local results on the liftings $\lambda_0$ and $\lambda$, as well as the full description of the global lifting $\lambda_0$ from $\mathbf{C}_0(\mathbb{A})$ to $\mathbf{H}(\mathbb{A})$ and the global lifting $\lambda$ from the image of $\lambda_0$ to the self-contragredient $\mathbf{G}(\mathbb{A})$-modules of type $I(2,2)$ (induced from the maximal parabolic of type (2,2)), we can complete the description of the lifting $\lambda$.

10.7 DEFINITION. The *stable* discrete spectrum of $L^2(\mathbf{H}(F)\backslash\mathbf{H}(\mathbb{A}))$ consists of all discrete spectrum representations $\pi_H$ of $\mathbf{H}(\mathbb{A})$ which are not in the image of the $\lambda_0$-lifting (thus there is no $\mathbf{C}_0(\mathbb{A})$-module $\pi_1 \times \pi_2$ such that $\pi_{Hv}$ is equivalent to $\lambda_0(\pi_{1v} \times \pi_{2v})$ for almost all $v$).

We proceed to describe the stable spectrum of $\mathbf{H}(\mathbb{A})$.

10.8 DEFINITION. Let $F$ be a global field. To define a (quasi-) packet $\{\pi\}$ of automorphic representations of $\mathbf{H}(\mathbb{A})$ we fix a (quasi-) packet $\{\pi_v\}$ of local representations for every place $v$ of $F$, such that $\{\pi_v\}$ contains an unramified representation $\pi_v^0$ for almost all $v$. The global (*quasi-*) *packet* $\{\pi\}$ which is determined by the local $\{\pi_v\}$ consists by definition of all products $\otimes\pi_v$ with $\pi_v$ in $\{\pi_v\}$ for all $v$ and $\pi_v = \pi_v^0$ for almost all $v$. Put in other words, the (quasi-) packet of an irreducible representation $\pi = \otimes\pi_v$ of $\mathbf{H}(\mathbb{A})$ consists of all products $\otimes\pi'_v$ where $\pi'_v$ is in the (quasi-) packet of $\pi_v$ and $\pi'_v = \pi_v$ for almost all $v$.

If a (quasi-) packet contains an automorphic member, its other members are not necessarily automorphic, as we saw in the case of $\lambda_0(\pi_1 \times \pi_2)$. Thus the multiplicity $m(\pi)$ of $\pi$ in the discrete spectrum of $\mathbf{H}(\mathbb{A})$ may fail to be constant over a (quasi-) packet.

10.9 THEOREM. *Every member of a* (*quasi-*) *packet of a stable discrete spectrum representation of* $\mathbf{H}(\mathbb{A})$ *is discrete spectrum* (*automorphic*) *representation, which occurs with multiplicity one in the discrete spectrum. Thus packets and quasi-packets partition the stable spectrum, and multiplicity one theorem holds for the discrete spectrum of* $\mathbf{H}(\mathbb{A})$ (*at least for those representations with at least three elliptic components*).

*Every stable packet which does not consist of a one dimensional representation* $\lambda$*-lifts to a* (*unique*) *cuspidal self-contragredient representation of* $\mathbf{G}(\mathbb{A})$.

*The quasi-packets in the stable spectrum of* $\mathbf{H}(\mathbb{A})$ *are all of the form* $\{L(\nu\xi, \nu^{-1/2}\pi_2)\}$, $\pi_2$ *cuspidal with central character* $\xi \neq 1 = \xi^2$.

*Every packet or quasi-packet in the discrete spectrum of* $\mathbf{H}(\mathbb{A})$ *with a local component which is one-dimensional or of the form* $L(\nu_v\xi_v, \nu_v^{-1/2}\pi_{2v})$, $\pi_{2v}$ *cuspidal with central character* $\xi_v \neq 1 = \xi_v^2$, *is globally so, and thus lies in the stable spectrum.*

In view of our global results we can write the remains of the trace formula identity as the equality of the sums

$$I' = \sum_{\pi}{}' \operatorname{tr} \pi(f \times \theta) \qquad \text{and} \qquad I'_H = \sum_{\pi_H}{}' m(\pi_H) \operatorname{tr} \pi_H(f_H).$$

The sum on the left, $I'$, ranges over all self-contragredient discrete spectrum representations of $\mathbf{G}(\mathbb{A})$ which are not $\lambda_1$-lifts from $\mathbf{C}(\mathbb{A})$. The sum on the right ranges over all discrete spectrum representations $\pi_H$ of $\mathbf{H}(\mathbb{A})$ which are not in packets or quasi-packets $\lambda_0$-lifted from $\mathbf{C}_0(\mathbb{A})$. Our test functions $f = \otimes f_v$ and $f_H = \otimes f_{Hv}$ have matching orbital integrals and at least at three places their components are elliptic (the orbital integrals vanish outside the elliptic set).

We first deal with the following residual case.

10.10 PROPOSITION. *For every cuspidal representation $\pi_2$ of $\mathrm{GL}(2, \mathbb{A})$ with central character $\xi \neq 1 = \xi^2$ (hence $\xi\pi_2 = \pi_2$) there exists a quasi-packet $\{L(\nu\xi, \nu^{-1/2}\pi_2)\}$ of representations of $\mathbf{H}(\mathbb{A})$ which $\lambda$-lifts to the residual (discrete spectrum but not cuspidal) self-contragredient representation*

$$J(\nu^{1/2}\pi_2, \nu^{-1/2}\pi_2) = \otimes_v J(\nu_v^{1/2}\pi_{2v}, \nu_v^{-1/2}\pi_{2v})$$

*of $\mathbf{G}(\mathbb{A})$.*

*Each irreducible in such a quasi-packet occurs in the discrete spectrum of $\mathbf{H}(\mathbb{A})$ with multiplicity one, and precisely one irreducible is residual, namely $\otimes_v L(\nu_v\xi_v, \nu_v^{-1/2}\pi_{2v})$.*

PROOF. If $\xi_v \neq 1$ and $\pi_{2v}$ is cuspidal, $J(\nu_v^{1/2}\pi_{2v}, \nu_v^{-1/2}\pi_{2v})$ is the $\lambda$-lift of

$$L(\nu_v\xi_v, \nu_v^{-1/2}\pi_{2v}).$$

If $\xi_v \neq 1$ and $\pi_{2v}$ is not cuspidal, $\pi_{2v}$ has the form $I(\mu_v, \mu_v\xi_v)$, $\mu_v^2 = 1$. If $\pi_{2v} = I(\mu_v, \mu_v\xi_v)$, $\mu_v^2 = 1$, $\xi_v^2 = 1$, then $J(\nu_v^{1/2}\pi_{2v}, \nu_v^{-1/2}\pi_{2v})$ is the quotient of the induced

$$I_G(\nu_v^{1/2}\mu_v, \nu_v^{-1/2}\mu_v, \nu_v^{1/2}\mu_v\xi_v, \nu^{-1/2}\mu_v\xi_v),$$

namely $I_G(\mu_v\mathbf{1}_2, \mu_v\xi_v\mathbf{1}_2)$. This is the $\lambda$-lift of the packet consisting of

$$L_v = L(\nu_v\xi_v, \xi_v \rtimes \mu_v\nu_v^{-1/2}) \quad \text{and} \quad X_v = X(\nu_v^{1/2}\xi_v \operatorname{sp}_{2v}, \xi_v\mu_v\nu_v^{-1/2}).$$

If $\xi_v = 1$ then $\pi_{2v}$ is induced with central character $\xi_v = 1$, thus $\pi_{2v} = I(\mu_v, \mu_v^{-1})$. We may assume that $\mu_v^2 \neq 1$ as the case of $\mu_v^2 = 1$ is dealt with in the previous paragraph, and that $1 > |\mu_v|^{-2} > |\nu_v|^{-1}$, since $\pi_{2v}$ is a component of a cuspidal $\pi_2$. Then $J(\nu_v^{1/2}\pi_{2v}, \nu_v^{-1/2}\pi_{2v})$ is the quotient $I_G(\mu_v \mathbf{1}_2, \mu_v^{-1}\mathbf{1}_2)$ of $I_G(\nu_v^{1/2}\pi_{2v}, \nu_v^{-1/2}\pi_{2v})$. It is the $\lambda$-lift of $\mu_v^{-2} \rtimes \mu_v \mathbf{1}_2$, which is irreducible since $\mu_v^{-2} \neq 1$, $\nu_v^{\pm 1}$, $\nu_v^{\pm 2}$ (Proposition V.2.1(b)). This $\mu_v^{-2} \rtimes \mu_v \mathbf{1}_2$ is the quotient of

$$\mu_v^{-2} \rtimes \mu_v I_2(\nu_v^{1/2}, \nu_v^{-1/2}) = \mu_v^{-2} \times \nu_v \rtimes \mu_v \nu_v^{-1/2}$$

$$= \nu_v \times \mu_v^{-2} \rtimes \mu_v \nu_v^{-1/2} = \nu_v \times \nu_v^{-1/2} I_2(\mu_v, \mu_v^{-1}).$$

So we write $L(\nu_v \xi_v, \nu_v^{-1/2}\pi_{2v})$ for $\mu_v^{-2} \rtimes \mu_v \mathbf{1}_2$ (when $\xi_v = 1$ and $\pi_{2v} = I(\mu_v, \mu_v^{-1})$); it $\lambda$-lifts to $J(\nu_v^{1/2}\pi_{2v}, \nu_v^{-1/2}\pi_{2v})$.

In summary, the quasi-packet of $L(\nu_v \xi_v, \nu_v^{-1/2}\pi_{2v})$ which $\lambda$-lifts to

$$J(\nu_v^{1/2}\pi_{2v}, \nu_v^{-1/2}\pi_{2v}),$$

$\pi_{2v}$ being a component of $\pi_2$ as in the proposition, consists of one irreducible, unless $\pi_v = I(\mu_v, \mu_v \xi_v)$, $\mu_v^2 = \xi_v^2 = 1$, when it consists of $L_v$ and $X_v$ (this includes all $v$ where $\xi_v \neq 1$ and $\pi_{2v}$ is not cuspidal).

Now to prove the proposition we apply the trace identity where the only entry on the side of $G$, having fixed almost all components, is the residual representation $J(\nu^{1/2}\pi_2, \nu^{-1/2}\pi_2)$. Generalized linear independence of characters on $\mathbf{H}(F_v)$ establishes the claim. Note that $L(\nu\xi, \nu^{-1/2}\pi_2) = \otimes_v L_v$ is residual ([Kim], Theorem 7.2), but any other irreducible in the packet is cuspidal. □

PROOF OF THEOREM. Since:

1. The one dimensional representations of $\mathbf{H}(\mathbb{A})$ $\lambda$-lift to the one dimensional representations of $\mathbf{G}(\mathbb{A})$; and
2. The discrete spectrum quasi-packet $\{L(\nu\xi, \nu^{-1/2}\pi_2)\}$ of $\mathbf{H}(\mathbb{A})$ $\lambda$-lifts to the residual representation $J(\nu^{1/2}\pi_2, \nu^{-1/2}\pi_2)$ (for every cuspidal representation $\pi_2$ of $\mathrm{GL}(2, \mathbb{A})$ with central character $\xi \neq 1 = \xi^2$, and $\xi\pi_2 = \pi_2$); and
3. The only other noncuspidal discrete spectrum self contragredient representations of $\mathbf{G}(\mathbb{A})$ are the residual $J(\nu^{1/2}\pi_2, \nu^{-1/2}\pi_2)$ where $\pi_2$ is a cuspidal representation of $\mathrm{PGL}(2, \mathbb{A})$, in which case this $J$ is the $\lambda_1$-lift of

$\mathbf{1}_2 \times \pi_2$ from $\mathbf{C}(\mathbb{A})$;
we may assume that $I'$ ranges only over cuspidal self contragredient representations of $\mathrm{PGL}(4, \mathbb{A})$.

We pass to the form of the identity where almost all components are fixed. If there is a global discrete spectrum $\pi_H$ with the prescribed local components then the sum $I'_H$ is nonzero, since the $m(\pi_H)$ are nonnegative, by generalized linear independence of characters. Hence $I' \neq 0$ and it consists of a single cuspidal $\pi$ by rigidity theorem for cuspidal representations of $\mathbf{G}(\mathbb{A})$. Each component $\pi_v$ of the self contragredient generic $\pi$ is a $\lambda$-lift of a packet $\{\pi_{Hv}\}$ of representations of $\mathbf{H}(F_v)$ (by our local results), hence our identity reads

$$\prod_{v \in V} \mathrm{tr}\{\pi_{Hv}\}(f_{Hv} \times \theta) = \sum_{\pi_H} m(\pi_H) \prod_{v \in V} \mathrm{tr}\, \pi_{Hv}(f_{Hv}).$$

Here $V$ is a finite set, its complement consists of finite places where unramified $\pi^0_{Hv}$ and $\pi^0_v$ with $\lambda(\pi^0_{Hv}) = \pi^0_v$ are fixed, and the sum ranges over the $\pi_H$ whose component at $v \notin V$ is $\pi_{Hv}$.

Generalized linear independence of characters then implies that the right side of our identity has the same form as the left, hence the multiplicity $m(\pi_H)$ is 1 and the $\pi_H$ which occur are precisely the members of the packet $\otimes_v\{\pi_{Hv}\}$, where $\{\pi^0_{Hv}\} = \pi^0_{Hv}$ for all $v$ outside $V$. □

Note that since we work with test functions which have at least three elliptic components, the only $\pi_H$ and $\pi$ which we see in our identity have three such components. The unconditional statement would follow once the unconditional identity of the trace formulae is established. As explained in 1G of the Introduction, "three" elliptic components can be reduced to "two", and even to "one real place", with available technology.

10.11 PROPOSITION. (1) *Every unstable packet* $\lambda_0(\pi_1 \times \pi_2)$ *of the group* $\mathrm{PGSp}(2, \mathbb{A})$, *where* $\pi_1$, $\pi_2$ *are cuspidal representations of* $\mathrm{PGL}(2, \mathbb{A})$, *contains precisely one generic representation. It is the only representation in the packet which is generic at all places. Every packet contains at most one generic representation.*
(2) *In a tempered packet* $\{\pi^+_H, \pi^-_H\}$ *of* $\mathrm{PGSp}(2, F)$, $F$ *local,* $\pi^+_H$ *is generic and* $\pi^-_H$ *is not.*
(3) *In a stable packet of* $\mathrm{PGSp}(2, \mathbb{A})$ *which lifts to a cuspidal representation*

*of* PGL(4, $\mathbb{A}$) *there is precisely one representation which is generic at each place.*

PROOF. (1) If $\pi_H^1$ and $\pi_H^2$ are generic, cuspidal, and lift to the same generic induced representation $I(\pi_1, \pi_2)$ of PGL(4, $\mathbb{A}$), namely they are in the same packet, then they are equivalent by [GRS]. The second claim follows from this and Lemma 7.3. The third claim follows from the rigidity theorem for generic representations of GSp(2), see [So], Theorem 1.5.

(2) Let $F$ be a global field such that at an odd number of places, say $v_1, \dots, v_5$, its completion is our local field. Construct cuspidal representations $\pi_1$, $\pi_2$ of PGL(2, $\mathbb{A}$) such that the set of places $v$ where both $\pi_{1v}$ and $\pi_{2v}$ are square integrable is precisely $v_1, \dots, v_5$, and such that $\lambda_0(\pi_{1v_i} \times \pi_{2v_i})$ is our local packet, now denoted $\{\pi_{Hv}^+, \pi_{Hv}^-\}$, $v = v_i$. In $\lambda_0(\pi_1 \times \pi_2)$ there is a unique cuspidal generic representation $\pi_H^0$, by [GRS]. By our multiplicity formula the cuspidal members of $\lambda_0(\pi_1 \times \pi_2)$ are those which have an even number of components $\pi_{Hv}^-$. Hence $\pi_H^0$ has a component $\pi_{Hv}^+$, so $\pi_{Hv}^+$ must be generic. If both $\{\pi_{Hv}^+, \pi_{Hv}^-\}$ were generic, Lemma 7.3 would imply that the packet of the cuspidal generic $\pi_H^0$ contains more than one generic cuspidal representation (in fact, $2^5$ of them), contradicting [GRS].

(3) Every irreducible in such a packet is in the discrete spectrum. The packet is the product of local packets. When the local packet consists of a single representation, it is generic. If the local packet has the form $\{\pi_{Hv}^+, \pi_{Hv}^-\}$, then $\pi_{Hv}^+$ is generic but $\pi_{Hv}^-$ is not. Hence the packet has precisely one irreducible which is everywhere locally generic. □

REMARK. Is the representation $\pi_H$ constructed in (3) above generic? By [GRS], it is, provided $L(S, \pi, \Lambda^2, s)$ has a pole at $s = 1$, where $\lambda(\pi_H) = \pi$. We do not know to rule out at present the possibility that there is a packet $\{\pi_H\}$ containing no generic member and $\lambda$-lifting to a cuspidal $\pi$, necessarily with $L(S, \pi, \Lambda^2, s)$ finite at $s = 1$. Note that the six-dimensional representation $\Lambda^2$ of the dual group Sp(2, $\mathbb{C}$) of PGSp(2) is the direct sum of the irreducible five-dimensional representation $\mathrm{id}_5 : \mathrm{Sp}(2, \mathbb{C}) \to \mathrm{SO}(5, \mathbb{C})$ (cf. Lemma 7.0) and the trivial representation (see [FH], Section 16.2, p. 245, for a formulation in terms of the Lie algebra of Sp(2, $\mathbb{C}$)). Hence $L(S, \pi_H, \Lambda^2, s) = L(S, \pi, \Lambda^2, s)$ has a pole at $s = 1$ provided $L(S, \pi_H, \mathrm{id}_5, s)$ is not zero at $s = 1$. This is guaranteed by [Sh1], Theorem 5.1 (as noted after Lemma 7.0) when $\pi_H$ is generic. Thus the locally generic $\pi_H$ of (3) is generic iff $L(S,\pi,\Lambda^2,s)$ has a pole at $s=1$, iff $L(S,\pi_H,\mathrm{id}_5,s)$ is not zero at $s=1$. An alternative approach is to consider

$$L(S,\pi\otimes\pi,s)=L(S,\pi,\Lambda^2,s)L(S,\pi,\mathrm{Sym}^2,s),$$

which has a simple pole at $s=1$ since $\pi\simeq\check{\pi}$. If $L(S,\pi,\Lambda^2,s)$ does not have a pole at $s=1$, $L(S,\pi,\mathrm{Sym}^2,s)$ has. One expects an analogue of [GRS] to show that $\pi$ is then a $\lambda_1$-lift from $\mathrm{SO}(4,\mathbb{A})$. We shall then conclude that $\pi$ is not a $\lambda$-lift from $\mathrm{PGSp}(2,\mathbb{A})$.

# 11. Representations of PGSp(2, ℝ)

The parametrization of the irreducible representations of the real symplectic group $\mathrm{PGSp}(2,\mathbb{R})$ is analogous to the $p$-adic case, but there are some differences. We review the listing next, starting with the case of $\mathrm{GL}(2,\mathbb{R})$. In particular we determine the cohomological representations, those which have Lie algebra $(\mathfrak{g},\mathrm{K})$-cohomology, with view for further applications.

## 11a. Representations of SL(2, ℝ)

Packets of representations of a real group $G$ are parametrized by maps of the Weil group $W_{\mathbb{R}}$ to the $L$-group ${}^LG$. Recall that

$$W_{\mathbb{R}}=\langle z,\sigma;z\in\mathbb{C}^\times,\sigma^2\in\mathbb{R}^\times-N_{\mathbb{C}/\mathbb{R}}\mathbb{C}^\times,\sigma z=\overline{z}\sigma\rangle$$

is

$$1\to W_{\mathbb{C}}\to W_{\mathbb{R}}\to\mathrm{Gal}(\mathbb{C}/\mathbb{R})\to 1$$

an extension of $\mathrm{Gal}(\mathbb{C}/\mathbb{R})$ by $W_{\mathbb{C}}=\mathbb{C}^\times$. It can also be viewed as the normalizer $\mathbb{C}^\times\cup\mathbb{C}^\times j$ of $\mathbb{C}^\times$ in $\mathbb{H}^\times$, where $\mathbb{H}=\mathbb{R}\langle 1,\,i,\,j,\,k\rangle$ is the Hamilton quaternions. The norm on $\mathbb{H}$ defines a norm on $W_{\mathbb{R}}$ by restriction ([D3], [Tt]). The discrete series (packets of) representations of $G$ are parametrized by the homomorphisms $\phi:W_{\mathbb{R}}\to\hat{G}\times W_{\mathbb{R}}$ whose projection to $W_{\mathbb{R}}$ is the identity and to the connected component $\hat{G}$ is bounded, and such that $C_\phi Z(\hat{G})/Z(\hat{G})$ is finite. Here $C_\phi$ is the centralizer $Z_{\hat{G}}(\phi(W_{\mathbb{R}}))$ in $\hat{G}$ of the image of $\phi$.

When $G = \mathrm{GL}(2,\mathbb{R})$ we have $\hat{G} = \mathrm{GL}(2,\mathbb{R})$, and these maps are $\phi_k$ ($k \geq 1$), defined by

$$W_{\mathbb{C}} = \mathbb{C}^\times \ni z \mapsto \begin{pmatrix} (z/|z|)^k & 0 \\ 0 & (|z|/z)^k \end{pmatrix} \times z, \qquad \sigma \mapsto \begin{pmatrix} 0 & 1 \\ \iota & 0 \end{pmatrix} \times \sigma.$$

Since $\sigma^2 = -1 \mapsto \begin{pmatrix} (-1)^k & 0 \\ 0 & (-1)^k \end{pmatrix} \times \sigma^2$, $\iota$ must be $(-1)^k$. Then $\det \phi_k(\sigma) = (-1)^{k+1}$, and so $k$ must be an odd integer ($= 1, 3, 5, \dots$) to get a discrete series (packet of) representation of $\mathrm{PGL}(2,\mathbb{R})$. In fact $\pi_1$ is the lowest discrete series representation, and $\phi_0$ parametrizes the so called limit of discrete series representations; it is tempered.

Even $k \geq 2$ and $\sigma \mapsto \begin{pmatrix} 0 & 1 \\ 1 & 0 \end{pmatrix} \times \sigma$ define discrete series representations of $\mathrm{GL}(2,\mathbb{R})$ with the quadratic nontrivial central character sgn. Packets for $\mathrm{GL}(2,\mathbb{R})$ and $\mathrm{PGL}(2,\mathbb{R})$ consist of a single discrete series irreducible representation $\pi_k$. Note that $\pi_k \otimes \mathrm{sgn} \simeq \pi_k$. Here $\mathrm{sgn} : \mathrm{GL}(2,\mathbb{R}) \to \{\pm 1\}$, $\mathrm{sgn}(g) = 1$ if $\det g > 0$, $= -1$ if $\det g < 0$.

The $\pi_k$ ($k > 0$) have the same central and infinitesimal character as the $k$th dimensional nonunitarizable representation

$$\mathrm{Sym}_0^{k-1}\,\mathbb{C}^2 = |\det g|^{-(k-1)/2}\,\mathrm{Sym}^{k-1}\,\mathbb{C}^2$$

into $\mathrm{SL}(k,\mathbb{C})^{\pm} = \{g \in \mathrm{GL}(2,\mathbb{C}); \det g \in \{\pm 1\}\}$. We have

$$\det \mathrm{Sym}^{k-1}(g) = \det g^{k(k-1)/2},$$

and the normalizing factor is $|\det \mathrm{Sym}^{k-1}|^{-1/k}$. Then $\mathrm{Sym}_0^{k-1}\begin{pmatrix} a & 0 \\ 0 & b \end{pmatrix}$

$$= \mathrm{diag}(\mathrm{sgn}(a)^{k-i}\mathrm{sgn}(b)^{i-1}|a|^{k-i-(k-1)/2}|b|^{i-1-(k-1)/2}; 1 \leq i \leq k).$$

In fact both $\pi_k$ and $\mathrm{Sym}_0^{k-1}\,\mathbb{C}^2$ are constituents of the normalizedly induced representation $I(\nu^{k/2}, \mathrm{sgn}^{k-1}\nu^{-k/2})$ whose infinitesimal character is $(\frac{k}{2}, -\frac{k}{2})$, where a basis for the lattice of characters of the diagonal torus in $\mathrm{SL}(2)$ is taken to be $(1,-1)$.

## 11b. Cohomological Representations

An irreducible admissible representation $\pi$ of $\mathbf{H}(\mathbb{A})$ which has nonzero Lie algebra cohomology $H^{ij}(\mathfrak{g}, K; \pi \otimes V)$ for some coefficients (finite dimensional representation) $V$ is called here *cohomological.* Discrete series representations are cohomological. The non discrete series representations which are cohomological are listed in [VZ]. They are nontempered. We proceed to list them here in our case of $\mathrm{PGSp}(2,\mathbb{R})$. We are interested in the $(\mathfrak{g},K)$-cohomology $H^{ij}(\mathrm{sp}(2,\mathbb{R}),\mathrm{U}(4);\pi\otimes V)$, so we need to compute

$$H^{ij}(\mathrm{sp}(2,\mathbb{R}),\mathrm{SU}(4);\pi\otimes V)$$

and observe that $\mathrm{U}(4)/\mathrm{SU}(4)$ acts trivially on the nonzero $H^{ij}$, which are $\mathbb{C}$. If $H^{ij}(\pi\otimes V)\neq 0$ then ([BW]) the infinitesimal character ([Kn]) of $\pi$ is equal to the sum of the highest weight ([FH]) of the self contragredient (in our case) $V$, and half the sum of the positive roots, $\delta$.

With the usual basis $(1,0)$, $(0,1)$ on $X^*(T_S^*)$, the positive roots are $(1,-1)$, $(0,2)$, $(1,1)$, $(2,0)$. Then $\delta=\frac{1}{2}\sum_{\alpha>0}\alpha$ is $(2,1)$.

Here $T_S^*$ denotes the diagonal subgroup $\{\mathrm{diag}(x,y,1/y,1/x)\}$ of the algebraic group $\mathrm{Sp}(2)$. Its lattice $X^*(T_S^*)$ of rational characters consists of

$$(a,b):\mathrm{diag}(x,y,1/y,1/x)\mapsto x^a y^b \qquad (a,b\in\mathbb{Z}).$$

The irreducible finite dimensional representations $V_{a,b}$ of $\mathrm{Sp}(2)$ are parametrized by the highest weight $(a,b)$ with $a\geq b\geq 0$ ([FH]). The central character of $V_{a,b}$ is $\zeta\mapsto\zeta^{a+b}$, $\zeta\in\{\pm 1\}$. It is trivial iff $a+b$ is even. Since $\mathrm{GSp}(2)=\mathrm{Sp}(2)\rtimes\{\mathrm{diag}(1,1,z,z)\}$, such $V_{a,b}$ extends to a representation of $\mathrm{PGSp}(2)$ by $(1,1,z,z)\mapsto z^{-(a+b)/2}$. This gives a representation of $H=\mathbf{H}(\mathbb{R})=\mathrm{PGSp}(2,\mathbb{R})$, extending its restriction to the index 2 connected subgroup $H^0=\mathrm{PSp}(2,\mathbb{R})$. Another – nonalgebraic – extension is $V'_{a,b}=V_{a,b}\otimes\mathrm{sgn}$, where $\mathrm{sgn}(1,1,z,z)=\mathrm{sgn}(z)$, $z\in\mathbb{R}^\times$. $V_{a,b}$ is self dual.

To list the irreducible admissible representations $\pi$ of $\mathrm{PGSp}(2,\mathbb{R})$ with nonzero Lie algebra cohomology $H^{i,j}(\mathrm{sp}(2,\mathbb{R}),\ \mathrm{SU}(4);\ \pi\otimes V_{a,b})$ for some $a\geq b\geq 0$ (the same results hold with $V_{a,b}$ replaced by $V'_{a,b}$), we first list the discrete series representations.

Packets of discrete series representations of the group $H=\mathrm{PGSp}(2,\mathbb{R})$ are parametrized by maps $\phi$ of $W_\mathbb{R}$ to ${}^LH=\hat{H}\times W_\mathbb{R}$ which are admissible ($\mathrm{pr}_2\circ\phi=\mathrm{id}$) and whose projection to $\hat{H}$ is bounded and $C_\phi Z(\hat{H})/Z(\hat{H})$ is finite. Here $C_\phi$ is $Z_{\hat{H}}(\phi(W_\mathbb{R}))$. They are parametrized $\phi=\phi_{k_1,k_2}$ by a pair $(k_1,k_2)$ of integers with odd $k_1>k_2>0$.

The homomorphism

$$\phi_{k_1,k_2}:W_\mathbb{R}\to{}^LG=\hat{G}\times W_\mathbb{R},\qquad \hat{G}=\mathrm{SL}(4,\mathbb{C}),$$

given by

$$z \mapsto \mathrm{diag}((z/|z|)^{k_1}, (z/|z|)^{k_2}, (|z|/z)^{k_2}, (|z|/z)^{k_1}) \times z$$

and

$$\sigma \mapsto \left(\begin{smallmatrix} 0 & w \\ -w & 0 \end{smallmatrix}\right) \times \sigma \qquad (\text{odd } k_1 > k_2 > 0)$$

or

$$\sigma \mapsto \left(\begin{smallmatrix} 0 & w \\ w & 0 \end{smallmatrix}\right) \times \sigma \qquad (\text{even } k_1 > k_2 > 0),$$

factorizes via (${}^LC_0 \to$) ${}^LH = \mathrm{Sp}(2,\mathbb{C}) \times W_{\mathbb{R}}$ precisely when $k_i$ are odd. When the $k_i$ are even it factorizes via ${}^LC = \mathrm{SO}(4,\mathbb{C}) \times W_{\mathbb{R}}$. When the $k_i$ are odd it parametrizes a packet $\{\pi^{\mathrm{Wh}}_{k_1,k_2}, \pi^{\mathrm{hol}}_{k_1,k_2}\}$ of discrete series representations of $\mathrm{PGSp}(2,\mathbb{R})$. Here $\pi^{\mathrm{Wh}}$ is generic and $\pi^{\mathrm{hol}}$ is holomorphic and antiholomorphic. Their restrictions to $H^0$ are reducible, consisting of $\pi^{\mathrm{Wh}}_{H^0}$ and $\pi^{\mathrm{Wh}}_{H^0} \circ \mathrm{Int}(\iota)$, $\pi^{\mathrm{hol}}_{H^0}$ and $\pi^{\mathrm{hol}}_{H^0} \circ \mathrm{Int}(\iota)$, $\iota = \mathrm{diag}(1,1,-1,-1)$, and $\pi^{\mathrm{Wh}} \otimes \mathrm{sgn} = \pi^{\mathrm{Wh}}$, $\pi^{\mathrm{hol}} \otimes \mathrm{sgn} = \pi^{\mathrm{hol}}$.

To compute the infinitesimal character of $\pi^*_{k_1,k_2}$ we note that

$$\pi_k \subset I(\nu^{k/2}, \mathrm{sgn}^{k-1}\nu^{-k/2})$$

(e.g. by [JL], Lemma I5.7 and Theorem I5.11) on $\mathrm{GL}(2,\mathbb{R})$. Via ${}^LC_0 \to {}^LH$ induced $I(\nu^{k_1/2}, \nu^{-k_1/2}) \times I(\nu^{k_2/2}, \nu^{-k_2/2})$ (in our case the $k_i$ are odd) lifts to the induced

$$I_H(\nu^{k_1/2}, \nu^{k_2/2}) = \nu^{(k_1+k_2)/2} \times \nu^{(k_1-k_2)/2} \rtimes \nu^{-k_2/2},$$

whose constituents (e.g. $\pi^*_{k_1,k_2}$, $* = \mathrm{Wh}, \mathrm{hol}$) have infinitesimal character

$$\left(\frac{k_1+k_2}{2}, \frac{k_1-k_2}{2}\right) = (2,1) + (a,b).$$

Here

$$a = \frac{k_1+k_2}{2} - 2 \geq b = \frac{k_1-k_2}{2} - 1 \geq 0$$

as $k_2 \geq 1$ and $k_1 > k_2$ and $k_1 - k_2$ is even. For these $a \geq b \geq 0$, thus $k_1 = a+b+3$, $k_2 = a-b+1$, we have

$$H^{ij}(\mathrm{sp}(2,\mathbb{R}), \mathrm{SU}(4); \pi^{\mathrm{Wh}}_{k_1,k_2} \otimes V_{a,b}) = \mathbb{C} \qquad \text{if} \quad (i,j) = (2,1),\ (1,2),$$

$$H^{ij}(\mathrm{sp}(2,\mathbb{R}), \mathrm{SU}(4); \pi^{\mathrm{hol}}_{k_1,k_2} \otimes V_{a,b}) = \mathbb{C} \qquad \text{if} \quad (i,j) = (3,0),\ (0,3).$$

Here $k_1 > k_2 > 0$ and $k_1$, $k_2$ are odd. In particular, the discrete series representations of $\mathrm{PGSp}(2,\mathbb{R})$ are endoscopic.

## 11c. Nontempered Representations

Quasi-packets including nontempered representations are parametrized by homomorphisms $\psi : W_{\mathbb{R}} \times \mathrm{SL}(2,\mathbb{R}) \to {}^L H$ and $\phi_\psi : W_{\mathbb{R}} \to {}^L H$ (see [A2]) defined by

$$\phi_\psi(\mathbf{w}) = \psi(\mathbf{w}, \begin{pmatrix} \|\mathbf{w}\|^{1/2} & 0 \\ 0 & \|\mathbf{w}\|^{-1/2} \end{pmatrix}).$$

The norm $\|.\| : W_{\mathbb{R}} \to \mathbb{R}^\times$ is defined by $\|z\| = z\bar{z}$ and $\|\sigma\| = 1$. Then $\phi_\psi(\sigma) = \psi(\sigma, I)$ and $\phi_\psi(z) = \psi(z, \mathrm{diag}(r, r^{-1}))$ if $z = re^{i\theta}$, $r > 0$. For example,

$$\psi : W_{\mathbb{R}} \times \mathrm{SL}(2,\mathbb{C}) \to \mathrm{SL}(2,\mathbb{C}), \quad \psi | W_{\mathbb{R}} : z\sigma^j \mapsto \xi(-1)^j, \quad \psi | \mathrm{SL}(2,\mathbb{C}) = \mathrm{id},$$

gives

$$\phi_\psi(z) = \begin{pmatrix} r & 0 \\ 0 & r^{-1} \end{pmatrix} \times z, \qquad \phi_\psi(\sigma) = \xi(-1) I_2 \times \sigma,$$

parametrizing the one dimensional representation

$$\xi_2 = J(\xi\nu^{1/2}, \xi\nu^{-1/2}) \quad \text{of} \quad \mathrm{PGL}(2,\mathbb{R}) \quad (\xi : \mathbb{R}^\times \to \{\pm 1\}, \quad \nu(z) = |z|).$$

Here $J$ denotes the Langlands quotient of the indicated induced representation, $I(\xi\nu^{1/2}, \xi\nu^{-1/2})$.

Similarly the one dimensional representation

$$\xi_4 = J(\xi\nu^{3/2}, \xi\nu^{1/2}, \xi\nu^{-1/2}, \xi\nu^{-3/2})$$

of $\mathrm{PGL}(4,\mathbb{R})$ is parametrized by $\psi : W_{\mathbb{R}} \times \mathrm{SL}(2,\mathbb{C}) \to \mathrm{SL}(4,\mathbb{C})$,

$$(\psi | W_{\mathbb{R}})(z\sigma^j) = \xi(-1)^j, \qquad \psi | \mathrm{SL}(2,\mathbb{C}) = \mathrm{Sym}_0^3,$$

thus

$$\phi_\psi(z) = \mathrm{diag}(r^3, r, r^{-1}, r^{-3}) \times z, \qquad \phi_\psi(\sigma) = \xi(-1) I_4 \times \sigma.$$

This parameter factorizes via $\psi : W_{\mathbb{R}} \times \mathrm{SL}(2,\mathbb{C}) \to \mathrm{Sp}(2,\mathbb{C})$, which parametrizes the one dimensional representation $\xi_H$ of $\mathrm{PGSp}(2,\mathbb{R})$, $h \mapsto \xi(\boldsymbol{\lambda}(h))$ where $\boldsymbol{\lambda}(h)$ denotes the factor of similitude of $h$, whose infinitesimal character is $(2,1) = \frac{1}{2}\sum_{\alpha>0} \alpha$. We have

$$H^{ij}(\mathrm{sp}(2,\mathbb{R}), \mathrm{SU}(4); \xi_H \otimes V_{0,0}) = \mathbb{C}$$

for $(i,j) = (0,0)$, $(1,1)$, $(2,2)$, $(3,3)$. Of course $1_H \neq \mathrm{sgn}_H$, and $\frac{1}{2}(1_H + \mathrm{sgn}_H)$ is the characteristic function of $H^0$ in $\mathrm{PGSp}(2,\mathbb{R})$. Moreover, the character of $\frac{1}{2}(1_H + \mathrm{sgn}_H) + \pi_{3,1}^{\mathrm{Wh}} + \pi_{3,1}^{\mathrm{hol}}$ vanishes on the regular elliptic set of $\mathrm{PGSp}(2,\mathbb{R})$, as $(\xi_H + \pi_{3,1}^{\mathrm{Wh}} + \pi_{3,1}^{\mathrm{hol}}) | H^0$ is a linear combination of properly induced ("standard") representations ([Vo]) in the Grothendieck group.

## 11d. The Nontempered: $L(\nu\,\mathrm{sgn}, \nu^{-1/2}\pi_{2k})$

The nontempered nonendoscopic representation $L(\nu\,\mathrm{sgn}, \nu^{-1/2}\pi_{2k})$ of the group $\mathrm{PGSp}(2,\mathbb{R})$ ($k \geq 1$) is the Langlands quotient of the representation $\nu\,\mathrm{sgn} \rtimes \nu^{-1/2}\pi_{2k}$ induced from the Heisenberg parabolic subgroup of $H$. It $\lambda$-lifts to $J(\nu^{1/2}\pi_{2k}, \nu^{-1/2}\pi_{2k})$, the Langlands quotient of the induced representation

$$I(\nu^{1/2}\pi_{2k}, \nu^{-1/2}\pi_{2k}) \quad \text{of} \quad \mathrm{PGL}(4,\mathbb{R}).$$

Note that the discrete series $\pi_{2k} \simeq \mathrm{sgn} \otimes \pi_{2k} \simeq \check{\pi}_{2k}$ has central character $\mathrm{sgn}$ ($\neq 1$). Now

$$\psi : W_{\mathbb{R}} \times \mathrm{SL}(2,\mathbb{C}) \to \mathrm{SL}(4,\mathbb{C}), \qquad \psi | W_{\mathbb{R}} : \mathbf{w} \mapsto \begin{pmatrix} \phi_{2k}(\mathbf{w}) & 0 \\ 0 & \phi_{2k}(\mathbf{w}) \end{pmatrix} \times \mathbf{w}$$

with

$$\phi_{2k}(z) = \begin{pmatrix} (z/|z|)^{2k} & 0 \\ 0 & (|z|/z)^{2k} \end{pmatrix} \times z, \qquad \phi_{2k}(\sigma) = w \times \sigma,$$

and $(\psi | \mathrm{SL}(2,\mathbb{C})) \begin{pmatrix} a & b \\ c & d \end{pmatrix} = \begin{pmatrix} aI & bI \\ cI & dI \end{pmatrix}$, defines

$$\phi_\psi(z) = \psi\left(z, \begin{pmatrix} |z| & 0 \\ 0 & |z|^{-1} \end{pmatrix}\right) = \begin{pmatrix} |z|\phi_{2k}(z) & 0 \\ 0 & |z|^{-1}\phi_{2k}(z) \end{pmatrix} \times z,$$

$$\phi_\psi(\sigma) = \psi(\sigma, I) = \begin{pmatrix} w & 0 \\ 0 & w \end{pmatrix}.$$

It factorizes via $\widehat{H} = \mathrm{Sp}(2,\mathbb{C}) \hookrightarrow \mathrm{SL}(4,\mathbb{C})$ and defines $L(\nu\,\mathrm{sgn}, \nu^{-1/2}\pi_{2k})$.

Note that when $2k$ is replaced by $2k+1$, $\phi_{2k+1}(\sigma) = \varepsilon w \times \sigma$, $\varepsilon = \mathrm{diag}(1,-1)$, then

$$\phi_\psi(\sigma) = \psi(\sigma, I) = \begin{pmatrix} \varepsilon w & 0 \\ 0 & \varepsilon w \end{pmatrix} = I \otimes \varepsilon w \in \hat{C},$$

$$\phi_\psi(z) = \begin{pmatrix} |z| & 0 \\ 0 & |z|^{-1} \end{pmatrix} \otimes \phi_{2k+1}(z) \in \hat{C},$$

thus $\phi_\psi$ defines a representation of $C(\mathbb{R})$ (which $\lambda_1$-lifts to the representation

$$J(\nu^{1/2}\pi_{2k+1}, \nu^{-1/2}\pi_{2k+1})$$

of $\mathrm{PGL}(4,\mathbb{R})$), but not a representation of $\mathrm{PGSp}(2,\mathbb{R})$.

As in [Ty] write $\pi^1_{2k,0}$ for $L(\mathrm{sgn}\nu, \nu^{-1/2}\pi_{2k+2})$. We have that $\pi^1_{2k,0} \simeq \mathrm{sgn}\otimes\pi^1_{2k,0}$, and $\pi^1_{2k,0}|H^0$ consists of two irreducibles. In the Grothendieck group the induced decomposes as

$$\nu\,\mathrm{sgn}\rtimes\nu^{-1/2}\pi_{2k} = L(\nu\,\mathrm{sgn}, \nu^{-1/2}\pi_{2k}) + \pi^{\mathrm{Wh}}_{2k+3,2k+1} + \pi^{\mathrm{hol}}_{2k+3,2k+1}, \qquad k \geq 1.$$

To compute the infinitesimal character of $\nu\,\mathrm{sgn}\rtimes\nu^{-1/2}\pi_{2k}$, note that it is a constituent of the induced

$$\nu\,\mathrm{sgn}\rtimes\nu^{-1/2}I(\nu^k, \mathrm{sgn}\nu^{-k}) \simeq \mathrm{sgn}\nu^{2k}\times\mathrm{sgn}\nu\rtimes\nu^{-k-1/2}\mathrm{sgn}$$

(using the Weyl group element (12)(34)), whose infinitesimal character is $(2k,1) = (2,1)+(a,0)$, with $a = 2k-2 \geq 0$ as $k \geq 1$. For $k \geq 1$ we have

$$H^{ij}(\mathrm{sp}(2,\mathbb{R}), \mathrm{SU}(4); \pi^1_{2k,0}\otimes V_{2k,0}) = \mathbb{C} \text{ if } (i,j) = (2,0),\ (0,2),\ (3,1),\ (1,3).$$

## 11e. The Nontempered: $L(\xi\nu^{1/2}\pi_{2k+1}, \xi\nu^{-1/2})$

The nontempered endoscopic representation $L(\xi\nu^{1/2}\pi_{2k+1}, \xi\nu^{-1/2})$ of the group $\mathrm{PGSp}(2,\mathbb{R})$ is the Langlands quotient of the induced representation $\xi\nu^{1/2}\pi_{2k+1}\rtimes\xi\nu^{-1/2}$ from the Siegel parabolic subgroup of $\mathrm{PGSp}(2,\mathbb{R})$. It is the $\lambda_0$-lift of $\pi_{2k+1}\times\xi_2$ and $\lambda$-lifts to the induced $I(\pi_{2k+1},\xi_2)$ of $\mathrm{PGL}(4,\mathbb{R})$. The central character of $\pi_{2k+1}$ is trivial, but that of $\pi_{2k}$ is sgn. Hence $I(\pi_{2k},\xi_2)$ defines a representation of $\mathrm{GL}(4,\mathbb{R})$ but not of $\mathrm{PGL}(4,\mathbb{R})$. The endoscopic map

$$\psi : W_{\mathbb{R}}\times\mathrm{SL}(2,\mathbb{C}) \to {}^LC_0 = \mathrm{SL}(2,\mathbb{C})\times\mathrm{SL}(2,\mathbb{C}) \overset{\lambda_0}{\to} \widehat{H},$$

$$\psi(z\sigma^j, s) = \lambda_0(\phi_{2k+1}(z\sigma^j), \xi(-1)^j s),$$

defines

$$\phi_\psi(z) = \psi\left(z, \begin{pmatrix} |z| & 0 \\ 0 & |z|^{-1}\end{pmatrix}\right) = \mathrm{diag}((z/|z|)^{2k+1}, |z|, |z|^{-1}, (|z|/z)^{2k+1})\times z,$$

$$\phi_\psi(\sigma) = \psi(\sigma, I) = \begin{pmatrix} & & & 1 \\ & & \xi(-1) & \\ & & & \xi(-1) \\ & (-1)^{2k+1} & & \end{pmatrix},$$

which lies in $\widehat{H}\subset\mathrm{SL}(4,\mathbb{C})$ since $2k+1$ is odd.

As in [Ty] we write $\pi^{2,\xi}_{k-1,k-1}$ for $L(\xi\nu^{1/2}\pi_{2k+1}, \xi\nu^{-1/2})$, $k \geq 0$.

Now $\xi\pi^{2,1} = \pi^{2,\xi}$ and $\pi^{2,\xi}|H^0$ is irreducible. In the Grothendieck group the induced decomposes as

$$\xi\nu^{1/2}\pi_{2k+1} \rtimes \xi\nu^{-1/2} = \pi^{2,\xi}_{k-1,k-1} + \pi^{\mathrm{Wh}}_{2k+1,1}.$$

Here $\pi^{\mathrm{Wh}}_{2k+1,1}$ is generic, discrete series if $k \geq 1$, tempered if $k = 0$.

Our $\xi\nu^{1/2}\pi_{2k+1} \rtimes \xi\nu^{-1/2}$ is a constituent of the induced

$$\xi\nu^{1/2} I(\nu^{(2k+1)/2}, \nu^{-(2k+1)/2}) \rtimes \xi\nu^{-1/2} = \xi\nu^{k+1} \times \xi\nu^{-k} \rtimes \xi\nu^{-1/2},$$

which is equivalent to $\xi\nu^{k+1} \times \xi\nu^{k} \rtimes \xi\nu^{-k-1/2}$ (using the Weyl group element (23)). Its infinitesimal character is $(k+1, k) = (2,1) + (k-1, k-1)$. We have

$$H^{ij}(\mathrm{sp}(2,\mathbb{R}), \mathrm{SU}(4); \pi^{2,\xi}_{k-1,k-1} \otimes V_{k-1,k-1}) = \mathbb{C} \qquad \text{if} \quad (i,j) = (1,1),\ (2,2).$$

In summary, $H^{ij}(\pi \otimes V_{a,b})$ is 0 except in the following four cases, where it is $\mathbb{C}$.

(1) One dimensional case: $(a,b) = (0,0)$ and $\pi$ is $\pi^{\mathrm{Wh}}_{3,1}$, $\pi^{\mathrm{hol}}_{3,1}$, $\xi_H$,

$$\pi^{1}_{0,0} = L(\nu\,\mathrm{sgn}, \nu^{-1/2}\pi_2), \quad \pi^{2,\xi}_{0,0} = L(\xi\nu^{1/2}\pi_3, \xi\nu^{-1/2}).$$

(2) Nontempered unstable case: $(a,b) = (k,k)$ $(k \geq 1)$ and $\pi$ is

$$\pi^{\mathrm{Wh}}_{2k+3,1}, \qquad \pi^{\mathrm{hol}}_{2k+3,1}, \qquad \pi^{2,\xi}_{k,k} = L(\xi\nu^{1/2}\pi_{2k+3}, \xi\nu^{-1/2}).$$

(3) Nontempered stable case: $(a,b) = (2k, 0)$ $(k \geq 1)$ and $\pi$ is

$$\pi^{\mathrm{Wh}}_{2k+3,2k+1}, \qquad \pi^{\mathrm{hol}}_{2k+3,2k+1}, \qquad \pi^{1}_{2k,0} = L(\nu\,\mathrm{sgn}, \nu^{-1/2}\pi_{2k+2}).$$

(4) Tempered case: any other $(a,b)$ with $a \geq b \geq 1$, $a+b$ even, and $\pi$ is $\pi^{\mathrm{Wh}}_{k_1,k_2}$, $\pi^{\mathrm{hol}}_{k_1,k_2}$. Here $k_1 = a+b+3 > k_2 = a-b+1 > 0$ are odd.

Applications of the classification above in the theory of Shimura varieties and their cohomology with arbitrary coefficients are discussed in [F7].

# VI. FUNDAMENTAL LEMMA

The following is a computation of the orbital integrals for GL(2), SL(2), and our GSp(2), for the characteristic function $1_K$ of $K$ in $G$, leading to a proof of the fundamental lemma for (PGSp(2), PGL(2)×PGL(2)), due to J.G.M. Mars (letter to me, 1997).

## 1. Case of SL(2)

**1.** Let $E/F$ be a (separable) quadratic extension of nonarchimedean local fields. Denote by $\mathcal{O}_E$ and $\mathcal{O}$ their rings of integers. Let $\boldsymbol{\pi} = \boldsymbol{\pi}_F$ be a generator of the maximal ideal in $\mathcal{O}$. Then $ef = 2$ where $e$ is the degree of ramification of $E$ over $F$. Let $V = E$, considered as a 2-dimensional vector space over $F$. Multiplication in $E$ gives an embedding $E \subset \mathrm{End}_R(V)$ and $E^\times \subset \mathrm{GL}(V)$. The ring of integers $\mathcal{O}_E$ is a lattice (free $\mathcal{O}$-module of maximal rank, namely which spans $V$ over $F$) in $V$ and $K = \mathrm{Stab}(\mathcal{O}_E)$ is a maximal compact subgroup of $\mathrm{GL}(V)$.

Let $\Lambda$ be a lattice in $V$. Then $R = R(\Lambda) = \{x \in E | x\Lambda \subset \Lambda\}$ is an order. The orders in $E$ are $R(m) = \mathcal{O} + \boldsymbol{\pi}^m \mathcal{O}_E$, $m \geq 0$ of $F$. This is well-known and easy to check. The quotient $R(m)/R(m+1)$ is a 1-dimensional vector space over $\mathcal{O}/\boldsymbol{\pi}$. If $R(\Lambda) = R(m)$, then $\Lambda = zR(m)$ for some $z \in E^\times$.

Choose a basis 1, $w$ of $E$ such that $\mathcal{O}_E = \mathcal{O} + \mathcal{O}w$. Define $d_m \in \mathrm{GL}(V)$ by $d_m(1) = 1$, $d_m(w) = \boldsymbol{\pi}^m w$. Then $R(m) = d_m \mathcal{O}_E$. It follows immediately that $\mathrm{GL}(V) = \bigcup_{m\geq 0} E^\times d_m K$, or, in coordinates with respect to 1, $w$:

$$\mathrm{GL}(2,F) = \bigcup_{m\geq 0} T \begin{pmatrix} 1 & 0 \\ 0 & \boldsymbol{\pi}^m \end{pmatrix} \mathrm{GL}(2,\mathcal{O}),$$

with $T = \left\{ \begin{pmatrix} a & \alpha b \\ b & a+\beta b \end{pmatrix};\ a,\ b \in F, \text{ not both } = 0 \right\}$, where $w^2 = \alpha + \beta w$, $\alpha$, $\beta \in \mathcal{O}$.

**2.** Put $G = \mathrm{GL}(V)$, $K = \mathrm{Stab}(\mathcal{O}_E)$. Choose the Haar measure $dg$ on $G$ such that $\int_K dg = 1$, and $dt$ on $E^\times$ such that $\int_{\mathcal{O}_E} dt = 1$. Choose $\gamma \in E^\times$, $\gamma \notin F^\times$. Let $1_K$ be the characteristic function of $K$ in $G$. Then

$$\int_{E^\times\backslash G} 1_K(g^{-1}\gamma g)\frac{dg}{dt} = \sum_{E^\times\backslash G/K} \frac{\mathrm{vol}(K)}{\mathrm{vol}(E^\times \cap gKg^{-1})} 1_K(g^{-1}\gamma g).$$

Now $E^\times\backslash G/K$ is the set of $E^\times$-orbits on the set of all lattices in $E$. Representatives are the lattices $R(m)$, $m \geq 0$. So our sum is

$$\sum_{m\geq 0,\gamma\in R(m)^\times} \frac{\mathrm{vol}(\mathcal{O}_E^\times)}{\mathrm{vol}(R(m)^\times)} = \sum_{m\geq 0,\gamma\in R(m)^\times} (\mathcal{O}_E^\times : R(m)^\times).$$

Note that $(\mathcal{O}_E^\times : R(m)^\times) = 1$ if $m = 0$, $\quad = q^{m+1-f}\frac{q^f-1}{q-1}$ if $m > 0$.

Put $M = \max\{m|\gamma \in R(m)^\times\}$. Then the integral equals

$$q^M\frac{q+1}{q-1} - \frac{2}{q-1} \quad \text{if} \quad e = 1, \qquad\qquad \frac{q^{M+1}-1}{q-1} \quad \text{if} \quad e = 2.$$

(If $\gamma \notin \mathcal{O}_E^\times$, then $\int = 0$). If $\gamma = a + bw \in \mathcal{O}_E^\times$, then $M = \mathfrak{v}_F(b)$, the order-valuation at $b$.

**3.** Let $G = \mathrm{SL}(V)$, $K = \mathrm{Stab}(\mathcal{O}_E) \cap G$, $E^1 = E^\times \cap G$. Choose the Haar measure $dg$ on $G$ such that $\int_K dg = 1$, and $dt$ on $E^1$ such that $\int_{E^1} dt = 1$. Let $\gamma \in E^1$, $\gamma \neq \pm 1$. Then

$$\int_{E^1\backslash G} 1_K(g^{-1}\gamma g)\frac{dg}{dt} = \int_G 1_K(g^{-1}\gamma g)dg = \sum_{G/K} 1_K(g^{-1}\gamma g)$$

is the number of lattices in the $G$-orbit of $\mathcal{O}_E$ fixed by $\gamma$.

Let $\Lambda$ be a lattice in $E$. If $R(\Lambda) = \mathcal{O}_E$, then $\Lambda \in G\cdot\mathcal{O}_E \Leftrightarrow \Lambda = \mathcal{O}_E$. And $\gamma\mathcal{O}_E = \mathcal{O}_E$ if $\gamma$ fixes $\Lambda$. If $R(\Lambda) = R(m)$ with $m > 0$, then $\Lambda = zR(m) \in G\cdot\mathcal{O}_E \Leftrightarrow N_{E/F}(z)\boldsymbol{\pi}^m \in \mathcal{O}^\times \Leftrightarrow f\mathfrak{v}_E(z) = -m$ and $\gamma\Lambda = \Lambda \Leftrightarrow \gamma \in R(m)^\times$.

Suppose $e = 1$. Then $m$ must be even and

$$\Lambda = \boldsymbol{\pi}^{-\frac{m}{2}}uR(m), \qquad u \in \mathcal{O}_E^\times \bmod R(m)^\times.$$

If $\gamma \in R(m)^\times$, this gives $(\mathcal{O}_E^\times : R(m)^\times) = q^{m-1}(q+1)$ lattices.

Suppose $e = 2$. Then $\Lambda = \pi_E^{-m} u R(m)$, $u \in \mathcal{O}_E^\times \bmod R(m)^\times$. If $\gamma \in R(m)^\times$ this gives $(\mathcal{O}_E^\times : R(m)^\times) = q^m$ lattices.

Put $N = \max\{m | \gamma \in R(m)^\times,\, m \equiv 0(f)\}$. Then the integral equals

$$\frac{q^{N+1} - 1}{q - 1}.$$

For $K = \mathrm{Stab}(R(1)) \cap G$ one find $\frac{q^{N'+1}-1}{q-1}$ with $N'$ defined as $N$, but with $m \equiv 1(f)$.

**4.** Notations as in 3. Choose $\pi = N_{E/F}(\pi_E)$ if $e = 2$. The description of the lattices in $G \cdot \mathcal{O}_E$ above gives the following decomposition for $\mathrm{SL}(2, F)$.

Choose a set $A_m$ of representatives for $N_{E/F}\mathcal{O}_E^\times / N_{E/F}R(m)^\times$ and for each $\varepsilon \in A_m$ choose $b_\varepsilon$ such that $N_{E/F}(b_\varepsilon) = \varepsilon$. For $m = 0$ we may take $A_0 = \{1\}$, $b_1 = 1$.

$$\mathrm{SL}(2, F) = \bigcup_{m \geq 0, \text{even}} \bigcup_{\varepsilon \in A_m} E^1 b_\varepsilon^{-1} \begin{pmatrix} 1 & 0 \\ 0 & \varepsilon \end{pmatrix} \begin{pmatrix} \pi^{-\frac{m}{2}} & 0 \\ 0 & \pi^{\frac{m}{2}} \end{pmatrix} K \qquad \text{if } e = 1$$

$$\mathrm{SL}(2, F) = \bigcup_{m \geq 0} \bigcup_{\varepsilon \in A_m} E^1 b_\varepsilon^{-1} \pi_E^{-m} \begin{pmatrix} 1 & 0 \\ 0 & \varepsilon \end{pmatrix} \begin{pmatrix} 1 & 0 \\ 0 & \pi^m \end{pmatrix} K \qquad \text{if } e = 2.$$

REMARK. If $e = 1$, $m > 0$, then

$$N_{E/F}\mathcal{O}_E^\times / N_{E/F}R(m)^\times = \mathcal{O}^\times / \mathcal{O}^{\times 2}(1 + \pi^m \mathcal{O})$$

(two elements, when $|2| = 1$). If $|2| = 1$ and $e = 2$, then

$$N_{E/F}R(m)^\times = N_{E/F}\mathcal{O}_E^\times$$

for all $m$.

# 2. Case of GSp(2)

## 2a. Preliminaries

**1.** Let $V$ be a symplectic vector space defined over a field $F$ of characteristic $\neq 2$. We have $G = \mathrm{Sp}(V) \subset \mathrm{GL}(V) \subset \mathrm{End}(V) = A$. Let $\gamma$ be a regular semisimple element of $G(F)$, $T$ the centralizer of $\gamma$ in $G$, $C$ the conjugacy class of $\gamma$ in $G$.

If $L$ denotes the centralizer of $\gamma$ in $A$, we have $L(F) = \prod L_i$, a direct product of separable field extensions of $F$. The space $V(F)$ is isomorphic to $L(F)$ as $L(F)$-module and $V(F) = \oplus V_i(F)$, where $V_i(F)$ is a 1-dimensional vector space over $L_i$.

We denote the symplectic form on $V$ by $\langle x, y\rangle$ and define the involution $\iota$ of $A(F)$ by

$$\langle ux, y\rangle = \langle x, {}^{\iota}uy\rangle \qquad (x, y \in V(F), \quad u \in A(F)).$$

From ${}^{\iota}\gamma = \gamma^{-1}$ it follows that $\iota$ stabilizes $L(F)$. The restriction $\sigma$ of $\iota$ to $L(F)$ is a $F$-automorphism of $L(F)$ of order 2. It may interchange two components $L_i$ and $L_j$ $(i \neq j)$ and it can leave a component $L_i$ fixed. If $T$ is $F$-anisotropic we have $\sigma(L_i) = L_i$ for all $i$. Note that $T(F)$ is the set of $u \in L(F)^{\times}$ such that $u\sigma(u) = 1$. If $\sigma(L_i) = L_i$, then $V_i \perp V_j$ for all $j \neq i$.

$$\{G(F)\text{-orbits in } C(F)\} \quad \leftrightarrow \quad G(F)\backslash\{h \in A(F)^{\times} | h\gamma h^{-1} \in C\}/L(F)^{\times}$$

$$h \mapsto {}^{\iota}hh \quad \downarrow \text{bij}$$

$$\{u \in L(F)^{\times} | \sigma(u) = u\}/\{u\sigma(u) | u \in L(F)^{\times}\}$$

**2.** If we take $G = \mathrm{GSp}(V)$ instead of $\mathrm{Sp}(V)$, we have ${}^{\iota}\gamma\gamma \in F^{\times}$ and $T(F)$ is the set of $u \in L(F)^{\times}$ such that $u\sigma(u) \in F^{\times}$. Now

$$\{G(F)\text{-orbits in } C(F)\} \quad \leftrightarrow \quad G(F)\backslash\{h \in A(F)^{\times} | h\gamma h^{-1} \in C\}/L(F)^{\times}$$

$$h \mapsto {}^{\iota}hh \quad \downarrow \text{bij}$$

$$\{u \in L(F)^{\times} | \sigma(u) = u\}/F^{\times}\{u\sigma(u) | u \in L(F)^{\times}\}$$

In this case consider $T$ such that $T/Z$ is $F$-anisotropic ($Z$ = center of $G$ and of $A^{\times}$).

Notation: $L' = \{u \in L(F) | \sigma(u) = u\}$, $Nu = u\sigma(u)$ if $u \in L(F)^\times$.

**3.** Assume $F$ is a nonarchimedean local field. If $\Lambda$ is a lattice in $V(F)$, the dual lattice is

$$\Lambda^* = \{x \in V(F) | \langle x, y\rangle \in \mathcal{O} \text{ for all } y \in \Lambda\} \simeq \mathrm{Hom}_{\mathcal{O}}(\Lambda, \mathcal{O}).$$

Properties:

$$(u\Lambda)^* = {}^t u^{-1}\Lambda^* \qquad (u \in \mathrm{GL}(V(F))),$$

in particular $(c\Lambda)^* = c^{-1}\Lambda^*$ if $c \in F^\times$ and $(g\lambda)^* = g\Lambda^*$ if $g \in \mathrm{Sp}(V(F))$. Further, $\Lambda^{**} = \Lambda$.

The lattices which are equal (resp. proportional by a factor in $F^\times$) to their dual form one orbit of $\mathrm{Sp}(V(F))$ (resp. $\mathrm{GSp}(V(F))$).

We want to compute the following numbers.

Orbital integral for $\mathrm{Sp}(V(F))$: $\mathrm{Card}\{\Lambda | \Lambda^* = \Lambda, \gamma\Lambda = \Lambda\}$.

Stable orbital integral for $\mathrm{Sp}(V(F))$:

$$\sum_{\nu \in L'^\times / N_{L(F)/L'} L(F)^\times} \mathrm{Card}\{\Lambda | \Lambda^* = \nu\Lambda, \gamma\Lambda = \Lambda\}.$$

Orbital integral for $\mathrm{GSp}(V(F))$:

$$\mathrm{Card}\{\Lambda | \Lambda^* \sim \Lambda, \gamma\Lambda = \Lambda\}/F^\times = \sum_{\alpha \in F^\times / F^{\times 2}\mathcal{O}^\times} \mathrm{Card}\{\Lambda | \Lambda^* = \alpha\Lambda, \gamma\Lambda = \Lambda\}.$$

Stable orbital integral for $\mathrm{GSp}(V(F))$:

$$\sum_{\nu \in L'^\times / F^\times N L(F)^\times} \ \sum_{\alpha \in F^\times / F^{\times 2}\mathcal{O}^\times} \mathrm{Card}\{\Lambda | \Lambda^* = \alpha\nu\Lambda, \gamma\Lambda = \Lambda\}$$

$$= \frac{2}{(F^\times : F^\times \cap NL(F)^\times)} \sum_{\nu \in L'^\times / NL(F)^\times} \mathrm{Card}\{\Lambda | \Lambda^* = \nu\Lambda, \gamma\Lambda = \Lambda\}.$$

So the stable orbital integrals for $\mathrm{Sp}(V(F))$ and $\mathrm{GSp}(V(F))$ differ by a factor, which is a power of 2, when $\gamma \in \mathrm{Sp}(V(F))$.

**4.** Let $L/F$ be a quadratic extension of nonarchimedean local fields. The orders of $L$ are $\mathcal{O}_L(n) = \mathcal{O}_F + \pi_F^n \mathcal{O}_L$ $(n \geq 0)$. We can find $w \in L$ such that $\mathcal{O}_L(n) = \mathcal{O}_F + \mathcal{O}_F \pi_F^n w$ for all $n \geq 0$. Any lattice in $L$ is of the form $z\mathcal{O}_L(n)$, $z \in L^\times$, $n \geq 0$.

Let a symplectic form on the $F$-vector space $L$ be given.

If $\langle 1, w\rangle \in \mathcal{O}_F^\times$, the lattice dual to $z\mathcal{O}_L(n)$ is $\overline{z}^{-1}\boldsymbol{\pi}_F^{-n}\mathcal{O}_L(n)$.

$$(\mathcal{O}_L^\times : \mathcal{O}_L(n)^\times) = 1 \quad (n = 0), \qquad q^{n-1}(q+1) \quad (n > 0),$$
$$\text{if } L/F \text{ is unramified,}$$
$$= q^n \quad (n \geq 0), \qquad \text{if } L/F \text{ is ramified.}$$

Here $q$ = number of elements of the residual field of $F$.

**5.** Let $V = V_1 \oplus V_2$ be a direct sum of two vector spaces over a nonarchimedean local field $F$. Let $\Lambda$ be a lattice in $V$. Put $M_i = \Lambda \cap V_i$ and $N_i = \mathrm{pr}_i(\Lambda)$. Then $M_i$ and $N_i$ are lattices in $V_i$, and $M_i \subset N_i$.

The set
$$\{(\nu_1 + M_1, \nu_2 + M_2) | \nu_1 + \nu_2 \in \Lambda\}$$
is the graph of an isomorphism between $N_1/M_1$ and $N_2/M_2$.

The lattices in $V$ correspond bijectively to the data: $M_1 \subset N_1$, lattices in $V_1$; $M_2 \subset N_2$, lattices in $V_2$; $N_1/M_1 \to N_2/M_2$, an isomorphism of (finite) $\mathcal{O}$-modules.

Assume a symplectic form is given on $V$ and $V = V_1 \oplus V_2$ is an orthogonal direct sum. If the lattice $\Lambda$ corresponds to the data

$$M_1 \subset N_1, \qquad M_2 \subset N_2, \qquad \varphi : N_1/M_1 \xrightarrow{\sim} N_2/M_2,$$

then the data of the dual lattice $\Lambda^*$ are:

$$N_1^* \subset M_1^*, \qquad N_2^* \subset M_2^*, \qquad -(\varphi^*)^{-1} : M_1^*/N_1^* \to M_2^*/N_2^*.$$

One may identify $M_i^*/N_i^*$ with $\mathrm{Hom}_{\mathcal{O}}(N_i/M_i, F/\mathcal{O})$ using $\langle \nu, \nu'\rangle$, $\nu \in N_i$, $\nu' \in M_i^*$. Then $\varphi^*$ is defined using this identification.

**6.** In the notation of section 1 assume that $L(F)$ is a field. For brevity write $L$ for this field. Let $L'$ be the field of fixed points of $\sigma$, so $[L{:}L'] = 2$. We identify $V(F)$ with $L(F) = L$ and have then $\langle x, y\rangle = \mathrm{tr}_{L/F}\,(a\sigma(x)y)$, with some $a \in L^\times$ such that $\sigma(a) = -a$. Put $\langle x, y\rangle' = \mathrm{tr}_{L/L'}(a\sigma(x)y)$. This is a symplectic form on $L$ over $L'$. We have $\langle x, y\rangle = \mathrm{tr}_{L'/F}(\langle x, y\rangle')$ and $\langle zx, y\rangle = \langle x, zy\rangle$ if $z \in L'$.

Assume now that $F$ is local, nonarchimedean. If $M$ is an $\mathcal{O}_{L'}$-lattice in $L$, then
$$M^* = \{x \in L | \langle x, y\rangle \in \mathcal{O} \quad \text{for all} \quad y \in M\}$$

is also an $\mathcal{O}_{L'}$-lattice. The dual

$$\widetilde{M} = \{x \in L | \langle x, y\rangle' \in \mathcal{O}_{L'} \quad \text{for all} \quad y \in M\}$$

of $M$ as an $\mathcal{O}_{L'}$-lattice is related to $M^*$ by the formula $\widetilde{M} = \mathcal{D}_{L'/F}M^*$, where $\mathcal{D}_{L'/F}$ is the different of $L'/F$.

We have $\widetilde{gM} = \det(g)^{-1}g\widetilde{M}$ if $g \in \mathrm{GL}_{L'}(L)$, in particular $\widetilde{uM} = \sigma(u)^{-1}\widetilde{M}$ if $u \in L^\times$.

In the remainder of this section we assume $\dim V = 4$.

The nonidentical automorphism of $L'/F$ is denoted by $\tau$ or by $z \mapsto \overline{z}$.

Let $\Lambda$ be an $\mathcal{O}$-lattice in $L$. Put $M = \mathcal{O}_{L'}\Lambda$, $N = \widetilde{\mathcal{O}_{L'}\Lambda^*}$. Then $M^* = \{x \in L | \mathcal{O}_{L'}x \subset \Lambda^*\}$ is the largest $\mathcal{O}_{L'}$-lattice contained in $\Lambda^*$ and $N = \mathcal{D}_{L'/F}\cdot$ largest $\mathcal{O}_{L'}$-lattice contained in $\Lambda$. We have $M \supset \Lambda \supset N$.

If $a_i \in \mathcal{O}_{L'}, x_i \in \Lambda$, then

$$\sum a_i x_i \in N \Rightarrow \sum \tau(a_i)x_i \in N.$$

Indeed, $u \in N \Leftrightarrow \langle u, y\rangle' \in \mathcal{O}_{L'}$ for all $y \in \Lambda^*$. If $\sum a_i x_i \in N$, then $\sum a_i\langle x_i y\rangle' \in \mathcal{O}_{L'}$ for all $y \in \Lambda^*$. Since

$$\langle x_i, y\rangle' + \tau(\langle x_i, y\rangle') = \langle x_i, y\rangle \in \mathcal{O},$$

it follows that $\sum \tau(a_i)\langle x_i, y\rangle' \in \mathcal{O}_{L'}$.

So we can define a homomorphism

$$\varphi : M/N \to M/N \quad \text{by} \quad \sum a_i x_i + N \mapsto \sum \tau(a_i)x_i + N,$$

whenever $a_i \in \mathcal{O}_{L'}, x_i \in \Lambda$.

The homomorphism $\varphi$ is $\mathcal{O}_{L'}$-semilinear and $\varphi^2 = \mathrm{id}$.

The set $\Lambda/N$ is the set of fixed points of $\varphi$. Indeed, if $\sum \tau(a_i)x_i - \sum a_i x_i \in N$, then

$$\sum \tau(a_i)\tau(\langle x_i, y\rangle') + \sum a_i\langle x_i, y\rangle' \in \mathcal{O}_{L'},$$

hence $\langle \sum a_i x_i, y\rangle \in \mathcal{O}$ for all $y \in \Lambda^*$, i.e. $\sum a_i x_i \in \Lambda$.

Conversely, let $M \supset N$ be two $\mathcal{O}_{L'}$-lattices in $L$ and $\varphi : M/N \to M/N$ an $\mathcal{O}_{L'}$-semilinear homomorphism.

Necessary conditions for $(M, N, \varphi)$ to correspond to a lattice $\Lambda$ are:

$$\varphi^2 = \mathrm{id}, \quad N \subset \mathcal{D}_{L'/F}M, \quad \varphi \equiv \mathrm{id} \bmod \mathcal{D}_{L'/F}M/N,$$

$$\varphi = \mathrm{id} \quad \text{on} \quad \mathcal{D}_{L'/F}^{-1}N/N.$$

These conditions are also sufficient when $L'/F$ is unramified (in which case the only condition is $\varphi^2 = \mathrm{id}$) and when $L'/F$ is tamely ramified. If $\Lambda$ exists, it is unique, since $\Lambda/N$ is the set of fixed points of $\varphi$.

The lattice $\widetilde{M}$ can be identified with $\mathrm{Hom}_{\mathcal{O}_{L'}}(M, \mathcal{O}_{L'})$ using $\langle m, \widetilde{m}\rangle'$ (this gives $M \xrightarrow{\sim} \widetilde{\widetilde{M}} = M$, $m \mapsto -m$). If $M \supset N$, then

$$\mathrm{Hom}_{\mathcal{O}_{L'}}(M/N, L'/\mathcal{O}_{L'}) = \widetilde{N}/\widetilde{M}.$$

If $\varphi : M/N \to M/N$ is $\mathcal{O}_{L'}$-semilinear, then $f \mapsto \tau f \varphi$ is a semilinear endomorphism $\widetilde{\varphi}$ of $\mathrm{Hom}_{\mathcal{O}_{L'}}(M/N, L'/\mathcal{O}_{L'})$, which on $\widetilde{N}/\widetilde{M}$ is given by

$$\langle \varphi(m), \widetilde{n}\rangle' \equiv \tau(\langle m, \widetilde{\varphi}(\widetilde{n})\rangle') \bmod \mathcal{O}_{L'}.$$

If $\Lambda \mapsto (M, N, \varphi)$ then $\Lambda^* \mapsto (\widetilde{N}, \widetilde{M}, -\widetilde{\varphi})$.

**7.** In the following computations $F$ is a nonarchimedean local field. Notations are as in section 1, $\dim V = 4$. We have either $L(F)$ is a field or $L(F)$ is the product of two quadratic fields.

## 2b. $L(F)$ is a Product

**1.** Assume $L(F) = L_1 \times L_2$, $[L_i{:}F] = 2$. Then $V(F) = V_1 \oplus V_2$, $V_i$ a 1-dimensional vector space over $L_i$, $V_1 \perp V_2$. We identify $V_i$ with $L_i$. Then

$$T(F) = \{(t_1, t_2) \in L_1^\times \times L_2^\times | N_{L_i/F}(t_i) = 1 \quad \text{for} \quad i = 1, 2\}.$$

We compute the number of lattices $\Lambda$ in $V(F)$ which satisfy $\Lambda^* = \nu\Lambda$ and $\gamma\Lambda = \Lambda$, for a given regular element $\gamma$ of $T(F)$ and a set of representatives $\nu$ of $F^\times/N_{L_1/F}L_1^\times \times F^\times/N_{L_2/F}L_2^\times$.

By section 2a.5 the lattice $\Lambda$ is given by lattices $M_i \subset N_i$ in $L_i$ $(i = 1, 2)$ and an isomorphism $\varphi : N_1/M_1 \to N_2/M_2$.

Let $\gamma = (t_1, t_2)$ and $\nu = (\nu_1, \nu_2)$.

The condition $\Lambda^* = \nu\Lambda$ means that $N_i = \nu_i^{-1}M_i^*$ $(i = 1, 2)$ and $\nu_2\varphi\nu_1^{-1} = -(\varphi^*)^{-1}$. Then $\gamma\Lambda = \Lambda$ is equivalent to $t_iM_i = M_i$ $(i = 1, 2)$ and $t_2\varphi t_1^{-1} = \varphi$.

Put $M_i = z_i\mathcal{O}_{L_i}(m_i)$ with $z_i \in L_i^\times, m_i \geq 0$.

Choose $w_i \in L_i$ such that $\mathcal{O}_{L_i} = \mathcal{O} + \mathcal{O}w_i$. On each $V_i = L_i$ there is only one symplectic form, up to a factor from $F^\times$, and in order to compute our four numbers we may assume that $\langle 1, w_1\rangle = \langle 1, w_2\rangle = 1$. Then $M_i^* = \overline{z}_i^{-1}\boldsymbol{\pi}^{m_i}\mathcal{O}_{L_i}(m_i)$ and

$$M_i \subset \nu_i^{-1}M_i^* \Leftrightarrow \nu_i N_{L_i/F}(z_i)\boldsymbol{\pi}^{m_i} \in \mathcal{O}.$$

Moreover, $\nu_1^{-1}M_1^*/M_1$ and $\nu_2^{-1}M_2^*/M_2$ have to be isomorphic. This means that $\mathfrak{v}(\nu_i N_{L_i/F}(z_i)\boldsymbol{\pi}^{m_i})$ must be independent of $i$. So put

$$m = \mathfrak{v}(\nu_1) + \mathfrak{v}(N_{L_i/F}(z_1)) + m_1 = \mathfrak{v}(\nu_2) + \mathfrak{v}(N_{L_2/F}(z_2)) + m_2 \geq 0.$$

Then

$$N_i/M_i = \nu_i^{-1}M_i^*/M_i \simeq \mathcal{O}_{L_i}(m_i)/\boldsymbol{\pi}^m\mathcal{O}_{L_i}(m_i).$$

With respect to the bases $1, \boldsymbol{\pi}^{m_i}w_i$ of $\mathcal{O}_{L_i}(m_i)$, the isomorphism $\varphi$ is given by a matrix $\varphi \in \mathrm{GL}(2, \mathcal{O}/\boldsymbol{\pi}^m\mathcal{O})$ satisfying (from $\nu_2\varphi\nu_1^{-1} = -(\varphi^*)^{-1}$)

$$\det(\varphi) = -\nu_2\nu_1^{-1}\boldsymbol{\pi}^{m_2-m_1}N_{L_2/F}(z_2)N_{L_1/F}(z_1)^{-1}(\mathrm{mod}\,\boldsymbol{\pi}^m\mathcal{O}).$$

The conditions with respect to $\gamma$ are: $\quad t_i \in \mathcal{O}_{L_i}(m_i), \quad t_2\varphi t_1^{-1} = \varphi.$

The number to compute is the sum over $m \geq 0$ of

$$\sum_{\substack{m_1,m_2\geq 0\\ t_i\in\mathcal{O}_{L_i}(m_i)}} \sum_{\substack{z_1\in L_1^\times, z_2\in L_2^\times\\ z_i \bmod \mathcal{O}_{L_i}(m_i)^\times\\ f_i\mathfrak{v}_{L_i}(z_i)=m-m_i-\mathfrak{v}(v_i)}} \mathrm{Card}\{\varphi \in \mathrm{GL}(2, \mathcal{O}/\boldsymbol{\pi}^m\mathcal{O})|\bullet\}$$

where $\bullet$ stands for $t_2\varphi = \varphi t_1, \det(\varphi) = u$ and

$$u = -\nu_2\nu_1^{-1}\boldsymbol{\pi}^{m_2-m_1}N_{L_2/F}(z_2)N_{L_1/F}(z_1)^{-1}.$$

Here Card is 1 when $m = 0$.

**2.** We now assume that $|2| = 1$ in $F$. Then $w_i$ can be so chosen that $w_i^2 = \alpha_i \in \mathcal{O}$. Put $t_i = a_i + b_i\boldsymbol{\pi}^{m_i}w_i$ with $a_i, b_i \in \mathcal{O}$.

Let $\varphi = \left(\begin{smallmatrix} x_1 & x_2 \\ x_3 & x_4 \end{smallmatrix}\right) \in \mathrm{GL}(2, \mathcal{O}/\boldsymbol{\pi}^m\mathcal{O})$. The matrix corresponding to $t_i$ is

$$\begin{pmatrix} a_i & \alpha_i b_i \boldsymbol{\pi}^{2m_i} \\ b_i & a_i \end{pmatrix}.$$

We have $a_1^2 - \alpha_i b_i^2 \boldsymbol{\pi}^{2m_i} = 1$ $(i = 1, 2)$.

Assume $m > 0$. The conditions on $\varphi$ are:

$$\left.\begin{aligned} (a_1 - a_2)x_1 + b_1 x_2 - \alpha_2 b_2 \boldsymbol{\pi}^{2m_2} x_3 &\equiv 0 \\ \alpha_1 b_1 \boldsymbol{\pi}^{2m_1} x_1 + (a_1 - a_2)x_2 - \alpha_2 b_2 \boldsymbol{\pi}^{2m_2} x_4 &\equiv 0 \\ b_2 x_1 + (a_2 - a_1)x_3 - b_1 x_4 &\equiv 0 \\ b_2 x_2 - \alpha_1 b_1 \boldsymbol{\pi}^{2m_1} x_3 + (a_2 - a_1)x_4 &\equiv 0 \\ x_1 x_4 - x_2 x_3 &\equiv u \end{aligned}\right\} \bmod \boldsymbol{\pi}^m$$

where $u$ is an element of $\mathcal{O}^\times$.

This system cannot be solvable unless $a_1 \equiv a_2(\boldsymbol{\pi}^m)$, since $t_2 = \varphi t_1 \varphi^{-1}$ implies $\mathrm{tr}(t_2) = \mathrm{tr}(t_1)$. Assume this. Then

$$\left.\begin{aligned} &(1) \quad b_2 x_1 \equiv b_1 x_4 \\ &(2) \quad \alpha_1 b_1 \boldsymbol{\pi}^{2m_1} x_1 \equiv \alpha_2 b_2 \boldsymbol{\pi}^{2m_2} x_4 \\ &(3) \quad b_1 x_2 \equiv \alpha_2 b_2 \boldsymbol{\pi}^{2m_2} x_3 \\ &(4) \quad b_2 x_2 \equiv \alpha_1 b_1 \boldsymbol{\pi}^{2m_1} x_3 \\ &(5) \quad x_1 x_4 - x_2 x_3 \equiv u \end{aligned}\right\} \bmod \boldsymbol{\pi}^m$$

This system is unsolvable when $\mathfrak{v}(b_2) > \mathfrak{v}(b_1)$ and $m > \mathfrak{v}(b_1)$, as (1) and (3) would imply that $x_4 \equiv x_2 \equiv 0(\boldsymbol{\pi})$; and also when $\mathfrak{v}(b_1) > \mathfrak{v}(b_2)$ and $m > \mathfrak{v}(b_2)$, as (1) and (4) would imply that $x_1 \equiv x_2 \equiv 0(\boldsymbol{\pi})$. It remains to consider: $m \le \mathfrak{v}(b_i)$ $(i = 1, 2)$ or $m > \mathfrak{v}(b_1) = \mathfrak{v}(b_2)$.

Suppose $m \le \mathfrak{v}(b_1)$ and $m \le \mathfrak{v}(b_2)$. Then $x_1 x_4 - x_2 x_3 \equiv u(\boldsymbol{\pi}^m)$ has $q^{3m-2}(q^2 - 1)$ solutions.

Suppose $m > \mathfrak{v}(b_1) = \mathfrak{v}(b_2)$. Put $k = \mathfrak{v}(b_1) = \mathfrak{v}(b_2)$. Put $c_i = \alpha_i b_i \boldsymbol{\pi}^{2m_i}$ $(i = 1, 2)$. Then (1)-(4) imply that

$$(b_1 c_1 - b_2 c_2) x_i \equiv 0 \bmod \boldsymbol{\pi}^{m+k}$$

for all $i$, so we must necessarily have $b_1 c_1 \equiv b_2 c_2 \bmod \boldsymbol{\pi}^{m+k}$. This is equivalent to $a_1^2 \equiv a_2^2 \bmod \boldsymbol{\pi}^{m+k}$ and implies that either $\mathfrak{v}(c_i) \ge m$ for $i = 1, 2$

or $\mathfrak{v}(c_1) = \mathfrak{v}(c_2) < m$. From (3) and (4) it follows that $x_2 \equiv 0(\boldsymbol{\pi})$, unless $\mathfrak{v}(c_1) = \mathfrak{v}(c_2) = k$, i.e. $\mathfrak{v}(\alpha_i) = 0$, $m_i = 0$ $(i = 1, 2)$. Assume we are not in this case. Then $\mathfrak{v}(c_i) > k$ for $i = 1, 2$. Also $x_2 \equiv 0(\boldsymbol{\pi})$, hence $x_1 \in \mathcal{O}^\times$ and (5) gives $x_4 \equiv x_1^{-1}x_2x_3 + x_1^{-1}u \bmod \boldsymbol{\pi}^m(*)$. (3) and (4) give $x_2 \equiv b_1^{-1}c_2x_3 \bmod \boldsymbol{\pi}^{m-k}(**)$. (2) is a consequence of (1). After substitution of $(*)$ and $(**)$ the congruence (1) reads

$$x_1^2 \equiv b_2^{-1}c_2x_3^2 + b_1b_2^{-1}u \bmod \boldsymbol{\pi}^{m-k}.$$

Here $b_2^{-1}c_2 \equiv 0(\boldsymbol{\pi})$.

We find $2q^{m+2k}$ solutions when $b_1b_2^{-1}u$ is a square in $F$, and otherwise no solution.

Now suppose that $\mathfrak{v}(\alpha_i) = 0$, $m_i = 0$ $(i = 1, 2)$. Then $L_1 = L_2 =$ the unramified quadratic extension of $F$. Take $\alpha_1 = \alpha_2 = \alpha$. We have $b_1^2 \equiv b_2^2 \bmod \boldsymbol{\pi}^{m+k}$. Now (1) and (2) are equivalent, (3) and (4) are equivalent. From (1), (3), (5) one deduces that $x_1^2 - \alpha x_3^2 \equiv b_1b_2^{-1}u \bmod \boldsymbol{\pi}^{m-k}$. This congruence has $q^{m-k-1}(q+1)$ solutions modulo $\boldsymbol{\pi}^{m-k}$. For each solution $x_1 \in \mathcal{O}^\times$ or $x_3 \in \mathcal{O}^\times$.

If $x_1 \in \mathcal{O}^\times$ we have

$$x_2 \equiv \alpha b_2 b_1^{-1} x_3 \ (\boldsymbol{\pi}^{m-k}), \qquad x_4 \equiv x_1^{-1}(x_2x_3 + u)\ (\boldsymbol{\pi}^m).$$

If $x_3 \in \mathcal{O}^\times$ we have $x_4 \equiv b_2b_1^{-1}x_1\ (\boldsymbol{\pi}^{m-k})$, $x_2 \equiv x_3^{-1}(x_1x_4 - u)\ (\boldsymbol{\pi}^m)$.
So there are $q^{m+2k-1}(q+1)$ solutions for the system in this case.

**3.** Recall that $\mathcal{O}_{L_i} = \mathcal{O} + \mathcal{O}w_i$, $w_i^2 = \alpha_i$ and $|2| = 1$. Let $t_i = a_i + b_iw_i$, $a_i^2 - \alpha_i b_i^2 = 1$ $(i = 1, 2)$. As $t = (t_1, t_2)$ is supposed to be regular, we have $b_1 \neq 0$, $b_2 \neq 0$, and in case $L_1 = L_2$, $a_1 \neq a_2$. Let us be given:

$\nu_1, \nu_2 \in F^\times \qquad (\nu_i \bmod N_{L_i/F}L_i^\times)$;
$m \geq 0$;
$m_1, m_2 \geq 0$ such that $t_i \in \mathcal{O}_{L_i}(m_i)$, i.e. $m_i \leq \mathfrak{v}(b_i)$;
$z_1 \in L_1^\times$, $z_2 \in L_2^\times$ $(z_i \bmod \mathcal{O}_{L_i}(m_i)^\times)$ with $f_i\mathfrak{v}_{L_i}(z_i) = m - m_i - \mathfrak{v}(v_i)$.

Put $u = -\nu_2\nu_1^{-1}\boldsymbol{\pi}^{m_2-m_1}N_{L_2/F}(z_2)N_{L_1/F}(z_1)^{-1}$. Then $u \in \mathcal{O}^\times$.
By section 2 we have that

$$\mathrm{Card}\{\varphi \in \mathrm{GL}(2, \mathcal{O}/\boldsymbol{\pi}^m\mathcal{O}) |\quad t_2\varphi = \varphi t_1, \quad \det(\varphi) = u\}$$

is given by

$1$ if $m = 0$;
$q^{3m-2}(q^2-1)$ if $m > 0$, $m + m_i \le \mathfrak{v}(b_i) (i = 1, 2)$, $a_1 \equiv a_2(\boldsymbol{\pi})$;
$2q^{m+2k}$ if $\mathfrak{v}(b_i) = m_i + k$, $0 \le k < m$, and
*either* $\mathfrak{v}(\alpha_i) + 2m_i + k \ge m$ $(i = 1, 2)$, $a_1 \equiv a_2(\boldsymbol{\pi})$, and $\bullet$,
where we put $\bullet$ for $-\nu_2\nu_1^{-1}b_2b_1^{-1}N(z_2)N(z_1)^{-1} \in F^{\times 2}$,
*or* $\alpha_1 = \alpha_2$, $m_1 = m_2$, $k < \mathfrak{v}(\alpha_i) + 2m_i + k < m$, $\bullet$, and $a_1 \equiv a_2(\boldsymbol{\pi}^{m+k})$;
$q^{m+2k-1}(q+1)$ if $\alpha_1 = \alpha_2 \in \mathcal{O}^\times$, $m_1 = m_2 = 0$, $0 \le \mathfrak{v}(b_1) = \mathfrak{v}(b_2) = k < m$, and $a_1 \equiv a_2(\boldsymbol{\pi}^{m+k})$.

It is zero in all other cases.

We are computing

$$\sum_{\nu_1,\nu_2} \sum_{m\ge 0} \sum_{\substack{0\le m_i \le \mathfrak{v}(b_i)\\ m_i \equiv m+\mathfrak{v}(\nu_i) \bmod f_i}} \sum_{z_i' \in \mathcal{O}_{L_i}^\times/\mathcal{O}_{L_i}(m_i)^\times} \mathrm{Card}\{\varphi\}.$$

We put $z_i = z_i'\boldsymbol{\pi}_{L_i}^{m-m_i-\mathfrak{v}(\nu_i)/f_i}$. The condition $\bullet$ becomes

$$N_{L_1/F}(z_1')N_{L_2/F}(z_2')^{-1}$$
$$\in -\nu_2\boldsymbol{\pi}_2^{-\mathfrak{v}(\nu_2)}\nu_1^{-1}\boldsymbol{\pi}_1^{\mathfrak{v}(\nu_1)}b_2\boldsymbol{\pi}_2^{-\mathfrak{v}(b_2)}b_1^{-1}\boldsymbol{\pi}_1^{\mathfrak{v}(b_1)}(\boldsymbol{\pi}_2\boldsymbol{\pi}_1^{-1})^{k+m}\mathcal{O}^{\times 2},$$

where $N_{L_i/F}(\boldsymbol{\pi}_{L_i}) = \boldsymbol{\pi}_i^{f_i}$.

Notice that $\mathrm{Card}\{\varphi\}$ is independent of $z_1'$, $z_2'$, except for the cases where the condition $\bullet$ plays a role. In those cases one has $m_i > 0$ when $L_i/F$ is unramified, so that $\mathcal{O}_{L_i}(m_i)^\times \subset \mathcal{O}_{L_i}^{\times 2}$. This is used in the following.

Our sum is the sum of the following sums.

*I)* $$\prod_{i=1,2} e_i \sum_{0\le k\le \mathfrak{v}(b_i)} (\mathcal{O}_{L_i}^\times : \mathcal{O}^{L_i}(k)^\times);$$

*II)* $$e_1e_2 \sum_{\substack{m>0, m_i\ge 0\\ m+m_i\le \mathfrak{v}(b_i)}} q^{3m-2}(q^2-1)AB$$

if $a_1 \equiv a_2(\boldsymbol{\pi})$; here $A = (\mathcal{O}_{L_1}^\times : \mathcal{O}_{L_1}(m_1)^\times)$, $B = (\mathcal{O}_{L_2}^\times : \mathcal{O}_{L_2}(m_2)^\times)$;

*III)* $$\frac{1}{2}e_1e_2 \sum_{\substack{0\le k<m, k\le \mathfrak{v}(b_i)\\ k+m\le \mathfrak{v}(\alpha_i)+2\mathfrak{v}(b_i)}} 2q^{m+2k}AB$$

if $a_1 \equiv a_2(\boldsymbol{\pi})$; here $A = (\mathcal{O}_{L_1}^\times : \mathcal{O}_{L_1}(\mathfrak{v}(b_1)-k)^\times)$, $(\mathcal{O}_{L_2}^\times : \mathcal{O}_{L_2}(\mathfrak{v}(b_2)-k)^\times)$.

If $\alpha_1 = \alpha_2$ and $\mathfrak{v}(b_1) = \mathfrak{v}(b_2)$, put $A = (\mathcal{O}_{L_1}^\times : \mathcal{O}_{L_1}(\mathfrak{v}(b_1) - k)^\times)^2$:

$$IV) \qquad \frac{1}{2} e_1 e_2 \sum_{\substack{0 \le k \le \mathfrak{v}(b_1), 2k < \mathfrak{v}(\alpha_1) + 2\mathfrak{v}(b_1) \\ \mathfrak{v}(\alpha_1) + 2\mathfrak{v}(b_1) < m + k \le \mathfrak{v}(a_1 - a_2)}} 2q^{m+2k} A.$$

If $\alpha_1 = \alpha_2 \in \mathcal{O}^\times$ and $\mathfrak{v}(b_1) = \mathfrak{v}(b_2)$:

$$V) \qquad \sum_{\mathfrak{v}(b_1) < m \le \mathfrak{v}(a_1 - a_2) - \mathfrak{v}(b_1)} q^{m + 2\mathfrak{v}(b_1) - 1}(q+1).$$

Put $M_i = \mathfrak{v}(b_i)$, $M = \max(M_1, M_2)$, $N = \min(M_1, M_2)$. The sub-sums are:

$$e_i \sum_{0 \le k \le M_i} (\mathcal{O}_{L_i}^\times : \mathcal{O}_{L_i}(k)^\times) = \frac{q^{M_i+1} + q^{M_i} - 2}{q-1} \qquad \text{if } e_i = 1,$$

$$I) \qquad = \frac{2(q^{M_i+1} - 1)}{q-1} \qquad \text{if } e_i = 2.$$

$$II) \quad \begin{cases} \frac{q+1}{q-1}\{q^{M+N-1}(q+1)^2 \frac{q^N-1}{q-1} & \text{if } e_1 = e_2 = 1 \\ \quad -2(q^M + q^N)(q+1)\frac{q^{2N}-1}{q^2-1} + 4q\frac{q^{3N}-1}{q^3-1}\}; & \\ 2\,\frac{q+1}{q-1}\{q^{M_1+M_2}(q+1)\frac{q^N-1}{q-1} & \text{if } e_1 = 1, e_2 = 2 \\ \quad -(q^{M_1+1} + q^{M_1} + 2q^{M_2+1})\frac{q^{2N}-1}{q^2-1} + 2q\frac{q^{3N}-1}{q^3-1}\}; & \\ 4\,\frac{q+1}{q-1}\{q^{M+N+1}\,\frac{q^N-1}{q-1} - (q^{M+1} + q^{N+1})\frac{q^{2N}-1}{q^2-1} + q\frac{q^{3N}-1}{q^3-1}\} & \\ & \text{if } e_1 = e_2 = 2; \end{cases}$$

if $a_1 \equiv a_2(\boldsymbol{\pi})$.

$$III) \quad \begin{cases} q^{M+N-1}(q+1)^2(q^N-1)(q^{N+1}-1)(q-1)^{-2} & \\ & \text{if } e_1 = e_2 = 1; \\ 2q^{M+N}(q+1)(q^N-1)(q^{N+1}-1)(q-1)^{-2} & \\ & \text{if } e_1 = 1, e_2 = 2, M_1 \le M_2; \\ 2q^{M+N}(q+1)(q^{N+1}-1)^2(q-1)^{-2} & \\ & \text{if } e_1 = 1, e_2 = 2, M_1 > M_2; \\ 4q^{M+N+1}(q^{N+1}-1)^2(q-1)^{-2} & \\ & \text{if } e_1 = e_2 = 2. \end{cases}$$

if $a_1 \equiv a_2(\boldsymbol{\pi})$.

$$IV)\begin{cases} q^{3N}(q+1)^2(q^N-1)(q^{\mathfrak{v}(a_1-a_2)-2N}-1)(q-1)^{-2} & \text{if } e_1=1,\\ 4q^{3N+2}(q^{N+1}-1)(q^{\mathfrak{v}(a_1-a_2)-2N-1}-1)(q-1)^{-2} & \text{if } e_2=2.\end{cases}$$

if $L_1 = L_2$, $a_1 \equiv a_2(\boldsymbol{\pi})$.

$$V) \qquad q^{3N}(q+1)(q^{\mathfrak{v}(a_1-a_2)-2N}-1)(q-1)^{-1}$$

if $L_1 = L_2$ is unramified and $a_1 \equiv a_2(\boldsymbol{\pi})$.

The formulas IV) and V) hold even when $\mathfrak{v}(b_1) \neq \mathfrak{v}(b_2)$, because we have then $\mathfrak{v}(a_1 - a_2) = 2N + \mathfrak{v}(\alpha_1)$.

## 2c. $L(F)$ is a Quartic Extension

**1.** Assume $L(F) = L$ is a field. We identify $V(F)$ with $L(F)$. A quadratic subfield $L'$ of $L$ is given and $T(F) = \{t \in L^\times | N_{L/L'}(t) = 1\}$. We compute the number of lattices $\Lambda$ in $L$ which satisfy $\Lambda^* = \nu\Lambda$ and $t\Lambda = \Lambda$, for a given regular element of $T(F)$ and a set of representatives $\nu$ of $L'^\times/N_{L/L'}L^\times$. That $t$ is regular means that $F(t) = L$.

We use the $L'$-bilinear alternating form $\langle x, y\rangle' = \mathrm{tr}_{L/L'}(a\sigma(x)y)$ introduced in section 6 of Case of SL(2) (we shall choose $a$ later). Assume that $L'/F$ is unramified or tamely ramified. A lattice $\Lambda$ is given by $\mathcal{O}_{L'}$-lattices $M \supset N$ and a map $\varphi : M/N \to M/N$ satisfying certain conditions (see section 6). Moreover, $\Lambda^* = \nu\Lambda$ is equivalent to $M = \nu^{-1}\widetilde{N}$ and $-\widetilde{\varphi} = \nu\varphi\nu^{-1}$. The identity $t\Lambda = \Lambda$ is equivalent to

$$tM = M, \qquad tN = N, \qquad t\varphi t^{-1} = \varphi.$$

Put $N = u\mathcal{O}_L(n)$, $u \in L^\times$, $n \geq 0$. Assuming that $\min\limits_{x,y\in\mathcal{O}_L} \mathfrak{v}_{L'}(\langle x, y\rangle') = 0$ (cf. section 4 of Case of SL(2); we come back to this later) we have

$$M = \nu^{-1}\widetilde{N} = \nu^{-1}\sigma(u)^{-1}\boldsymbol{\pi}_{L'}^{-n}\mathcal{O}_L(n).$$

Now

$$M \supset N \Leftrightarrow \nu N_{L/L'}(u)\boldsymbol{\pi}_{L'}^{n} \in \mathcal{O}_{L'}.$$

Put

$$m = \mathfrak{v}_{L'}(\nu) + f_{L/L'}\mathfrak{v}_L(u) + n \geq 0.$$

Then

$$M/N \simeq \mathcal{O}_L(n)/\boldsymbol{\pi}_{L'}^m \mathcal{O}_L(n)$$

and we consider $\varphi$ as a semilinear endomorphism of this $\mathcal{O}_{L'}$-module. We choose $\boldsymbol{\pi}_{L'} = \boldsymbol{\pi}$ when $L'/F$ is unramified, $\boldsymbol{\pi}_{L'}^2 \in F$ when $L'/F$ is tamely ramified. In any case $\varphi$ must satisfy $\varphi^2 = \mathrm{id}$. When $L'/F$ is tamely ramified ($\mathcal{D}_{L'/F} = \boldsymbol{\pi}_{L'}\mathcal{O}_{L'}$), there are more conditions, namely:
1) $N \subset \mathcal{D}_{L'/F}M$, i.e. $m \geq 1$;
2) $\varphi = \mathrm{id} \bmod \boldsymbol{\pi}_{L'}$;
3) $\varphi = \mathrm{id}$ on $\boldsymbol{\pi}_{L'}^{m-1}\mathcal{O}_L(n)/\boldsymbol{\pi}_{L'}^m\mathcal{O}_L(n)$.

When 2) holds, condition 3) means that $m$ is odd.

The condition $-\widetilde{\varphi} = \nu\varphi\nu^{-1}$ translates to:

$$* \quad -c\langle\varphi(x), y\rangle' \equiv \overline{\langle x, \varphi(y)\rangle'} \bmod \boldsymbol{\pi}_{L'}^{m+n}\mathcal{O}_{L'} \qquad \text{for all } x, y \in \mathcal{O}_L(n),$$

where

$$c = \nu N_{L/L'}(u)/\overline{\nu}\,\overline{N_{L/L'}(u)}.$$

Write

$$\nu N_{L/L'}(u) = c_1\boldsymbol{\pi}_{L'}^{m-2}, \qquad c_1 \in \mathcal{O}_{L'}^\times.$$

Then $c = c_1/\overline{c}_1$ when $L'/F$ is unramified or $n$ is odd and $c = -c_1/\overline{c}_1$, when $L'/F$ is ramified and $n$ is even. Now $*$ is:

$$** \quad \langle x, c_1\varphi(y)\rangle' \equiv \pm\overline{\langle c_1\varphi(x), y\rangle'} \bmod \boldsymbol{\pi}_{L'}^{m+n}\mathcal{O}_{L'} \qquad \text{for all } x, y \in \mathcal{O}_L(n)$$

(+ when $L'/F$ is ramified and $n$ even, – otherwise).

Choose $w_L \in L$ such that $\mathcal{O}_L = \mathcal{O}_{L'} + \mathcal{O}_{L'}w_L$. Then $\mathcal{O}_L(n) = \mathcal{O}_{L'} + \mathcal{O}_{L'}\boldsymbol{\pi}_{L'}^n w_L$. In $\langle x, y\rangle' = \mathrm{tr}_{L/L'}(a\sigma(x)y)$ the element $a$ is such that $\sigma(a) = -a$. We may take $a = a_1(w_L - \sigma(w_L))^{-1}$ with any $a_1 \in L'^\times$. Note that $\langle 1, \boldsymbol{\pi}_{L'}^n w_L\rangle' = a_1\boldsymbol{\pi}_{L'}^n$. So, when we take a unit for $a_1$, we have $\langle 1, w_L\rangle' \in \mathcal{O}_{L'}^\times$, which was used above. A possible choice is: $a_1 = 1$ if $L'/F$ is ramified, $a_1 \in \mathcal{O}_{L'}^\times$ such that $\overline{a}_1 = -a_1$ if $L'/F$ is unramified. Then $\langle 1, \boldsymbol{\pi}_{L'}^n w_L\rangle'/\overline{\langle 1, \boldsymbol{\pi}_{L'}^n w_L\rangle'}$ is just the sign in $**$.

With respect to the basis $\{1, \boldsymbol{\pi}_{L'}^n w_L\}$ the map $c_1\varphi$ is given by a matrix $Z$ in

$$\mathrm{GL}(2, \mathcal{O}_{L'}/\boldsymbol{\pi}_{L'}^m\mathcal{O}_{L'})$$

satisfying

$$\begin{cases} {}^t\overline{Z}J = JZ, \qquad J = \left(\begin{smallmatrix} 0 & 1 \\ -1 & 0 \end{smallmatrix}\right), \\ Z\overline{Z} = c_1\overline{c_1}, \\ tZ = Z\overline{t}, \\ m \text{ is odd and } Z \equiv c_1 \bmod \boldsymbol{\pi}_{L'} \qquad \text{if } L'/F \quad \text{is (tamely) ramified.} \end{cases}$$

It is perhaps better to say that the map is given by $Z\tau$: if $Z = \left(\begin{smallmatrix} z_1 & z_2 \\ z_3 & z_4 \end{smallmatrix}\right)$ and $x = \left(\begin{smallmatrix} x_1 \\ x_2 \end{smallmatrix}\right)$, then

$$c_1\varphi(x) = \left(\begin{smallmatrix} z_1 & z_2 \\ z_3 & z_4 \end{smallmatrix}\right)\begin{pmatrix} \overline{x}_1 \\ \overline{x}_2 \end{pmatrix}.$$

**2.** We now assume that $|2| = 1$ in $F$. Then we can take $w_L$ such that $w_L^2 \in \mathcal{O}_{L'}$. Suppose $t \in \mathcal{O}_L(n)$. Put $t = t_1 + t_2\boldsymbol{\pi}_{L'}^n w_L$ with $t_1, t_2 \in \mathcal{O}_{L'}$. The matrix corresponding to multiplication by $t$ is $\left(\begin{smallmatrix} t_1 & \lambda t_2 \\ t_2 & t_1 \end{smallmatrix}\right)$ with $\lambda = \boldsymbol{\pi}_{L'}^{2n} w_L^2 \in \mathcal{O}_{L'}$.

We have $t_1^2 - \lambda t_2^2 = 1$. Let

$$Z = \left(\begin{smallmatrix} z_1 & z_2 \\ z_3 & z_4 \end{smallmatrix}\right) \in \mathrm{GL}(2, \mathcal{O}_{L'}/\boldsymbol{\pi}_{L'}^m\mathcal{O}_{L'}), \qquad m > 0.$$

The conditions on $Z$ are (all $\equiv \bmod \boldsymbol{\pi}_{L'}^m$):

$$(1)\begin{cases} z_2 + \overline{z}_2 \equiv 0 \\ z_3 + \overline{z}_3 \equiv 0 \\ z_4 \equiv \overline{z}_1 \\ z_1\overline{z}_1 + z_2\overline{z}_3 \equiv c_1\overline{c}_1 \end{cases} \qquad (2)\begin{cases} (t_1 - \overline{t}_1)z_1 - \overline{t}_2 z_2 + \lambda t_2 z_3 \equiv 0 \\ -\overline{\lambda}\,\overline{t}_2 z_1 + (t_1 - \overline{t}_1)z_2 + \lambda t_2 z_4 \equiv 0 \\ t_2 z_1 + (t_1 - \overline{t}_1)z_3 - \overline{t}_2 z_4 \equiv 0 \\ t_2 z_2 - \overline{\lambda}\,\overline{t}_2 z_3 + (t_1 - \overline{t}_1)z_4 \equiv 0 \end{cases}$$

and if $L'/F$ is ramified: $z_1 \equiv c_1 \bmod \boldsymbol{\pi}_{L'}$. It follows from (1), in this case, that

$$z_2 \equiv z_3 \equiv 0(\boldsymbol{\pi}_{L'}), \qquad z_1 \equiv \pm c_1(\boldsymbol{\pi}_{L'}).$$

When (1) holds, (2) is equivalent to

$$(t_1 - \overline{t}_1)z_i \equiv 0 \quad (\text{all } i), \quad \overline{t}_2 z_2 \equiv \lambda t_2 z_3, \quad t_2 z_1 \equiv \overline{t}_2\overline{z}_1, \quad \lambda t_2 \overline{z}_1 \equiv \overline{\lambda}\,\overline{t}_2 z_1.$$

Necessary for solvability of the system is that $t_1 \equiv \overline{t}_1(\boldsymbol{\pi}_{L'}^m)$.

We treat the cases $L'/F$ ramified, resp. unramified, separately.

Suppose $m > \mathfrak{v}_{L'}(t_2)$. Put $t_2 = \boldsymbol{\pi}_{L'}^k b_1$, $b_1 \in \mathcal{O}_{L'}^\times$, $0 \le k < m$. The congruences (2) are $\overline{b}_1 z_2 \equiv (-1)^k \lambda b_1 z_3$,

$$b_1 z_1 \equiv (-1)^k \overline{b}_1 \overline{z}_1, \quad \lambda b_1 \overline{z}_1 \equiv (-1)^k \overline{\lambda}\, \overline{b}_1 z_1 \operatorname{mod} \boldsymbol{\pi}_{L'}^{m-k} \mathcal{O}_{L'}.$$

Introduce the new variable $z_1' = b_1 z_1$. Then

$$z_1' \equiv (-1)^k \overline{z}', \qquad \lambda b_1^2 \overline{z}_1' \equiv (-1)^k \overline{\lambda}\, \overline{b}_1^2 z_1' \operatorname{mod} \boldsymbol{\pi}_{L'}^{m-k} \mathcal{O}_{L'}.$$

And from (1):

$$z_2 + \overline{z}_2 \equiv 0, \quad z_3 + \overline{z}_3 \equiv 0, \quad z_1' \overline{z}_1 - b_1 \overline{b}_1 z_2 z_3 \equiv b_1 \overline{b}_1 c_1 \overline{c}_1 \operatorname{mod} \boldsymbol{\pi}_{L'}^m \mathcal{O}_{L'}$$

and $z_1' \equiv b_1 c_1 \operatorname{mod} \boldsymbol{\pi}_{L'} \mathcal{O}_{L'}$.

From the last congruence and $z_1' \equiv (-1)^k \overline{z}_1'$, we see that $k$ must be even. Then

$$z_1' \equiv x_1 + \boldsymbol{\pi}^{\frac{m-k-1}{2}} y_1 \boldsymbol{\pi}_{L'}, \quad z_2 \equiv y_2 \boldsymbol{\pi}_{L'}, \quad z_3 \equiv y_3 \boldsymbol{\pi}_{L'} \operatorname{mod} \boldsymbol{\pi}_{L'}^m \mathcal{O}_{L'}$$

with $x_1, y_i \in \mathcal{O}$ and

(5) $b_1 \overline{b}_1 y_2 \equiv \lambda b_1^2 y_3 \operatorname{mod} \boldsymbol{\pi}_{L'}^{m-k-1} \mathcal{O}_{L'}$, $\qquad \lambda b_1^2 x_1 \equiv \overline{\lambda}\, \overline{b}_1^2 x_1 \operatorname{mod} \boldsymbol{\pi}_{L'}^{m-k} \mathcal{O}_{L'}$,

(6) $x_1^2 - \boldsymbol{\pi}^{m-k} y_1^2 - b_1 \overline{b}_1 \boldsymbol{\pi} y_2 y_3 \equiv b_1 \overline{b}_1 c_1 \overline{c}_1 \operatorname{mod} \boldsymbol{\pi}^{\frac{m+1}{2}} \mathcal{O}$,

$$x_1 \equiv \tfrac{b_1 c_1 + \overline{b}_1 \overline{c}_1}{2} \operatorname{mod} \boldsymbol{\pi} \mathcal{O}.$$

Here $x_1$ has to be taken $\operatorname{mod} \boldsymbol{\pi}^{\frac{m+1}{2}}$, $y_1 \operatorname{mod} \boldsymbol{\pi}^{\frac{k}{2}}$, $y_2$ and $y_3 \operatorname{mod} \boldsymbol{\pi}^{\frac{m-1}{2}}$.

The congruences (5) are equivalent to

$$(\lambda b_1^2 - \overline{\lambda}\, \overline{b}_1^2) x_1 \equiv 0 \operatorname{mod} \boldsymbol{\pi}_{L'}^{m-k} \mathcal{O}_{L'}, \qquad (\lambda b_1^2 - \overline{\lambda}\, \overline{b}_1^2) y_3 \equiv 0 \operatorname{mod} \boldsymbol{\pi}_{L'}^{m-k-1} \mathcal{O}_{L'},$$

$$2 b_1 \overline{b}_1 y_2 \equiv y_3 \operatorname{tr}_{L'/F}(\lambda b_1^2) \operatorname{mod} \boldsymbol{\pi}^{\frac{m-k-1}{2}} \mathcal{O}.$$

We must necessarily have $\lambda b_1^2 \equiv \overline{\lambda}\, \overline{b}_1^2 \operatorname{mod} \boldsymbol{\pi}_{L'}^{m-k} \mathcal{O}_{L'}$, since $x_1 \in \mathcal{O}^\times$ by (6). Then there are $q^{\frac{m-1}{2}+k}$ solutions ($b_1 \overline{b}_1 c_1 \overline{c}_1$ is always a square in $F$, because $L'/F$ is ramified).

REMARK. The condition $\lambda b_1^2 \equiv \overline{\lambda}\, \overline{b}_1^2 \operatorname{mod} \boldsymbol{\pi}_{L'}^{m-k} \mathcal{O}_{L'}$, is equivalent to

$$t_1^2 \equiv \overline{t}_1^2 \operatorname{mod} \boldsymbol{\pi}_{L'}^{m+k} \mathcal{O}_{L'}$$

in both cases ($L'/F$ ramified or not).

**3.** Recall that

$$\mathcal{O}_L = \mathcal{O}_{L'} + \mathcal{O}_{L'} w_L, \quad w_L^2 \in \mathcal{O}_{L'}. \quad \text{Let} \quad t = t_1 + t_2 w_L, \quad t_1^2 - t_2^2 w_L^2 = 1.$$

As $t$ is regular, we have $t_2 \neq 0$ and $t_1 \neq \bar{t}_1$. Let us be given:

$$\left| \begin{array}{l} \nu \in L'^{\times} \qquad (\nu \bmod N_{L/L'} L^{\times}), \\ m \geq 0, \\ n \geq 0 \text{ such that } t \in \mathcal{O}_L(n), \text{ i.e. } n \leq \mathfrak{v}_{L'}(t_2), \\ u \in L^{\times} (u \bmod \mathcal{O}_L(n)^{\times}) \text{ such that } f_{L/L'} \mathfrak{v}_L(u) = m - n - \mathfrak{v}_{L'}(\nu). \end{array} \right.$$

By section 2 the number of corresponding $\varphi$ is:
If $L'/F$ is unramified:

$$\begin{cases} 1 & \text{if} \quad m = 0; \\ q^{3m-2}(q^2+1) & \text{if} \quad m > 0, \quad m + n \leq \mathfrak{v}_{L'}(t_2), \quad t_1 \equiv \bar{t}_1 \bmod \boldsymbol{\pi}; \\ 2q^{m+2k} & \text{if} \quad \mathfrak{v}_{L'}(t_2) \equiv n + k, \quad 0 \leq k < m, \\ & \text{and} \quad t_1 \equiv \bar{t}_1 \bmod \boldsymbol{\pi}^{m+k} \mathcal{O}_{L'}, \quad \nu \bar{\nu} N_{L/k}(u) t_2 \bar{t}_2 \in F^{\times 2}. \end{cases}$$

If $L'/F$ is ramified and $m$ odd

$$\begin{cases} q^{\frac{3(m-1)}{2}} & \text{if} \quad m + n \leq \mathfrak{v}_{L'}(t_2), \quad t_1 \equiv \bar{t}_1 \bmod \boldsymbol{\pi}_{L'}; \\ q^{\frac{m-1}{2}+k} & \text{if} \quad \mathfrak{v}_{L'}(t_2) = n + k, \quad 0 \leq k < m, \\ & \text{and} \quad k \text{ even}, \quad t_1 \equiv \bar{t}_1 \bmod \boldsymbol{\pi}_{L'}^{m+k} \mathcal{O}_{L'}. \end{cases}$$

It is 0 in all other cases.

First, we consider the case where $L'/F$ is unramified. We have

$$\mathcal{O}_{L'} = \mathcal{O} + \mathcal{O} w', \quad w'^2 = \alpha \in \mathcal{O}^{\times}, \quad \boldsymbol{\pi}_{L'} = \boldsymbol{\pi}.$$

Suppose $m \leq \mathfrak{v}_{L'}(t_2)$. Only the congruences (1) are left. We have

$$z_1 \equiv x_1 + y_1 w', \quad z_2 \equiv y_2 w', \quad z_3 \equiv y_3 w'$$

with $x_1, y_1, y_2, y_3 \in \mathcal{O}(\bmod \boldsymbol{\pi}^m)$. Further

$$x_1^2 - \alpha y_1^2 - \alpha y_2 y_3 \equiv c_1 \bar{c}_1 \bmod \boldsymbol{\pi}^m.$$

There are $q^{3m}(1+q^{-2})$ solutions.

Suppose $m > \mathfrak{v}_{L'}(t_2)$. Put $\quad t_2 = \boldsymbol{\pi}^k b_1, \quad b_1 \in \mathcal{O}_{L'}^{\times}, \quad 0 \leq k < m.$

The congruences (2) become

$$\overline{b}_1 z_2 \equiv \lambda b_1 z_3, \quad b_1 z_1 \equiv \overline{b}_1 \overline{z}_1, \quad \lambda b_1 \overline{z}_1 \equiv \overline{\lambda}\, \overline{b}_1 z_1 \bmod \boldsymbol{\pi}^{m-k} \mathcal{O}_{L'}.$$

Introduce the new variable $z_1' = b_1 z_1$. Then

$$z_1' \equiv \overline{z}_1', \qquad \lambda b_1^2 \overline{z}_1' \equiv \overline{\lambda}\, \overline{b}_1^2 z_1' \bmod \boldsymbol{\pi}^{m-k} \mathcal{O}_{L'}.$$

Moreover we have, from (1):

$$\left.\begin{array}{l} z_2 + \overline{z}_2 \equiv 0, \quad z_3 + \overline{z}_3 \equiv 0 \\ z_1' \overline{z}_1' - b_1 \overline{b}_1 z_2 z_3 \equiv b_1 \overline{b}_1 c_1 \overline{c}_1 \end{array}\right\} \bmod \boldsymbol{\pi}^m \mathcal{O}_{L'}$$

$z_1' \equiv x_1 + \boldsymbol{\pi}^{m-k} y_1 w', \quad z_2 \equiv y_2 w', \quad z_3 \equiv y_3 w' \bmod \boldsymbol{\pi}^m \mathcal{O}_{L'}$ with $x_1$, $y_1$, $y_2$, $y_3 \in \mathcal{O}$ and

(3) $\qquad b_1 \overline{b}_1 y_2 \equiv \lambda b_1^2 y_3, \quad \lambda b_1^2 x_1 \equiv \overline{\lambda}\, \overline{b}_1^2 x_1 \bmod \boldsymbol{\pi}^{m-k} \mathcal{O}_{L'},$

(4) $\qquad x_1^2 - \alpha \boldsymbol{\pi}^{2m-2k} y_1^2 - \alpha b_1 \overline{b}_1 y_2 y_3 \equiv b_1 \overline{b}_1 c_1 \overline{c}_1 \bmod \boldsymbol{\pi}^m \mathcal{O}.$

The elements $x_1$, $y_2$, $y_3$ are to be taken modulo $\boldsymbol{\pi}^m$ and $y_1$ modulo $\boldsymbol{\pi}^k$.

The congruences (3) are equivalent to

$$(\lambda b_1^2 - \overline{\lambda}\, \overline{b}_1^2) x_1 \equiv 0, \quad (\lambda b_1^2 - \overline{\lambda}\, \overline{b}_1^2) y_3 \equiv 0,$$

$$2 b_1 \overline{b}_1 y_2 \equiv y_3 \operatorname{tr}_{L'/F}(\lambda b_1^2) \bmod \boldsymbol{\pi}^{m-k} \mathcal{O}_{L'}.$$

We must necessarily have $\lambda b_1^2 \equiv \overline{\lambda}\, \overline{b}_1^2 \bmod \boldsymbol{\pi}^{m-k} \mathcal{O}_{L'}$, since $x_1$ and $y_3$ cannot be both $\equiv 0(\boldsymbol{\pi})$ because of (4).

It follows from $\lambda b_1^2 \equiv \overline{\lambda}\, \overline{b}_1^2 \bmod \boldsymbol{\pi}^{m-k} \mathcal{O}_{L'}$ that $\lambda b_1^2$ is congruent to an element of $\mathcal{O}$, which must be in $\boldsymbol{\pi}\mathcal{O}$, for otherwise $\lambda$ would be a square in $L'$. So $\lambda \in \boldsymbol{\pi} \mathcal{O}_{L'}$ and $y_2 \in \boldsymbol{\pi}\mathcal{O}$. Hence, for (4) to be solvable, $b_1 \overline{b}_1 c_1 \overline{c}_1$ must be a square in $F$. The number of solutions is

$$2q^{m+2k} \quad \text{if} \quad \lambda b_1^2 \equiv \overline{\lambda}\, \overline{b}_1^2 \bmod \boldsymbol{\pi}^{m-k} \mathcal{O}_{L'} \quad \text{and} \quad b_1 \overline{b}_1 c_1 \overline{c}_1 \in F^{\times 2},$$

and 0 otherwise.

Next, consider the case where $L'/F$ is ramified.

We have $\mathcal{O}_{L'} = \mathcal{O} + \mathcal{O}\pi_{L'}$, $\pi_{L'}^2 = \pi$, a uniformizing element of $F$. Now $m$ is odd and

$$\pi_{L'}^m \mathcal{O}_{L'} = \mathcal{O}\pi^{\frac{m+1}{2}} + \mathcal{O}\pi^{\frac{m-1}{2}}\pi_{L'}.$$

Suppose $m \leq \mathfrak{v}_{L'}(t_2)$. Then

$$z_1 \equiv x_1 + y_1\pi_{L'}, \qquad z_2 \equiv y_2\pi_{L'}, \qquad z_3 \equiv y_3\pi_{L'} \bmod \pi_{L'}^m \mathcal{O}_{L'}$$

with $x_1, y_i \in \mathcal{O}$. Further

$$x_1^2 - \pi y_1^2 - \pi y_2 y_3 \equiv c_1\bar{c}_1 \bmod \pi^{\frac{m+1}{2}}\mathcal{O}, \qquad x_1 \equiv \frac{c_1 + \bar{c}_1}{2} \bmod \pi\mathcal{O}.$$

Here $x_1$ is to be taken modulo $\pi^{\frac{m+1}{2}}$ and the $y_i$ modulo $\pi^{\frac{m-1}{2}}$.

There are $q^{\frac{3(m-1)}{2}}$ solutions.

We compute

$$\sum_{\nu} \sum_{m\geq 0} \sum_{\substack{0\leq n\leq \mathfrak{v}_{L'}(t_2) \\ n\equiv m-\mathfrak{v}_{L'}(\nu) \bmod f_{L/L'}}} \sum_{u_1 \in \mathcal{O}_L^\times/\mathcal{O}_L(n)^\times} \mathrm{Card}\{\varphi\},$$

where we put

$$u = u_1 \pi_L^{m-n-\mathfrak{v}_{L'}(\nu)/f_{L/L'}}.$$

The following observations can be used to handle the sum over $u_1$.

a) If $L/L'$ is unramified, $\mathfrak{v}_{L'}$ induces a bijection $L'^\times/N_{L/L'}L^\times \to \mathbb{Z}/2\mathbb{Z}$. If $L/L'$ is ramified,

$$\mathcal{O}_{L'}^\times/\mathcal{O}_{L'}^{\times 2} = \mathcal{O}_{L'}^\times/N_{L/L'}\mathcal{O}_L^\times \xrightarrow{\sim} L'^\times/N_{L/L'}L^\times.$$

b) Assume $L'/F$ unramified. Then $N_{L/F}\mathcal{O}_L^\times = \mathcal{O}^\times$ if $L/L'$ is unramified, $= \mathcal{O}^{\times 2}$ if $L/L'$ is ramified.

c) Assume $L/F$ unramified. Then $N_{L/F}$: $\mathcal{O}_L^\times/\mathcal{O}_L^{\times 2} \xrightarrow{\sim} \mathcal{O}^\times/\mathcal{O}^{\times 2}$. Moreover, in the case where $\mathfrak{v}_{L'}(t_2) < m+n$, we have $n > 0$, so $\mathcal{O}_L(n)^\times \subset \mathcal{O}_L^{\times 2}$.

Our sum is the sum of the following sums.

If $L'/F$ is unramified:

I) $e_{L/L'} \sum_{0 \le n \le \mathfrak{v}_{L'}(t_2)} (\mathcal{O}_L^\times : \mathcal{O}_L(n)^\times)$.

II) $e_{L/L'} \sum_{\substack{m>0, n\ge 0 \\ m+n \le \mathfrak{v}_{L'}(t_2)}} q^{3m-2}(q^2+1)(\mathcal{O}_L^\times : \mathcal{O}_L(n)^\times)$,

if $t_1 \equiv \bar{t}_1 \bmod \pi\mathcal{O}_{L'}$.

III) $\frac{1}{2} e_{L/L'} \sum_{\substack{0 \le k < m, k \le \mathfrak{v}_{L'}(t_2) \\ m+k \le \mathfrak{v}_{L'}(t_1 - \bar{t}_1)}} 2q^{m+2k}(\mathcal{O}_L^\times : \mathcal{O}_L(\mathfrak{v}_{L'}(t_2) - k)^\times)$.

If $L/F$ is ramified:

IV) $e_{L/L'} \sum_{\substack{m>0, n \ge 0, m \text{ odd} \\ m+n \le \mathfrak{v}_{L'}(t_2)}} q^{\frac{3(m-1)}{2}} (\mathcal{O}_L^\times : \mathcal{O}_L(n)^\times)$,

if $t_1 \equiv \bar{t}_1 \bmod \pi_{L'}\mathcal{O}_{L'}$.

V) $e_{L/L'} \sum_{\substack{0 \le k < m, m \text{ odd}, \ k \text{ even} \\ k \le \mathfrak{v}_{L'}(t_2), m+k \le \mathfrak{v}_{L'}(t_1 - \bar{t}_1)}} q^{\frac{m-1}{2}+k} (\mathcal{O}_L^\times : \mathcal{O}_L(\mathfrak{v}_{L'}(t_2) - k)^\times)$.

Put $A = \mathfrak{v}_{L'}(t_1 - \bar{t}_1)$, $B = \mathfrak{v}_{L'}(t_2)$. We have $t_1^2 - \delta t_2^2 = 1$, with $\delta = w_L^2$.

LEMMA. a) $A \ge 2B + \mathfrak{v}_{L'}(\delta)$.
b) $A = 2B + \mathfrak{v}_{L'}(\delta)$ *except for the cases where $L/F$ is the noncyclic Galois extension.*

PROOF. a) follows from $t_1^2 - \bar{t}_1^2 = \delta \bar{t}_2^2 (t_2^2 \bar{t}_2^{-2} - \delta^{-1}\bar{\delta})$. Note that $t_1 + \bar{t}_1 \in \mathcal{O}_{L'}^\times$ if $2B + \mathfrak{v}_{L'}(\delta) > 0$.

b) If $A > 2B + \mathfrak{v}_{L'}(\delta)$, then $t_2^2 \bar{t}_2^{-2} \equiv \delta^{-1}\bar{\delta} \bmod \pi_{L'}$. One checks case-by-case that this is impossible when $L/F$ is not the composite of the three quadratic extensions of $F$. □

The sums (I)-(V) are:

$$I)\quad \frac{q^{2B+2}+q^{2B}-2}{q^2-1} \qquad (L/L' \text{ unramified}),$$

$$\frac{2(q^{2B+2}-1)}{q^2-1} \qquad (L/L' \text{ ramified});$$

$$II)\quad q^{2B-1}\frac{(q^2+1)^2}{q^2-1}\frac{q^B-1}{q-1}-2q\frac{q^2+1}{q^2-1}\frac{q^{3B}-1}{q^3-1} \qquad (L/L' \text{ unramified});$$

$$2q\frac{q^2+1}{q^2-1}\left\{q^{2B}\frac{q^B-1}{q-1}-\frac{q^{3B}-1}{q^3-1}\right\} \qquad (L/L' \text{ ramified});$$

$$III)\quad \frac{q^{2B+1}(q^2+1)(q^B-1)(q^{B+1}-1)}{(q-1)^2} \qquad (L/L' \text{ unramified});$$

$$\frac{2q^{2B+1}(q^{B+1}-1)(q^{A-B}-1)}{(q-1)^2} \qquad (L/L' \text{ ramified});$$

Here $A = 2B+1$ if $L/F$ is cyclic.

$$IV)\quad \frac{q^{B-1}(q+1)(q^{\left[\frac{B+1}{2}\right]}-1)}{(q-1)^2}-\frac{2(q^{3\left[\frac{B+1}{2}\right]}-1)}{(q-1)(q^3-1)} \qquad (L/L' \text{ unramified})$$

$$\frac{2q^B(q^{\left[\frac{B+1}{2}\right]}-1}{(q-1)^2}-\frac{2(q^{3\left[\frac{B+1}{2}\right]}-1)}{(q-1)(q^3-1)} \qquad (L/L' \text{ ramified})$$

$$V)\quad q^{\frac{3B}{2}}\frac{q^{\frac{A+1}{2}-B}-1}{q-1}\delta(B,2\left[\frac{B}{2}\right])$$

$$+\frac{q^{B-1}(q+1)(q^{\left[\frac{B+1}{2}\right]}-1)(q^{\frac{A+1}{2}-\left[\frac{B-1}{2}\right]}-1)}{(q-1)^2} \qquad (L/L' \text{ unramified});$$

$$\frac{2q^B(q^{\left[\frac{B}{2}\right]+1}+1)(q^{\left[\frac{B+3}{2}\right]}-1)}{(q-1)^2} \qquad (L/L' \text{ ramified}).$$

# PART 2. ZETA FUNCTIONS OF SHIMURA VARIETIES OF PGSp(2)

# I. PRELIMINARIES

## 1. Introduction

Eichler expressed the Hasse-Weil Zeta function of a modular curve as a product of $L$-functions of modular forms in 1954, and, a few years later, Shimura introduced the theory of canonical models and used it to similarly compute the Zeta functions of the quaternionic Shimura curves. Both authors based their work on congruence relations.

Ihara introduced (1967) a new technique, based on comparison of the number of points on the Shimura variety over various finite fields with the Selberg trace formula. He used this to study forms of higher weight. Deligne [D1] explained Shimura's theory of canonical models in group theoretical terms, and obtained Ramanujan's conjecture for some cusp forms on $\mathrm{GL}(2, \mathbb{A}_{\mathbb{Q}})$, namely that their Hecke eigenvalues are algebraic and all of their conjugates have absolute value 1 in $\mathbb{C}^\times$, for almost all components.

Langlands [L3-5] developed Ihara's approach to predict the contribution of the tempered automorphic representations to the Zeta function of arbitrary Shimura varieties, introducing in the process the theory of endoscopic groups. He carried out the computations in [L5] for subgroups of the multiplicative groups of nonsplit quaternionic algebras.

Using Arthur's conjectural description [A2-4] of the automorphic nontempered representations, Kottwitz [K3] developed Langlands' conjectural description of the Zeta function to include nontempered representations. In [K4] he associated Galois representations to automorphic representations which occur in the cohomology of unitary groups associated to division algebras. In this anisotropic case the trace formula simplifies.

To deal with isotropic cases, where the Shimura variety is not proper and one has continuous spectrum on the automorphic side, Deligne conjectured that the Lefschetz fixed point formula for a correspondence on a variety over a finite field remains valid if the correspondence is twisted by a sufficiently high power of the Frobenius.

Deligne's conjecture was used with Kazhdan in [FK3] to decompose the cohomology with compact supports of the Drinfeld moduli scheme of elliptic modules, and relate Galois representations and automorphic representations of $\mathrm{GL}(n)$ over function fields of curves over finite fields.

Deligne's conjecture was proven in some cases by Zink [Zi], Pink [P], Shpiz [Sh], and in general by Fujiwara [Fu]. See Varshavsky [Va] for a recent simple proof. We use it here to express the Zeta function of the Shimura varieties of the projective symplectic group of similitudes $H = \mathrm{PGSp}(2)$ of rank 2 over any totally real field $F$ and with any coefficients, in terms of automorphic representations of this group and of its unique proper elliptic endoscopic group, $C_0 = \mathrm{PGL}(2) \times \mathrm{PGL}(2)$.

Moreover we decompose the cohomology (étale, with compact supports) of the Shimura variety (with coefficients in a finite dimensional representation of $H$), thus associating a Galois representation to any "cohomological" automorphic representation of $H(\mathbb{A})$. Here $\mathbb{A} = \mathbb{A}_F$ denotes the ring of adèles of $F$, and $\mathbb{A}_\mathbb{Q}$ of $\mathbb{Q}$. Our results are consistent with the conjectures of Langlands and Kottwitz [Ko4]. We make extensive use of the results of [Ko4], expressing the Zeta function in terms of stable trace formulae of $\mathrm{PGSp}(2)$ and its endoscopic group $C_0$, also for twisted coefficients. We use the fundamental lemma proven in this case in [F5] and assumed in [Ko4] in general.

Using congruence relations Taylor [Ty] associated Galois representations to automorphic representations of $\mathrm{GSp}(2, \mathbb{A}_\mathbb{Q})$ which occur in the cohomology of the Shimura three-fold, in the case of $F = \mathbb{Q}$. Laumon [Ln] used the Arthur-Selberg trace formula and Deligne's conjecture to get more precise results on such representations again for the case $F = \mathbb{Q}$ where the Shimura variety is a 3-fold, and with trivial coefficients. Similar results were obtained by Weissauer [W] (unpublished) using the topological trace formula of Harder and Goresky-MacPherson.

However, a description of the automorphic representations of the group $\mathrm{PGSp}(2, \mathbb{A}_F)$ has recently become available [F6]. We use this, together with the fundamental lemma [F5] and Deligne's conjecture [Fu], [Va], to decompose the $\overline{\mathbb{Q}}_\ell$-adic cohomology with compact supports and describe all of its constituents. This permits us to compute the Zeta function, in addition to describing the Galois representation associated to each automorphic representation occurring in the cohomology. To use [F6] when

$F = \mathbb{Q}$ we work only with automorphic representations which have an elliptic component at a finite place. There is no restriction when $F \neq \mathbb{Q}$.

We work with any coefficients, and with any totally real base field $F$. In the case $F \neq \mathbb{Q}$ the Galois representations which occur are related to the interesting "twisted tensor" representation of the dual group. Using Deligne's "mixed purity" theorem [D6] we conclude that for all good primes $p$ the Hecke eigenvalues of any automorphic representation $\pi = \otimes \pi_p$ occurring in the cohomology are algebraic and all of their conjugates lie on the unit circle for $\pi$ which lift ([F6]) to representations on PGL(4) induced from cuspidal ones, or are related by lifting – in a way which we make explicit – to automorphic representations of GL(2) with such a property. This is known as the "generalized" Ramanujan conjecture (for PGSp(2)).

# 2. Statement of Results

To describe our results we briefly introduce the subjects of study; more detailed account is given in the body of the work. Let $F$ be a totally real number field, $H = \mathrm{GSp}(2)$ the group of symplectic similitudes (whose Borel subgroup is the group of upper triangular matrices), $H' = \mathrm{R}_{F/\mathbb{Q}}\,\mathrm{GSp}(2)$ the $\mathbb{Q}$-group obtained by restriction of scalars, $\mathbb{A}_{\mathbb{Q}}$ and $\mathbb{A}_{\mathbb{Q}f}$ the rings of adèles and finite adèles of $\mathbb{Q}$, $K_f$ an open compact subgroup of $H'(\mathbb{A}_{\mathbb{Q}f})$ of the form $\prod_{p<\infty} K_p$, $K_p$ open compact in $H'(\mathbb{Z}_p)$ for all $p$ with equality for almost all primes $p$, $h : \mathrm{R}_{\mathbb{C}/\mathbb{R}}\mathbb{G}_m \to H'_{\mathbb{R}}$ an $\mathbb{R}$-homomorphism satisfying the axioms of [D5] and $\mathcal{S}_{K_f}$ the associated Shimura variety, defined over its reflex field $\mathbb{E}$, which is $\mathbb{Q}$.

The finite dimensional irreducible algebraic representations of $H$ are parametrized by their highest weights $(a, b; c)$ : $\mathrm{diag}(x, y, z/y, z/x) \mapsto x^a y^b z^c$, where $a$, $b$, $c \in \mathbb{Z}$ and $a \geq b \geq 0$. Those with trivial central character have $a + b = -2c$ even, and we denote them by $(\rho_{a,b}, V_{a,b})$. For each rational prime $\ell$, the representation

$$(\rho_{\mathbf{a},\mathbf{b}} = \otimes_{v\in S}\rho_{a_v,b_v}, V_{\mathbf{a},\mathbf{b}} = \otimes_{v\in S} V_{a_v,b_v})$$

of $H'$ over $F$ ($S$ is the set of embeddings of $F$ in $\mathbb{R}$) defines a smooth $\overline{\mathbb{Q}}_\ell$-adic sheaf $\mathbb{V}_{\mathbf{a},\mathbf{b};\ell}$ on $\mathcal{S}_{K_f}$. We are concerned with the decomposition of the

$\overline{\mathbb{Q}}_\ell$-adic vector space $H^i_c(\mathcal{S}_{K_f} \otimes_{\mathbb{Q}} \overline{\mathbb{Q}}, \mathbb{V}_{\mathbf{a},\mathbf{b};\ell})$ as a $C_c(K_f \backslash H'(\mathbb{A}_{\mathbb{Q}f})/K_f, \overline{\mathbb{Q}}_\ell) \times \mathrm{Gal}(\overline{\mathbb{Q}}/\mathbb{Q})$-module, or more precisely the virtual bi-module

$$H^*_c = \oplus(-1)^i H^i_c, \qquad 0 \le i \le 2\dim \mathcal{S}_{K_f}.$$

We fix an isomorphism of fields from $\overline{\mathbb{Q}}_\ell$ to $\mathbb{C}$. Write $H^*_c(\pi_{Hf})$ for

$$\mathrm{Hom}_{\mathbb{H}_{K_f}}(\pi_f, H^*_c(\mathcal{S}_{K_f} \otimes_{\mathbb{E}} \overline{\mathbb{Q}}, \mathbb{V}_{\mathbf{a},\mathbf{b};\ell})).$$

We are concerned only with $a_v \ge b_v \ge 0$ with even $a_v - b_v$, and we consider only the part of $H^*_c$ isotypic under $Z'(\mathbb{A}_{\mathbb{Q}f})$. Thus we work with functions in the Hecke convolution algebra of compactly supported modulo the center $Z'(\mathbb{A}_{\mathbb{Q}f})$ of $H'(\mathbb{A}_{\mathbb{Q}f})$, $K_f$-biinvariant functions on $H'(\mathbb{A}_{\mathbb{Q}f})$ which transform trivially under $Z'(\mathbb{A}_{\mathbb{Q}f})$. Alternatively we take our group $H$ to be the projective symplectic group of similitudes. We make this restriction since this is the case studied in [F6]. The fundamental lemma is established in [F5] for any central character. Thus from now on $H' = \mathrm{R}_{F/\mathbb{Q}}H$, $H = \mathrm{PGSp}(2)$. In the next line, $C_c$ is $C_c(K_f \backslash H'(\mathbb{A}_{\mathbb{Q}f})/K_f)$.

THEOREM 1. *The irreducible $C_c \times \mathrm{Gal}(\overline{\mathbb{Q}}/\mathbb{Q})$-modules which occur nontrivially in $H^*_c(\mathcal{S}_{K_f} \otimes_{\mathbb{Q}} \overline{\mathbb{Q}}, \mathbb{V}_{\mathbf{a},\mathbf{b};\ell})$ are of the form $\pi_{Hf}^{K_f} \otimes H^*_c(\pi_{Hf})$, where $\pi_{Hf}$ is the finite component $\otimes_{p<\infty}\pi_{Hp}$ of a discrete spectrum automorphic representation $\pi_H$ of $H'(\mathbb{A}_{\mathbb{Q}f})$, and $\pi_{Hf}^{K_f}$ denotes its subspace of $K_f$-fixed vectors. The archimedean component $\pi_{H\infty} = \otimes_{v\in S}\pi_{Hv}$ of $\pi$, $S = \{F \hookrightarrow \mathbb{R}\}$ and $H'(\mathbb{R}) = \prod_{v\in S} H(F \otimes_{F,v} \mathbb{R})$, has components $\pi_{Hv}$ whose infinitesimal character is $(a_v, b_v) + (2,1)$. Here $(2,1)$ is half the sum of the positive roots.*

*Conversely, any discrete spectrum representation $\pi_H$ of $H'(\mathbb{A}_{\mathbb{Q}f})$ whose archimedean component $\pi_{H\infty} = \otimes_{v\in S}\pi_{Hv}$ is such that the infinitesimal character of $\pi_{Hv}$ is $(a_v, b_v) + (2,1)$, $a_v \ge b_v \ge 0$, even $a_v - b_v$, for each $v \in S$ (we call such representations $\pi_H$* **cohomological***), and $\pi_{Hf}^{K_f} \ne \{0\}$, occurs in $H^*_c(\mathcal{S}_{K_f} \otimes_{\mathbb{Q}} \overline{\mathbb{Q}}, \mathbb{V}_{\mathbf{a},\mathbf{b};\ell})$ with multiplicity one as $\pi_{Hf}^{K_f} \otimes H^*_c(\pi_{Hf})$.*

The main point here is that the $\pi_H$ which occur in $H^*_c$ are automorphic, in fact discrete spectrum with the prescribed behavior at $\infty$ and ramification controlled by $K_f$. Each cohomological $\pi_H$ occurs for some $K_f$ depending on $\pi_H$. The first statement here is known for $IH$ by [BC].

We proceed to describe the semisimplification of the Galois representation $H_c^*(\pi_{Hf})$ attached to $\pi_{Hf}$. For this purpose we first need to list the cohomological $\pi_H$. Note that $H'(\mathbb{Q}) = H(F)$ and $H'(\mathbb{A}_{\mathbb{Q}}) = H(\mathbb{A}_F)$.

The $\pi_H$ are described in [F6] in terms of packets and quasi-packets, and liftings $\lambda : \hat{H} = \mathrm{Sp}(2,\mathbb{C}) \to \hat{G} = \mathrm{SL}(4,\mathbb{C})$ (natural embedding), and $\lambda_0 : \mathrm{SL}(2,\mathbb{C}) \times \mathrm{SL}(2,\mathbb{C}) = \hat{C}_0 \hookrightarrow \hat{H}$, $\hat{C}_0$ is viewed as the centralizer of $\mathrm{diag}(1,-1,-1,1)$ in $\hat{H}$. A detailed account of the lifting theorems of [F6] is given in the text below, as are the definitions of [F6] of packets and quasi-packets. Quasi-packets refer to nontempered representations. We distinguish five types of cohomological representations $\pi_H$ of $\mathrm{PGSp}(2,\mathbb{A}_F)$.

(1) $\pi_H$ in a (stable) packet which $\lambda$-lifts to a cuspidal representation of $G(\mathbb{A}_F)$, $G = \mathrm{PGL}(4)$; the components $\pi_{Hv} (v \in S)$ are discrete series with infinitesimal characters $(a_v, b_v) + (2,1)$.

(2) $\pi_H$ in a (stable) quasi-packet of the form $\{L(\xi\nu, \nu^{-1/2}\pi^2)\}$ which $\lambda$-lifts to the residual noncuspidal representation $J(\nu^{1/2}\pi^2, \nu^{-1/2}\pi^2)$ of $\mathrm{PGL}(4,\mathbb{A}_F)$. Here $\pi^2$ is a cuspidal representation of $\mathrm{GL}(2,\mathbb{A}_F)$ with quadratic central character $\xi \neq 1$ with $\xi\pi^2 = \pi^2$, and discrete series components $\pi_v^2 = \pi_{2k_v+2}$, $k_v \geq 0$ for all $v \in S$. Here $(a_v, b_v) = (2k_v, 0)$.

(3) One dimensional representation $\pi_H(g) = \xi(\boldsymbol{\lambda}(g))$ of $H(\mathbb{A}_F)$. Here $\boldsymbol{\lambda}(g)$ is the factor of similitude of $g$, $\xi$ is a character $\mathbb{A}_F^\times / F^\times \mathbb{A}_F^{\times 2} \to \{\pm 1\}$, and $(a_v, b_v) = (0,0)$.

(4) $\pi_H$ in a packet which is the $\lambda_0$-lift of $\pi^1 \times \pi^2$, where $\pi^1$ and $\pi^2$ are distinct cuspidal representations of $\mathrm{PGL}(2,\mathbb{A}_F)$ such that $\{\pi_v^1, \pi_v^2\} = \{\pi_{k_{1v}}, \pi_{k_{2v}}\}$, $k_{1v} > k_{2v} > 0$ odd integers for all $v \in S$. This packet $\lambda$-lifts to the (normalizedly) induced representation $I(\pi^1, \pi^2)$ of $\mathrm{PGL}(4,\mathbb{A}_F)$. Here $(a_v, b_v) = (\frac{1}{2}(k_{1v} + k_{2v}) - 2, \frac{1}{2}(k_{1v} - k_{2v}) - 1)$.

(5) $\pi_H$ is in a quasi-packet $\{L(\xi\nu^{1/2}\pi^2, \xi\nu^{-1/2})\}$ which is the $\lambda_0$-lift of $\xi \times \pi^2$, where $\xi$ is a character $\mathbb{A}_F^\times / F^\times \mathbb{A}_F^{\times 2} \to \{\pm 1\}$ and $\pi^2$ is a cuspidal representation of $\mathrm{PGL}(2,\mathbb{A}_F)$ with $\pi_v^2 = \pi_{2k_v+3}$, $k_v \geq 0$, $v \in S$. Here $(a_v, b_v) = (k_v, k_v)$.

A global (quasi-)packet is the restricted product of local (quasi-)packets, which are sets of one or two irreducibles, pointed by the property of being unramified (the local (quasi-) packets contains a single unramified representation at almost all places). The packets (1) and (3) and the quasi-packet (2) are stable: each member is automorphic and occurs in the discrete spectrum with multiplicity one. The packets (4) and quasi-packets

(5) are not stable, their members occur in the discrete spectrum with multiplicity one or zero, according to a formula of [F6] recalled below.

We now describe the semisimplification $H_c^*(\pi_{Hf})^{ss}$ of the representation $H_c^*(\pi_{Hf})$ of $\mathrm{Gal}(\overline{\mathbb{Q}}/\mathbb{Q})$ associated to each of these $\pi_{Hf}$. From now on, we write $H_c^*(\pi_{Hf})$ for $H_c^*(\pi_{Hf})^{ss}$. The Chebotarev's density theorem asserts that the Frobenius elements $\mathrm{Fr}_p$ for almost all $p$ make a dense subgroup of $\mathrm{Gal}(\overline{\mathbb{Q}}/\mathbb{Q})$. Hence it suffices to specify the conjugacy class of $H_c^*(\pi_{Hf})(\mathrm{Fr}_p)$ for almost all $p$. This makes sense since $H_c^*(\pi_{Hf})$ is unramified at almost all $p$, trivial on the inertia subgroup $I_p$ of the decomposition group $D_p = \mathrm{Gal}(\overline{\mathbb{Q}}_p/\mathbb{Q}_p)$ of $\mathrm{Gal}(\overline{\mathbb{Q}}/\mathbb{Q})$, and $D_p/I_p$ is (topologically) generated by $\mathrm{Fr}_p$. The conjugacy class $H_c^*(\pi_{Hf})(\mathrm{Fr}_p)$ is determined by its trace, and since $H_c^*(\pi_{Hf})(\mathrm{Fr}_p)$ is semisimple it is determined by $H_c^*(\pi_{Hf})(\mathrm{Fr}_p^j)$ for all sufficiently large $j$. We consider only $p$ which are unramified in $F$, thus the residual cardinality $q_u$ of $F_u$ at any place $u$ of $F$ over $p$ is $p^{n_u}$, $n_u = [F_u : \mathbb{Q}_p]$. Further we use only $p$ with $K_f = K_p K^p$, where $K_p = H'(\mathbb{Z}_p)$ is the standard maximal compact, thus $\mathcal{S}_{K_f}$ has good reduction at $p$. Note that $\dim \mathcal{S}_{K_f} = 3[F : \mathbb{Q}]$.

Part of the data defining the Shimura variety is the $\mathbb{R}$-homomorphism $h : \mathrm{R}_{\mathbb{C}/\mathbb{R}}\mathbb{G}_m \to H' = \mathrm{R}_{F/\mathbb{Q}}H$. Over $\mathbb{C}$ the one-parameter subgroup $\mu : \mathbb{C}^\times \to H'(\mathbb{C})$, $\mu(z) = h(z,1)$ factorizes through any maximal $\mathbb{C}$-torus $T'_H(\mathbb{C}) \subset H'(\mathbb{C})$. The $H'(\mathbb{C})$-conjugacy class of $\mu$ defines then a Weyl group $W_{\mathbb{C}}$-orbit $\mu = \prod_\tau \mu_\tau$ in $X_*(T'_H) = X^*(\widehat{T}'_H)$. The dual torus $\widehat{T}'_H = \prod_\sigma \widehat{T}_H$ in $\widehat{H}' = \prod_\sigma \widehat{H}$, $\sigma \in \mathrm{Emb}(F, \mathbb{R})$, can be taken to be the diagonal subgroup, and $X^*(\widehat{T}_H) = \mathbb{Z}^2$. See section 10 for more detail. We choose $\mu_\tau$ to be the character $(1,0) : \mathrm{diag}(a, b, b^{-1}, a^{-1}) \mapsto a$ of $\widehat{T}_H$. Thus the $H(\mathbb{C})$-orbit of the coweight $\mu_\tau$ determines a $W_{\mathbb{C}}$-orbit of a character – again denoted by $\mu_\tau$ – of $\widehat{T}_H$, which is the highest weight of the standard representation $r_\tau^0 = \mathrm{st}$ of $\mathrm{Sp}(2, \mathbb{C})$. Put $r_\mu^0 = \otimes_\tau r_\tau^0$. It is a representation of $\widehat{H}'$.

The Galois group $\mathrm{Gal}(\overline{\mathbb{Q}}/\mathbb{Q})$ acts on $\mathrm{Emb}(F, \overline{\mathbb{Q}})$. The stabilizer of $\mu$, $\mathrm{Gal}(\overline{\mathbb{Q}}/\mathbb{E})$, defines the reflex field $\mathbb{E}$. In our case $\mathbb{E} = \mathbb{Q}$.

An irreducible admissible representation $\pi_{Hp}$ of $H(F \otimes \mathbb{Q}_p) = H'(\mathbb{Q}_p) = \prod_{u|p} H(F_u)$ has the form $\otimes_u \pi_{Hu}$. Suppose it is unramified. Then $\pi_{Hu}$ has the form $\pi_H(\mu_{1u}, \mu_{2u})$, a subquotient of the normalizedly induced representation $I(\mu_{1u}, \mu_{2u})$ of $H(F_u) = \mathrm{PGSp}(2, F_u)$, where $\mu_{iu}$ are unramified characters of $F_u^\times$. Write $\mu_{mu}$ for the value $\mu_{mu}(\boldsymbol{\pi}_u)$ at any uniformizing parameter $\boldsymbol{\pi}_u$ of $F_u^\times$. Put $t_u = t(\pi_{Hu}) = \mathrm{diag}(\mu_{1u}, \mu_{2u}, \mu_{2u}^{-1}, \mu_{1u}^{-1})$. Note that $\mathrm{tr}[t_u^j] = \mu_{1u}^j + \mu_{2u}^j + \mu_{2u}^{-j} + \mu_{1u}^{-j}$.

The representation $\pi_p$ is parametrized by the conjugacy class of $\mathbf{t}(\pi_p) = \mathbf{t}_p \times \mathrm{Fr}_p$ in the unramified dual group

$$^L H'_p = \widehat{H}^{[F:\mathbb{Q}]} \rtimes \langle \mathrm{Fr}_p \rangle.$$

Here $\mathbf{t}_p$ is the $[F:\mathbb{Q}]$-tuple $(\mathbf{t}_u; u|p)$ of diagonal matrices in $\widehat{H} = \mathrm{Sp}(2,\mathbb{C})$, where each $\mathbf{t}_u = (t_{u1}, \dots, t_{un_u})$ is any $n_u = [F_u:\mathbb{Q}_p]$-tuple with $\prod_i t_{ui} = t_u$. The Frobenius $\mathrm{Fr}_p$ acts on each $\mathbf{t}_u$ by permutation to the left: $\mathrm{Fr}_p(\mathbf{t}_u) = (t_{u2}, \dots, t_{un_u}, t_{u1})$. Each $\pi_u$ is parametrized by the conjugacy class of $\mathbf{t}(\pi_{Hu}) = \mathbf{t}_u \times \mathrm{Fr}_p$ in the unramified dual group $^L H'_u = \widehat{H}^{n_u} \times \langle \mathrm{Fr}_p \rangle$, or alternatively by the conjugacy class of $t_u \times \mathrm{Fr}_u$ in $^L H_u = \widehat{H} \times \langle \mathrm{Fr}_u \rangle$, where $\mathrm{Fr}_u = \mathrm{Fr}_p^{n_u}$.

Our determination of the Galois representation attached to $\pi_{Hf}$ is in terms of the traces of the representation $r_\mu^0$ of the dual group $^L H'_{\mathbb{E}} = \widehat{H}' \rtimes W_{\mathbb{E}}$ at the positive powers of the $n_\wp$th powers of the classes $\mathbf{t}(\pi_{Hp}) = (\mathbf{t}(\pi_u); u|p)$ parametrizing the unramified components $\pi_{Hp} = \otimes_{u|p} \pi_{Hu}$. The representation $H_c^*(\pi_{Hf})$ is determined by $\mathrm{tr}[\mathrm{Fr}_\wp^j | H_c^*(\pi_{Hf})]$ for every integer $j \geq 0$, prime $p$ unramified in $E$, and $\mathbb{E}$-prime $\wp$ dividing $p$. As $\mathbb{E} = \mathbb{Q}$ here, $\wp = p$ and $n_\wp = 1$.

The following very detailed statement describes the Galois representation $H_c^*(\pi_{Hf})(\pi_H)$ attached to the cohomological $\pi_H$.

THEOREM 2. (1) *Fix $\pi_H$ of type (1) which occurs in the cohomology with coefficients in $\mathbb{V}_{\mathbf{a},\mathbf{b}}$, $a_v \geq b_v \geq 0$, even $a_v - b_v$. Thus $\pi_H$ has archimedean components $\pi^*_{k_{1v},k_{2v}}$, $* = \mathrm{Wh}$ or $\mathrm{hol}$, $k_{1v} = a_v + b_v + 3 > k_{2v} = a_v - b_v + 1 > 1$ are odd. It contributes to the cohomology only in dimension $3[F:\mathbb{Q}]$. Denote by $\pi_{Hu} = \pi_H(\mu_{1u}, \mu_{2u})$ the component of the representation $\pi_H$ of $H(\mathbb{A}_F)$ at a place $u$ of $F$ above $p$. It is parametrized by the conjugacy class $t(\pi_{Hu}) = \mathrm{diag}(t_{1u}, t_{2u}, t_{2u}^{-1}, t_{1u}^{-1})$ in $\hat{H} = \mathrm{Sp}(2,\mathbb{C})$, where $t_{mu} = \mu_{mu}(\boldsymbol{\pi}_u)$, $m = 1, 2$. Then $H_c^*(\pi_{Hf})$ is $4^{[F:\mathbb{Q}]}$-dimensional, and with $j_u = (j, n_u)$,*

$$\mathrm{tr}[\mathrm{Fr}_p^j | H_c^*(\pi_{Hf})] = p^{\frac{j}{2} \dim S_{K_f}} \,\mathrm{tr}\, r_\mu^0[(\mathbf{t}(\pi_p) \times \mathrm{Fr}_p)^j] = \prod_{u|p} (\mathrm{tr}[t_u^{j/j_u}])^{j_u}.$$

*Namely $H_c^*(\pi_{Hf})(\mathrm{Fr}_p)$ is $\otimes_{u|p} \nu_u^{-1/2} r_u(\mathrm{Fr}_u)$, where $r_u(\mathrm{Fr}_u)$ acts on the twisted tensor representation $(r_u, (\mathbb{C}^4)^{n_u})$ as*

$$\mathbf{t}(\pi_{Hu}) \times \mathrm{Fr}_u, \qquad \mathbf{t}(\pi_{Hu}) = (t_1, \dots, t_{n_u}),$$

*$t_m$ diagonal with $\prod_{1\le m\le n_u} t_m = t(\pi_{Hu})$. Here $\nu_u$ is the unramified character of $\mathrm{Gal}(\overline{\mathbb{Q}}_u/\mathbb{Q}_u)$ with $\nu_u(\mathrm{Fr}_u) = q_u^{-1}$. The eigenvalues of $H_c^*(\pi_{Hf})(\mathrm{Fr}_p)$ are $p^{\frac{3}{2}[F:\mathbb{Q}]}\prod_{u|p} t_{m(u),u}^{\iota(u)}$, $m(u) \in \{1,2\}$, $\iota(u) \in \{\pm1\}$. In our stable case (1), the representation $H_c^*(\pi_{Hf})$ depends only on the packet of $\pi_H$, and not on $\pi_H$ itself.*

*The Hecke eigenvalues $t_{1u}$, $t_{2u}$ are algebraic and each of their conjugates has complex absolute value one.*

*(2) Representations $\pi_H$ in a quasi-packet $\{L(\xi\nu, \nu^{-1/2}\pi^2)\}$ of type (2) occur in the cohomology with coefficients in $\mathbb{V}_{2k_v,0}$, $k_v \ge 0$, thus $\pi_H$ has archimedean components $\pi_{Hv} = L(\nu\ \mathrm{sgn}, \nu^{-1/2}\pi_{2k_v+2})$, $v \in S$. At a place $u$ of $F$ over $p$ the component $\pi_u^2$ of $\pi^2$ is unramified of the form $\pi^2(z_{1u}, z_{2u})$. The Hecke eigenvalues $z_{1u}$, $z_{2u}$ satisfy $z_{1u}z_{2u} = \xi_u(\boldsymbol{\pi}_u) \in \{\pm1\}$. The component $\pi_{Hu}$ has parameter*

$$t_u = \mathrm{diag}(q_u^{1/2}z_{1u}, q_u^{1/2}z_{2u}, q_u^{-1/2}z_{2u}^{-1}, q_u^{-1/2}z_{1u}^{-1})$$

*in $\hat{H} = \mathrm{Sp}(2,\mathbb{C})$. The associated representation $H_c^*(\pi_{Hf})$ has dimension $4^{[F:\mathbb{Q}]}$ and $H_c^*(\pi_{Hf})(\mathrm{Fr}_p)$ is the same as in case (1) but with $t_{1u} = q_u^{1/2}z_{1u}$, $t_{2u} = q_u^{1/2}z_{2u}$. The $z_{1u}$, $z_{2u}$ are algebraic, all their conjugates lie on the unit circle in $\mathbb{C}$.*

*(3) The case of type (3) of the one dimensional representation $\pi_H = \xi\circ\boldsymbol{\lambda}$, $\xi^2 = 1$, occurs in the cohomology with coefficients in $\mathbb{V}_{\mathbf{0},\mathbf{0}}$ only. The parameter $t(\pi_{Hu})$ is*

$$\mathrm{diag}(\xi_u q_u^{3/2}, \xi_u q_u^{1/2}, \xi_u q_u^{-1/2}, \xi_u q_u^{-3/2}).$$

*The associated representation $H_c^*(\pi_{Hf})$ is again $4^{[F:\mathbb{Q}]}$-dimensional and $H_c^*(\pi_{Hf})(\mathrm{Fr}_p)$ is the same as in case (1) but with $t_{1u} = \xi_u q_u^{3/2}$, $t_{2u} = \xi_u q_u^{1/2}$, $\xi_u \in \{\pm1\}$.*

*(4) The $\pi_H$ of the unstable tempered case (4) occur in the cohomology with coefficients in $\mathbb{V}_{\mathbf{a},\mathbf{b}}$, $a_v \ge b_v \ge 0$, even $a_v - b_v$. Thus the archimedean components $\pi_{Hv}$ are in $\{\pi^{\mathrm{Wh}}_{k_{1v},k_{2v}}, \pi^{\mathrm{hol}}_{k_{1v},k_{2v}}\}$, $k_{1v} = a_v + b_v + 3 > k_{2v} = a_v - b_v + 1 > 0$ are odd. The component $\pi_{Hu}$ of $\pi_H$ at a place $u$ of $F$ over $p$ is unramified of the form $\pi_{Hu} = \pi_H(\mu_{1u}, \mu_{2u})$, parametrized by $t_u = \mathrm{diag}(t_{1u}, t_{2u}, t_{2u}^{-1}, t_{1u}^{-1})$, $t_{mu} = \mu_{mu}(\boldsymbol{\pi}_u)$, $m = 1, 2$, in $\hat{H}$. The packet $\{\pi_H\}$ of $\pi_H$ is the $\lambda_0$-lift of $\pi^1 \times \pi^2$, where $\pi^1$, $\pi^2$ are cuspidal representations of $\mathrm{PGL}(2, \mathbb{A}_F)$. It is defined by means of local packets*

*$\{\pi_{Hw}\}$, which are singletons unless $\pi_w^1$ and $\pi_w^2$ are discrete series, in which case $\{\pi_{Hw}\} = \{\pi_{Hw}^+, \pi_{Hw}^-\}$, with + indicating generic and – nongeneric. If $\{\pi_{Hw}\}$ consists of a single term, it is $\pi_{Hw}^+$, and we put $\pi_{Hw}^- = 0$. We say that $\pi_{Hf}$ lies in $\{\pi_{Hf}\}^+$ if it has an even number of components $\pi_{Hw}^-$ ($w < \infty$), and in $\{\pi_{Hf}\}^-$ otherwise. Write $n(\pi^1 \times \pi^2)$ for the number of archimedean places $v \in V$ with $(\pi_v^1, \pi_v^2) = (\pi_{k_{2v}}, \pi_{k_{1v}})$ (recall: $\{\pi_{Hv}\} = \lambda_0(\pi_{k_{1v}} \times \pi_{k_{2v}})$, $k_{1v} > k_{2v} > 0$). Then the dimension of $H_c^*(\pi_{Hf})$ is $\frac{1}{2} \cdot 4^{[F:\mathbb{Q}]}$ and the trace of $H_c^*(\pi_{Hf})(\mathrm{Fr}_p^j)$ is $\frac{1}{2}p^{\frac{j}{2}\dim S_{K_f}}$ times*

$$\operatorname{tr} r_\mu^0[(\mathbf{t}(\pi_p) \times \mathrm{Fr}_p)^j] \pm (-1)^{n(\pi^1\times\pi^2)} \operatorname{tr} r_\mu^0[s(\mathbf{t}(\pi_p) \times \mathrm{Fr}_p)^j]$$

$$= \prod_{u|p}(t_{1u}^{j/j_u} + t_{1u}^{-j/j_u} + t_{2u}^{j/j_u} + t_{2u}^{-j/j_u})^{j_u}$$

$$\pm(-1)^{n(\pi^1\times\pi^2)} \prod_{u|p}(t_{1u}^{j/j_u} + t_{1u}^{-j/j_u} - (t_{2u}^{j/j_u} + t_{2u}^{-j/j_u}))^{j_u}$$

*if $\pi_{Hf} \in \{\pi_{Hf}\}^\pm$. The $t_{1u}$, $t_{2u}$ are algebraic and their conjugates lie on the unit circle.*

*Thus $H_c^*(\pi_{Hf})(\mathrm{Fr}_p^j)$ is*

$$\frac{1}{2}[\otimes_{u|p}\nu_u^{-1/2} r_u^+(\mathrm{Fr}_u^j) \pm (-1)^{n(\pi^1\times\pi^2)} \otimes_{u|p} \nu_u^{-1/2} r_u^-(\mathrm{Fr}_u^j)],$$

*where $r_u^\pm(\mathrm{Fr}_u)$ acts on the twisted tensor representation $(r_u, (\mathbb{C}^4)^{[F_u:\mathbb{Q}_u]})$ as $s^\pm \mathbf{t}(\pi_{Hu}) \times \mathrm{Fr}_u$ where $s^+ = I$ and $s^- = (s, I, \dots, I)$, $s = \mathrm{diag}(1, -1, -1, 1)$.*

*(5) The $\pi_H$ of the unstable nontempered case (5) occur in the cohomology with coefficients in $\mathbb{V}_{\boldsymbol{k},\boldsymbol{k}}$, $\boldsymbol{k} = (k_v)$, $k_v \geq 0$. Its archimedean components $\pi_{Hv}$ are $\pi_{2k_v+3,1}^{\mathrm{Wh}}$, $\pi_{2k_v+3,1}^{\mathrm{hol}}$ or the nontempered $L(\xi\nu^{1/2}\pi_{2k_v+3}, \xi\nu^{-1/2})$, $\xi = 1$ or sgn. It lies in a quasi-packet $\{L(\xi\nu^{1/2}\pi^2, \xi\nu^{-1/2})\}$, $\pi^2$ cuspidal representation of $\mathrm{PGL}(2, \mathbb{A}_F)$, whose real components are $\pi_{2k_v+3}$, and $\xi$ is a character $\mathbb{A}_F^\times / F^\times \mathbb{A}_F^{\times 2} \to \{\pm 1\}$. The unramified components $\pi_{Hu}$ are $\pi_{Hu}^\times = L(\xi_u\nu_u^{1/2}\pi_u^2, \xi_u\nu_u^{-1/2})$, $\pi_u^2 = \pi_u^2(z_{1u}, z_{2u})$, $z_{1u}z_{2u} = 1$, parametrized by $t_u = \mathrm{diag}(t_{1u}, t_{2u}, t_{2u}^{-1}, t_{2u}^{-1})$, $t_{1u} = \xi_u q_u^{1/2} z_{1u}$, $t_{2u} = \xi q_u^{1/2} z_{1u}^{-1}$. The quasi-packet $\{\pi_H\}$ is the $\lambda_0$-lift of $\pi^2 \times \xi 1_2$, defined using the local quasi-packets $\{\pi_{Hw}^\times, \pi_{Hw}^-\}$, $\pi_{Hw}^\times = L(\xi_w\nu_w^{1/2}\pi_w^2, \xi_w\nu_w^{-1/2})$, $\pi_{Hw}^-$ is 0 unless $\pi_w^2$ is square integrable in which case $\pi_{Hw}^-$ is square integrable (in the real case $\pi_{Hv}^-$ is $\pi_{2k_v+3,1}^{\mathrm{hol}}$).*

*We write $\pi_{Hf} \in \{\pi_{Hf}\}^\times$ if the number of components $\pi^-_{Hw}(w < \infty)$ of $\pi_{Hf}$ is even, and $\pi_{Hf} \in \{\pi_{Hf}\}^-$ if this number is odd. Then the dimension of $H^*_c(\pi_{Hf})$ is $\frac{1}{2}\cdot 4^{[F:\mathbb{Q}]}$ and the trace of $H^*_c(\pi_{Hf})(\mathrm{Fr}^j_p)$ is $\frac{1}{2}p^{\frac{j}{2}\dim \mathcal{S}_{K_f}}$ times*

$$\varepsilon(\xi\pi^2, \frac{1}{2}) \prod_{u|p}(t_{1u}^{j/j_u} + t_{1u}^{-j/j_u} + t_{2u}^{j/j_u} + t_{2u}^{-j/j_u})^{j_u}$$

$$\pm \prod_{u|p}(t_{1u}^{j/j_u} + t_{1u}^{-j/j_u} - (t_{2u}^{j/j_u} + t_{2u})^{-j/j_u})^{j_u}$$

*with + if $\pi_{Hf} \in \{\pi_{Hf}\}^\times$ and – if $\pi_{Hf} \in \{\pi_{Hf}\}^-$. Here $z_{1u}$ is algebraic, its conjugates are all on the complex unit circle. Thus $H^*_c(\pi_{Hf})(\mathrm{Fr}_p)$ has the same description as in case (4), except for the values of $t_{1u}$ and $t_{2u}$.*

Note that the Hodge types of $\pi_{Hv}$ for each $v \in S$ are (1,2), (2,1), (0,3), (3,0) in types (1) and (4); (2,0), (0,2), (1,3), (3,1) in type (2); (1,1) and (2,2) in type (5); and (0,0), (1,1), (2,2), (3,3) in type (3), specifying in which $H^{i,j}_c(\mathcal{S}_{K_f} \otimes_{\mathbb{Q}} \overline{\mathbb{Q}}, \mathbb{V}_{\mathbf{a},\mathbf{b}})$ each $\pi_{Hf}$ may occur.

In particular $H^{ij}_c$ is 0 if $i \neq j$ and $i + j < 2[F : \mathbb{Q}]$ or $i + j > 4[F : \mathbb{Q}]$; $H^{ii}_c$ has contributions only from one dimensional representations $\pi_H$ (of type (3)) if $i < [F : \mathbb{Q}]$ or $i > 2[F : \mathbb{Q}]$; $H^{2[F:\mathbb{Q}]}_c$ (and $H^{4[F:\mathbb{Q}]}_c$) has contributions only from representations of type (2), (3), (5). For example, $H^{2[F:\mathbb{Q}],0}_c$ (and $H^{0,2[F:\mathbb{Q}]}_c$) has contributions only from representations of type (2). The representations of types (2) and (5) are parametrized only by certain representations of GL(2) (and quadratic characters); these have smaller parametrizing set than the representations of type (4) (two copies of PGL(2)) or of type (1) (representations of PGL(4)).

In stating Theorem 2 we implicitly made a choice of a square root of $p$.

For unitary groups defined using division algebras endoscopy does not show and Kottwitz [K4] used the trace formula in this anisotropic case to associate Galois representations $H^*_c(\pi_{Hf})$ to some automorphic $\pi_H$ and obtain some of their properties. However, in this case the classification of automorphic representations and their packets is not yet known.

For GSp(2), in the case of $F = \mathbb{Q}$ and trivial coefficients $a_\infty = b_\infty = 0$, in particular trivial central character (PGSp(2)), Laumon [Ln], Thm 7.5, gave a list of possibilities for the trace of $H^*_c(\pi_{Hf})$ at $\mathrm{Fr}_p$ for $\pi_H$ in the stable spectrum, removing Eisensteinian contributions, see [Ln], (6.1). His

Thm 7.5 (1), (2) says $\pi_H$ might be Eisensteinian (our cases (2), (3), (5)) or endoscopic (our cases (4), (5)), his (3) corresponds to our case (1), but our cases (2), (5) are included again as a possibility in his Thm 7.5 (4). That is, by [F6] the $\pi_H$ in his Thm 7.5 (4) are already included in his (1) and (2).

Using the results of [F6], namely classification of and multiplicity one for the automorphic representations of the symplectic group, as well as the fundamental lemma of [F5] and Deligne's conjecture of [Fu], [Va], makes it possible for us to obtain more precise results, namely specify the $H_c^*(\pi_{Hf})$ such that $\pi_{Hf} \otimes H_c^*(\pi_{Hf})$ occurs in $H_c^*$, for all $\pi_H$, and list the $\pi_H$ which occur. Also, knowing the structure of packets and quasi-packets from [F6] lets us state and deal with the general case of $F \neq \mathbb{Q}$.

Laumon [Ln] works with $F = \mathbb{Q}$ and uses very extensively Arthur's deep analysis of the distributions occurring in the trace formula, together with the ideas of the simple trace formula of [FK3], [FK2] (the test function has an elliptic ([Ln], p. 301: "very cuspidal") real component and a regular component). This lets him put no restriction on the test function, but leads to very involved usage of the spectral side of the Arthur trace formula.

A simple trace formula (for a test measure with no cuspidal components) is available for comparisons in cases of $F$-rank 1 (see [F2;I], [F3;VI], [F4;III]) but not yet in $F$-rank 2. Hence in [F6] we use instead the trace formula with 3 discrete components (in fact 2 suffice, as explained in [F6], 1G). Using the results of [F6] leads us to the restriction (elliptic component at a finite place) we made here when $F = \mathbb{Q}$. This can be removed, to get unconditional result also for $F = \mathbb{Q}$, on using Arthur's deep analysis of the distributions occurring in the trace formula, as [Ln] explains.

Results similar to [Ln] have been obtained by Weissauer [W] (unpublished), who used the topological trace formula of Harder and Goresky-MacPherson. This trace formula applies to "geometric" representations only, namely those with elliptic (in fact cohomological) components at the real places. Previously some results (for a dense set of places) were derived by Taylor [Ty] from the congruence relations.

# 3. The Zeta Function

The Zeta function $Z$ of the Shimura variety is a product over the rational primes $p$ of local factors $Z_p$ each of which is a product over the primes $\wp$ of the reflex field $\mathbb{E}$ which divide $p$ of local factors $Z_\wp$. In our case $\mathbb{E} = \mathbb{Q}$ and $\wp = p$ but we keep using the symbol $\wp$ to suggest the general form. Write $q = q_\wp$ for the cardinality of the residue field $\mathbb{F} = R_p/\wp R_\wp$ ($R_\wp$ denotes the ring of integers of $\mathbb{E}_\wp$; $q$ is $p$ in our case). We work only with "good" $p$, thus $K_f = K_p K_f^p$, $K_p = H'(\mathbb{Z}_p)$, $\mathcal{S}_{K_f}$ is defined over $R_\wp$ and has good reduction mod $\wp$.

A general form of the Zeta function is for a correspondence, namely for a $K_f$-biinvariant $\overline{\mathbb{Q}}_\ell$-valued function $f_H^p$ on $H(\mathbb{A}_f^p)$, ($\mathbb{A}$ is $\mathbb{A}_F$ and we fix a field isomorphism $\overline{\mathbb{Q}}_\ell \xrightarrow{\sim} \mathbb{C}$), and with coefficients in the smooth $\overline{\mathbb{Q}}_\ell$-sheaf $\mathbb{V}_{\mathbf{a},\mathbf{b};\ell}$ constructed from an absolutely irreducible algebraic finite dimensional representation $V_{\mathbf{a},\mathbf{b}} = \otimes_{v\in S} V_{a_v,b_v}$ of $H'$ over $F$, each $V_{a_v,b_v}$ with highest weight $(a_v, b_v)$, $a_v \geq b_v \geq 0$, even $a_v - b_v$.

The standard form of the Zeta function is stated for $f_H^p = 1_{H(\mathbb{A}_f^p)}$, and for the trivial coefficient system ($(a_v, b_v) = (0,0)$ for all $v$). In this case the coefficients of the Zeta function store the number of points of the Shimura variety over finite residue fields. Thus the Zeta function, or rather its natural logarithm, is defined by

$$\ln Z_\wp(s, \mathcal{S}_{K_f}, f_H^p, \mathbb{V}_{\mathbf{a},\mathbf{b};\ell})_c$$

$$= \sum_{j=1}^{\infty} \frac{1}{jq_\wp^{js}} \sum_{i=0}^{2\dim \mathcal{S}_{K_f}} (-1)^i \operatorname{tr}[\mathrm{Fr}_\wp^j \circ f_H^p; H_c^i(\mathcal{S}_{K_f} \otimes_{\mathbb{E}} \overline{\mathbb{Q}}, \mathbb{V}_{\mathbf{a},\mathbf{b};\ell})].$$

The subscript $c$ on the left emphasizes that we work with $H_c$ rather than $H$ or $IH$; we drop it from now on. One can add a superscript $i$ on the left to isolate the contribution from $H_c^i$.

Our results decompose the alternating sum of the traces on the cohomology for a correspondence $f_H^p$ projecting on the subspace parametrized by those representations $\pi_H$ of $H(\mathbb{A}_F)$ with at least 2 discrete series components. We make this assumption from now on. The coefficient of $1/jq_\wp^{js}$ is then equal to the sum of 5 types of terms. The first 3, stable, terms, are of the form

$$\sum_{\{\pi_H\}} \sum_{\pi_H \in \{\pi_H\}} \operatorname{tr}\{\pi^p_{Hf}\}(f^p_H) \cdot q_\wp^{\frac{j}{2}\dim \mathcal{S}_{K_f}} \operatorname{tr}[\bigotimes_{u|p} r^0_u(\mathbf{t}(\pi_{Hu}))^j].$$

The 4th, unstable tempered term, is the sum of

$$\sum_{\{\pi_H\}} \sum_{\pi_H \in \{\pi_H\}} [\operatorname{tr}\{\pi^p_{Hf}\}^+(f^p_H) + \operatorname{tr}\{\pi^p_{Hf}\}^-(f^p_H)] \cdot q_\wp^{\frac{j}{2}\dim \mathcal{S}_{K_f}} \operatorname{tr}[\bigotimes_{u|p} r^0_u(\mathbf{t}(\pi_{Hu}))^j]$$

and

$$\sum_{\pi_1\times\pi_2} (-1)^{n(\pi_1\times\pi_2)} [\operatorname{tr}\{\pi^p_{Hf}\}^+(f^p_H) - \operatorname{tr}\{\pi^p_{Hf}\}^-(f^p_H)]$$

$$\cdot q_\wp^{\frac{j}{2}\dim \mathcal{S}_{K_f}} \operatorname{tr}[\bigotimes_{u|p} r^0_u(\mathbf{s}\ \mathbf{t}(\pi_{Hu}))^j].$$

The 5th, unstable nontempered term, is the sum of

$$\varepsilon(\xi\pi^2, \frac{1}{2}) \sum_{\{\pi_H\}} \sum_{\pi_H \in \{\pi_H\}} [\operatorname{tr}\{\pi^p_{Hf}\}^\times(f^p_H) - \operatorname{tr}\{\pi^p_{Hf}\}^-(f^p_H)]$$

$$\cdot q_\wp^{\frac{j}{2}\dim \mathcal{S}_{K_f}} \operatorname{tr}[\bigotimes_{u|p} r^0_u(\mathbf{t}(\pi_{Hu}))^j]$$

and

$$\sum_{\pi_1\times\pi_2} [\operatorname{tr}\{\pi^p_{Hf}\}^\times(f^p_H) + \operatorname{tr}\{\pi^p_{Hf}\}^-(f^p_H)] \cdot q_\wp^{\frac{j}{2}\dim \mathcal{S}_{K_f}} \operatorname{tr}[\bigotimes_{u|p} r^0_u(\mathbf{s}\ \mathbf{t}(\pi_{Hu}))^j].$$

The representation $(r^0_u, (\mathbb{C}^4)^{[F_u:\mathbb{Q}_p]})$ is the twisted tensor representation of ${}^L(\mathrm{R}_{F_u/\mathbb{Q}_p} H) = \hat{H}^{[F_u:\mathbb{Q}_p]} \rtimes \operatorname{Gal}(F_u/\mathbb{Q}_p)$. Here $\mathbb{C}^4$ is the standard representation of $\hat{H} \subset \operatorname{GL}(4,\mathbb{C})$ and the generator $\operatorname{Fr}_u$ of $\operatorname{Gal}(F_u/\mathbb{Q}_p)$ acts by permutation $\operatorname{Fr}_u(x_1 \otimes x_2 \otimes \cdots \otimes x_{n_u}) = x_2 \otimes \cdots \otimes x_{n_u} \otimes x_1$, $n_u = [F_u : \mathbb{Q}_p]$. The class $\mathbf{t}(\pi_{Hu})$ is $(t_1, \ldots, t_{n_u})$, $t_m$ is diagonal in $\hat{H}$ with $\prod_{1\le m\le n_u} t_m = t(\pi_{Hu})$ being the Satake parameter of the unramified component $\pi_{Hu}$. Further, $\mathbf{s} = (s, I, \ldots, I)$, $s = \operatorname{diag}(1,-1,-1,1)$.

The three stable contributions to the first sum are parametrized by:

(1) Stable packets $\{\pi_H\}$. These $\lambda$-lift to cuspidal $\theta$-invariant representations $\pi$ of $G(\mathbb{A}_F)$, $G = \operatorname{PGL}(4)$. The infinitesimal character of each

archimedean component $\pi_{Hv}(v \in S)$ is $(a_v, b_v) + (2,1)$, determined by $(\mathbf{a}, \mathbf{b})$. The components $\pi_{Hu}$ for each place $u$ of $F$ over $p$ are unramified and tempered. In fact the 4 nonzero, namely diagonal, entries of $t(\pi_{Hu})$ are algebraic, all conjugates lie on the complex unit circle.

(2) Stable quasi-packets $\{\pi_H = L(\xi\nu, \nu^{-1/2}\pi^2)\}$, which $\lambda$-lift to the quotient $J$ of the induced $I(\nu^{1/2}\pi^2, \nu^{-1/2}\pi^2)$ of $\mathrm{PGL}(4, \mathbb{A}_F)$. Here $\pi^2$ is a cuspidal representation of $\mathrm{GL}(2, \mathbb{A}_F)$ with central character $\xi \neq 1 = \xi^2$, archimedean components $\pi_v^2 = \pi_{2k_v}$, $k_v \geq 1$, $v \in S$, and unramified components $\pi_u^2$, $u|p$. The infinitesimal character of $\pi_{Hv}$ is $(2k_v, 1) = (2,1) + (a_v, 0)$, thus these contributions occur only when all $b_v$ are 0. The diagonal entries of $t(\pi_{Hu})$ are $(q_u^{1/2} z_{mu})^{\pm 1}$, $m = 1, 2$, where the Satake eigenvalues $z_{mu}$ of $\pi_u^2$ are algebraic all of whose conjugates are on the complex unit circle.

(3) One dimensional representations $\pi_H = \xi \circ \boldsymbol{\lambda}$, $\boldsymbol{\lambda}$ denotes the factor of similitude, $\xi$ a character $\mathbb{A}_F^\times / F^\times \mathbb{A}_F^{\times 2} \to \{\pm 1\}$. This case occurs only when $(a_v, b_v) = (0, 0)$ for all $v \in S$, and we have

$$t(\pi_{Hu}) = \mathrm{diag}(\xi_u q_u^{3/2}, \xi_u q_u^{1/2}, \xi_u q_u^{-1/2}, \xi_u q_u^{-3/2}),$$

$\xi_u \in \{\pm 1\}$ indicates the value at $\boldsymbol{\pi}_u$ of the $u$-component of $\xi$.

The two unstable contributions are parametrized by:

(4) Unordered pairs $\pi^1 \times \pi^2$ of cuspidal representations of $\mathrm{PGL}(2, \mathbb{A}_F)$, $\pi^1 \neq \pi^2$, with discrete series archimedean components $\pi_v^1 = \pi_{k_{1v}}$, $\pi_v^2 = \pi_{k_{2v}}$, $k_{1v} > k_{2v} > 0$ odd, specify the packet $\{\pi_H\} = \lambda_0(\pi^1 \times \pi^2)$. This occurs when $a_v = \frac{1}{2}(k_{1v} + k_{2v}) - 2$, $b_v = \frac{1}{2}(k_{1v} - k_{2v}) - 1$ for all $v \in S$. In this case the $\mathbf{t}(\pi_{Hu})$ are as in (1). The number of $v \in S$ with $(\pi_v^1, \pi_v^2) = (\pi_{k_{2v}}, \pi_{k_{1v}})$, $k_{1v} > k_{2v}$, is denoted by $n(\pi_1 \times \pi_2)$.

(5) Pairs $\pi^2 \times \xi \mathbf{1}_2$, where $\pi^2$ is a cuspidal representation of $\mathrm{PGL}(2, \mathbb{A}_F)$ with discrete series archimedean components $\pi_v^2 = \pi_{2k_v+3}$, $k_v \geq 0$, all $v \in S$, unramified components $\pi_u^2$, $u|p$, with Satake parameters $z_u^{\pm 1}$, and character $\xi : \mathbb{A}_F^\times / F^\times \mathbb{A}_F^{\times 2} \to \{\pm 1\}$. Such pair specifies the quasi-packet of $\pi_H^\times = L(\xi\nu^{1/2}\pi^2, \xi\nu^{-1/2}) = \lambda_0(\pi^2 \times \xi \mathbf{1}_2)$, whose archimedean components have infinitesimal characters $(2,1)$ plus $(a_v, b_v) = (k_v, k_v)$ for all $v \in S$. Thus this case occurs only for $(\mathbf{a}, \mathbf{b})$ with $a_v = b_v$ for all $v \in S$. The diagonal entries of $t(\pi_{Hu})$, $u|p$, are $(\xi_u q_u^{1/2} z_u^{\pm 1})^{\pm 1}$. The $z_u$ are algebraic, all its conjugates have absolute value one. The terms in the first sum are multiplied by $\varepsilon(\xi\pi^2, \frac{1}{2})$.

To express the Zeta function as a product of $L$-functions, recall that

$$\ln L_p(s,\pi_H,r) = \ln\det(1-p^{-s}r(\mathbf{t}_p(\pi_{Hp})))^{-1} = \sum_{j=1}^{\infty}\frac{1}{jp^{js}}\operatorname{tr} r(\mathbf{t}_p(\pi_{Hp})^j),$$

where $r$ is a representation of ${}^L H' = \hat{H}' \times \mathrm{Gal}(\overline{\mathbb{Q}}_p/\mathbb{Q}_p)$ and $\pi_{Hp}$ is unramified. For general $\mathbb{E}$, $r = \mathrm{Ind}(r^0; W_{\mathbb{Q}}, W_{\mathbb{E}})$.

We now continue with $\mathbb{E} = \mathbb{Q}$, $\wp = (p)$, $q = p$.

THEOREM 3. *The Zeta function is equal to the product over the* $\{\pi_H\}$ *and the* $\pi_H$ *in* $\{\pi_H\}$ *of*

$$L_p(s - \frac{1}{2}\dim \mathcal{S}_{K_f}, \pi_H, r)^{\operatorname{tr}\{\pi^p_{Hf}\}(f^p_H)}$$

*if* $\pi_H$ *is stable (of type* (1), (2) *or* (3)), *or*

$$L_p(s - \frac{1}{2}\dim \mathcal{S}_{K_f}, \pi_H, r + (-1)^{n(\pi_1\times\pi_2)} r\circ s)^{\operatorname{tr}\{\pi^p_{Hf}\}^+(f^p_H)}$$

*times*

$$L_p(s - \frac{1}{2}\dim \mathcal{S}_{K_f}, \pi_H, r - (-1)^{n(\pi_1\times\pi_2)} r\circ s)^{\operatorname{tr}\{\pi^p_{Hf}\}^-(f^p_H)}$$

*if* $\pi_H$ *is (unstable and tempered) of type* (4), *or*

$$L_p(s - \frac{1}{2}\dim \mathcal{S}_{K_f}, \pi_H, \varepsilon(\xi\pi^2, \frac{1}{2})r + r\circ s)^{\operatorname{tr}\{\pi^p_{Hf}\}^\times(f^p_H)}$$

*times*

$$L_p(s - \frac{1}{2}\dim \mathcal{S}_{K_f}, \pi_H, -\varepsilon(\xi\pi^2, \frac{1}{2})r + r\circ s)^{\operatorname{tr}\{\pi^p_{Hf}\}^-(f^p_H)}$$

*if* $\pi_H$ *is (unstable and nontempered) of type* (5). *Here*

$$r(\mathbf{t}_p(\pi_{Hp})) \quad \textit{is} \quad \otimes_{u|p}\, r_u(\mathbf{t}(\pi_{Hu}))$$

*and*

$$(r\circ s)(\mathbf{t}_p(\pi_{Hp})) = \otimes_{u|p} r(\mathbf{s}\ \mathbf{t}(\pi_{Hu})).$$

In the case of Shimura varieties associated with subgroups of GL(2), a similar statement is obtained in Langlands [L5]. In general, our result is predicted by Langlands [L3-5] and more precisely by Kottwitz [Ko4].

# 4. The Shimura Variety

Let $G$ be a connected reductive group over the field $\mathbb{Q}$ of rational numbers. Suppose that there exists a homomorphism $h : \mathrm{R}_{\mathbb{C}/\mathbb{R}}\mathbb{G}_m \to G_{\mathbb{R}}$ of algebraic groups over the field $\mathbb{R}$ of real numbers which satisfies the conditions (2.1.1.1-3) of Deligne [D5]. The $G(\mathbb{R})$-conjugacy class $X_\infty = \mathrm{Int}(G(\mathbb{R}))(h)$ of $h$ is isomorphic to $G(\mathbb{R})/K_\infty$, where $K_\infty$ is the fixer of $h$ in $G(\mathbb{R})$; it carries a natural structure of a Hermitian symmetric domain. Let $K_f$ be an open compact subgroup of $G(\mathbb{A}_{\mathbb{Q}f})$, where $\mathbb{A}_{\mathbb{Q}f}$ is the ring of adèles of $\mathbb{Q}$ without the real component, sufficiently small so that the set

$$\mathcal{S}_{K_f}(\mathbb{C}) = G(\mathbb{Q})\backslash[X_\infty \times (G(\mathbb{A}_{\mathbb{Q}f})/K_f)] = G(\mathbb{Q})\backslash G(\mathbb{A}_{\mathbb{Q}})/K_\infty K_f$$

is a smooth complex variety (manifold).

The group $\mathrm{R}_{\mathbb{C}/\mathbb{R}}\mathbb{G}_m$ obtained from the multiplicative group $\mathbb{G}_m$ on restricting scalars from the field $\mathbb{C}$ of complex numbers to $\mathbb{R}$ is defined over $\mathbb{R}$. Its group $(\mathrm{R}_{\mathbb{C}/\mathbb{R}}\mathbb{G}_m)(\mathbb{R})$ of real points can be realized as $\{(z,\overline{z}); z \in \mathbb{C}^\times\}$ in $(\mathrm{R}_{\mathbb{C}/\mathbb{R}}\mathbb{G}_m)(\mathbb{C}) = \mathbb{C}^\times \times \mathbb{C}^\times$. The $G(\mathbb{C})$-conjugacy class $\mathrm{Int}(G(\mathbb{C}))\mu_h$ of the homomorphism $\mu_h : \mathbb{G}_{m,\mathbb{C}} \to G_{\mathbb{C}}$, $z \mapsto h(z,1)$, is acted upon by the Galois group $\mathrm{Gal}(\mathbb{C}/\mathbb{Q})$. The subgroup which fixes $\mathrm{Int}(G(\mathbb{C}))(\mu_h)$ has the form $\mathrm{Gal}(\mathbb{C}/\mathbb{E})$, where $\mathbb{E}$ is a number field, named the reflex field. There is a smooth variety over $\mathbb{E}$ determined by the structure of its special points (see [D5]), named the canonical model $\mathcal{S}_{K_f}$ of the Shimura variety associated with $(G,h,K_f)$, whose set of complex points is $\mathcal{S}_{K_f}(\mathbb{C})$ displayed above.

Let $L$ be a number field, and let $\rho$ be an absolutely irreducible finite dimensional representation of $G$ on an $L$-vector space $V_\rho$. Denote by $p$ the natural projection $G(\mathbb{A}_{\mathbb{Q}})/K_\infty K_f \to \mathcal{S}_{K_f}(\mathbb{C})$. The sheaf $\mathbb{V} : U \mapsto V_\rho(L) \underset{\rho,G(\mathbb{Q})}{\times} p^{-1}U$ of $L$-vector spaces over $\mathcal{S}_{K_f}(\mathbb{C})$ is locally free of rank $\dim_L V_\rho$. For any finite place $\lambda$ of $L$ the local system $\mathbb{V} \otimes_L L_\lambda : U \to V_\rho(L_\lambda) \underset{\rho,G(\mathbb{Q})}{\times} p^{-1}U$ defines a smooth $L_\lambda$-sheaf $\mathbb{V}_\lambda$ on $\mathcal{S}_{K_f}$ over $\mathbb{E}$.

The Baily-Borel-Satake compactification $\mathcal{S}'_{K_f}$ of $\mathcal{S}_{K_f}$ has a canonical model over $\mathbb{E}$ as does $\mathcal{S}_{K_f}$. The Hecke convolution algebra $\mathbb{H}_{K_f,L}$ of compactly supported bi-$K_f$-invariant $L$-valued functions on $G(\mathbb{A}_{\mathbb{Q}f})$ is generated by the characteristic functions of the double cosets $K_f \cdot g \cdot K_f$ in $G(\mathbb{A}_{\mathbb{Q}f})$. It acts on the cohomology spaces $H^i(\mathcal{S}_{K_f}(\mathbb{C}),\mathbb{V})$, the cohomology with compact supports $H^i_c(\mathcal{S}_{K_f}(\mathbb{C}),\mathbb{V})$, and on the intersection cohomology $L$-spaces $IH^i(\mathcal{S}'_{K_f}(\mathbb{C}),\mathbb{V})$. These modules are related by maps: $H_c^i \to IH^i \to H^i$. The action is compatible with the isomorphism $H_c^i(\mathcal{S}_{K_f}(\mathbb{C}),\mathbb{V}) \otimes_L L_\lambda \simeq H_c^i(\mathcal{S}_{K_f} \otimes_{\mathbb{E}} \overline{\mathbb{Q}}, \mathbb{V}_\lambda)$, (same for $H^i$ and for $IH^i(\mathcal{S}')$), but the last étale cohomology spaces have in addition an action of the absolute Galois group $\mathrm{Gal}(\overline{\mathbb{Q}}/\mathbb{E})$, which commutes with the action of the Hecke algebra ($X \otimes_{\mathbb{E}} \overline{\mathbb{Q}}$ abbreviates $X \times_{\mathrm{Spec}\,\mathbb{E}} \mathrm{Spec}\,\overline{\mathbb{Q}}$).

# 5. Decomposition of Cohomology

Of interest is the decomposition of the finite dimensional $L_\lambda$-vector spaces $IH^i$, $H^i$ and $H_c^i$ as $\mathbb{H}_{K_f,L_\lambda} \times \mathrm{Gal}(\overline{\mathbb{Q}}/\mathbb{E})$-modules. They vanish unless $0 \le i \le 2\dim \mathcal{S}_{K_f}$. Thus

$$H_c^i(\mathcal{S}_{K_f} \otimes_{\mathbb{E}} \overline{\mathbb{Q}}, \mathbb{V}_\lambda) = \oplus\, \pi_{f,L_\lambda}^{K_f} \otimes H_c^i(\pi_{f,L_\lambda}^{K_f}). \tag{1}$$

The (finite) sum ranges over the inequivalent irreducible $\mathbb{H}_{K_f,L_\lambda}$-modules $\pi_{f,L_\lambda}^{K_f}$. The $H_c^i(\pi_{f,L_\lambda}^{K_f})$ are finite dimensional representations of $\mathrm{Gal}(\overline{\mathbb{Q}}/\mathbb{E})$ over $L_\lambda$. Similar decomposition holds for $H^i$ and $IH^i(\mathcal{S}')$.

In the case of $IH$, the Zucker conjecture [Zu], proved by Looijenga and Saper-Stern, asserts that the intersection cohomology of $\mathcal{S}'_{K_f}$ is isomorphic to the $L^2$-cohomology of $\mathcal{S}_{K_f}$. The isomorphism commutes with the action of the Hecke algebra. The $L^2$-cohomology with coefficients in the sheaf $\mathbb{V}_{\mathbb{C}} : U \mapsto V_\rho(\mathbb{C}) \underset{\rho, G(\mathbb{Q})}{\times} p^{-1}(U)$ of $\mathbb{C}$-vector spaces, $H_{(2)}^i(\mathcal{S}_{K_f}(\mathbb{C}), \mathbb{V}_{\mathbb{C}})$, has a ("Matsushima-Murakami") decomposition (see Borel-Casselman [BC]) in terms of discrete spectrum automorphic representations. Thus

$$H_{(2)}^i(\mathcal{S}_{K_f}(\mathbb{C}), \mathbb{V}_{\mathbb{C}}) = \bigoplus_{\pi} m(\pi)\pi_f^{K_f} \otimes H^i(\mathfrak{g}, K_\infty; \pi_\infty \otimes V_\rho(\mathbb{C})).$$

Here $\pi$ ranges over the equivalence classes of the discrete spectrum (irreducible) automorphic representations of $G(\mathbb{A}_{\mathbb{Q}})$ in

$$L_d^2 = L_d^2(G(\mathbb{Q})\backslash G(\mathbb{A}_{\mathbb{Q}}), \mathbb{C})$$

and $m(\pi)$ denotes the multiplicity of $\pi$ in $L_d^2$. Write $\pi = \pi_f \otimes \pi_\infty$ as a product of irreducible representations $\pi_f$ of $G(\mathbb{A}_{\mathbb{Q}f})$ and $\pi_\infty$ of $G(\mathbb{R})$, according to $\mathbb{A}_{\mathbb{Q}} = \mathbb{A}_{\mathbb{Q}f}\mathbb{R}$, and $\pi_f^{K_f}$ for the space of $K_f$-fixed vectors in

$\pi_f$. Then $\pi_f^{K_f}$ is a finite dimensional complex space on which $\mathbb{H}_{K_f} = \mathbb{H}_{K_f,L} \otimes_L \mathbb{C}$ acts irreducibly. The representation $\pi_\infty$ is viewed as a $(\mathfrak{g}, K_\infty)$-module, where $\mathfrak{g}$ denotes the Lie algebra of $G(\mathbb{R})$, and

$$H^i(\mathfrak{g}, K_\infty; \pi_\infty \otimes \rho_{\mathbb{C}}) = H^i(\mathfrak{g}, K_\infty; \pi_\infty \otimes V_\rho(\mathbb{C})), \quad \rho_{\mathbb{C}} = \rho \otimes_L \mathbb{C},$$

denotes the Lie-algebra cohomology of $\pi_\infty$ twisted by the finite dimensional representation $\rho_{\mathbb{C}}$ of $G(\mathbb{R})$. Then the finite dimensional complex space $H^i(\mathfrak{g}, K_\infty; \pi_\infty \otimes \rho_{\mathbb{C}})$ vanishes unless the central character $\omega_{\pi_\infty}$ and the infinitesimal character $\inf(\pi_\infty)$ are equal to those $\omega_{\check{\rho}_{\mathbb{C}}}$, $\inf(\check{\rho}_{\mathbb{C}})$ of the contragredient $\check{\rho}_{\mathbb{C}}$ of $\rho_{\mathbb{C}}$; see Borel-Wallach [BW].

There are only finitely many equivalence classes of $\pi$ in $L_d^2$ with fixed central and infinitesimal character, and a nonzero $K_f$-fixed vector ($\pi_f^{K_f} \neq 0$). The multiplicities $m(\pi)$ are finite. Hence $H^i_{(2)}(\mathcal{S}_{K_f}(\mathbb{C}), \mathbb{V}_{\mathbb{C}})$ is finite dimensional. The Zucker isomorphism (for a fixed embedding of $L_\lambda$ in $\mathbb{C}$) of $\mathbb{H}_{K_f,L} \otimes_L \mathbb{C} = \mathbb{H}_{K_f}$-modules

$$IH^i(\mathcal{S}'_{K_f} \times_{\mathbb{E}} \overline{\mathbb{Q}}, \mathbb{V}_\lambda) \otimes_{L_\lambda} \mathbb{C} \xrightarrow{\sim} H^i_{(2)}(\mathcal{S}_{K_f}(\mathbb{C}), \mathbb{V}_{\mathbb{C}})$$

then implies that the decomposition (1) ranges over the finite set of equivalence classes of irreducible $\pi$ in $L_d^2$ with $\pi_f^{K_f} \neq 0$ and $\pi_\infty$ with central and infinitesimal characters equal to those of $\check{\rho}_{\mathbb{C}}$. Further, $\pi_{f,L_\lambda}^{K_f}$ of (1) is an irreducible $\mathbb{H}_{K_f,L_\lambda}$-module with $\pi_{f,L_\lambda}^{K_f} \otimes_{L_\lambda} \mathbb{C} = \pi_f^{K_f}$ for such a discrete spectrum $\pi = \pi_f \otimes \pi_\infty$, and

$$\dim_{L_\lambda} IH^i(\pi_f^{K_f}) = \sum_{\pi_\infty} m(\pi_f \otimes \pi_\infty) \dim_{\mathbb{C}} H^i(\mathfrak{g}, K_\infty; \pi_\infty \otimes \check{\rho}_{\mathbb{C}}).$$

Moreover, each discrete spectrum $\pi = \pi_f \otimes \pi_\infty$ such that the central and infinitesimal characters of $\pi_\infty$ coincide with those of $\check{\rho}_{\mathbb{C}}$ (where $\rho$ is an absolutely irreducible representation of $G$ on a finite dimensional vector space over $L$) has the property that for some open compact subgroup $K_f \subset G(\mathbb{A}_{\mathbb{Q}f})$ for which $\pi_f^{K_f} \neq \{0\}$, there is an $L$-model $\pi_{f,L}^{K_f}$ of $\pi_f^{K_f}$.

It is also known that the cuspidal cohomology in $H_c^i$, that is, its part which is indexed by the cuspidal $\pi$, makes an orthogonal direct summand in $H_c^i \otimes_{L_\lambda} \mathbb{C}$, and also in $IH^i \otimes_{L_\lambda} \mathbb{C}$ (and $H^i \otimes_{L_\lambda} \mathbb{C}$). When we study the $\pi_f$-isotypic component of $H_c^i \otimes_{L_\lambda} \mathbb{C}$ for the finite component $\pi_f$ of a cuspidal representation $\pi$, we shall then be able to view it as such a component of $IH^i$.

Our aim is then to recall the classification of automorphic representations of PGSp(2) given in [F6], in particular list the possible $\pi_H = \pi_{Hf} \otimes \pi_{H\infty}$ in the cuspidal and discrete spectrum. This means listing the possible $\pi_{Hf}$, then the $\pi_{H\infty}$ which make $\pi_{Hf} \otimes \pi_{H\infty}$ occur in the cuspidal or discrete spectrum. Further we list the cohomological $\pi_{H\infty}$, those for which $H^i(\mathfrak{h}, K; \pi_{H\infty} \otimes \rho_{\mathbb{C}})$ is nonzero, and describe these spaces. In particular we can then compute the dimension of the contribution of $\pi_{Hf}$ to $IH^*$. Then we describe the trace of $\mathrm{Fr}_p$ acting on the Galois representation $H_c^*(\pi_{Hf})$ attached to $\pi_{Hf}$ in terms of the Satake parameters of $\pi_{Hp}$, in fact any sufficiently large power of $\mathrm{Fr}_p$. This determines uniquely the Galois representation $H_c^*(\pi_{Hf})$, of $\mathrm{Gal}(\overline{\mathbb{Q}}/\mathbb{Q})$, and in particular its dimension. The displayed formula of "Matsushima-Murakami" type will be used to estimate the absolute values of the eigenvalues of the action of the Frobenius on $H_c^*(\pi_{Hf})$.

# 6. Galois Representations

The decomposition (1) for $IH$ then defines a map $\pi_f \mapsto IH^i(\pi_f)$ from the set of irreducible representations $\pi_f$ of $G(\mathbb{A}_{\mathbb{Q}f})$ for which there exists an irreducible representation $\pi_\infty$ of $G(\mathbb{A}_{\mathbb{Q}})$ with central and infinitesimal characters equal to those of $\check{\rho}_{\mathbb{C}}$ such that $\pi_\infty \otimes \pi_f$ is discrete spectrum, to the set of finite dimensional representations of $\mathrm{Gal}(\overline{\mathbb{Q}}/\mathbb{E})$. We wish to determine the representation $IH^i(\pi_f)$ associated with $\pi_f$, namely its restriction to the decomposition groups at almost all primes.

However, the cohomology with which we work in this paper is $H_c^i$ and not $IH^i(\mathcal{S}')$.

Let $p$ be a rational prime. Assume that $G$ is unramified at $p$, thus it is quasi-split over $\mathbb{Q}_p$ and splits over an unramified extension of $\mathbb{Q}_p$. Assume that $K_f$ is unramified at $p$, thus it is of the form $K_f^p K_p$ where $K_f^p$ is a compact open subgroup of $G(\mathbb{A}_{\mathbb{Q}f}^p)$ and $K_p = G(\mathbb{Z}_p)$. Then $\mathbb{E}$ is unramified at $p$. Let $\wp$ be a place of $\mathbb{E}$ lying over $p$ and $\lambda$ a place of $L$ such that $p$ is a unit in $L_\lambda$. Let $f = f^p f_{K_p}$ be a function in the Hecke algebra $\mathbb{H}_{K_f,L}$, where $f^p$ is a function on $G(\mathbb{A}_{\mathbb{Q}f}^p)$ and $f_{K_p}$ is the quotient of the characteristic function of $K_p$ in $G(\mathbb{Q}_p)$ by the volume of $K_p$. Denote by $\mathrm{Fr}_\wp$ a geometric Frobenius element of the decomposition group $\mathrm{Gal}(\overline{\mathbb{Q}}_p/\mathbb{E}_\wp)$.

Choose models of $\mathcal{S}_{K_f}$ and of $\mathcal{S}'_{K_f}$ over the ring of integers of $\mathbb{E}$. For almost all primes $p$ of $\mathbb{Q}$, for each prime $\wp$ of $\mathbb{E}$ over $p$, the representation $H^i_c(\mathcal{S}_{K_f} \otimes_{\mathbb{E}} \overline{\mathbb{Q}}, \mathbb{V}_\lambda)$ of $\mathrm{Gal}(\overline{\mathbb{Q}}/\mathbb{E})$ is unramified at $\wp$, thus its restriction to $\mathrm{Gal}(\overline{\mathbb{Q}}_p/\mathbb{E}_\wp)$ factorizes through the quotient $\mathrm{Gal}(\mathbb{Q}_p^{\mathrm{ur}}/\mathbb{E}_\wp) \simeq \mathrm{Gal}(\overline{\mathbb{F}}/\mathbb{F})$ which is (topologically) generated by $\mathrm{Fr}_\wp$; here $\mathbb{Q}_p^{\mathrm{ur}}$ is the maximal unramified extension of $\mathbb{Q}_p$ in the algebraic closure $\overline{\mathbb{Q}}_p$, $\mathbb{F}$ is the residue field of $\mathbb{E}_\wp$ and $\overline{\mathbb{F}}$ an algebraic closure of $\mathbb{F}$. Denote the cardinality of $\mathbb{F}$ by $q_\wp$; it is a power of $p$. As a $\mathrm{Gal}(\overline{\mathbb{F}}/\mathbb{F})$-module $H^i_c(\mathcal{S}_{K_f} \otimes_{\mathbb{E}} \overline{\mathbb{Q}}, \mathbb{V}_\lambda)$ is isomorphic to $H^i_c(\mathcal{S}_{K_f} \otimes_{\mathbb{E}} \overline{\mathbb{F}}, \mathbb{V}_\lambda)$.

Deligne's conjecture proven by Zink [Zi] for surfaces, by Pink [P] and Shpiz [Sh] for varieties $X$ (such as $\mathcal{S}_{K_f}$) which have smooth compactification $\overline{X}$ which differs from $X$ by a divisor with normal crossings, and unconditionally by Fujiwara [Fu], and recently Varshavsky [Va], implies that for each correspondence $f^p$ there exists an integer $j_0 \geq 0$ such that for any $j \geq j_0$ the trace of $f^p \cdot \mathrm{Fr}_\wp^j$ on

$$\bigoplus_{i=0}^{2\dim \mathcal{S}_{K_f}} (-1)^i \; H^i_c(\mathcal{S}_{K_f} \otimes_{\mathbb{E}} \overline{\mathbb{F}}, \mathbb{V}_\lambda)$$

has contributions only from the variety $\mathcal{S}_{K_f}$ and not from any boundary component of $\mathcal{S}'_{K_f}$. The trace is the same in this case as if the scheme $\mathcal{S}_{K_f} \otimes_{\mathbb{E}} \overline{\mathbb{F}}$ were proper over $\overline{\mathbb{F}}$, and it is given by the usual formula of the Lefschetz fixed point formula. This is the reason why we work with $H^i_c$ in this paper, and not with $IH^i(\mathcal{S}')$.

# II. AUTOMORPHIC REPRESENTATIONS

## 1. Stabilization and the Test Function

Kottwitz computed the trace of $f^p \cdot \mathrm{Fr}_{\wp}^j$ on this alternating sum (see [Ko6], and [Ko4], chapter III, for $\rho = 1$) at least in the case considered here. The result, stated in [Ko4], (3.1) as a conjecture, is a certain sum

$$\sum_{\gamma_0} \sum_{(\gamma,\delta)} c(\gamma_0;\gamma,\delta) \cdot O(\gamma, f^p) \cdot TO(\delta,\phi_j) \cdot \mathrm{tr}\,\rho(\gamma_0),$$

rewritten in [Ko4], (4.2) in the form

$$\tau(G) \sum_{\gamma_0} \sum_{\kappa} \sum_{(\gamma,\delta)} \langle \alpha(\gamma_0;\gamma,\delta),\kappa\rangle \cdot e(\gamma,\delta) \cdot O(\gamma, f_{\mathbb{C}}^p)$$

$$\cdot TO(\delta,\phi_j) \cdot \frac{\mathrm{tr}\,\rho_{\mathbb{C}}(\gamma_0)}{|I(\infty)(\mathbb{R})/A_G(\mathbb{R})^0|},$$

where $O$ and $TO$ are orbital and twisted orbital integrals and $\phi_j$ is a spherical ($K_p = G(\mathbb{Z}_p)$-biinvariant) function on $G(\mathbb{Q}_p)$. Theorem 7.2 of [Ko4] expresses this as a sum

$$\sum \iota(G,H)\,\mathrm{STF}_e^{\mathrm{reg}}(f_H^{j,s,\rho})$$

over a set of representatives for the isomorphism classes of the elliptic endoscopic triples $(H, s, \eta_0 : \hat{H} \to \hat{G})$ for $G$. The $\mathrm{STF}_e^{\mathrm{reg}}(f_H^{j,s,\rho})$ indicates the $(G,H)$-regular $\mathbb{Q}$-elliptic part of the stable trace formula for a function $f_H^{j,s,\rho}$ on $H(\mathbb{A}_{\mathbb{Q}})$. The function $f_H^{j,s,\rho}$, denoted simply by $h$ in [Ko4], is constructed in [Ko4], Section 7 assuming the "fundamental lemma" and "matching orbital integrals", both known in our case by [F5] and [W].

Thus $f_H^{j,s,\rho}$ is the product of the functions $f_H^p$ on $H(\mathbb{A}_{\mathbb{Q}f}^p)$ which are obtained from $f_{\mathbb{C}}^p$ by matching of orbital integrals, $f_{Hp}^{js}$ on $H(\mathbb{Q}_p)$ which is a spherical function obtained by the fundamental lemma from the spherical function $\phi_j$, and $f_{H\infty}^{s\rho}$ on $H(\mathbb{R})$ which is constructed from pseudo-coefficients of discrete series representations of $H(\mathbb{R})$ which lift to discrete series representations of $G(\mathbb{R})$ whose central and infinitesimal characters coincide with those of $\check{\rho}_{\mathbb{C}}$. We denote by $f_H^{j,s,\rho} = f_H^p f_{Hp}^{js} f_{H\infty}^{s\rho}$ Kottwitz's function $h = h^p h_p h_\infty$, so that functions on the adèle groups are denoted by $f$, and the notation does not conflict with that of $h : \mathrm{R}_{\mathbb{C}/\mathbb{R}}\mathbb{G}_m \to G$.

The factor $\langle \alpha^p(\gamma_0; \gamma), s\rangle$ is missing on the right side of [Ko4], (7.1). Here

$$\alpha^p = \prod_{v \neq p,\infty} \alpha_v, \quad \text{where} \quad \alpha_v(\gamma_0; \gamma_v) \in X^*(Z(\hat{I}_0)^{\Gamma(v)}/Z(\hat{I}_0)^{\Gamma(v),0}Z(\hat{G}^{\Gamma(v)}))$$

as defined in [Ko4], p. 166, bottom paragraph

We need to compare the elliptic regular part $\mathrm{STF}_e^{\mathrm{reg}}(f_H^{j,s,\rho})$ of the stable trace formula with the spectral side. To simplify matters we shall work only with a special class of test functions $f^p = \otimes_{v\neq p,\infty} f_v$ for which the complicated parts of the trace formulae vanish. Thus we choose a place $v_0$ where $G$ is quasi-split, and a maximal split torus $A$ of $G$ over $\mathbb{Q}_{v_0}$, and require that the component $f_{v_0}$ of $f^p$ be in the span of the functions on $G(\mathbb{Q}_{v_0})$ which are bi-invariant under an Iwahori subgroup $I_{v_0}$ and supported on a double coset $I_{v_0} a I_{v_0}$, where $a \in A(\mathbb{Q}_{v_0})$ has $|\alpha(a)| \neq 1$ for all roots $\alpha$ of $A$. The orbital integrals of such a function $f_{v_0}$ vanish on the singular set, and the matching functions $f_{Hv_0}$ on $H(\mathbb{Q}_{v_0})$ have the same property. This would permit us to deal only with regular conjugacy classes in the elliptic part of the stable trace formulae $\mathrm{STF}_e^{\mathrm{reg}}(f_H^{j,s,\rho})$, and would restrict no applicability.

To avoid dealing with weighted orbital integrals and the continuous spectrum, we note that these vanish if two components of the test function $f_H^{j,s,\rho}$ are discrete, by which we mean that they have orbital integrals which are zero on the regular nonelliptic set. The component $f_{\infty H}^{s\rho}$ has this property. If $G = \mathrm{R}_{F/\mathbb{Q}}G_1$ is obtained by restriction of scalars from a group $G_1$ defined over a totally real field $F$, then $G(\mathbb{Q}) = G_1(F)$ and $G(\mathbb{R}) = G_1(F\otimes\mathbb{R}) = \prod G_1(\mathbb{R})$; the last product has $[F : \mathbb{Q}]$ factors. Correspondingly the function $f_{H\infty}^{s\rho}$ is a product of $[F : \mathbb{Q}]$ discrete factors. This gives the equality of the elliptic regular part of the stable trace formula with the discrete spectral side when $F \neq \mathbb{Q}$. If $F = \mathbb{Q}$, and in general, we may take some of the components $f_w$ of $f^p$ to be discrete, for example pseudo-coefficients of discrete series representations, to achieve this vanishing of the weighted terms in the trace formula. Such a choice of course will limit our results to only those automorphic representations with the specified (by the $f_w$) elliptic components.

# 2. Automorphic Representations of PGSp(2)

We then need to describe the stable trace formulae. This we can do only in the special case, studied in [F6]. We then use the notations of [F6] from now on, and in particular the group denoted so far by $G$ will be denoted from now on by $H' = \mathrm{R}_{F/\mathbb{Q}}H$, where $F$ is a totally real number field and $H$ is the projective group PGSp(2) of symplectic similitudes over $F$. When describing the automorphic representations of $H'$, note that $H'(\mathbb{Q}) = H(F)$ and $H'(\mathbb{A}_{\mathbb{Q}}) = H(\mathbb{A})$, where $\mathbb{A}$ denotes now the ring of adèles of $F$ (and $\mathbb{A}_{\mathbb{Q}}$ of $\mathbb{Q}$). It is more convenient to describe the automorphic representations of $H/F$. Working with PGSp(2) is the same as working with GSp(2) and functions transforming trivially under the center.

A detailed description of the automorphic representations of $H/F$ is given in [F6]. We recall here only the most essential facts. The group $H = \mathrm{PGSp}(2)$ is the quotient of (we put $J = \begin{pmatrix} 0 & w \\ -w & 0 \end{pmatrix}$, $w = \begin{pmatrix} 0 & 1 \\ 1 & 0 \end{pmatrix}$)

$$\mathrm{GSp}(2) = \{(g, \boldsymbol{\lambda}) \in \mathrm{GL}(4) \times \mathbb{G}_m;\ {}^t g J g = \boldsymbol{\lambda} J\}$$

by its center $\{(\boldsymbol{\lambda}, \boldsymbol{\lambda}^2);\ \boldsymbol{\lambda} \in \mathbb{G}_m\}$. It has a single proper elliptic endoscopic group $C_0 = \mathrm{PGL}(2) \times \mathrm{PGL}(2)$ over $F$. The group $H$ itself is one of the two elliptic endoscopic groups of $G = \mathrm{PGL}(4)$ with respect to the involution $\theta$, $\theta(g) = J^{-1}\, {}^t g^{-1} J$. The other $\theta$-twisted elliptic endoscopic group of $G$ is

$$C = \text{“}\mathrm{SO}(4)/F\text{”} = \{(g_1, g_2) \in \mathrm{GL}(2) \times \mathrm{GL}(2);\ \det g_1 = \det g_2\}/\mathbb{G}_m.$$

The automorphic representations of $H$ are described in [F6] in terms of liftings, defined by means of the natural embeddings of $L$-groups. The groups $G$, $H$, $C$, $C_0$ are split. Hence their $L$-groups (${}^L G, \dots$) are the direct product of the connected component of the identity ($\hat{G}, \dots$) with the Weil group. Let $\hat{\theta}$ be the involution on $\hat{G}$ defined by the formula which defines $\theta$. Writing $Z_{\hat{G}}(\hat{s}\hat{\theta})$ for the group of $g$ in $\hat{G}$ with $\hat{s}\hat{\theta}(g)\hat{s}^{-1} = g$, the $L$-group homomorphisms are

$$\lambda : \hat{H} = \mathrm{Sp}(2, \mathbb{C}) = Z_{\hat{G}}(\theta) \hookrightarrow \hat{G} = \mathrm{SL}(4, \mathbb{C}),$$

$$\lambda_1 : \hat{C} = \text{“}\mathrm{SO}(4, \mathbb{C})\text{”} = Z_{\hat{G}}(\hat{s}\hat{\theta}) \hookrightarrow \hat{G}, \quad \lambda_0 : \hat{C}_0 = Z_{\hat{H}}(\hat{s}_0) \hookrightarrow \hat{H}.$$

Here $\hat{s}_0 = \text{diag}(1,-1,-1,1)$, $\hat{s} = \text{diag}(-1,1,-1,1)$, and $\hat{C}$ consists of the $A\otimes B = \begin{pmatrix} aB & bB \\ cB & dB \end{pmatrix}$, where $\left(A = \begin{pmatrix} a & b \\ c & d \end{pmatrix}, B\right)$ ranges over $\text{GL}(2,\mathbb{C})\times\text{GL}(2,\mathbb{C})$ with $\det A\cdot\det B = 1$, modulo $(z, z^{-1})$, $z \in \mathbb{C}^\times$. These homomorphisms of complex groups define liftings of unramified representations via the Satake transform. They are extended in [F6] to ramified representation by character relations involving packets and quasi-packets (which are introduced in [F6]).

The packets and quasi-packets define a partition of the discrete spectrum of $H(\mathbb{A})$. To define a global (quasi-) packet $P = \{\pi\}$, fix a local (quasi-) packet $P_v = \{\pi_v\}$ at every place $v$ of $F$, such that $P_v = \{\pi_v\}$ contains an unramified representation $\pi_v^0$ at almost all places. Then $P = \{\pi\}$ consists of all products $\otimes\pi_v$ over all $v$, where $\pi_v \in P_v = \{\pi_v\}$ for every $v$ and $\pi_v = \pi_v^0$ for almost all $v$.

Before we recall the definition of local packets, we state that the discrete spectrum of $H(\mathbb{A})$ is the disjoint union of what we call the stable and unstable spectra. The lifting $\lambda$ defines a bijection from the set of packets and quasi-packets of discrete spectrum representations in the stable spectrum to the set of self contragredient discrete spectrum (cuspidal or residual) representations of $G(\mathbb{A})$ which are not in the image of $\lambda_1$.

In particular, $\lambda$ maps one dimensional representations of $H(\mathbb{A})$ to one dimensional representations of $G(\mathbb{A})$, stable non one-dimensional packets of $H(\mathbb{A})$ to cuspidal self contragredient representations of $G(\mathbb{A})$, and the quasi-packets in the stable discrete spectrum of $H(\mathbb{A})$, each of which has the form $\{L(\xi\nu, \nu^{-1/2}\pi^2)\}$, to $J(\nu^{1/2}\pi^2, \nu^{-1/2}\pi^2)$, residual representations of $G(\mathbb{A})$.

Here $L(\xi\nu, \nu^{-1/2}\pi^2)$ is the unique quotient of the representation of $H(\mathbb{A})$ normalizedly induced from the "Heisenberg" maximal parabolic subgroup (whose unipotent radical is a (nonabelian) Heisenberg group) and the indicated representation on the Levi subgroup $\mathbb{A}^\times \times \text{GL}(2,\mathbb{A})$: $\pi^2$ is a cuspidal irreducible automorphic representation of $\text{GL}(2,\mathbb{A})$ with central character $\xi \neq 1$ of order two and $\xi\pi^2 = \pi^2$.

The $J(\nu^{1/2}\pi^2, \nu^{-1/2}\pi^2)$ is the unique quotient of the representation $I(\nu^{1/2}\pi^2, \nu^{-1/2}\pi^2)$ normalizedly induced from the parabolic subgroup of type (2,2) and the indicated representation of the Levi factor, where $\nu(x) = |x|$ and $\pi^2$ is a cuspidal automorphic representation of $\text{GL}(2,\mathbb{A})$ with central character $\xi \neq 1$ of order two and $\xi\pi^2 = \pi^2$.

In particular, the image of $\lambda$ in the discrete spectrum self-contragredient representations of $\mathrm{PGL}(4,\mathbb{A})$ is precisely the complement of the lifting $\lambda_1$ from $C(\mathbb{A})$.

Similarly, the lifting $\lambda_0$ defines a bijection to the set of packets and quasi-packets in the unstable spectrum of $H(\mathbb{A})$ from the set of unordered pairs $\{\pi^1 \times \pi^2, \pi^2 \times \pi^1; \pi^1 \neq \pi^2\}$ of discrete spectrum automorphic representations of $\mathrm{PGL}(2,\mathbb{A})$. This last set is bijected by $\lambda$ with the set of automorphic (irreducible) representations $I(\pi^1, \pi^2)$ normalizedly induced from the representation $\pi^1 \otimes \pi^2$ on the Levi subgroup of $G(\mathbb{A})$ of type (2,2), where $\pi^1$, $\pi^2$ are discrete spectrum on $\mathrm{PGL}(2,\mathbb{A})$ with $\pi^1 \neq \pi^2$. In fact if $\pi^1 \times \pi^2$ is cuspidal it is mapped by $\lambda_0$ to a packet, while if not, that is when $\pi^1$ or $\pi^2$ are one-dimensional, $\lambda_0(\pi^1 \times \pi^2)$ is a quasi-packet.

To repeat, the global liftings are defined by the $L$-group homomorphisms for almost all components, which are unramified, and it is a theorem that the liftings extend to all places in terms of packets and quasi-packets, and have the properties listed above.

The stable part of the discrete spectrum, defined above by means of the bijection $\lambda$, has the property that the multiplicity in the discrete spectrum of $H(\mathbb{A})$ is stable, namely constant over each packet. Thus each member $\otimes_v \pi_v$ of a packet $\{\pi\}$ which $\lambda$-lifts to a discrete spectrum representation $\pi \simeq \check{\pi}$ of $\mathrm{PGL}(4,\mathbb{A})$ occurs in the discrete spectrum of $H(\mathbb{A})$ with multiplicity one. The same is true for the stable quasi-packets, each of which is of the form $\{L(\xi\nu, \nu^{-1/2}\pi^2)\}$.

# 3. Local Packets

The multiplicity is not constant on the unstable packets, but it is bounded by one. It is possible that a member in an unstable packet will not occur in the discrete spectrum of $H(\mathbb{A})$. Then its multiplicity is zero. To specify the multiplicity, we need to describe the local packets. For this purpose we recall the main local theorem of [F6]. It has 4 parts.

Let $F$ be a local field.

(1) For any unordered pair $\pi^1$, $\pi^2$ of irreducible square integrable representations of $\mathrm{PGL}(2,F)$ there exists a unique pair $\pi_H^+$, $\pi_H^-$ of tempered

(square integrable if $\pi^1 \neq \pi^2$, cuspidal if $\pi^1 \neq \pi^2$ are cuspidal) representations of $H(F)$, $\pi_H^+$ is generic, $\pi_H^-$ is not, with

$$\operatorname{tr}(\pi^1 \times \pi^2)(f_{C_0}) = \operatorname{tr}\pi_H^+(f_H) - \operatorname{tr}\pi_H^-(f_H)$$
$$\operatorname{tr} I_G(\pi^1, \pi^2; f \times \theta) = \operatorname{tr}\pi_H^+(f_H) + \operatorname{tr}\pi_H^-(f_H)$$

for every triple of matching functions $f$, $f_H$, $f_{C_0}$, for a suitable choice of an operator $\pi(\theta)$, $\pi = I_G(\pi^1, \pi^2)$, intertwining $\pi$ with ${}^\theta\pi$, and having order 2.

We define the packet of $\pi_H^+$ and of $\pi_H^-$ to be $\{\pi_H^+, \pi_H^-\}$. The packet of any other irreducible representations of $H(F)$ is defined to be a singleton. More details are known.

If $\pi^1 = \pi^2$ is cuspidal, $\pi_H^+$ and $\pi_H^-$ are the two inequivalent constituents of the induced representation $1 \rtimes \pi^1$ from the Heisenberg parabolic subgroup, $\pi_H^+$ is the generic constituent.

If $\pi^1 = \pi^2 = \sigma\,\mathrm{sp}_2$ where $\sigma$ is a character of $F^\times$ with $\sigma^2 = 1$, then $\pi_H^+$ and $\pi_H^-$ are the two tempered inequivalent constituents $\tau(\nu^{1/2}\,\mathrm{sp}_2, \sigma\nu^{-1/2})$ and $\tau(\nu^{1/2}\mathbf{1}_2, \sigma\nu^{-1/2})$ of $1 \rtimes \sigma\,\mathrm{sp}_2$.

If $\pi^1 = \sigma\,\mathrm{sp}_2$, $\sigma^2 = 1$, and $\pi^2$ is cuspidal, then $\pi_H^+$ is the square integrable constituent $\delta(\sigma\nu^{1/2}\pi^2, \sigma\nu^{-1/2})$ of the induced $\sigma\nu^{1/2}\pi^2 \rtimes \sigma\nu^{-1/2}$ from the Siegel maximal parabolic subgroup of $H(F)$ (with abelian unipotent radical). The $\pi_H^-$ is cuspidal, denote by $\delta^-(\sigma\nu^{1/2}\pi^2, \sigma\nu^{-1/2})$.

If $\pi^1 = \sigma\,\mathrm{sp}_2$ and $\pi^2 = \xi\sigma\,\mathrm{sp}_2$, $\xi$ ($\neq 1 = \xi^2$) and $\sigma$ ($\sigma^2 = 1$) are characters of $F^\times$, then $\pi_H^+$ is the square integrable constituent

$$\delta(\xi\nu^{1/2}\,\mathrm{sp}_2, \sigma\nu^{-1/2})$$

of the induced $\xi\nu^{1/2}\,\mathrm{sp}_2 \rtimes \sigma\nu^{-1/2}$. The $\pi_H^-$ is cuspidal, denoted by

$$\delta^-(\xi\nu^{1/2}\,\mathrm{sp}_2, \sigma\nu^{-1/2}).$$

(2) For every character $\sigma$ of $F^\times/F^{\times 2}$ and square integrable $\pi^2$ there exists a nontempered representation $\pi_H^\times$ of $H(F)$ such that

$$\operatorname{tr}(\pi^2 \times \sigma\mathbf{1}_2)(f_{C_0}) = \operatorname{tr}\pi_H^\times(f_H) + \operatorname{tr}\pi_H^-(f_H)$$
$$\operatorname{tr} I_G(\pi^2, \sigma\mathbf{1}_2; f \times \theta) = \operatorname{tr}\pi_H^\times(f_H) - \operatorname{tr}\pi_H^-(f_H)$$

for every triple $(f, f_H, f_{C_0})$ of matching functions. Here

$$\pi_H^- = \pi_H^-(\sigma\,\mathrm{sp}_2 \times \pi^2) \quad \text{and} \quad \pi_H^\times = L(\sigma\nu^{1/2}\pi^2, \sigma\nu^{-1/2}).$$

(3) For any characters $\xi$, $\sigma$ of $F^\times/F^{\times 2}$ and matching $f$, $f_H$, $f_{C_0}$ we have

$$\begin{aligned}&\mathrm{tr}(\sigma\xi\mathbf{1}_2 \times \sigma\mathbf{1}_2)(f_{C_0})\\&\quad = \mathrm{tr}\, L(\nu\xi, \xi \rtimes \sigma\nu^{-1/2})(f_H) - \mathrm{tr}\, X(\xi\nu^{1/2}\,\mathrm{sp}_2, \xi\sigma\nu^{-1/2})(f_H),\\&\mathrm{tr} I_G(\sigma\xi\mathbf{1}_2, \sigma\mathbf{1}_2; f \times \theta)\\&\quad = \mathrm{tr}\, L(\nu\xi, \xi \rtimes \sigma\nu^{-1/2})(f_H) + \mathrm{tr}\, X(\xi\nu^{1/2}\,\mathrm{sp}_2, \xi\sigma\nu^{-1/2})(f_H).\end{aligned}$$

Here $X = \delta^-$ if $\xi \neq 1$ and $X = L$ if $\xi = 1$.

(4) Any $\theta$-invariant irreducible square integrable representation $\pi$ of $G$ which is not a $\lambda_1$-lift is a $\lambda$-lift of an irreducible square integrable representation $\pi_H$ of $H$, thus $\mathrm{tr}\,\pi(f \times \theta) = \mathrm{tr}\,\pi_H(f_H)$ for all matching $f$, $f_H$. In particular, the square integrable (resp. nontempered) constituent $\delta(\xi\nu, \nu^{-1/2}\pi^2)$ (resp. $L(\xi\nu, \nu^{-1/2}\pi^2)$) of the induced representation $\xi\nu \rtimes \nu^{-1/2}\pi^2$ of $H$, where $\pi^2$ is a cuspidal (irreducible) representation of $\mathrm{GL}(2, F)$ with central character $\xi \neq 1 = \xi^2$ and $\xi\pi^2 = \pi^2$, $\lambda$-lifts to the square integrable (resp. nontempered) constituent

$$S(\nu^{1/2}\pi^2, \nu^{-1/2}\pi^2) \qquad (\text{resp.} \qquad J(\nu^{1/2}\pi^2, \nu^{-1/2}\pi^2))$$

of the induced representation $I_G(\nu^{1/2}\pi^2, \nu^{-1/2}\pi^2)$ of $G = \mathrm{PGL}(4, F)$.

We define a *quasi-packet* only for the nontempered irreducible representations $\pi_H^\times$, and $L = L(\nu\xi, \xi \rtimes \sigma\nu^{-1/2})$, to consist of $\{\pi_H^\times, \pi_H^-\}$, and of $\{L, X\}$, $X = X(\xi\nu^{1/2}\,\mathrm{sp}_2, \xi\sigma\nu^{-1/2})$.

Using the notations of sections IV.1-IV.5 below, we state the analogue of these results in the real case: $F = \mathbb{R}$. In (1), $\pi^1 = \pi_{k_1}$ and $\pi^2 = \pi_{k_2}$, $k_1 \geq k_2 > 0$ and $k_1$, $k_2$ are odd, are discrete series representations of $\mathrm{PGL}(2, \mathbb{R})$, and $\pi_H^+$ is the generic $\pi_{k_1,k_2}^{\mathrm{Wh}}$, $\pi_H^-$ is the holomorphic $\pi_{k_1,k_2}^{\mathrm{hol}}$, which are discrete series when $k_1 > k_2$. When $k_1 = k_2$, $\pi_H^+$ is the generic and $\pi_H^-$ is the nongeneric constituents of the induced $1 \rtimes \pi_{2k_1+1}$. There is no special or Steinberg representation of $\mathrm{GL}(2, \mathbb{R})$; the analogue is the lowest discrete series $\pi^1$. It is self invariant under twist with sgn. In (2) with $\pi^2 = \pi_{2k+3}$ ($k \geq 0$), $\pi_H^\times$ is $L(\sigma\nu^{1/2}\pi_{2k+3}, \sigma\nu^{-1/2})$, $\pi_H^-$ is $\pi_{2k+3,1}^{\mathrm{hol}}$. In (3), if $\xi = \mathrm{sgn}$ then $X$ is $\pi_H^- \subset 1 \rtimes \pi^1$, if $\xi = 1$ then $X$ is $L(\nu^{1/2}\pi^1, \sigma\nu^{-1/2})$, but both of these $X$, as well as $L(\nu\xi, \xi \rtimes \sigma\nu^{-1/2})$, are not cohomological, and will not concern us in this work.

# 4. Multiplicities

We are now ready to describe the multiplicities of the representations in the packets and quasi-packets in the unstable spectrum of $H(\mathbb{A})$.

Each member of a stable packet occurs in the discrete spectrum of $\mathrm{PGSp}(2,\mathbb{A})$ with multiplicity one. The multiplicity $m(\pi_H)$ of a member $\pi_H = \otimes\pi_{Hv}$ in an unstable [quasi-] packet $\lambda_0(\pi^1 \times \pi^2)$ ($\pi^1 \neq \pi^2$) is not ("stable", namely) constant over the [quasi-] packet.

If $\pi^1 \times \pi^2$ is cuspidal then

$$m(\pi_H) = \frac{1}{2}(1 + (-1)^{n(\pi_H)}) \quad (\in \{0,1\}).$$

Here $n(\pi_H)$ is the number of components $\pi_{Hv}^-$ of $\pi_H$ ($n(\pi_H)$ is bounded by the number of places $v$ where both $\pi_v^1$ and $\pi_v^2$ are square integrable). If $m(\pi_H) = 1$ then $\pi_H$ is cuspidal.

If $\pi^2$ is a cuspidal representation of $\mathrm{PGL}(2,\mathbb{A})$ and $\sigma$ is a character of $\mathbb{A}^\times/F^\times\mathbb{A}^{\times 2}$, the multiplicity $m(\pi_H)$ of $\pi_H = \otimes\pi_{Hv}$ in a quasi-packet $\lambda_0(\pi^2 \times \sigma\mathbf{1}_2)$ is

$$\frac{1}{2}\left(1 + \varepsilon(\sigma\pi^2, \frac{1}{2})(-1)^{n(\pi_H)}\right) \quad (= 0 \text{ or } 1),$$

where $n(\pi_H)$ is the number of components $\pi_{Hv}^-$ of $\pi_H$, and $\varepsilon = \varepsilon(\sigma\pi^2, \frac{1}{2})$ is 1 or $-1$, being the value at $\frac{1}{2}$ of the $\varepsilon$-factor occurring in the functional equation of the $L$-function $L(\sigma\pi^2, s)$ of $\sigma\pi^2$. This $\varepsilon$ is 1 if and only if $\pi_H^\times = \otimes\pi_{Hv}^\times$ ($n(\pi_H) = 0$) is discrete series.

Finally we have $m(\pi_H) = \frac{1}{2}(1+(-1)^{n(\pi_H)})$ for $\pi_H = \otimes\pi_{Hv}$ in $\lambda_0(\sigma\xi\mathbf{1}_2 \times \sigma\mathbf{1}_2)$ with $n(\pi_H)$ components $\pi_{Hv} = X_v$. Here $\pi_H = \otimes L_v$ ($n(\pi_H) = 0$) is residual.

# 5. Spectral Side of the Stable Trace Formula

We are now in a position to describe the spectral side of the stable trace formula for a test function $f_H = \otimes f_{Hv}$ with at least two discrete components, on $H(\mathbb{A})$. Thus $\mathrm{STF}_H(f_H)$ is the sum of five parts: $I(H,1), \ldots, I(H,5)$.

The first, $I(H,1)$, is the sum of three subterms: $I(H,1)_i$, $i=1,2,3$, each of which is a sum of products

$$\prod_v \operatorname{tr}\{\pi_{Hv}\}(f_{Hv}),$$

where $\operatorname{tr}\{\pi_{Hv}\}$ indicates the sum of $\operatorname{tr}\pi_{Hv}$ over all $\pi_{Hv}$ in a packet or quasi-packet $\{\pi_{Hv}\}$, over all packets and quasi-packets in the stable spectrum.

$I(H,1)_1$ ranges over the packets $\{\pi_H\}$ which $\lambda$-lift to cuspidal self contragredient representations $\pi$ of $\mathrm{PGL}(4,\mathbb{A})$ not in the image of $\lambda_1$.

$I(H,1)_2$ ranges over the discrete series quasi-packets $\{L(\xi\nu,\nu^{-1/2}\pi^2)\}$ (which $\lambda$-lift to the residual $J(\nu^{1/2}\pi^2,\nu^{-1/2}\pi^2)$, cuspidal $\pi^2$ with quadratic central character $\xi\neq 1$ with $\xi\pi^2=\pi^2$).

$I(H,1)_3$ is a sum over the one dimensional representations $\pi_H$ of $H(\mathbb{A})$.

The second part, $I(H,2)$, of $\mathrm{STF}_H(f_H)$, is the sum of

$$\frac{1}{2}\prod_v\{\operatorname{tr}\pi^+_{Hv}(f_{Hv})+\operatorname{tr}\pi^-_{Hv}(f_{Hv})\}$$

over all unordered pairs $(\pi^1,\pi^2)$ of distinct cuspidal representations of $\mathrm{PGL}(2,\mathbb{A})$. Here $\{\pi_H\}$ is the $\lambda_0$-lift of $\pi^1\times\pi^2$, that is $\lambda_0(\pi^1_v\times\pi^2_v)=\{\pi^+_{Hv},\pi^-_{Hv}\}$ for all $v$, and $\pi^-_{Hv}$ is zero if $\pi^1_v$ and $\pi^2_v$ are not both discrete series.

The third part, $I(H,3)$, is the sum of

$$\frac{\varepsilon(\sigma\pi^2,\frac{1}{2})}{2}\prod_v\{\operatorname{tr}\pi^\times_{Hv}(f_{Hv})-\operatorname{tr}\pi^-_{Hv}(f_{Hv})\}$$

over all pairs $(\sigma,\pi^2)$, where $\pi^2$ is a cuspidal representation of $\mathrm{PGL}(2,\mathbb{A})$ and $\sigma$ is a character of $\mathbb{A}^\times/F^\times\mathbb{A}^{\times 2}$. For each $v$ the pair $\{\pi^\times_{Hv},\pi^-_{Hv}\}$ is the quasi-packet $\lambda_0(\pi^2_v\times\sigma_v\mathbf{1}_2)$ when $\pi^2_v$ is discrete series, while it consists only of $\pi^\times_{Hv}$ (and $\pi^-_{Hv}$ is zero) when $\pi^2_v$ is not discrete series.

The fourth part, $I(H,4)$, is the sum of

$$\frac{1}{2}\prod_v\{\operatorname{tr}L_{Hv}(f_{Hv})+\operatorname{tr}X_{Hv}(f_{Hv})\}$$

over all unordered pairs $(\sigma\xi,\sigma)$ of characters of $\mathbb{A}^\times/F^\times\mathbb{A}^{\times 2}$ with $\xi\neq 1$. For each $v$ the pair $\{L_{Hv},X_{Hv}\}$ is the $\lambda$-lift of $\sigma_v\xi_v\mathbf{1}_2\times\sigma_v\mathbf{1}_2$.

The fifth part, $I(H,5)$, is the sum over all discrete spectrum representations $\pi^2$ of $\mathrm{PGL}(2,\mathbb{A})$ of the terms

$$\frac{1}{4}\prod_v \operatorname{tr} R_v \circ (1 \rtimes \pi_v^2)(f_{Hv}).$$

At each place $v$ where $\pi_v^2$ is properly induced (hence irreducible), the normalized intertwining operator $R_v$ is the scalar 1, and $\operatorname{tr}(1 \rtimes \pi_v^2)(f_{Hv}) = \operatorname{tr}(\pi_v^2 \times \pi_v^2)(f_{C_0v})$ for a matching function $f_{C_0v}$ on $C_0(F_v)$. If $\pi_v^2$ is square integrable (or one dimensional), our local results assert that the two constituents of the composition series of $1 \rtimes \pi_v^2$ can be labeled $\pi_{Hv}^+$ (or $\pi_{Hv}^\times$) and $\pi_{Hv}^-$, such that for matching functions $\operatorname{tr}(\pi_v^2 \times \pi_v^2)(f_{C_0v})$ is $\operatorname{tr}\pi_{Hv}^+(f_{Hv}) - \operatorname{tr}\pi_{Hv}^-(f_{Hv})$ (or $\operatorname{tr}\pi_{Hv}^\times(f_{Hv}) + \operatorname{tr}\pi_{Hv}^-(f_{Hv})$). Moreover, $R_v$ acts on $\pi_{Hv}^+$ as 1 and on $\pi_{Hv}^-$ as $-1$ when $\pi_v^2$ is square integrable, and as 1 on both $\pi_{Hv}^+$ and $\pi_{Hv}^-$ when $\pi_v^2$ is one dimensional.

# 6. Proper Endoscopic Group

The spectral side of the other trace formula which we need is for a function $f_{C_0} = \otimes f_{C_0v}$ on $C_0(\mathbb{A}) = \mathrm{PGL}(2,\mathbb{A}) \times \mathrm{PGL}(2,\mathbb{A})$. It comes multiplied by the coefficient $\frac{1}{4}$. Since PGL(2) has no proper elliptic endoscopic groups, this trace formula is already stable. Thus $\mathrm{STF}_{C_0}(f_0) = \mathrm{TF}_{C_0}(f_0)$. It is a sum of three sums, $I(C_0,i), i = 1,2,3$. The first, $I(C_0,1)$, is a sum of

$$\prod_v \operatorname{tr}(\pi_v^1 \times \pi_v^2)(f_{C_0v})$$

over all ordered pairs $(\pi^1,\pi^2)$ of cuspidal representations of $\mathrm{PGL}(2,\mathbb{A})$. The second part, $I(C_0,2)$, is a sum of

$$\prod_v \operatorname{tr}(\pi_v^2 \times \sigma_v \mathbf{1}_2)(f_{C_0v}) + \prod_v \operatorname{tr}(\sigma_v \mathbf{1}_2 \times \pi_v^2)(f_{C_0v})$$

over all pairs $(\sigma,\pi^2)$, where $\pi^2$ is a cuspidal representation of $\mathrm{PGL}(2,\mathbb{A})$ and $\sigma$ is a character of $\mathbb{A}^\times/F^\times\mathbb{A}^{\times 2}$. The third, $I(C_0,3)$, is the sum over all ordered pairs $(\sigma,\xi\sigma)$ of characters of $\mathbb{A}^\times/F^\times\mathbb{A}^{\times 2}$ of the products

$$\prod_v \operatorname{tr}(\sigma_v\xi_v\mathbf{1}_2 \times \sigma_v\mathbf{1}_2)(f_{C_0v}).$$

At all places $v \neq p, \infty$ the component $f_{C_0 v}$ is matching $f_{Hv}$, so the local factor indexed by $v$ in each of the 3 cases can be replaced by

$$\operatorname{tr} \pi^+_{Hv}(f_{Hv}) - \operatorname{tr} \pi^-_{Hv}(f_{Hv}),$$
$$\operatorname{tr} \pi^\times_{Hv}(f_{Hv}) + \operatorname{tr} \pi^-_{Hv}(f_{Hv}),$$
$$\operatorname{tr} L_{Hv}(f_{Hv}) - \operatorname{tr} X_{Hv}(f_{Hv}).$$

# III. LOCAL TERMS

## 1. Representations of the Dual Group

Part of the data which is used to define the Shimura variety is the $G(\mathbb{C})$-conjugacy class $\mathrm{Int}(G(\mathbb{C}))(\mu_h)$ of the homomorphism $\mu_h : \mathbb{G}_m \to G$ over $\mathbb{C}$. Let $C_k$ denote the set of conjugacy classes of homomorphisms $\mu : \mathbb{G}_m \to G$ over a field $k$. The embedding $\overline{\mathbb{Q}} \to \mathbb{C}$ induces an $\mathrm{Aut}(\mathbb{C}/\mathbb{Q})$-equivariant map $C_{\overline{\mathbb{Q}}} \to C_{\mathbb{C}}$. This map is bijective. Indeed, choose a maximal torus $\overline{T}$ of $G$ defined over $\overline{\mathbb{Q}}$. Then $\mathrm{Hom}_{\overline{\mathbb{Q}}}(\mathbb{G}_m, \overline{T})/W \to C_{\overline{\mathbb{Q}}}$ is a bijection, where $W$ is the Weyl group of $\overline{T}$ in $G(\overline{\mathbb{Q}})$. Similarly $\mathrm{Hom}_{\mathbb{C}}(\mathbb{G}_m, \overline{T})/W \to C_{\mathbb{C}}$ is a bijection. Since $\mathrm{Hom}_{\overline{\mathbb{Q}}}(\mathbb{G}_m, \overline{T}) \to \mathrm{Hom}_{\mathbb{C}}(\mathbb{G}_m, \overline{T})$ is an $\mathrm{Aut}(\mathbb{C}/\mathbb{Q})$-equivariant bijection, so is $C_{\overline{\mathbb{Q}}} \to C_{\mathbb{C}}$. The conjugacy class of $\mu_h$ over $\mathbb{C}$ is then a point in $\mathrm{Hom}_{\overline{\mathbb{Q}}}(\mathbb{G}_m, \overline{T})/W$. The subgroup of $\mathrm{Gal}(\overline{\mathbb{Q}}/\mathbb{Q})$ which fixes it has the form $\mathrm{Gal}(\overline{\mathbb{Q}}/\mathbb{E})$, where $\mathbb{E}$ is a number field, named the reflex field. It is contained in any field $\mathbb{E}_1$ over which $G$ splits, since $\overline{T}$ can be chosen to split over $\mathbb{E}_1$.

In our case ($G$ is) $H' = \mathrm{R}_{F/\mathbb{Q}} H$, where $H$ is $\mathrm{PGSp}(2)$ over a totally real field $F$. Thus $H'$ is split over $\mathbb{Q}$, and $\mathbb{E} = \mathbb{Q}$. Note that $H'(\mathbb{Q}) = H(F)$ and $H'(\mathbb{R}) = H(\mathbb{R}) \times \cdots \times H(\mathbb{R})$ ($[F : \mathbb{Q}]$ times). The dimension of the corresponding Shimura variety is $3[F : \mathbb{Q}]$, where 3 is half the real dimension of the symmetric space $H(\mathbb{R})/K_{H(\mathbb{R})}$.

Let $(r_\mu^0, V_\mu)$ be the representation of ${}^L H'_{\mathbb{E}} = \hat{H}' \rtimes W_{\mathbb{E}}$ determined by $\mathrm{Int}(H'(\mathbb{C}))\mu_h$ (see [L5] and section 1). It is determined by two properties. (1) The restriction of $r_\mu^0$ to $\hat{H}'$ is irreducible with extreme weight $-\mu$. Here $\mu = \mu_h \in X^*(\hat{T}) = X_*(T)$ is a character of a maximal torus $\hat{T}$ of $\hat{H}'$, uniquely determined up to the action of the Weyl group. (2) Let $y$ be a splitting ([Ko3], Section 1) of $\hat{H}'$. Assume that $y$ is fixed by the Weil group $W_{\mathbb{E}}$ of $\mathbb{E}$. Then $W_{\mathbb{E}} \subset {}^L H'_{\mathbb{E}}$ acts trivially on the highest weight space of $V_\mu$ corresponding to $y$. Put $r = r_\mu$ for the representation induced from $r_\mu^0$ on $\hat{H}' \rtimes W_{\mathbb{E}}$ to $\hat{H}' \rtimes W_{\mathbb{Q}}$.

We proceed to specify this representation explicitly in our case, as the twisted tensor $4^{[F:\mathbb{Q}]}$-dimensional representation of $\mathrm{Sp}(2,\mathbb{C})^{[F:\mathbb{Q}]} \rtimes W_{\mathbb{E}}$, $\mathbb{E} =$

$\mathbb{Q}$. In particular, $r = r^0$.

Consider first $H = \mathrm{PGSp}(2)/\mathbb{Q}$. Take $h : \mathrm{R}_{\mathbb{C}/\mathbb{R}}\mathbb{G}_m \to H_\mathbb{R}$ to be defined by $h(a+bi) = \begin{pmatrix} aI & bI \\ -bI & aI \end{pmatrix}, I = I_2$. Over $\mathbb{C}$, the homomorphism $h$ can be diagonalized to $(z,w) \mapsto \mathrm{diag}(zI, wI)$. We claim that the representation $r$ of $\hat{H} = \mathrm{Sp}(2,\mathbb{C})$ is its natural embedding in $\mathrm{GL}(4,\mathbb{C})$. Let $T_H^*$ be the diagonal torus in $H$, and $\hat{T}_H$ the diagonal torus in $\hat{H}$. Then $X_*(\hat{T}_H) = \{(a,b,-b,-a);\, a,b \in \mathbb{Z}\}$ and

$$X^*(\hat{T}_H) = \{(x,y,z,t) \bmod (n,m,m,n); x,y,z,t \in \mathbb{Z}\}.$$

Here $(x,y,z,t)$ takes $\mathrm{diag}(a,b,b^{-1},a^{-1})$ in $\hat{T}_H$ to $a^{x-t}b^{y-z}$. The isomorphism $u : X^*(\hat{T}_H) \xrightarrow{\sim} X_*(T_H^*)$

$$= \{(\alpha,\beta,\gamma,\delta) \bmod (\epsilon,\epsilon,\epsilon,\epsilon); \alpha,\beta,\gamma,\delta,\epsilon \in \mathbb{Z}, \alpha+\delta = \beta+\gamma\}$$

is given by $u : (x,y,z,t) \mapsto (x+y, x+z, y+t, z+t)$, with inverse $u^{-1} : (\alpha,\beta,\gamma,\delta) \mapsto (\alpha-\gamma, \alpha-\beta, 0, 0)$. Now $X_*(T_H^*)$ is spanned by the cocharacters $\alpha_0 = (0,0,1,1) : x \mapsto \mathrm{diag}(1,1,x,x)$,

$$\alpha_1 = (1,0,0,-1) : \ x \mapsto \mathrm{diag}(x,1,1,x^{-1}),$$
$$\alpha_2 = (0,1,-1,0) : \ x \mapsto \mathrm{diag}(1,x,x^{-1},1).$$

An extremal weight of $r$ is $\alpha_0$, viewed as a character of $\hat{T}_H$, thus $u^{-1}(\alpha_0) = (-1,0,0,0)$.

The orbit under the Weyl group $W = \langle (23), (12)(34) \rangle$ of $\alpha_0$ is

$$\alpha_0, \qquad (23)\alpha_0 = \alpha_0 + \alpha_2 = (0,1,0,1), \qquad (14)\alpha_0 = \alpha_0 + \alpha_1 = (1,0,1,0),$$

$(23)(14)\ \alpha_0 = \alpha_0 + \alpha_1 + \alpha_2 = (1,1,0,0)$. Their images under $u^{-1}$ are $(-1,0,0,0)$, $(0,-1,0,0)$ (equivalently (0,0,0,1), (0,0,1,0)), (0,1,0,0), and (1,0,0,0).

The representation $r$ with these weights is the natural embedding $r : \hat{H} = \mathrm{Sp}(2,\mathbb{C}) \hookrightarrow \mathrm{GL}(4,\mathbb{C})$.

The unramified representation $\pi_H(\mu_1,\mu_2)$ of $H = \mathrm{PGSp}(2,\mathbb{Q}_p)$ contained in the composition series of the representation normalizedly induced from the character $n \cdot \mathrm{diag}(\alpha,\beta,\gamma,\delta) \mapsto \mu_1(\alpha/\gamma)\mu_2(\alpha/\beta)$ of the upper triangular subgroup is parametrized by the conjugacy class of $t \times \mathrm{Fr}_p$ in

${}^L H = \hat{H} \times W(\mathbb{Q}_p^{\mathrm{ur}}/\mathbb{Q}_p)$. Here $\mathrm{Fr}_p$ is the Frobenius element, which generates the unramified Weil group $W(\mathbb{Q}_p^{\mathrm{ur}}/\mathbb{Q}_p)$. Further $t = t(\pi_H(\mu_1, \mu_2)) = \mathrm{diag}(\mu_1, \mu_2, \mu_2^{-1}, \mu_1^{-1})$ in $\hat{H} = \mathrm{Sp}(2, \mathbb{C})$ (where we write here $\mu_i$ for $\mu_i(\boldsymbol{\pi})$). The matrix $r(t(\pi_H(\mu_1, \mu_2)))$ has the eigenvalues $\mu_1, \mu_2, \mu_2^{-1}, \mu_1^{-1}$, the values of the weights (1,0,0,0), (0,1,0,0), ... at $t = \mathrm{diag}(\mu_1, \mu_2, \mu_2^{-1}, \mu_1^{-1})$.

Let now $F$ be a totally real number field, $H = \mathrm{PGSp}(2)$ over $F$, and $H' = \mathrm{R}_{F/\mathbb{Q}} H$. Fix an embedding $\iota : F \hookrightarrow \overline{\mathbb{Q}} \cap \mathbb{R}$. Then the set $S$ of archimedean places of $F$ can be identified with the coset space $\mathrm{Gal}(\overline{\mathbb{Q}}/\mathbb{Q})/\mathrm{Gal}(\overline{\mathbb{Q}}/F)$ by $\tau \mapsto \iota_\tau$, where $\iota_\tau : F \hookrightarrow \overline{\mathbb{Q}}$ is $x \mapsto \tau\iota(x)$. Then $H'(\mathbb{Q}) = H(F)$ and $H'(\mathbb{R}) = \prod_S H(\mathbb{R})$. The connected dual group $\hat{H}'$ is $\prod_S \hat{H}$, $\hat{H} = \mathrm{Sp}(2, \mathbb{C})$, and the $L$-group is the semidirect product ${}^L H' = \hat{H}' \rtimes W_{\mathbb{Q}}$ where the Weil group $W_{\mathbb{Q}}$ acts by translation of the factors via its projection to $\mathrm{Gal}(F/\mathbb{Q})$. The homomorphism $h : \mathrm{R}_{\mathbb{C}/\mathbb{R}} \mathbb{G}_m \to H'_{\mathbb{R}}$ is taken to be

$$h(a + bi) = \left( \begin{pmatrix} aI & bI \\ -bI & aI \end{pmatrix}, \dots, \begin{pmatrix} aI & bI \\ -bI & aI \end{pmatrix} \right)$$

($[F : \mathbb{Q}]$ copies on the right). Up to conjugacy by the Weyl group, the weight $\mu : \hat{T}_{H'} \to \mathbb{C}^\times$, where $\hat{T}_{H'}$ is the diagonal torus in $\hat{H}'$ (product of the $|S| = [F : \mathbb{Q}]$ diagonal tori in $\hat{H}$), has the form

$$\mu(\prod_v \mathrm{diag}(a_v, b_v, b_v^{-1}, a_v^{-1})) = \prod_{v \in S} a_v.$$

The group $\mathrm{Gal}(\overline{\mathbb{Q}}/\mathbb{Q})$ stabilizes $\mu$, thus the reflex field $\mathbb{E}$ is $\mathbb{Q}$. Let $r_1$ be the natural embedding of $\hat{H} = \mathrm{Sp}(2, \mathbb{C})$ in $\mathrm{GL}(4, \mathbb{C})$. Then the representation $r = r_\mu$ of ${}^L H'_{\mathbb{E}} = \hat{H}' \rtimes W_{\mathbb{E}}$ is defined on the $[F : \mathbb{Q}]$-fold tensor product $\otimes_S \mathbb{C}^4$, as follows. For $h = (h_v; v \in S) \in \hat{H}'$ we have $r_\mu(h) = \otimes_{v \in S} r_1(h_v)$. The Weil group $W_{\mathbb{E}}$ acts by permuting the factors, thus by left multiplication on $S$. Then $\dim r_\mu = 4^{[F:\mathbb{Q}]}$ and $r_\mu$ is the twisted tensor representation.

## 2. Local Terms at $p$

Let $p$ be a rational prime which is unramified in $F$. The $\mathbb{Q}$-group $H' = \mathrm{R}_{F/\mathbb{Q}} H$ is $\mathbb{Q}_p$-isomorphic to $\prod_{u|p} H'_u$, where $H'_u = \mathrm{R}_{F_u/\mathbb{Q}_p} H$ and $u$ ranges over the primes of $F$ over $\mathbb{Q}$. The set $S$ of embeddings $\iota$ of $F$ into $\overline{\mathbb{Q}}$ (or $\mathbb{R}$, $\mathbb{C}$

or $\overline{\mathbb{Q}}_p$) is parametrized by the homogeneous space $\mathrm{Gal}(\overline{\mathbb{Q}}/\mathbb{Q})/\mathrm{Gal}(\overline{\mathbb{Q}}/F)$, once we fix such an embedding. The Galois group $\mathrm{Gal}(\overline{\mathbb{Q}}_p/\mathbb{Q}_p)$ acts on the left. If $p$ is unramified in $F$ this action factorizes via its quotient $\mathrm{Gal}(\mathbb{Q}_p^{\mathrm{ur}}/\mathbb{Q}_p)$ by the inertia subgroup. The orbits of the Frobenius generator $\mathrm{Fr}_p$ are the places $u$ of $F$ over $\mathbb{Q}$. The group of $\mathbb{Q}_p$-points of $H'$ is $H'(\mathbb{Q}_p) = \prod_{u|p} H'_u(\mathbb{Q}_p) = \prod_{u|p} H(F_u)$.

An irreducible admissible representation $\pi_{Hp}$ of $H'(\mathbb{Q}_p)$ has the form $\otimes_u \pi_{Hu}$. If $\pi_{Hp}$ is unramified then each $\pi_{Hu}$ has the form $\pi_H(\mu_{1u}, \mu_{2u})$, where $\mu_{1u}$, $\mu_{2u}$ are unramified characters of $F_u^\times$. We write $\mu_{mu}$ $(m = 1, 2)$ also for its value $\mu_{mu}(\boldsymbol{\pi}_u)$ at any uniformizing parameter $\boldsymbol{\pi}_u$ of $F_u^\times$. Put $t_u = t(\pi_{Hu}) = \mathrm{diag}(\mu_{1u}, \mu_{2u}, \mu_{2u}^{-1}, \mu_{1u}^{-1})$.

The representation $\pi_{Hp}$ is parametrized in the unramified dual group ${}^L H'_p = \hat{H}^{[F:\mathbb{Q}]} \rtimes \langle \mathrm{Fr}_p \rangle$ by the conjugacy class of $\mathbf{t}_p \times \mathrm{Fr}_p$. Here $\mathbf{t}_p$ is the $[F : \mathbb{Q}]$-tuple $(\mathbf{t}_u; u|p)$ of diagonal matrices in $\hat{H} = \mathrm{Sp}(2, \mathbb{C})$, each $\mathbf{t}_u = (t_{u1}, \dots, t_{un_u})$ is any $n_u = [F_u : \mathbb{Q}_p]$-tuple with $\prod_i t_{ui} = t_u$. The Frobenius $\mathrm{Fr}_p$ acts on each $\mathbf{t}_u$ by permutation to the left: $\mathrm{Fr}_p(\mathbf{t}_u) = (t_{u2}, \dots, t_{un_u}, t_{u1})$. Each $\pi_{Hu}$ is parametrized by the conjugacy class of $\mathbf{t}_u \times \mathrm{Fr}_p$ in the unramified dual group ${}^L H'_u = \hat{H}^{[F_u:\mathbb{Q}_p]} \rtimes \langle \mathrm{Fr}_p \rangle$, or alternatively by the conjugacy class of $t_u \times \mathrm{Fr}_u$ in ${}^L H_u = \hat{H} \times \langle \mathrm{Fr}_u \rangle$, where $\mathrm{Fr}_u = \mathrm{Fr}_p^{[F_u:\mathbb{Q}_p]}$.

The representation $r = \otimes_S r_\iota$ of ${}^L H'_p$ can be written as the product $\otimes_{u|p} r_u$, where $r_u = \otimes_{\iota \in u} r_\iota$. A basis for $r$ is given by $\otimes_S e_\iota$, where $e_\iota$ lies in the standard basis $\{e_1, e_2, e_3, e_4\}$ of $\mathbb{C}^4$. A basis for $r_u$ is given by $\otimes_{\iota \in u} e_\iota$. The representation $r_u$ is called the twisted tensor representation. The vectors fixed by $\mathrm{Fr}_p$ are those which are homogeneous on each orbit of $S$, in the sense that $e_\iota = e_i$ for a fixed $i = i(u)$ for all $\iota \in u$. In particular

$$\mathrm{tr}\, r(\mathbf{t}_p \times \mathrm{Fr}_p) = \prod_{u|p} \mathrm{tr}\, r_u(\mathbf{t}_u \times \mathrm{Fr}_p) = \prod_{u|p} \mathrm{tr}(t_u) = \prod_{u|p} (\mu_{1u} + \mu_{2u} + \mu_{2u}^{-1} + \mu_{1u}^{-1}).$$

More generally let us compute the trace

$$\mathrm{tr}\, r_\mu[(\mathbf{t}_p \times \mathrm{Fr}_p)^j] = \prod_{u|p} \mathrm{tr}\, r_u[(\mathbf{t}_u \times \mathrm{Fr}_p)^j].$$

We proceed to describe the action of $\mathrm{Fr}_p$ on $\mathrm{Emb}(F, \mathbb{R})$.

Fixing a $\sigma_0 : F \hookrightarrow \overline{\mathbb{Q}} \cap \mathbb{R}$ $(\subset \mathbb{R})$, we identify $\mathrm{Gal}(\overline{\mathbb{Q}}/\mathbb{Q})/\mathrm{Gal}(\overline{\mathbb{Q}}/F)$ with $\mathrm{Emb}(F, \overline{\mathbb{Q}} \cap \mathbb{R}) = \{\sigma_1, \dots, \sigma_n\}$. The decomposition group of $\mathbb{Q}$ at

$p$, $\mathrm{Gal}(\overline{\mathbb{Q}}_p/\mathbb{Q}_p)$, acts by left multiplication. Suppose $p$ is unramified in $F$. Then $\mathrm{Fr}_p$ acts, and the $\mathrm{Fr}_p$-orbits in $\mathrm{Emb}(F,\mathbb{R})$ are in bijection with the places $u_1,\dots,u_r$ of $F$ over $p$.

The Frobenius $\mathrm{Fr}_p$ acts transitively on its orbit $u = \mathrm{Emb}(F_u,\overline{\mathbb{Q}}_p)$. The smallest positive power of $\mathrm{Fr}_p$ which fixes each $\sigma \in u$ is $n_u$. The action of $\mathrm{Fr}_p$ on $\widehat{G}'_u = \widehat{G}^{n_u}$ is by $\mathrm{Fr}_p(\mathbf{t}_u) = (t_{u2},\dots,t_{un_u},t_{u1})$, where $\mathbf{t}_u = (t_{u1},\dots,t_{un_u})$. Then $\mathrm{Fr}_p^{n_u}(\mathbf{t}_u)$ is $\mathbf{t}_u$,

$$(\mathbf{t}_u \times \mathrm{Fr}_p)^{n_u} = \Big(\prod_{1\le i\le n_u} t_{ui},\dots,\prod_{1\le i\le n_u} t_{ui}\Big) \times \mathrm{Fr}_p^{n_u},$$

and

$$(\mathbf{t}_u \times \mathrm{Fr}_p)^j = (\dots,t_{u,i}t_{u,i+1}\dots t_{u,i+j-1},\dots;1\le i\le n_u)\times \mathrm{Fr}_p^j.$$

A basis for the $4^{n_u}$-dimensional representation $r_u = \otimes_\sigma r_\sigma$, $\sigma \in u$, is given by $\otimes_{\sigma\in u} e^\sigma_{\ell(\sigma)}$, where $e_{\ell(\sigma)}$ lies in the standard basis $\{e_1,e_2,e_3,e_4\}$ of $\mathbb{C}^4$ for each $\sigma$. To compute the action of $\mathrm{Fr}_p^j$ on these vectors it is convenient to enumerate the $\sigma$ so that the vectors become

$$\otimes_{1\le i\le n_u} e^i_{\ell(i)} = e^1_{\ell(1)}\otimes e^2_{\ell(2)}\otimes\cdots\otimes e^{n_u}_{\ell(n_u)},$$

and $\mathrm{Fr}_p$ acts by sending this vector to

$$\otimes_i e^{i-1}_{\ell(i)} = \otimes_i e^i_{\ell(i+1)} = e^1_{\ell(2)}\otimes e^2_{\ell(3)}\otimes\cdots\otimes e^{n_u}_{\ell(1)}.$$

Then $\mathrm{Fr}_p^{n_u}$ fixes each vector, and a vector is fixed by $\mathrm{Fr}_p^j$ iff it is fixed by $\mathrm{Fr}_p^{j_0}$, $0\le j_0<n_u$, $j\equiv j_0(\mathrm{mod}\, n_u)$. A vector $\otimes_i e^i_{\ell(i)}$ is fixed by $\mathrm{Fr}_p^j$ iff it is equal to $\otimes_i e^{i-j}_{\ell(i)} \equiv \otimes_i e^{i-j_0}_{\ell(i)}$, thus $\ell(i)$ depends only on $i \bmod j$ (and $i \bmod n_u$), namely only on $i \bmod j_u$, where $j_u=(j,n_u)$. Then

$$(\mathbf{t}_u\times \mathrm{Fr}_p)^{j_u} = (\dots,\prod_{0\le k<j_u} t_{u,i+k},\dots)\times \mathrm{Fr}_p^{j_u}.$$

This is

$$(t_{u1}t_{u2}\dots t_{u,j_u},\ t_{u2}t_{u3}\dots t_{u,j_u+1},\ \dots,t_{u,j_u}t_{u,j_u+1}\dots t_{u,2j_u-1};$$
$$t_{u,j_u+1}\dots t_{u,2j_u},\dots)\times \mathrm{Fr}_p^{j_u}.$$

It acts on vectors of the form

$$(e^1_{u,\ell(1)} \otimes e^2_{u,\ell(2)} \otimes \cdots \otimes e^{j_u}_{u,\ell(j_u)}) \otimes (e^1_{u,\ell(1)} \otimes e^2_{u,\ell(2)} \otimes \cdots \otimes e^{j_u}_{u,\ell(j_u)}) \otimes \cdots.$$

The product of the first $j_u$ vectors is repeated $n_u/j_u$ times.

On the vectors with superscript 1 the class $(\mathbf{t}_u \times \mathrm{Fr}_p)^{j_u}$ acts as

$$t_{u,1}t_{u,2}\ldots t_{u,j_u} \cdot t_{u,j_u+1}\ldots t_{u,2j_u} \cdot \ldots \cdot t_{u,(\frac{n_u}{j_u}-1)j_u+1}\ldots t_{u,\frac{n_u}{j_u}j_u}$$

$$= \prod_{1\leq i\leq n_u} t_{u,i} = t_u = \mathrm{diag}(\mu_{1u}, \mu_{2u}, \mu_{2u}^{-1}, \mu_{1u}^{-1}),$$

and so $(\mathbf{t}_u \times \mathrm{Fr}_p)^j$ acts as $t_u^{j/j_u}$. The trace is then $\mu_{1u}^{j/j_u} + \mu_{2u}^{j/j_u} + \mu_{2u}^{-j/j_u} + \mu_{1u}^{-j/j_u}$. The same holds for each superscript, so we get the product of $j_u$ such factors. Put $j_u = (j, n_u)$. We then have

$$\mathrm{tr}\, r_u[(\mathbf{t}_u \times \mathrm{Fr}_p)^j] = (\mu_{1u}^{\frac{j}{j_u}} + \mu_{2u}^{\frac{j}{j_u}} + \mu_{2u}^{\frac{-j}{j_u}} + \mu_{1u}^{\frac{-j}{j_u}})^{j_u}.$$

The spherical function $f^{sj}_{C_0p}$ is defined by means of $L$-group homomorphisms ${}^LC_0' \to {}^LH' \to {}^LH_j'$, where $H_j' = \mathrm{R}_{\mathbb{Q}_j/\mathbb{Q}_p}H'$ and $\mathbb{Q}_j$ denotes the unramified extension of $\mathbb{Q}_p$ in $\overline{\mathbb{Q}}_p$ of degree $j$. Since the groups $C_0'$ and $H'$ are products of groups $C_{0u}' = \mathrm{R}_{F_u/\mathbb{Q}_p}C_0$ and $H_u' = \mathrm{R}_{F_u/\mathbb{Q}_p}H$, it suffices to work with these groups. Thus $H_j' = \prod_{u|p} H_{uj}'$, where $H_{uj}' = \mathrm{R}_{\mathbb{Q}_j/\mathbb{Q}_p}H_u'$. The function $f^{sj}_{C_0p}$ will be $\otimes f^{sj}_{C_0u}$, for analogously defined $f^{sj}_{C_0u}$.

Now

$${}^LH_j' = (\hat{H}')^j \rtimes \langle \mathrm{Fr}_p\rangle = \prod_{u|p}(\hat{H}_u')^j \rtimes \langle \mathrm{Fr}_p\rangle, \qquad \hat{H}' = \hat{H}^{[F:\mathbb{Q}]}, \qquad \hat{H}_u' = \hat{H}^{n_u},$$

and $\mathrm{Fr}_p$ acts on

$$\mathbf{x} = (\mathbf{x}_u), \qquad \mathbf{x}_u = (\mathbf{x}_{u1}, \ldots, \mathbf{x}_{uj}), \qquad \mathbf{x}_{ui} \in \hat{H}_u' = \hat{H}^{n_u},$$

by

$$\mathrm{Fr}_p(\mathbf{x}) = (\mathrm{Fr}_p(\mathbf{x}_u)), \qquad \mathrm{Fr}_p(\mathbf{x}_u) = (\mathrm{Fr}_p(\mathbf{x}_{u2}), \ldots, \mathrm{Fr}_p(\mathbf{x}_{uj}), \mathrm{Fr}_p(\mathbf{x}_{u1})).$$

It suffices to work with ${}^LH_{uj}' = (\hat{H}_u')^j \rtimes \langle \mathrm{Fr}_p\rangle$.

Let $\mathbf{s}_1, \dots, \mathbf{s}_j$ be $\mathrm{Fr}_p$-fixed elements in $Z(\hat{C}'_{0u})$, thus $\mathbf{s}_i = (s_i, \dots, s_i)$, $s_i \in Z(\hat{C}_0) = \{\pm I_2\} \times \{\pm I_2\}$ repeated $n_u$ times, with $\mathbf{s}_1 \dots \mathbf{s}_j = \mathbf{s} = (s, \dots, s)$, $s = \operatorname{diag}(1, -1, -1, 1)$. Define

$$\tilde{\eta}_j : {}^L C'_{0u} = \hat{C}_0^{n_u} \times \langle \mathrm{Fr}_p \rangle \to {}^L H'_{uj} = (\hat{H}'_u)^j \rtimes \langle \mathrm{Fr}_p \rangle$$

by

$$\mathbf{t} \mapsto (\mathbf{t}, \dots, \mathbf{t}), \qquad \mathrm{Fr}_p \mapsto (\mathbf{s}_1, \mathbf{s}_2, \dots, \mathbf{s}_j) \times \mathrm{Fr}_p,$$

thus

$$\mathrm{Fr}_p^i \mapsto (\mathbf{s}_1 \mathbf{s}_2 \dots \mathbf{s}_i, \mathbf{s}_2 \dots \mathbf{s}_{i+1}, \dots, \mathbf{s}_j \mathbf{s}_1 \dots \mathbf{s}_{i-1}) \times \mathrm{Fr}_p^i .$$

The diagonal map $H'_u \to H'_{uj}$ defines ${}^L H'_{uj} \to {}^L H'_u$, $(\mathbf{t}_1, \dots, \mathbf{t}_j) \times \mathrm{Fr}_p^i \mapsto \mathbf{t}_1 \dots \mathbf{t}_j \times \mathrm{Fr}_p^i$. The composition $\eta_j : {}^L C'_{0u} \to {}^L H'_u$ gives

$$\mathbf{t} \times \mathrm{Fr}_p^i \mapsto \mathbf{t}^j \mathbf{s}^i \times \mathrm{Fr}_p^i .$$

The homomorphism $\tilde{\eta}_j$ defines a dual homomorphism

$$\mathbb{H}(K_{uj} \backslash H_{uj} / K_{uj}) \to \mathbb{H}(K_{0u} \backslash C_{0u} / K_{0u})$$

of Hecke algebras. The function $f_{C_0 u}^{sj}$ is defined to be the image by the relation

$$\operatorname{tr} \pi_{Hu}(\tilde{\eta}_j(t))(\phi_{ju}) = \operatorname{tr} \pi_{C_0 u}(t)(f_{C_0 u}^{sj})$$

of the function $\phi_j$ of [Ko4], p. 173, or rather the $u$-component $\phi_{uj}$ of $\phi_j$, which is the characteristic function of $K_{uj} \cdot \mu_{F_j}(p^{-1}) \cdot K_{uj}$. Theorem 2.1.3 of [Ko3] (see also [Ko4], p. 193) asserts that the product over $u|p$ in $F$ of these traces is the product of $p^{\frac{j}{2} \dim S_{K_f}}$ with the product over $u|p$ of

$$\operatorname{tr} r_u(\mathbf{st}(\pi_{Hu})^j \times \mathrm{Fr}_p) = \operatorname{tr}(st(\pi_{Hu})^j) = \mu_{1u}^j + \mu_{1u}^{-j} - \mu_{2u}^j - \mu_{2u}^{-j}.$$

Similarly for $s = I$ we have that the analogous factor (with $C_0$ replaced by $H$) is the product with factors

$$\operatorname{tr} r_u(\mathbf{t}(\pi_{Hu})^j \times \mathrm{Fr}_p) = \operatorname{tr}(t(\pi_{Hu})^j) = \mu_{1u}^j + \mu_{1u}^{-j} + \mu_{2u}^j + \mu_{2u}^{-j}.$$

# 3. The Eigenvalues at p

We proceed to describe the eigenvalues $\mu_{1u}, \mu_{2u}$ for the various terms in the formula, beginning with $\mathrm{STF}_H(f_H)$, according to the five parts which make it. Note that

$$\lambda(\pi_H(\mu_{1u}, \mu_{2u})) = \pi_G(\mu_{1u}, \mu_{2u}, \mu_{2u}^{-1}, \mu_{1u}^{-1}),$$

where $G = \mathrm{PGL}(4, F_u)$. We choose the complex numbers $\mu_{mu}$ to have $|\mu_{mu}| \geq 1$ (otherwise we replace as we may $\mu_{mu}$ by $\mu_{mu}^{-1}$; $m = 1, 2$). Write $t_u$ for $\mathrm{diag}(\mu_{1u}, \mu_{2u}, \mu_{2u}^{-1}, \mu_{1u}^{-1})$.

The first part of $\mathrm{STF}_H(f_H)$ describes the stable spectrum. It has 3 types of terms.

(1) For the packets $\{\pi_H\}$ which $\lambda$-lift to cuspidal $\pi_G \simeq \check{\pi}_G$ ($\notin \mathrm{Im}\,\lambda_1$), $t_u$ is

$$\mathrm{diag}(\mu_{1u}, \mu_{2u}, \mu_{2u}^{-1}, \mu_{1u}^{-1}) \qquad \text{with} \qquad q_u^{-1/2} < |\mu_{mu}| < q_u^{1/2},$$

since $\pi_G$ is unitary and so its component $\pi_{Gu}$ is unitarizable. Note that the unramified component $\pi_{Gu}$ is generic (since $\pi_G$ is), hence fully induced.

(2) For the quasi-packets $\{L(\xi\nu, \nu^{-1/2}\pi^2)\}$, they $\lambda$-lift to the residual $J(\nu^{1/2}\pi^2,\ \nu^{-1/2}\pi^2)$, $\pi^2$ cuspidal with central character $\xi \neq 1 = \xi^2$ satisfying $\xi\pi^2 = \pi^2$. The component $\pi_u^2$ of $\pi^2$ at $u$ is unramified of the form $\pi^2(z_{1u},\ z_{2u})$. This is an unramified generic representation of $\mathrm{GL}(2, F_u)$, hence fully induced, normalizedly from the character $\begin{pmatrix} a & b \\ 0 & c \end{pmatrix} \mapsto z_{1u}^{\mathrm{val}(a)} z_{2u}^{\mathrm{val}(c)}$. We have that $z_{1u}z_{2u} = \xi_u(\boldsymbol{\pi}_u)$ has square 1. If $\xi_u \neq 1$ then $\{z_{1u}, z_{2u}\} = \{1, -1\}$. If $\xi_u = 1$, since $\pi^2$ is unitary its component $\pi_u^2$ is unitarizable, and so $q_u^{-1/2} < |z_{mu}| < q_u^{1/2}$. In both cases we have

$$t_u = \mathrm{diag}(q_u^{1/2} z_{1u}, q_u^{1/2} z_{2u}, q_u^{-1/2} z_{2u}^{-1}, q_u^{-1/2} z_{1u}^{-1}).$$

Better estimates are known for the $|z_{mu}|$ (the exponent $1/2$ can be reduced to $1/4$ by the theory of the symmetric square lifting), but for our $\pi^2$ we shall show below that $|z_{mu}| = 1$.

(3) For one dimensional representations $\pi_H$, $\lambda(\pi_H) = \pi_G$ is one dimensional representation $g \mapsto \chi(\det g)$, where $\chi$ is a character of order 2, and $t(\pi_{Hu}) = \mathrm{diag}(\mu_{1u}, \mu_{2u}, \mu_{2u}^{-1}, \mu_{1u}^{-1})$ is

$$\mathrm{diag}(\chi_u q_u^{3/2}, \chi_u q_u^{1/2}, \chi_u q_u^{-1/2}, \chi_u q_u^{-3/2}),$$

where $\chi_u = \chi(\boldsymbol{\pi}_u)$ has square 1. Since $\pi_H$ is a quadratic character we have that $\mu_{1u} = \pm q_u^{3/2}$, $\mu_{2u} = \pm q_u^{1/2}$.

The second part of $\mathrm{STF}_H(f_H)$ is a sum of terms indexed by $\{\pi_H\} = \lambda_0(\pi^1 \times \pi^2)$. Here $\pi^1$, $\pi^2$ are cuspidal representations of $\mathrm{PGL}(2, \mathbb{A})$, and $\operatorname{tr} \pi_{Hu}^-(f_{Hu}) = 0$ as $f_{Hu}$ is spherical. Then the component of $\pi^m$ $(m = 1, 2)$ at $u$ is the unramified generic thus fully induced $\pi_u^m = \pi^2(z_{mu}, z_{mu}^{-1})$, and $t_u = \operatorname{diag}(z_{1u}, z_{2u}, z_{2u}^{-1}, z_{1u}^{-1})$, where $|z_{mu}|^{\pm 1} \leq q_u^{1/2}$.

The terms in the third part of $\mathrm{STF}_H(f_H)$ correspond to $\lambda_0(\pi^2 \times \sigma \mathbf{1}_2)$, where $\pi^2$ is a cuspidal representation of $\mathrm{PGL}(2, \mathbb{A})$ and $\sigma$ is a character of $\mathbb{A}^\times / F^\times \mathbb{A}^{\times 2}$. The factors at $u|p$ of $\pi^2$ are $\pi_u^2(z_u, z_u^{-1})$, $q_u^{-1/2} < |z_i| < q_u^{1/2}$. So $t_u = \operatorname{diag}(\sigma_u q_u^{1/2}, z_u, z_u^{-1}, \sigma_u q_u^{-1/2})$, where $\sigma_u = \sigma_u(\boldsymbol{\pi}_u)$ has square 1.

The terms in the fourth part correspond to $\lambda_0(\sigma\xi \mathbf{1}_2 \times \sigma \mathbf{1}_2)$, where $\sigma$, $\xi$ are characters of $\mathbb{A}^\times / F^\times \mathbb{A}^{\times 2}$ with $\xi \neq 1$. Put $\sigma_u = \sigma_u(\boldsymbol{\pi}_u)$. Then

$$t_u = \operatorname{diag}(\sigma_u \xi_u q_u^{1/2}, \sigma_u q_u^{1/2}, \sigma_u q_u^{-1/2}, \sigma_u \xi_u q_u^{-1/2}), \quad \xi_u = \xi_u(\boldsymbol{\pi}_u).$$

The fifth part consists of terms indexed by $\pi_H = 1 \rtimes \pi^2$ where $\pi^2$ is a cuspidal representation of $\mathrm{PGL}(2, \mathbb{A})$. At $u$ the factor $\pi_u^2 = \pi^2(z_u, z_u^{-1})$ is fully induced with $|z_u|^{\pm 1} < q_u^{1/2}$ and $\lambda(1 \rtimes \pi_u^2) = I(\pi_u^2, \pi_u^2)$ so that $t_u = \operatorname{diag}(z_u, z_u^{-1}, z_u, z_u^{-1})$.

In summary, as noted in the last section, the factor at $p$ of each of the summands in $\mathrm{STF}_H(f_H)$ has the form (where $j_u = (n_u, j)$)

$$p^{\frac{j}{2} \dim \mathcal{S}_{K_f}} \operatorname{tr} r_\mu[(\mathbf{t}(\pi_{Hp}) \times \mathrm{Fr}_p)^j] = p^{\frac{j}{2} \dim \mathcal{S}_{K_f}} \prod_{u|p} (\operatorname{tr}[t_u \times \mathrm{Fr}_p]^j)$$

$$= p^{\frac{j}{2} \dim \mathcal{S}_{K_f}} \prod_{u|p} (t_{1u}^{j/j_u} + t_{1u}^{-j/j_u} + t_{2u}^{j/j_u} + t_{2u}^{-j/j_u})^{j_u}.$$

REMARK. As $p$ splits in $F$ into a product of primes $u$ with $F_u/\mathbb{Q}_p$ unramified with $[F : \mathbb{Q}] = \sum_{u|p} [F_u : \mathbb{Q}_p]$, and the dimension of the symmetric space $H(\mathbb{R})/K_{H(\mathbb{R})}$ is 3, we note that

$$\dim \mathcal{S}_{K_f} = 3[F : \mathbb{Q}] = 3 \sum_{u|p} [F_u : \mathbb{Q}_p].$$

# 4. Terms at $p$ for the Endoscopic Group

The other trace formula which contributes is that of $C_0(\mathbb{A}) = \mathrm{PGL}(2,\mathbb{A}) \times \mathrm{PGL}(2,\mathbb{A})$. The factors at $p$ of the various summands have the form

$$p^{\frac{j}{2}\dim \mathcal{S}_{K_f}} \prod_{u|p} \mathrm{tr}(s[t_u \times \mathrm{Fr}_p]^j)$$

$$= p^{\frac{j}{2}\dim \mathcal{S}_{K_f}} \prod_{u|p} (t_{1u}^{j/j_u} + t_{1u}^{-j/j_u} - t_{2u}^{j/j_u} - t_{2u}^{-j/j_u})^{j_u},$$

where $s = \mathrm{diag}(1,-1,-1,1)$ is the element in $\hat{H} = \mathrm{Sp}(2,\mathbb{C})$ whose centralizer is $\hat{C}_0 = \mathrm{SL}(2,\mathbb{C}) \times \mathrm{SL}(2,\mathbb{C})$. We need to specify the 4-tuples $t_u$ again, according to the three parts of $\mathrm{STF}_{C_0}(f_0)$. For the first part, where the summands are indexed by pairs $\pi^1 \times \pi^2$ of cuspidal representations of $\mathrm{PGL}(2,\mathbb{A})$, the $t_u$ is the same as in the second part of $\mathrm{STF}_H(f_H)$ if $\pi^1 \neq \pi^2$, and as in the fifth part if $\pi^1 = \pi^2$. For the second part of $\mathrm{STF}_{C_0}(f_0)$, the $t_u$ for the term indexed by $(\sigma, \pi^2)$ is the same as for the third part of $\mathrm{STF}_H(f_H)$. For the third part of $\mathrm{STF}_{C_0}(f_0)$, the $t_u$ for the term indexed by $(\sigma, \xi\sigma)$ is the same as for the fourth part of $\mathrm{STF}_H(f_H)$ if $\xi \neq 1$ or the fifth part when $\xi = 1$.

# IV. REAL REPRESENTATIONS

## 1. Representations of SL(2, ℝ)

Packets of representations of a real group $G$ are parametrized by maps of the Weil group $W_{\mathbb{R}}$ to the $L$-group ${}^LG$. Recall that $W_{\mathbb{R}} = \langle z, \sigma; z \in \mathbb{C}^\times, \sigma^2 \in \mathbb{R}^\times - N_{\mathbb{C}/\mathbb{R}}\mathbb{C}^\times, \sigma z = \overline{z}\sigma \rangle$ is

$$1 \to W_{\mathbb{C}} \to W_{\mathbb{R}} \to \mathrm{Gal}(\mathbb{C}/\mathbb{R}) \to 1$$

an extension of $\mathrm{Gal}(\mathbb{C}/\mathbb{R})$ by $W_{\mathbb{C}} = \mathbb{C}^\times$. It can also be viewed as the normalizer $\mathbb{C}^\times \cup \mathbb{C}^\times j$ of $\mathbb{C}^\times$ in $\mathbb{H}^\times$, where $\mathbb{H} = \mathbb{R}\langle 1, i, j, k \rangle$ is the Hamilton quaternions. The norm on $\mathbb{H}$ defines a norm on $W_{\mathbb{R}}$ by restriction ([D3], [Tt]). The discrete series (packets of) representations of $G$ are parametrized by the homomorphisms $\phi : W_{\mathbb{R}} \to \hat{G} \times W_{\mathbb{R}}$ whose projection to $W_{\mathbb{R}}$ is the identity and to the connected component $\hat{G}$ is bounded, and such that $C_\phi Z(\hat{G})/Z(\hat{G})$ is finite. Here $C_\phi$ is the centralizer $Z_{\hat{G}}(\phi(W_{\mathbb{R}}))$ in $\hat{G}$ of the image of $\phi$.

When $G = \mathrm{GL}(2, \mathbb{R})$ we have $\hat{G} = \mathrm{GL}(2, \mathbb{R})$, and these maps are $\phi_k$ ($k \geq 1$), defined by

$$W_{\mathbb{C}} = \mathbb{C}^\times \ni z \mapsto \begin{pmatrix} (z/|z|)^k & 0 \\ 0 & (|z|/z)^k \end{pmatrix} \times z, \qquad \sigma \mapsto \begin{pmatrix} 0 & 1 \\ \iota & 0 \end{pmatrix} \times \sigma.$$

Since $\sigma^2 = -1 \mapsto \begin{pmatrix} (-1)^k & 0 \\ 0 & (-1)^k \end{pmatrix} \times \sigma^2$, $\iota$ must be $(-1)^k$. Then $\det \phi_k(\sigma) = (-1)^{k+1}$, and so $k$ must be an odd integer ($= 1, 3, 5, \dots$) to get a discrete series (packet of) representation of $\mathrm{PGL}(2, \mathbb{R})$. In fact $\pi_1$ is the lowest discrete series representation, and $\phi_0$ parametrizes the so called limit of discrete series representations; it is tempered. Even $k \geq 2$ and $\sigma \mapsto \begin{pmatrix} 0 & 1 \\ 1 & 0 \end{pmatrix} \times \sigma$ define discrete series representations of $\mathrm{GL}(2, \mathbb{R})$ with the quadratic nontrivial central character sgn. Packets for $\mathrm{GL}(2, \mathbb{R})$ and $\mathrm{PGL}(2, \mathbb{R})$ consist of a single discrete series irreducible representation $\pi_k$. Note that $\pi_k \otimes \mathrm{sgn} \simeq \pi_k$. Here $\mathrm{sgn} : \mathrm{GL}(2, \mathbb{R}) \to \{\pm 1\}$, $\mathrm{sgn}(g) = 1$ if $\det g > 0$, $= -1$ if $\det g < 0$.

The $\pi_k$ $(k > 0)$ have the same central and infinitesimal character as the $k$th dimensional nonunitarizable representation

$$\mathrm{Sym}_0^{k-1}\,\mathbb{C}^2 = |\det g|^{-(k-1)/2}\,\mathrm{Sym}^{k-1}\,\mathbb{C}^2$$

into

$$\mathrm{SL}(k,\mathbb{C})^{\pm} = \{g \in \mathrm{GL}(2,\mathbb{C}); \det g \in \{\pm 1\}\}.$$

Note that $\det \mathrm{Sym}^{k-1}(g) = \det g^{k(k-1)/2}$. The normalizing factor is

$$|\det \mathrm{Sym}^{k-1}|^{-1/k}. \qquad \text{Then} \qquad \mathrm{Sym}_0^{k-1}\left(\begin{smallmatrix} a & 0 \\ 0 & b \end{smallmatrix}\right)$$

$$= \mathrm{diag}(\mathrm{sgn}(a)^{k-i}\,\mathrm{sgn}(b)^{i-1}|a|^{k-i-(k-1)/2}|b|^{i-1-(k-1)/2}; 1 \leq i \leq k).$$

In fact both $\pi_k$ and $\mathrm{Sym}_0^{k-1}\,\mathbb{C}^2$ are constituents of the normalizedly induced representation $I(\nu^{k/2}, \mathrm{sgn}^{k-1}\,\nu^{-k/2})$ whose infinitesimal character is $(\frac{k}{2}, -\frac{k}{2})$, where a basis for the lattice of characters of the diagonal torus in SL(2) is taken to be $(1,-1)$.

## 2. Cohomological Representations

An irreducible admissible representation $\pi$ of $H(\mathbb{A})$ which has nonzero Lie algebra cohomology $H^{ij}(\mathfrak{g}, K; \pi \otimes V)$ for some coefficients (finite dimensional representation) $V$ is called here *cohomological*. Discrete series representations are cohomological. The non discrete series representations which are cohomological are listed in [VZ]. They are nontempered. We proceed to list them here in our case of PGSp(2,$\mathbb{R}$). We are interested in the $(\mathfrak{g}, K)$-cohomology $H^{ij}(\mathrm{sp}(2,\mathbb{R}), \mathrm{U}(4); \pi \otimes V)$, so we need to compute $H^{ij}(\mathrm{sp}(2,\mathbb{R}), \mathrm{SU}(4); \pi \otimes V)$ and observe that $\mathrm{U}(4)/\mathrm{SU}(4)$ acts trivially on the nonzero $H^{ij}$, which are $\mathbb{C}$. If $H^{ij}(\pi \otimes V) \neq 0$ then ([BW]) the infinitesimal character ([Kn]) of $\pi$ is equal to the sum of the highest weight ([FH]) of the self contragredient (in our case) $V$, and half the sum of the positive roots, $\delta$. With the usual basis $(1,0)$, $(0,1)$ on $X^*(T_S^*)$, the positive roots are $(1,-1)$, $(0,2)$, $(1,1)$, $(2,0)$. Then $\delta = \frac{1}{2}\sum_{\alpha>0}\alpha$ is $(2,1)$.

Here $T_S^*$ denotes the diagonal subgroup $\{\mathrm{diag}(x, y, 1/y, 1/x)\}$ of the algebraic group Sp(2). Its lattice $X^*(T_S^*)$ of rational characters consists of

$$(a,b) : \mathrm{diag}(x, y, 1/y, 1/x) \mapsto x^a y^b \qquad (a, b \in \mathbb{Z}).$$

The irreducible finite dimensional representations of Sp(2) are $V_{a,b}$, parametrized by the highest weight $(a,b)$ with $a \geq b \geq 0$ ([FH]). The central character of $V_{a,b}$ is $\zeta \mapsto \zeta^{a+b}$, $\zeta \in \{\pm 1\}$. It is trivial iff $a+b$ is even. Since $\mathrm{GSp}(2) = \mathrm{Sp}(2) \rtimes \{\mathrm{diag}(1,1,z,z)\}$, such $V_{a,b}$ extends to a representation of PGSp(2) by $(1,1,z,z) \mapsto z^{-(a+b)/2}$. This gives a representation of $H(\mathbb{R}) = \mathrm{PGSp}(2,\mathbb{R})$, extending its restriction to the index 2 connected subgroup $H^0 = \mathrm{PSp}(2,\mathbb{R})$. Another – nonalgebraic – extension is $V'_{a,b} = V_{a,b} \otimes \mathrm{sgn}$, where $\mathrm{sgn}(1,1,z,z) = \mathrm{sgn}(z)$, $z \in \mathbb{R}^\times$. $V_{a,b}$ is self dual.

To list the irreducible admissible representations $\pi$ of $\mathrm{PGSp}(2,\mathbb{R})$ with nonzero Lie algebra cohomology $H^{i,j}(\mathrm{sp}(2,\mathbb{R}), \mathrm{SU}(4); \pi \otimes V_{a,b})$ for some $a \geq b \geq 0$ (the same results hold with $V_{a,b}$ replaced by $V'_{a,b}$), we first list the discrete series representations.

Packets of discrete series representations of $H(\mathbb{R}) = \mathrm{PGSp}(2,\mathbb{R})$ are parametrized by maps $\phi$ of $W_\mathbb{R}$ to ${}^LH = \hat{H} \times W_\mathbb{R}$ which are admissible ($\mathrm{pr}_2 \circ \phi = \mathrm{id}$) and whose projection to $\hat{H}$ is bounded and $C_\phi Z(\hat{H})/Z(\hat{H})$ is finite. Here $C_\phi$ is $Z_{\hat{H}}(\phi(W_\mathbb{R}))$. They are parametrized $\phi = \phi_{k_1,k_2}$ by a pair $(k_1,k_2)$ of integers with $k_1 > k_2 > 0$ and odd $k_1$, $k_2$.

The homomorphism $\phi_{k_1,k_2} : W_\mathbb{R} \to {}^LG = \hat{G} \times W_\mathbb{R}$, $\hat{G} = \mathrm{SL}(4,\mathbb{C})$, given by

$$z \mapsto \mathrm{diag}((z/|z|)^{k_1}, (z/|z|)^{k_2}, (|z|/z)^{k_2}, (|z|/z)^{k_1}) \times z$$

and

$$\sigma \mapsto \begin{pmatrix} 0 & w \\ -w & 0 \end{pmatrix} \times \sigma \qquad (\text{odd } k_1 > k_2 > 0)$$

or

$$\sigma \mapsto \begin{pmatrix} 0 & w \\ w & 0 \end{pmatrix} \times \sigma \qquad (\text{even } k_1 > k_2 > 0),$$

factorizes via $({}^LC_0 \to)\ {}^LH = \mathrm{Sp}(2,\mathbb{C}) \times W_\mathbb{R}$ precisely when $k_i$ are odd. When the $k_i$ are even it factorizes via ${}^LC = \mathrm{SO}(4,\mathbb{C}) \times W_\mathbb{R}$. When the $k_i$ are odd it parametrizes a packet $\{\pi^{\mathrm{Wh}}_{k_1,k_2}, \pi^{\mathrm{hol}}_{k_1,k_2}\}$ of discrete series representations of $H(\mathbb{R})$. Here $\pi^{\mathrm{Wh}}$ is generic and $\pi^{\mathrm{hol}}$ is holomorphic and antiholomorphic. Their restrictions to $H^0$ are reducible, consisting of $\pi^{\mathrm{Wh}}_{H^0}$ and $\pi^{\mathrm{Wh}}_{H^0} \circ \mathrm{Int}(\iota)$, $\pi^{\mathrm{hol}}_{H^0}$ and $\pi^{\mathrm{hol}}_{H^0} \circ \mathrm{Int}(\iota)$, $\iota = \mathrm{diag}(1,1,-1,-1)$, and $\pi^{\mathrm{Wh}} \otimes \mathrm{sgn} = \pi^{\mathrm{Wh}}$, $\pi^{\mathrm{hol}} \otimes \mathrm{sgn} = \pi^{\mathrm{hol}}$.

To compute the infinitesimal character of $\pi^*_{k_1,k_2}$ we note that $\pi_k \subset I(\nu^{k/2}, \mathrm{sgn}^{k-1} \nu^{-k/2})$ (e.g. by [JL], I5.7 and I5.11) on $\mathrm{GL}(2,\mathbb{R})$. Via ${}^LC_0 \to {}^LH$ induced $I(\nu^{k_1/2}, \nu^{-k_1/2}) \times I(\nu^{k_2/2}, \nu^{-k_2/2})$ (in our case the

$k_i$ are odd) lifts to the induced

$$I_H(\nu^{k_1/2}, \nu^{k_2/2}) = \nu^{(k_1+k_2)/2} \times \nu^{(k_1-k_2)/2} \rtimes \nu^{-k_2/2},$$

whose constituents (e.g. $\pi^*_{k_1,k_2}$, $* = \text{Wh, hol}$) have infinitesimal character $(\frac{k_1+k_2}{2}, \frac{k_1-k_2}{2}) = (2,1) + (a,b)$. Here $a = \frac{k_1+k_2}{2} - 2 \geq b = \frac{k_1-k_2}{2} - 1 \geq 0$ as $k_2 \geq 1$ and $k_1 > k_2$ and $k_1 - k_2$ is even. For these $a \geq b \geq 0$, thus $k_1 = a+b+3$, $k_2 = a-b+1$, we have

$$H^{ij}(\text{sp}(2,\mathbb{R}), \text{SU}(4); \pi^{\text{Wh}}_{k_1,k_2} \otimes V_{a,b}) = \mathbb{C} \qquad \text{if} \quad (i,j) = (2,1), (1,2),$$

$$H^{ij}(\text{sp}(2,\mathbb{R}), \text{SU}(4); \pi^{\text{hol}}_{k_1,k_2} \otimes V_{a,b}) = \mathbb{C} \qquad \text{if} \quad (i,j) = (3,0), (0,3).$$

Here $k_1 > k_2 > 0$ and $k_1$, $k_2$ are odd. In particular, the discrete series representations of $\text{PGSp}(2,\mathbb{R})$ are endoscopic.

# 3. Nontempered Representations

Quasi-packets including nontempered representations are parametrized by homomorphisms $\psi : W_{\mathbb{R}} \times \text{SL}(2,\mathbb{R}) \to {}^L H$ and $\phi_\psi : W_{\mathbb{R}} \to {}^L H$ defined ([A]) by

$$\phi_\psi(\mathbf{w}) = \psi(\mathbf{w}, \begin{pmatrix} \|\mathbf{w}\|^{1/2} & 0 \\ 0 & \|\mathbf{w}\|^{-1/2} \end{pmatrix}).$$

The norm $\|.\| : W_{\mathbb{R}} \to \mathbb{R}^\times$ is defined by $\|z\| = z\overline{z}$ and $\|\sigma\| = 1$. Then $\phi_\psi(\sigma) = \psi(\sigma, I)$ and $\phi_\psi(z) = \psi(z, \text{diag}(r, r^{-1}))$ if $z = re^{i\theta}$, $r > 0$. For example, $\psi : W_{\mathbb{R}} \times \text{SL}(2,\mathbb{C}) \to \text{SL}(2,\mathbb{C})$,

$$\psi | W_{\mathbb{R}} : z\sigma^j \mapsto \xi(-1)^j, \qquad \psi | \text{SL}(2,\mathbb{C}) = \text{id},$$

gives

$$\phi_\psi(z) = \begin{pmatrix} r & 0 \\ 0 & r^{-1} \end{pmatrix} \times z, \qquad \phi_\psi(\sigma) = \xi(-1) I_2 \times \sigma,$$

parametrizing the one dimensional representation $\xi_2 = J(\xi\nu^{1/2}, \xi\nu^{-1/2})$ of $\text{PGL}(2,\mathbb{R})$ ($\xi : \mathbb{R}^\times \to \{\pm 1\}$, $\nu(z) = |z|$). Here $J$ denotes the Langlands quotient of the indicated induced representation, $I(\xi\nu^{1/2}, \xi\nu^{-1/2})$.

Similarly the one dimensional representation

$$\xi_4 = J(\xi\nu^{3/2}, \xi\nu^{1/2}, \xi\nu^{-1/2}, \xi\nu^{-3/2})$$

of $\mathrm{PGL}(4,\mathbb{R})$ is parametrized by $\psi : W_{\mathbb{R}} \times \mathrm{SL}(2,\mathbb{C}) \to \mathrm{SL}(4,\mathbb{C})$,

$$(\psi|W_{\mathbb{R}})(z\sigma^j) = \xi(-1)^j, \qquad \psi|\,\mathrm{SL}(2,\mathbb{C}) = \mathrm{Sym}_0^3,$$

thus

$$\phi_\psi(z) = \mathrm{diag}(r^3, r, r^{-1}, r^{-3}) \times z, \qquad \phi_\psi(\sigma) = \xi(-1)I_4 \times \sigma.$$

This parameter factorizes via $\psi : W_{\mathbb{R}} \times \mathrm{SL}(2,\mathbb{C}) \to \mathrm{Sp}(2,\mathbb{C})$, which parametrizes the one dimensional representation $\xi_H$ of $H(\mathbb{R})$, $h \mapsto \xi(\boldsymbol{\lambda}(h))$ where $\boldsymbol{\lambda}(h)$ denotes the factor of similitude of $h$, whose infinitesimal character is $(2,1) = \frac{1}{2}\sum_{\alpha>0}\alpha$. We have

$$H^{ij}(\mathrm{sp}(2,\mathbb{R}), \mathrm{SU}(4); \xi_H \otimes V_{0,0}) = \mathbb{C}$$

for $(i,j) = (0,0)$, $(1,1)$, $(2,2)$, $(3,3)$. Of course $1_H \neq \mathrm{sgn}_H$, and $\frac{1}{2}(1_H + \mathrm{sgn}_H)$ is the characteristic function of $H^0$ in $H(\mathbb{R})$. Moreover, the character of $\frac{1}{2}(1_H + \mathrm{sgn}_H) + \pi_{3,1}^{\mathrm{Wh}} + \pi_{3,1}^{\mathrm{hol}}$ vanishes on the regular elliptic set of $H(\mathbb{R})$, as $(\xi_H + \pi_{3,1}^{\mathrm{Wh}} + \pi_{3,1}^{\mathrm{hol}})|H^0$ is a linear combination of properly induced ("standard") representations ([Vo], [Ln]) in the Grothendieck group.

## 4. The Cohomological $L(\nu\,\mathrm{sgn}, \nu^{-1/2}\pi_{2k})$

The nontempered nonendoscopic representation $L(\nu\,\mathrm{sgn}, \nu^{-1/2}\pi_{2k})$ of the group $H(\mathbb{R})$ $(k \geq 1)$ is the Langlands quotient of the representation $\nu\,\mathrm{sgn} \rtimes \nu^{-1/2}\pi_{2k}$ induced from the Heisenberg parabolic subgroup of $H$. It $\lambda$-lifts to

$$J(\nu^{1/2}\pi_{2k}, \nu^{-1/2}\pi_{2k}),$$

the Langlands quotient of the induced representation $I(\nu^{1/2}\pi_{2k}, \nu^{-1/2}\pi_{2k})$ of $\mathrm{PGL}(4,\mathbb{R})$. Note that the discrete series $\pi_{2k} \simeq \mathrm{sgn} \otimes \pi_{2k} \simeq \check{\pi}_{2k}$ has central character $\mathrm{sgn}(\neq 1)$. Now

$$\psi : W_{\mathbb{R}} \times \mathrm{SL}(2,\mathbb{C}) \to \mathrm{SL}(4,\mathbb{C}), \qquad \psi|W_{\mathbb{R}} : \mathbf{w} \mapsto \begin{pmatrix} \phi_{2k}(\mathbf{w}) & 0 \\ 0 & \phi_{2k}(\mathbf{w}) \end{pmatrix} \times \mathbf{w}$$

with

$$\phi_{2k}(z) = \begin{pmatrix} (z/|z|)^{2k} & 0 \\ 0 & (|z|/z)^{2k} \end{pmatrix} \times z, \qquad \phi_{2k}(\sigma) = w \times \sigma,$$

and $(\psi|\,\mathrm{SL}(2,\mathbb{C}))\begin{pmatrix} a & b \\ c & d \end{pmatrix} = \begin{pmatrix} aI & bI \\ cI & dI \end{pmatrix}$, defines

$$\phi_\psi(z) = \psi\left(z, \begin{pmatrix} |z| & 0 \\ 0 & |z|^{-1} \end{pmatrix}\right) = \begin{pmatrix} |z|\phi_{2k}(z) & 0 \\ 0 & |z|^{-1}\phi_{2k}(z) \end{pmatrix} \times z,$$

$$\phi_\psi(\sigma) = \psi(\sigma, I) = \begin{pmatrix} w & 0 \\ 0 & w \end{pmatrix},$$

which factorizes via $\widehat{H} = \mathrm{Sp}(2,\mathbb{C}) \hookrightarrow \mathrm{SL}(4,\mathbb{C})$ and parametrizes

$$L(\nu\ \mathrm{sgn}, \nu^{-1/2}\pi_{2k}).$$

Note that when $2k$ is replaced by $2k+1$, $\phi_{2k+1}(\sigma) = \varepsilon w \times \sigma$, $\varepsilon = \mathrm{diag}(1,-1)$, then

$$\phi_\psi(\sigma) = \psi(\sigma, I) = \begin{pmatrix} \varepsilon w & 0 \\ 0 & \varepsilon w \end{pmatrix} = I \otimes \varepsilon w \in \hat{C},$$

$$\phi_\psi(z) = \begin{pmatrix} |z| & 0 \\ 0 & |z|^{-1} \end{pmatrix} \otimes \phi_{2k+1}(z) \in \hat{C},$$

thus $\phi_\psi$ defines a representation of $C(\mathbb{R})$ (which $\lambda_1$-lifts to the representation

$$J(\nu^{1/2}\pi_{2k+1}, \nu^{-1/2}\pi_{2k+1})$$

of $\mathrm{PGL}(4,\mathbb{R})$), but not a representation of $H(\mathbb{R})$.

As in [Ty] write $\pi^1_{2k,0}$ for $L(\mathrm{sgn}\,\nu, \nu^{-1/2}\pi_{2k+2})$. We have that $\pi^1_{2k,0} \simeq \mathrm{sgn} \otimes \pi^1_{2k,0}$, and $\pi^1_{2k,0}|H^0$ consists of two irreducibles. In the Grothendieck group the induced decomposes as

$$\nu\ \mathrm{sgn} \rtimes \nu^{-1/2}\pi_{2k} = L(\nu\ \mathrm{sgn}, \nu^{-1/2}\pi_{2k}) + \pi^{\mathrm{Wh}}_{2k+3,2k+1} + \pi^{\mathrm{hol}}_{2k+3,2k+1} \qquad k \geq 1.$$

To compute the infinitesimal character of $\nu\ \mathrm{sgn} \rtimes \nu^{-1/2}\pi_{2k}$, note that it is a constituent of the induced $\nu\ \mathrm{sgn} \rtimes \nu^{-1/2} I(\nu^k, \mathrm{sgn}\,\nu^{-k}) \simeq \mathrm{sgn}\,\nu^{2k} \times \mathrm{sgn}\,\nu \rtimes \nu^{-k-1/2}\,\mathrm{sgn}$ (using the Weyl group element $(12)(34)$), whose infinitesimal character is $(2k,1) = (2,1) + (a,0)$, with $a = 2k-2 \geq 0$ as $k \geq 1$. For $k \geq 1$ we have $H^{ij}(\mathrm{sp}(2,\mathbb{R}), \mathrm{SU}(4); \pi^1_{2k,0} \otimes V_{2k,0}) = \mathbb{C}$ if $(i,j) = (2,0)$, $(0,2)$, $(3,1)$, $(1,3)$.

# 5. The Cohomological $L(\xi\nu^{1/2}\pi_{2k+1}, \xi\nu^{-1/2})$

The nontempered endoscopic representation $L(\xi\nu^{1/2}\pi_{2k+1}, \xi\nu^{-1/2})$ of the group $H(\mathbb{R})$ is the Langlands quotient of the representation $\xi\nu^{1/2}\pi_{2k+1} \rtimes \xi\nu^{-1/2}$ induced from the Siegel parabolic subgroup of $H(\mathbb{R})$. It is the $\lambda_0$-lift of $\pi_{2k+1} \times \xi\mathbf{1}_2$ and $\lambda$-lifts to the induced $I(\pi_{2k+1}, \xi\mathbf{1}_2)$ of $\mathrm{PGL}(4,\mathbb{R})$. The central character of $\pi_{2k+1}$ is trivial, but that of $\pi_{2k}$ is sgn. Hence $I(\pi_{2k}, \xi_2)$ defines a representation of $\mathrm{GL}(4,\mathbb{R})$ but not of $\mathrm{PGL}(4,\mathbb{R})$. The endoscopic map

$$\psi : W_{\mathbb{R}} \times \mathrm{SL}(2,\mathbb{C}) \to {}^L C_0 = \mathrm{SL}(2,\mathbb{C}) \times \mathrm{SL}(2,\mathbb{C}) \overset{\lambda_0}{\to} \widehat{H},$$

$$\psi(z\sigma^j, s) = \lambda_0(\phi_{2k+1}(z\sigma^j), \xi(-1)^j s),$$

defines

$$\phi_\psi(z) = \psi\left(z, \begin{pmatrix} |z| & 0 \\ 0 & |z|^{-1} \end{pmatrix}\right) = \mathrm{diag}((z/|z|)^{2k+1}, |z|, |z|^{-1}, (|z|/z)^{2k+1}) \times z,$$

$$\phi_\psi(\sigma) = \psi(\sigma, I) = \begin{pmatrix} & & & 1 \\ & \xi(-1) & & \\ & & \xi(-1) & \\ (-1)^{2k+1} & & & \end{pmatrix},$$

which lies in $\widehat{H} \subset \mathrm{SL}(4,\mathbb{C})$ since $2k+1$ is odd.

As in [Ty] we write $\pi^{2,\xi}_{k-1,k-1}$ for $L(\xi\nu^{1/2}\pi_{2k+1}, \xi\nu^{-1/2})$, $k \geq 0$. Now $\xi\pi^{2,1} = \pi^{2,\xi}$ and $\pi^{2,\xi}|H^0$ is irreducible. In the Grothendieck group the induced decomposes as

$$\xi\nu^{1/2}\pi_{2k+1} \rtimes \xi\nu^{-1/2} = \pi^{2,\xi}_{k-1,k-1} + \pi^{\mathrm{Wh}}_{2k+1,1}.$$

Here $\pi^{\mathrm{Wh}}_{2k+1,1}$ is generic, discrete series if $k \geq 1$, tempered if $k = 0$. Our $\xi\nu^{1/2}\pi_{2k+1} \rtimes \xi\nu^{-1/2}$ is a constituent of the induced

$$\xi\nu^{1/2} I(\nu^{(2k+1)/2}, \nu^{-(2k+1)/2}) \rtimes \xi\nu^{-1/2} = \xi\nu^{k+1} \times \xi\nu^{-k} \rtimes \xi\nu^{-1/2},$$

which is equivalent to $\xi\nu^{k+1} \times \xi\nu^{k} \rtimes \xi\nu^{-k-1/2}$ (using the Weyl group element (23)). Its infinitesimal character is $(k+1,k) = (2,1) + (k-1,k-1)$. We have

$$H^{ij}(\mathrm{sp}(2,\mathbb{R}), \mathrm{SU}(4); \pi^{2,\xi}_{k-1,k-1} \otimes V_{k-1,k-1}) = \mathbb{C} \qquad \text{if} \quad (i,j) = (1,1),\ (2,2).$$

In summary, $H^{ij}(\pi \otimes V_{a,b})$ is 0 except in the following four cases, where it is $\mathbb{C}$.

(1) One dimensional case: $(a,b) = (0,0)$ and $\pi$ is $\pi^{\rm Wh}_{3,1}$, $\pi^{\rm hol}_{3,1}$, $\xi_H$,

$$\pi^1_{0,0} = L(\nu \text{ sgn}, \nu^{-1/2}\pi_2), \qquad \pi^{2,\xi}_{0,0} = L(\xi\nu^{1/2}\pi_3, \xi\nu^{-1/2}).$$

(2) Unstable nontempered case: $(a,b) = (k,k)$ $(k \geq 1)$ and $\pi$ is

$$\pi^{\rm Wh}_{2k+3,1}, \qquad \pi^{\rm hol}_{2k+3,1}, \qquad \pi^{2,\xi}_{k,k} = L(\xi\nu^{1/2}\pi_{2k+3}, \xi\nu^{-1/2}).$$

(3) Stable nontempered case: $(a,b) = (2k,0)$ $(k \geq 1)$ and $\pi$ is

$$\pi^{\rm Wh}_{2k+3,2k+1}, \qquad \pi^{\rm hol}_{2k+3,2k+1}, \qquad \pi^1_{2k,0} = L(\nu \text{ sgn}, \nu^{-1/2}\pi_{2k+2}).$$

(4) Tempered case: any other $(a,b)$, thus $a > b \geq 1$, $a+b$ even, and $\pi$ is $\pi^{\rm Wh}_{k_1,k_2}$, $\pi^{\rm hol}_{k_1,k_2}$. Here $k_1 = a+b+3 > k_2 = a-b+1 > 0$ are odd.

## 6. Finite Dimensional Representations

The $\mathbb{Q}$-rational representation $(\rho, V)$ of $H' = \mathrm{R}_{F/\mathbb{Q}}H$ has the form $(h_\iota) \mapsto \otimes\rho_\iota(h_\iota)$, where $H' = \prod_\iota H_\iota$, $H_\iota = H$, over $\overline{\mathbb{Q}}$, and $\rho_\iota$ is a representation (irreducible and finite dimensional) of $H_\iota$. Here $\iota$ ranges over $S = \mathrm{Gal}(\overline{\mathbb{Q}}/\mathbb{Q})/\mathrm{Gal}(\overline{\mathbb{Q}}/F)$, $= \mathrm{Hom}(F,\mathbb{R})$ and so $H' = \{(h_\iota);\ h_\iota \in H\}$. The Galois group $\mathrm{Gal}(\overline{\mathbb{Q}}/\mathbb{Q})$ acts by $\tau((h_\iota)) = ((\tau h_\iota)_{\tau\iota}) = ((\tau h_{\tau^{-1}\iota})_\iota)$. The fixed points are the the $(h_\iota)$ with $h_\iota = \iota h_1$, where $h_1$ ranges over $H(F)$ (the "1" indicates the fixed embedding $F \hookrightarrow \mathbb{R}$). Thus $H'(\mathbb{Q}) = H(F)$ and $H'(\mathbb{R}) = \prod_S H(\mathbb{R})$ with $|S| = [F:\mathbb{Q}]$ since $F$ is totally real; $S$ is the set of embeddings $F \hookrightarrow \mathbb{R}$. Now the representation $\rho$ is defined over $\mathbb{Q}$, namely fixed under the action of $\mathrm{Gal}(\overline{\mathbb{Q}}/\mathbb{Q})$. Thus $\otimes_\iota\rho_\iota(h_\iota) = \otimes_\iota\rho_{\tau\iota}(\tau h_\iota)$. The element $h = (h_1, 1, \ldots, 1)$ (thus $h_\iota = 1$ for all $\iota \neq 1$) is mapped by $\tau$ to $(1,\ldots,1,\tau h_1, 1, \ldots, 1)$ (the entry $\tau h_1$ is at the place parametrized by $\tau$). Hence $\rho_1(h_1)$ equals $\rho_\tau(\tau h_1)$ (both are equal to $\rho(h)(=\rho(\tau h))$). Hence $\rho_\tau = {}^\tau\rho_1(: h_1 \mapsto \rho_1(\tau^{-1}h_1))$, and the components $\rho_\tau$ of $\rho$ are all translates of the same representation $\rho_1$. For $(h_\iota) = (\iota h_1)$ in $H'(\mathbb{Q}) = H(F)$, $\rho((h_\iota)) = \otimes_\iota\rho_\iota(\iota h_1) = \otimes_\iota\rho_1(h_1) = \rho_1(h_1) \otimes \cdots \otimes \rho_1(h_1)$ ($[F:\mathbb{Q}]$ times).

However, over $F$ we have $H' \simeq \prod_{v\in S} H_v$ with $H_v = H$. An irreducible representation $(\boldsymbol{\rho}, \mathbf{V})$ of $H'$ over $F$ has the form $(\rho_{\mathbf{a},\mathbf{b}} = \otimes_{v\in S}\rho_{a_v,b_v}, V_{\mathbf{a},\mathbf{b}} = \otimes_{v\in S}V_{a_v,b_v})$, where $a_v \geq b_v \geq 0$, even $a_v - b_v$ for all $v \in S$.

# 7. Local Terms at $\infty$

Next we wish to compute the factors at $\infty$ of each of the terms in the trace formulae $\mathrm{STF}_H(f_H)$ and $\mathrm{TF}_{C_0}(f_{C_0})$. The functions $f_{H,\infty}$ ($= h_\infty$ of [Ko4]) and $f_{C_0,\infty}$ are products $\otimes f_{Hv}$ and $\otimes f_{C_0 v}$ over $v$ in $S$. We fixed a finite dimensional representation

$$(\boldsymbol{\rho}, V_{\boldsymbol{\rho}}) = (\rho_{\mathbf{a},\mathbf{b}} = \otimes_{v\in S}\rho_{a_v,b_v}, V_{\mathbf{a},\mathbf{b}} = \otimes_{v\in S}V_{a_v,b_v}), \qquad a_v \geq b_v \geq 0,$$

even $a_v - b_v$ for all $v \in S$, over $F$ of the group $H'$ over $\mathbb{Q}$. Denote by $\{{}_\rho\pi_{Hv}\}$ the packet of discrete series representations of $H(\mathbb{R})$ with infinitesimal character $(a_v, b_v) + (2,1)$.

For any $(\rho_v, V_{a_v,b_v})$, the packet $\{{}_\rho\pi_{Hv}\}$ consists of two representations, ${}_\rho\pi^+_{Hv} = \pi^{\mathrm{Wh}}_{k_{1v},k_{2v}}$ and ${}_\rho\pi^-_{Hv} = \pi^{\mathrm{hol}}_{k_{1v},k_{2v}}$, where $k_{1v} = a_v + b_v + 3 > k_{2v} = a_v - b_v + 1 > 0$ are odd. It is the $\lambda_0$-lift of the representations $\pi_{k_{1v}} \times \pi_{k_{2v}}$ and $\pi_{k_{2v}} \times \pi_{k_{1v}}$ of $C_0(\mathbb{R}) = \mathrm{PGL}(2,\mathbb{R}) \times \mathrm{PGL}(2,\mathbb{R})$. Denote by $h(\pi^{\mathrm{Wh}}_{k_{1v},k_{2v}})$, $h(\pi^{\mathrm{hol}}_{k_{1v},k_{2v}})$ a pseudo coefficient of the indicated representation. Then

$$f_{H',\infty} = {}_\rho f_{H',\infty} = \prod_{v\in S} h_{H,v},$$

$$h_{H,v} = h_{H,v}(\{{}_\rho\pi^{\pm}_{Hv}\}) = \frac{(-1)^{q(H)}}{2}[h({}_\rho\pi^+_{Hv}) + h({}_\rho\pi^-_{Hv})].$$

Put

$$f_{C_0',\infty} = {}_\rho f_{C_0',\infty} = \prod_{v\in S} h_{C_0,v}, \qquad C_0' = \mathrm{R}_{F/\mathbb{Q}}C_0,$$

$$h_{C_0,v} = h_{C_0,v}(\pi_{k_{1v}} \times \pi_{k_{2v}}) = (-1)^{q(H)}[h(\pi_{k_{1v}} \times \pi_{k_{2v}}) - h(\pi_{k_{2v}} \times \pi_{k_{1v}})].$$

Note that if $\pi_{k_1} = \pi_{k_{2v}}$ then $\lambda_0(\pi_{k_1} \times \pi_{k_{2v}}) = 1 \rtimes \pi_{k_1}$ which is not discrete series but properly induced. In particular, the fifth term $I(H,5)$ of $\mathrm{STF}_H(f_H)$, and the corresponding terms of $I(C_0,2)$ in $\mathrm{TF}_{C_0}(f_{C_0})$ – those which are parametrized by $\pi^2 \times \sigma\mathbf{1}_2$ where $\pi^2$ is cuspidal whose components at $\infty$ are $\pi_1$, vanish for our functions $f_H$, $f_{C_0}$. Moreover, as explained at the end of section 6, $I(H,4)$ and $I(C_0,3)$ are 0 for our $f_H$, $f_{C_0}$.

Note that $q(H') = [F:\mathbb{Q}]q(H)$ is half the real dimension of the symmetric space attached to $H'(\mathbb{R})$, and $q(H)$ is that of $H(\mathbb{R})$, thus $q(H) = 3$ in our case.

Then $\operatorname{tr}\pi^{\mathrm{Wh}}_{k_{1v},k_{2v}}(h_{H,v}) = \operatorname{tr}\pi^{\mathrm{hol}}_{k_{1v},k_{2v}}(h_{H,v}) = \frac{1}{2}(-1)^{q(H)} = -\frac{1}{2}$. When $(a,b) = (2k,0)$, $k \geq 0$, we have in addition $\operatorname{tr} L(\nu \text{ sgn}, \nu^{-1/2}\pi_{2k+2})(h_{H,v}) = 1$. When $(a,b) = (k,k)$, $k \geq 0$, we have $\operatorname{tr} L(\xi\nu^{1/2}\pi_{2k+3}, \xi\nu^{-1/2})(h_{H,v}) = \frac{1}{2}$. When $(a,b) = (0,0)$, we have in addition $\operatorname{tr}\xi_H(h_{H,v}) = 1$, $\xi^2 = 1$.

Note that if $\pi_H$ contributes to $I(H,1)_2$ then its archimedean components $\pi_{Hv}$ have infinitesimal characters of the form $(2k_v, 0)$, $k_v \geq 0$, for all $v \in S$.

If $\pi_H$ contributes to $I(H,3)$ then there is a contribution to $I(C_0,2)$, and the archimedean components $\pi_{Hv}$ have infinitesimal characters of the form $(k_v, k_v)$, $k_v \geq 0$, for all $v \in S$.

If $\pi_H$ contributes to $I(H,1)_3$ the infinitesimal characters of its archimedean components are $(0,0)$.

# V. GALOIS REPRESENTATIONS

## 1. Tempered Case

We apply the Lefschetz formula in Deligne's conjecture form to the étale cohomology

$$H_c^*(\mathcal{S}_{K_f} \otimes_{\mathbb{E}} \overline{\mathbb{F}}, \mathbb{V}_{\mathbf{a},\mathbf{b};\lambda})$$

with compact supports and coefficients in the representation $(\rho_{\mathbf{a},\mathbf{b}}, V_{\mathbf{a},\mathbf{b}})$, even $a_v - b_v$, for all $v \in S$.

Suppose $\pi_H$ occurs in the stable spectrum, namely in $I(H,1)_1$.

The choice of the function ${}_{\rho}f_{\infty H}$ guarantees that the components $\pi_{Hv}$ lie in the packet $\{\pi^{\mathrm{Wh}}_{k_{1v},k_{2v}}, \pi^{\mathrm{hol}}_{k_{1v},k_{2v}}\}$, $k_{1v} = a_v + b_v + 3$, $k_{2v} = a_v - b_v + 1$, at each archimedean place $v \in S$.

We start by fixing a cuspidal representation $\pi_H$ with $\pi_{Hv}$ in the set $\{\pi^{\mathrm{Wh}}_{k_{1v},k_{2v}}, \pi^{\mathrm{hol}}_{k_{1v},k_{2v}}\}$ for all $v$ in $S$ and with $\pi_{Hf}^{K_f} \neq 0$. In particular the component at $p$ of such $\pi_H$ is unramified, of the form $\otimes_{u|p}\pi_H(\mu_{1u}, \mu_{2u})$.

We use a correspondence $f_H^p$, which is a $K_f^p$-biinvariant function on $H(\mathbb{A}_f^p)$. Since there are only finitely many discrete series representations of $H(\mathbb{A})$ with a given infinitesimal character (determined by $\boldsymbol{\rho}$) and a nonzero $K_f$-fixed vector, we can choose $f_H^p$ to be a projection onto $\{\pi_{Hf}^{K_f}\}$. Writing $t_{mu}$ for $\mu_{mu}(\pi_u)$, $m = 1, 2$, the trace of the action of $\mathrm{Fr}_p^j$ on the $\{\pi_{Hf}^{K_f}\}$-isotypic component of $H_c^*(\mathcal{S}_{K_f} \otimes_{\mathbb{F}} \overline{\mathbb{F}}, \mathbb{V}_{\boldsymbol{\rho}})$ (which vanishes outside the middle dimension $3[F:\mathbb{Q}]$), is multiplication by (we put $j_u = (j, n_u)$)

$$p^{\frac{j}{2}\dim \mathcal{S}_{K_f}} \prod_{u|p} (t_{1u}^{j/j_u} + t_{2u}^{j/j_u} + t_{2u}^{-j/j_u} + t_{1u}^{-j/j_u})^{j_u}.$$

Note that $H_c^{3[F:\mathbb{Q}]}$ occurs in the alternating sum $H_c^*$ with coefficient $(-1)^{3[F:\mathbb{Q}]}$. This sign is canceled by the sign $(-1)^{q(H')}$ of the definition of the functions $f_{H',\infty} = \otimes_{v \in S} h_{Hv}$.

Thus the $\{\pi_{Hf}^{K_f}\}$-isotypic part of $H_c^{3[F:\mathbb{Q}]}$ (namely the $\pi_{Hf}^{K_f}$-isotypic part for each member of the packet) is of the form $\{\pi_{Hf}^{K_f}\} \otimes \rho(\{\pi_H\})$. Here

$\rho(\{\pi_H\})$ is a $4^{[F:\mathbb{Q}]}$-dimensional representation of $\mathrm{Gal}(\overline{\mathbb{Q}}/\mathbb{Q})$. The $4^{\#\{u|p\}}$ nonzero eigenvalues $t$ of the action of $\mathrm{Fr}_p$ are $p^{\frac{3}{2}[F:\mathbb{Q}]}\prod_{u|p} t_{m(u),u}^{\iota(u)}$, $m(u) \in \{1,2\}$, $\iota(u) \in \{\pm 1\}$. This we see first for sufficiently large $j$ by Deligne's conjecture, but then for all $j \geq 0$, by multiplicativity.

Deligne's "Weil conjecture" purity theorem asserts that the Frobenius eigenvalues are algebraic numbers and all their conjugates have equal complex absolute values of the form $q_\wp^{i/2}$ $(0 \leq i \leq 2\dim \mathcal{S})$. This is also referred to as "mixed purity". The eigenvalues of $\mathrm{Fr}_\wp$ on $IH^i$ have complex absolute values equal $q_\wp^{i/2}$, by a variant of the purity theorem due to Gabber. We shall use this to show that the absolute values in our case are all equal to $q_\wp^{\frac{1}{2}\dim \mathcal{S}}$. In our case $\mathbb{E} = \mathbb{Q}$, the ideal $\wp$ is $(p)$, the residual cardinality $q_\wp$ is $p$, and $n_\wp = [\mathbb{E}_\wp : \mathbb{Q}_p]$ is 1.

Note that the cuspidal $\pi$ define part not only of the cohomology

$$H_c^i(\mathcal{S}_{K_f} \otimes_{\mathbb{E}} \overline{\mathbb{Q}}, \mathbb{V})$$

but also part of the intersection cohomology $\mathrm{IH}^i(\mathcal{S}'_{K_f} \otimes_{\mathbb{E}} \overline{\mathbb{Q}}, \mathbb{V})$. By the Zucker isomorphism it defines a contribution to the $L^2$-cohomology, which is of the form $\pi_f^{K_f} \otimes H^i(\mathfrak{g}, K_\infty; \pi_\infty \otimes V_\xi(\mathbb{C}))$. We shall compute this $(\mathfrak{g}, K_\infty)$-cohomology space to know for which $i$ there is nonzero contribution corresponding to our $\pi_f$. We shall then be able to evaluate the absolute values of the conjugates of the Frobenius eigenvalues using Deligne's "Weil conjecture" theorem.

The space $H^{i,j}(\mathfrak{g}, K; \pi \otimes V_{a,b})$ is 0 for $\pi = \pi^*_{k_1,k_2}$, $* =$ Wh or hol, $k_1 > k_2 > 0$ are odd (indexed by $a \geq b \geq 0$) except when $(i,j) = (2,1)$, $(1,2)$, $(3,0)$, $(0,3)$ (respectively), when this space is $\mathbb{C}$. From the "Matsushima-Murakami" decomposition of section 2, first for the $L^2$-cohomology $H_{(2)}$ but then by Zucker's conjecture also for $IH^*$, and using the Künneth formula, we conclude that $IH^i(\pi_f)$ is zero unless $i$ is equal to $\dim \mathcal{S}_{K_f} = 3[F:\mathbb{Q}]$, and there $\dim IH^{2[F:\mathbb{Q}]}(\pi_f)$ is $4^{[F:\mathbb{Q}]}$ (as there are $[F:\mathbb{Q}]$ real places of $F$). Since $\pi_f$ is the finite component of cuspidal representations only, $\pi_f$ contributes also to the cohomology $H_c^i(\mathcal{S}_{K_f} \otimes_{\mathbb{E}} \overline{\mathbb{Q}}, \mathbb{V}_{\mathbf{a},\mathbf{b};\lambda})$ only in dimension $i = 3[F:\mathbb{Q}]$, and $\dim H_c^{3[F:\mathbb{Q}]}(\pi_f) = 4^{[F:\mathbb{Q}]}$. This space depends only on the packet of $\pi_f$ and not on $\pi_f$ itself.

Deligne's theorem [D6] (in fact its $IH$-version due to Gabber) asserts that the eigenvalues $t$ of the Frobenius $\mathrm{Fr}_\wp$ acting on the $\ell$-adic intersection cohomology $IH^i$ of a variety over a finite field of $q_\wp$ elements are algebraic and "pure", namely all conjugates have the same complex absolute value, of the form $q_\wp^{i/2}$. In our case $i = \dim \mathcal{S}_{K_f} = 3[F:\mathbb{Q}]$, hence the eigenvalues of the Frobenius are algebraic and each of their conjugates is $q_\wp^{\frac{3}{2}[F:\mathbb{Q}]}$ in absolute value. Consequently the eigenvalues $\mu_{1u}$, $\mu_{2u}$ are algebraic, and all of their conjugates have complex absolute value 1.

Note that we could not use only "mixed-purity" (that the eigenvalues are powers of $q_\wp^{1/2}$ in absolute value) and the unitarity estimates $|\mu_{mu}|^{\pm 1} < q_u^{1/2}$ on the Hecke eigenvalues, since the estimate (less than $(\sqrt{q_\wp})^{\frac{1}{2}\dim \mathcal{S}}$ away from $(\sqrt{q_\wp})^{\dim \mathcal{S}}$) does not define the absolute value $((\sqrt{q_\wp})^{\dim \mathcal{S}})$ uniquely. This estimate does suffice to show unitarity when $\dim \mathcal{S} = 1$.

In summary, the representation $\rho = \rho(\pi_{Hf})$ of $\mathrm{Gal}(\overline{\mathbb{Q}}/\mathbb{Q})$ attached to $\pi_{Hf}$ depends only on the packet of $\pi_{Hf}$, its dimension is $4^{[F:\mathbb{Q}]}$. Its restriction to $\mathrm{Gal}(\overline{\mathbb{Q}}_p/\mathbb{Q}_p)$ is unramified, and the trace of $\rho(\mathrm{Fr}_p)$ on the $\{\pi_{Hf}^{K_f}\}$-isotypic part of $H_c^{3[F:\mathbb{Q}]}$ is equal to the trace of $\otimes \nu_u^{-1/2} r_u(\mathbf{t}(\pi_{Hu}) \times \mathrm{Fr}_u)$. Here $(r_u, (\mathbb{C}^4)^{[F_u:\mathbb{Q}_p]})$ denotes the twisted tensor representation of ${}^L\mathrm{R}_{F_u/\mathbb{Q}_p} H = \hat{H}^{[F_u:\mathbb{Q}_p]} \rtimes \mathrm{Gal}(F_u/\mathbb{Q}_p)$, $\mathrm{Fr}_u$ is $\mathrm{Fr}_p^{[F_u:\mathbb{Q}_p]}$, and $\nu_u$ is the character of ${}^L\mathrm{R}_{F_u/\mathbb{Q}_p} H$ which is trivial on the connected component of the identity and whose value at $\mathrm{Fr}_u$ is $q_u^{-1}$, where $q_u = p^{[F_u:\mathbb{Q}_p]}$. The eigenvalues of $\mathbf{t}(\pi_{Hu})$ and all of their conjugates, lie on the complex unit circle.

We continue by fixing a cuspidal representation $\pi_H$ with $\pi_{Hv}$ in the set $\{\pi^{\mathrm{Wh}}_{k_{1v},k_{2v}}, \pi^{\mathrm{hol}}_{k_{1v},k_{2v}}\}$ for all $v$ in $S$ and with $\pi_H^{K_f} \neq 0$. But now we assume it occurs in the unstable spectrum, namely in $I(H,2)$. We fix a correspondence $f_H^p$ which projects to the packet $\{\pi_{Hf}^p\}$. Since the function $f_{C_0}^p$ is chosen to be matching $f_H^p$, by [F6] the contributions to $I(C_0,1)$ are precisely those parametrized by $\pi^1 \times \pi^2$ and $\pi^2 \times \pi^1$, where $\pi^m$ are cuspidal representations of $\mathrm{PGL}(2,\mathbb{A})$ whose real components are $\{\pi_v^1, \pi_v^2\} = \{\pi_{k_{1v}}, \pi_{k_{2v}}\}$, a set of cardinality two.

Write $\{\pi_{Hf}\}^+$ for the set of $\pi_{Hf} = \otimes_w \pi_{Hw}$, $w < \infty$, which are the finite part of an irreducible $\pi_H$ in our packet $\{\pi_H\}$, such that $\pi_{Hw}$ is $\pi_{Hw}^-$ for an even number of places $w < \infty$. Similarly define $\{\pi_{Hf}\}^-$ by replacing "even" with "odd". The contribution of $\{\pi_H\}$ to $I(H,2)$ is

$$\frac{1}{2} \prod_{v|\infty} \mathrm{tr}\{\pi_{Hv}\}(h_{Hv}) \cdot [\mathrm{tr}\{\pi_{Hf}\}^+(f_H^p) + \mathrm{tr}\{\pi_{Hf}\}^-(f_H^p)]$$

$$\cdot p^{\frac{j}{2}\dim \mathcal{S}_{K_f}} \prod_{u|p} (t_{1u}^{j/j_u} + t_{1u}^{-j/j_u} + t_{2u}^{j/j_u} + t_{2u}^{-j/j_u})^{j_u}.$$

Here and below $f_H^p$ indicates – as suitable – its product with the unit element of the $H'(\mathbb{Z}_p)$-Hecke algebra of $H'(\mathbb{Q}_p)$.

The corresponding contribution to $I(C_0, 1)$ is twice (from $\pi^1 \times \pi^2$ and $\pi^2 \times \pi^1$)

$$\frac{1}{4} \prod_{v|\infty} \mathrm{tr}\{\pi_v^1 \times \pi_v^2\}(h_{C_0 v}) \cdot \mathrm{tr}(\pi_f^1 \times \pi_f^2)(f_{C_0}^p)$$

$$\cdot p^{\frac{j}{2}\dim \mathcal{S}_{K_f}} \prod_{u|p} (t_{1u}^{j/j_u} + t_{1u}^{-j/j_u} - t_{2u}^{j/j_u} - t_{2u}^{-j/j_u})^{j_u}.$$

By choice of $f_{C_0}^p$ we have that $\mathrm{tr}(\pi_f^1 \times \pi_f^2)(f_{C_0}^p) = \mathrm{tr}\{\pi_{Hf}\}^+(f_H^p) - \mathrm{tr}\{\pi_{Hf}\}^-(f_H^p)$. The choice of $h_{Hv}$ is such that $\mathrm{tr}\{\pi_{Hv}\}(h_{Hv}) = (-1)^{q(H)}$, and $\mathrm{tr}(\pi_v^1 \times \pi_v^2)(h_{C_0 v})$ is $(-1)^{q(H)}$ if $\pi_v^1 \times \pi_v^2$ is $\pi_{k_{1v}} \times \pi_{k_{2v}}$, and $-(-1)^{q(H)}$ if it is $\pi_{k_{2v}} \times \pi_{k_{1v}}$.

We conclude that for each irreducible $\pi_{Hf} \in \{\pi_{Hf}\}^+$, the $\pi_{Hf}^{K_f}$-isotypic part of $H_c^i$ is zero unless $i = 3[F:\mathbb{Q}]$ (middle dimension), in which case it is $\pi_{Hf}^{K_f} \otimes \rho(\{\pi_{Hf}\}^+)$, and $\mathrm{Fr}_p^j$ acts on $\rho(\{\pi_{Hf}\}^+)$ with trace

$$\frac{1}{2} p^{\frac{j}{2}\dim \mathcal{S}_{K_f}} [\prod_{u|p} (t_{1u}^{j/j_u} + t_{1u}^{-j/j_u} + t_{2u}^{j/j_u} + t_{2u}^{-j/j_u})^{j_u}$$

$$+(-1)^{n(\pi^1\times\pi^2)} \prod_{u|p} (t_{1u}^{j/j_u} + t_{1u}^{-j/j_u} - (t_{2u}^{j/j_u} + t_{2u}^{-j/j_u}))^{j_u}].$$

We write $n(\pi^1 \times \pi^2)$ for the number of archimedean places $v$ of $F$ with $(\pi_v^1, \pi_v^2) = (\pi_{k_{2v}}, \pi_{k_{1v}})$.

Similarly, for each irreducible $\pi_{Hf} \in \{\pi_{Hf}\}^-$, the $\pi_{Hf}^{K_f}$-isotypic part of $H_c^i$ is zero unless $i = 3[F:\mathbb{Q}]$ (middle dimension), in which case it is $\pi_{Hf}^{K_f} \otimes \rho(\{\pi_{Hf}\}^-)$, and $\mathrm{Fr}_p^j$ acts on $\rho(\{\pi_{Hf}\}^-)$ with trace

$$\frac{1}{2} p^{\frac{j}{2}\dim \mathcal{S}_{K_f}} [\prod_{u|p} (t_{1u}^{j/j_u} + t_{1u}^{-j/j_u} + t_{2u}^{j/j_u} + t_{2u}^{-j/j_u})^{j_u}$$

$$-(-1)^{n(\pi^1\times\pi^2)} \prod_{u|p} (t_{1u}^{j/j_u} + t_{1u}^{-j/j_u} - (t_{2u}^{j/j_u} + t_{2u}^{-j/j_u}))^{j_u}].$$

As usual, we conclude from Deligne's mixed purity [D6] that the Hecke eigenvalues $t_{mu}$ are algebraic and their conjugates all lie in the unit circle in $\mathbb{C}$.

## 2. Nontempered Case

Next we deal with the case of $\pi_H$ which occurs in $I(H,1)_2$, namely in a quasi packet $\{L(\xi\nu, \nu^{-1/2}\pi^2)\}$ which $\lambda$-lifts to the residual representation $J(\nu^{1/2}\pi^2, \nu^{-1/2}\pi^2)$ of $G(\mathbb{A})$, $G = \mathrm{PGL}(4)$. Here $\pi^2$ is a cuspidal representation of $\mathrm{GL}(2,\mathbb{A})$ with quadratic central character $\xi \neq 1$ and $\xi\pi^2 = \pi^2$, and $\pi_v^2 = \pi_{2k_v+2}$ at each $v \in S$. The infinitesimal character of $\pi_{Hv}$ is $(2k_v, 0) + (2,1)$, $k_v \geq 0$, for all $v \in S$. Choosing $f_H^p$ to project on $\pi_{Hf}^{K_f}$ for such $\pi_H$, we note that there are no contributions from the endoscopic group $C_0$, thus $I(C_0, i)$ are zero. Namely the contributions to $I(H,1)_2$ are stable. The result for Shimura varieties associated with GL(2) assures us that the Hecke eigenvalues, or Satake parameters, of each component $\pi_u^2$ of $\pi^2$ at $u|p$ are algebraic and their conjugates have complex absolute value one. Alternatively we can conclude that the components above $p$ are all – unramified – of the form $L(\xi_u\nu_u, \nu_u^{-1/2}\pi_u^2)$ where $\pi_u^2 = \pi(\mu_{1u}, \xi_u/\mu_{1u})$, $\xi_u^2 = 1$, $\mu_{1u}$ unramified and equals 1 or $-1$ if $\xi_u = -1$.

As noted in section 12,

$$t(\pi_{Hu}) = \mathrm{diag}(q_u^{1/2} z_{1u}, q_u^{1/2} z_{2u}, q_u^{-1/2} z_{2u}^{-1}, q_u^{-1/2} z_{1u}^{-1})$$

with $z_{1u} = \mu_{1u}$, $z_{2u} = \xi_u/\mu_{1u}$, where we write $\mu_{1u}$ and $\xi_u$ also for their values at $\boldsymbol{\pi}_u$. Using the estimate $q_u^{-1/2} < |z_{mu}| < q_u^{1/2}$ we conclude from Deligne's theorem [D6] that $\mu_{1u}$ is algebraic and the complex absolute value of each of its conjugates is equal to one. On the $\pi_{Hf}^{K_f}$-isotypic part of the cohomology, $\mathrm{Fr}_p$ acts with the $4^{\#\{u|p\}}$ eigenvalues $p^{\frac{1}{2}\dim S_{K_f}} \prod_{u|p} a_u$, where

$$a_u \in \{q_u^{1/2}\mu_{1u}, \quad q_u^{1/2}\xi_u\mu_{1u}^{-1}, \quad q_u^{-1/2}\xi_u\mu_{1u}, \quad q_u^{-1/2}\mu_{1u}^{-1}\}.$$

Note that $\pi_{Hv} = L(\mathrm{sgn}\,\nu_v, \nu_v^{-1/2}\pi_{2k_v+2})$ has $H^{ij}(\pi_{Hv} \otimes V_{a_v,b_v}) \neq 0$ only when $a_v = 2k_v$, $b_v = 0$, and $(i,j) = (2,0)$, $(0,2)$, $(3,1)$, $(1,3)$.

Next we deal with the case of $\pi_H$ which occurs in $I(H,3)$, namely in a quasi-packet $\lambda_0(\pi^2 \times \xi\mathbf{1}_2)$, where $\pi^2$ is a cuspidal representation of $\mathrm{PGL}(2,\mathbb{A})$ whose real components are $\pi_{2k_v+3}$, and $\xi$ is a character of $\mathbb{A}^\times/F^\times\mathbb{A}^{\times 2}$. There are corresponding contributions in $I(C_0,2)$ from $\pi^2 \times \xi\mathbf{1}_2$ and from $\xi\mathbf{1}_2 \times \pi^2$. The infinitesimal character of $\pi_{Hv}$ is $(k_v,k_v)$, $k_v \geq 0$, for all $v \in S$.

We fix a correspondence $f_H^p$ which projects to the packet $\{\pi_{Hf}^p\}$. Since the function $f_{C_0}^p$ is chosen to be matching $f_H^p$, by [F6] the contributions to $I(C_0,1)$ are precisely those parametrized by $\pi^2 \times \xi\mathbf{1}_2$ and $\xi\mathbf{1}_2 \times \pi^2$.

Write $\{\pi_{Hf}\}^\times$ for the set of $\pi_{Hf} = \otimes_w \pi_{Hw}$, $w<\infty$, which are the finite part of an irreducible $\pi_H$ in our quasi-packet $\{\pi_H\}$, such that $\pi_{Hw}$ is $\pi_{Hw}^-$ for an even number of places $w<\infty$. Similarly define $\{\pi_{Hf}\}^-$ by replacing "even" with "odd". The contribution of $\{\pi_H\}$ to $I(H,3)$ is

$$\frac{\varepsilon(\xi\pi^2,\frac{1}{2})}{2}\prod_{v|\infty}\mathrm{tr}(\pi_{Hv}^\times - \pi_{Hv}^-)(h_{Hv})\cdot[\mathrm{tr}\{\pi_{Hf}\}^\times(f_H^p) - \mathrm{tr}\{\pi_{Hf}\}^-(f_H^p)]$$

$$\cdot p^{\frac{j}{2}\dim S_{K_f}}\prod_{u|p}[(\xi_u q_u^{1/2}\mu_u)^{j/j_u} + (\xi_u q_u^{1/2}\mu_u)^{-j/j_u}$$

$$+(\xi_u q_u^{1/2}\mu_u^{-1})^{j/j_u} + (\xi_u q_u^{1/2}\mu_u^{-1})^{-j/j_u}]^{j_u}.$$

Here $\pi_u^2 = I(\mu_u,\mu_u^{-1})$, and we abbreviate $\mu_u(\boldsymbol{\pi}_u)$ to $\mu_u$ and $\xi_u(\boldsymbol{\pi}_u)$ to $\xi_u$.

The corresponding contribution to $I(C_0,2)$ is twice (from $\pi^2 \times \xi\mathbf{1}_2$ and from $\xi\mathbf{1}_2 \times \pi^2$)

$$\frac{1}{4}\prod_{v|\infty}\mathrm{tr}(\pi_v^2 \times \xi\mathbf{1}_2)(h_{C_0v})\cdot \mathrm{tr}(\pi_f^2 \times \xi_f\mathbf{1}_2)(f_{C_0}^p)$$

$$\cdot p^{\frac{j}{2}\dim S_{K_f}}\prod_{u|p}[(\xi_u q_u^{1/2}\mu_u)^{j/j_u} + (\xi_u q_u^{1/2}\mu_u)^{-j/j_u}$$

$$-(\xi_u q_u^{1/2}\mu_u^{-1})^{j/j_u} - (\xi_u q_u^{1/2}\mu_u^{-1})^{-j/j_u}]^{j_u}.$$

By choice of $f_{C_0}^p$ we have that $\mathrm{tr}(\pi_f^2 \times \xi_f\mathbf{1}_2)(f_{C_0}^p) = \mathrm{tr}\{\pi_{Hf}\}^\times(f_H^p) + \mathrm{tr}\{\pi_{Hf}\}^-(f_H^p)$. The choice of $h_{Hv}$ is such that $\mathrm{tr}\,\pi_{Hv}^-(h_{Hv}) = \frac{1}{2}(-1)^{q(H)} = -\frac{1}{2}$, $\mathrm{tr}\,\pi_{Hv}^\times(h_{Hv}) = \frac{1}{2}$, and $\mathrm{tr}(\pi_v^1 \times \pi_v^2)(h_{C_0v})$ is $(-1)^{q(H)}$ if $\pi_v^1 \times \pi_v^2$ is $\pi_v^2 \times \xi_v\mathbf{1}_2$ and $-(-1)^{q(H)}$ if it is $\xi_v\mathbf{1}_2 \times \pi_v^2$.

We conclude that $\mathrm{Fr}_p^j$ acts on the $\pi_{Hf}^{K_f}$-isotypic past of $H_c^*$, for each irreducible $\pi_{Hf} \in \{\pi_{Hf}\}^\times$, with trace $\frac{1}{2}p^{\frac{j}{2}\dim S_{K_f}}$ times

$$\varepsilon(\xi\pi^2, \frac{1}{2}) \prod_{u|p} [(\xi_u q_u^{1/2}\mu_u)^{j/j_u} + (\xi_u q_u^{1/2}\mu_u)^{-j/j_u} + (\xi_u q_u^{1/2}\mu_u^{-1})^{j/j_u} + (\xi_u q_u^{1/2}\mu_u^{-1})^{-j/j_u}]^{j_u}$$
$$+ \prod_{u|p} [(\xi_u q_u^{1/2}\mu_u)^{j/j_u} + (\xi_u q_u^{1/2}\mu_u)^{-j/j_u} - (\xi_u q_u^{1/2}\mu_u^{-1})^{j/j_u} - (\xi_u q_u^{1/2}\mu_u^{-1})^{-j/j_u}]^{j_u}.$$

Similarly, $\mathrm{Fr}_p^j$ acts on the $\pi_{Hf}^{K_f}$-isotypic past of $H_c^*$, for each irreducible $\pi_{Hf} \in \{\pi_{Hf}\}^-$, with trace $\frac{1}{2}p^{\frac{j}{2}\dim S_{K_f}}$ times

$$\varepsilon(\xi\pi^2, \frac{1}{2}) \prod_{u|p} [(\xi_u q_u^{1/2}\mu_u)^{j/j_u} + (\xi_u q_u^{1/2}\mu_u)^{-j/j_u} + (\xi_u q_u^{1/2}\mu_u^{-1})^{j/j_u} + (\xi_u q_u^{1/2}\mu_u^{-1})^{-j/j_u}]^{j_u}$$
$$- \prod_{u|p} [(\xi_u q_u^{1/2}\mu_u)^{j/j_u} + (\xi_u q_u^{1/2}\mu_u)^{-j/j_u} - (\xi_u q_u^{1/2}\mu_u^{-1})^{j/j_u} - (\xi_u q_u^{1/2}\mu_u^{-1})^{-j/j_u}]^{j_u}.$$

Note that $\pi_{Hv} = L(\xi_v \nu_v^{1/2}\pi_{2k+3}, \xi_v \nu_v^{-1/2})$ has $H^{ij}(\pi_{Hv} \otimes V_{a_v,b_v}) \neq 0$ only when $a_v = k_v$, $b_v = k_v$, and $(i,j) = (1,1)$ or $(2,2)$.

As usual, we conclude from Deligne's mixed purity [D6] that the $\mu_{mu}$ are algebraic and their conjugates all lie in the unit circle in $\mathbb{C}$.

Finally we deal with the case of a one dimensional representation $\pi_H = \xi_H$, which occurs in $I(H,1)_3$. We can choose $f_H^p$ to factorize through a projection onto this one dimensional representation $\pi_H = \xi$ such that $\pi_{Hf}^{K_f} \neq 0$. The infinitesimal character of $\pi_{Hv}$ is $(0,0)$ for all $v \in S$. In particular the component at $p$ of such $\pi_H$ is unramified, and the trace of the action of $\mathrm{Fr}_p^j$ on the $\pi_{Hf}$-isotypic component of $H_c^*(\mathcal{S}_{K_f} \otimes_{\mathbb{F}} \overline{\mathbb{F}}, \mathbb{V}_\rho)$ is

$$p^{\frac{j}{2}\dim S_{K_f}} \prod_{u|p} [(\xi_u q_u^{3/2})^{j/j_u} + (\xi_u q_u^{1/2})^{j/j_u} + (\xi_u q_u^{-1/2})^{j/j_u} + (\xi_u q_u^{-3/2})^{j/j_u}]^{j_u}.$$

Note that $H_c^{ij}(\mathrm{sp}(2,\mathbb{R}),\mathrm{SU}(4);\mathbb{C})$ is $\mathbb{C}$ for $(i,j)=(0,0)$, $(1,1)$, $(2,2)$, $(3,3)$ and $\{0\}$ otherwise. Thus $\pi_H=\xi_H$ contributes only to the (even) part

$$\bigoplus_{0\leq m\leq \dim \mathcal{S}_{K_f}} H_c^{2m}(\mathcal{S}_{K_f}\otimes_{\mathbb{F}}\overline{\mathbb{F}},\mathbf{1}).$$

Note that the functions $f_{H',\infty}=\otimes_{v\in S}h_{Hv}$ satisfy

$$\mathrm{tr}(\xi_{Hv})(h_{Hv})=-(-1)^{q(H)}=1.$$

We conclude that the representation $\rho(\pi_{Hf})$ of $\mathrm{Gal}(\overline{\mathbb{Q}}/\mathbb{Q})$ on $H_c^*$ attached to $\pi_{Hf}$ is $4^{[F:\mathbb{Q}]}$-dimensional. Its restriction to $\mathrm{Gal}(\overline{\mathbb{Q}}_p/\mathbb{Q}_p)$ is unramified. Its trace is equal to the trace of $\otimes_{u|p}\nu_u^{-1/2}r_u(\mathrm{Fr}_u)$. Here $r_u(\mathrm{Fr}_u)$ acts on the twisted tensor representation $(r_u,(\mathbb{C}^4)^{[F_u:\mathbb{Q}_p]})$ as $\mathbf{t}(\xi_u)\times\mathrm{Fr}_u$, $\mathbf{t}(\xi_u)=(t_1,\ldots,t_{n_u})$, $t_m$ diagonal with

$$\prod_{1\leq m\leq n_u} t_m=\mathrm{diag}(\xi_u q_u^{3/2},\xi_u q_u^{1/2},\xi_u q_u^{-1/2},\xi_u q_u^{-3/2}).$$

# PART 3. BACKGROUND

# I. ON AUTOMORPHIC FORMS

## 1. Class Field Theory

Underlying the discipline of Automorphic Representations is a hypothetical reciprocity law that would generalize to the context of connected reductive groups $G$ over local or global fields $F$ the deep Class Field Theory, which simply asserts that $W_F^{\mathrm{ab}} \simeq C_F$, and is to be viewed as the special case of $\mathrm{GL}(1) = \mathbb{G}_m$. We review some of the key notions here, starting with basics. Key topics are in bold letters, and new terms are in italics.

Number Theory concerns number fields $F$, finite extensions of the field $\mathbb{Q}$ of rational numbers. The completion of $F$ at each of its valuations, $v$, is denoted by $F_v$. It is the field $\mathbb{C}$ of complex numbers or the field $\mathbb{R}$ of real numbers if $v$ is archimedean ($|x+y|_v \leq |x|_v + |y|_v$), or a finite extension of $\mathbb{Q}_p$ for a prime $p$ if $v$ is nonarchimedean ($|x+y|_v \leq \max(|x|_v, |y|_v)$). There is a positive characteristic analogue, where $F$ is the function field of a curve $C$ over a finite field $\mathbb{F}_q$, the places $v$ are the closed points of $C$ and $F_v$ is $\mathbb{F}_{q^r}((t))$, the field of power series over a finite extension of $\mathbb{F}_q$. In the nonarchimedean case denote by $R_v$ the ring of integers of $F_v$ (defined by $|x|_v \leq 1$).

The ring of $F$-**adèles**, denoted $\mathbb{A}_F$ or simply $\mathbb{A}$, is the union over all finite sets $S$ of valuations of $F$ containing the archimedean ones, of the products $\prod_{v\in S} F_v \times \prod_{v\notin S} R_v$. Thus an adèle is a tuple $(x_v)$, $x_v \in F_v$ for all $v$ and $x_v \in R_v$ for almost all $v$ (finite number of exceptions). The field $F$ embeds diagonally ($x_v = x$ for all $v$) in $\mathbb{A}$ as a discrete subgroup, and $\mathbb{A} \bmod F$ is compact. By $\mathbb{A}_{F,f}$, or $\mathbb{A}_f$, we denote the ring of adèles without archimedean components. Thus $\mathbb{A} = \mathbb{A}_f \prod_{v\in\infty} F_v$ where $\infty$ is the set of archimedean places of $F$.

The multiplicative group of $\mathbb{A}$ is the group of idèles, $\mathbb{A}^\times$, consisting of $(x_v)$ with $x_v \in F_v^\times$ for all $v$, $x_v \in R_v^\times$ for almost all $v$, where the multiplicative group $R_v^\times$ of $R_v$ is the group of units, defined by $|x|_v = 1$. Thus $\mathbb{A}^\times = \cup_S \prod_{v\in S} F_v^\times \times \prod_{v\notin S} R_v^\times$. The multiplicative group $F^\times$ embeds diagonally as a discrete subgroup in $\mathbb{A}^\times$, and $\mathbb{A}^1/F^\times$ is compact, where

$\mathbb{A}^1$ consists of the $(x_v)$ in $\mathbb{A}^\times$ with $\prod_v |x_v|_v = 1$. The product formula $\prod_v |x|_v = 1$ for $x$ in $F^\times$ implies that $F^\times \subset \mathbb{A}^1$. Denote by $\boldsymbol{\pi}_v$ a generator of the maximal ideal $R_v - R_v^\times$ of the local ring $R_v$, when $v$ is nonarchimedean. The field $R_v/(\boldsymbol{\pi}_v)$ is finite, of cardinality $q_v$ and residual characteristic $p_v$.

The quotient space $\mathbb{A}^\times/F^\times$ is called the idèle class group and is denoted by $C_F$. When $F$ is a local field put $C_F = F^\times$. For more on valuations, adèle, idèles, see, e.g., Platonov-Rapinchuk [PR].

**Class Field Theory** can be stated as providing a bijection between the set of characters $\chi$ of finite order of the profinite Galois group $\text{Gal}(\overline{F}/F)$ of $F$ ($\overline{F}$ denotes a separable algebraic closure of $F$), and the set of characters $\pi$ of finite order of $C_F$.

When $F$ is global, the bijection is defined as follows.

The decomposition group $D_v$ of $v$, which consists of the $g$ in $\text{Gal}(\overline{F}/F)$ which fix an extension $\overline{v}$ of $v$ to $\overline{F}$, is isomorphic to $\text{Gal}(\overline{F}_v/F_v)$. (In fact $D_v$ depends on $\overline{v}$. Replacing $\overline{v}$ by $\overline{v}'$ leads to a subgroup $D_{\overline{v}'}$ conjugate to $D_{\overline{v}}$. Thus $D_v$ is determined by $v$ only up to conjugacy). Its inertia subgroup $I_v$ consists of the $g \in D_v$ which induce the identity on $R_{\overline{v}}$ modulo its maximal ideal. The quotient group $D_v/I_v$ is $\text{Gal}(\overline{\mathbb{F}}_{q_v}/\mathbb{F}_{q_v})$. Any element of $D_v$ which maps to the generator $x \mapsto x^{q_v}$ of the Galois group of $\mathbb{F}_{q_v}$ is called a *Frobenius* at $v$, denoted $\text{Fr}_v$. Now $\chi$ is unramified at almost all $v$, which means that its restriction to $D_v$ is trivial on $I_v$. It is then determined by its value $\chi(\text{Fr}_v)$ at $\text{Fr}_v$. *Chebotarev's density theorem* asserts that $\chi$ is uniquely determined by $\chi(\text{Fr}_v)$ at almost all $v$.

On the other hand, the character $\pi$ of $\mathbb{A}^\times$ is the product $\otimes_v \pi_v$, where $\pi_v$ is the restriction of $\pi$ to $F_v^\times$ ($F_v^\times$ is embedded in $\mathbb{A}^\times$ as $(x_w)$, $x_w = 1$ if $w \neq v$). Since $\pi$ is continuous, almost all components $\pi_v$ are unramified, namely trivial on $R_v^\times$. Thus they are determined by their value $\pi_v(\boldsymbol{\pi}_v)$ at the generator $\boldsymbol{\pi}_v$ of the maximal ideal $R_v - R_v^\times$ in the local ring $R_v$. By the *Chinese Remainder Theorem* $F^\times \cdot \prod_{v \notin S} F_v^\times$ is dense in $\mathbb{A}^\times$. Hence the character $\pi$ of $\mathbb{A}^\times/F^\times$ is uniquely determined by $\pi_v(\boldsymbol{\pi}_v)$ for almost all $v$.

The bijection of *global Class Field Theory* is $\chi \leftrightarrow \pi$ if $\chi(\text{Fr}_v) = \pi(\boldsymbol{\pi}_v)$ for almost all $v$.

The bijection of local class field theory can be derived from this on embedding a local situation in a global one, thus starting from $\chi_v$ or $\pi_v$ one can construct global $\chi$ and $\pi$ with components $\chi_v$ and $\pi_v$ at $v$, when $\chi_v$ or $\pi_v$ are ramified.

In fact, CFT provides a homomorphism $C_F \to \mathrm{Gal}(\overline{F}/F)^{\mathrm{ab}}$, named the *reciprocity law*, where the maximal abelian quotient $\mathrm{Gal}(\overline{F}/F)^{\mathrm{ab}}$ of $\mathrm{Gal}(\overline{F}/F)$ is the inverse limit of $\mathrm{Gal}(E/F)$ over all abelian extensions $E$ of $F$ in $\overline{F}$ (if $G$ is a topological group, $G^{\mathrm{ab}}$ is its quotient by the closure $G^c$ of its commutator subgroup).

However, in this form the statement is unsatisfactory, as it applies only to characters of finite order, and indeed these are all the continuous characters of the compact, profinite group $\mathrm{Gal}(\overline{F}/F)$. However $\mathbb{A}^\times/F^\times$ is not compact, and has characters of infinite order, e.g. $x \mapsto \|x\| = \prod_v |x_v|_v$. To extend CFT to characters of $C_F$ of any order, Weil introduced the group $W_F$ that we describe next, following Deligne [D2] and Tate [Tt].

To introduce **Weil groups**, note that a *Weil datum* for $\overline{F}/F$, $F$ local or global and $\overline{F}$ a separable algebraic closure, is a triple $(W_F, \varphi, \{r_E\})$. Here $W_F$ is a topological group and $\varphi : W_F \to \mathrm{Gal}(\overline{F}/F)$ is a continuous homomorphism with dense image; $E$ ranges over all finite extensions of $F$ in $\overline{F}$. Put $W_E = \varphi^{-1}(\mathrm{Gal}(\overline{F}/E))$. It is open in $W_F$ for each $E$ since $\varphi$ is continuous and $\{\mathrm{Gal}(\overline{F}/E)\}_E$ makes a basis of the topology of $\mathrm{Gal}(\overline{F}/F)$. As $\mathrm{Im}\,\varphi$ is dense in $\mathrm{Gal}(\overline{F}/F)$, $\varphi$ induces a bijection of homogeneous spaces

$$W_F/W_E \xrightarrow{\sim} \mathrm{Gal}(\overline{F}/F)/\mathrm{Gal}(\overline{F}/E) \simeq \mathrm{Hom}_F(E, \overline{F})$$

for each $E$, and a group isomorphism $W_F/W_E \xrightarrow{\sim} \mathrm{Gal}(E/F)$ when $E/F$ is Galois. The $r_E : C_E \xrightarrow{\sim} W_E^{\mathrm{ab}}$ are isomorphisms. A Weil datum is called a *Weil group* if

$(W_1)$ For each $E$, the composition $\quad C_E \overset{r_E}{\underset{\sim}{\to}} W_E^{\mathrm{ab}} \xrightarrow{\varphi} \mathrm{Gal}(\overline{F}/E)^{\mathrm{ab}} \quad$ is the reciprocity law homomorphism of CFT.

$(W_2)$ For each $w \in W_F$ and any $E$, commutative is the square

$$\begin{array}{ccc} C_E & \xrightarrow{r_E} & W_E^{\mathrm{ab}} \\ {\scriptstyle \varphi(w)}\downarrow & & \downarrow{\scriptstyle \mathrm{Int}(w)} \\ C_{\varphi(w)E} & \xrightarrow[r_{\varphi(w)E}]{} & W_{\varphi(w)E}^{\mathrm{ab}} \end{array}.$$

$(W_3)$ If $E' \subset E$, commutative is the square

$$\begin{array}{ccc} C_{E'} & \xrightarrow{r_{E'}} & W_{E'}^{\text{ab}} \\ {\scriptstyle E'\subset E}\big\downarrow & & \big\downarrow{\scriptstyle \text{tr}} \\ C_E & \xrightarrow[r_E]{} & W_E^{\text{ab}}. \end{array}$$

The transfer map on the right is defined as follows. Suppose $H$ is a closed subgroup of finite index in a topological group $G$, $s : H\backslash G \to G$ a section. For any $g \in G$, $x \in H\backslash G$, define $h_{g,x} \in H$ by $s(x)g = h_{g,x}s(xg)$, and $\text{tr}(gG^c) = \prod_{x\in H\backslash G} h_{g,x} (\text{mod}\, H^c)$. Then $\text{tr} : G^{\text{ab}} \to H^{\text{ab}}$ is a homomorphism.

$(W_4)$ Put $W_{E/F}$ for $W_F/W_E^c$. The natural map $W_F \to \varprojlim W_{E/F}$ is an isomorphism of topological groups.

It follows that if $(W_F, \varphi, \{r_E\})$ is a Weil group for $\overline{F}/F$ and $E$ is a finite extension of $F$ in $\overline{F}$, then $(W_E, \varphi|W_E, \{r_{E_1}\}_{E_1\supset E})$ is a Weil group for $\overline{F}/E$. We usually abbreviate the triple to $W_F$. Note that via $r_E$, the norm $N_{E/E_1} : C_E \to C_{E_1}$ (for $F \subset E_1 \subset E$) becomes the map $W_E^{\text{ab}} \to W_{E_1}^{\text{ab}}$ induced by the inclusion $W_E \subset W_{E_1}$. Note also the exactness of

$$1 \to C_E \to W_{E/F} \to \text{Gal}(E/F) \to 1$$

whenever $E/F$ is a Galois extension.

When $F$ is local archimedean, if $F = \mathbb{C}$ we take $W_F = \mathbb{C}^\times$, $\varphi : \mathbb{C}^\times \to \{1\}$, $r_F = \text{id}$.

If $F = \mathbb{R}$ we take $W_\mathbb{R}$ to be the subgroup $\mathbb{C}^\times \cup j\mathbb{C}^\times$ of $\mathbb{H}^\times$, where $\mathbb{H}$ is the Hamiltonian quaternions. It is $\langle z, j; z \in \mathbb{C}^\times, j^2 = -1, jz = \overline{z}j\rangle$ where $\overline{z}$ is the complex conjugate of $z$. Then $\varphi : W_\mathbb{R} \to \text{Gal}(\mathbb{C}/\mathbb{R})$ takes $\mathbb{C}^\times$ to 1 and $j\mathbb{C}^\times$ to the nontrivial element in $\text{Gal}(\mathbb{C}/\mathbb{R})$. Further $r_\mathbb{C} = 1$ and $r_\mathbb{R} : \mathbb{R}^\times \to W_\mathbb{R}^{\text{ab}}$ is $x \mapsto \sqrt{x}W_\mathbb{R}^c$ if $x > 0$, and $-1 \mapsto jW_\mathbb{R}^c$, where $W_\mathbb{R}^c$ is the unit circle $\mathbb{C}^1 = \{z/\overline{z}; z \in \mathbb{C}^\times\} = \ker N_{\mathbb{C}/\mathbb{R}}$. The norm map $N : \mathbb{H}^\times \to \mathbb{R}_{>0}^\times$ induces a norm $z_1 + jz_2 \mapsto z_1\overline{z}_1 + z_2\overline{z}_2$ on $W_\mathbb{R}$.

When $F$ is local nonarchimedean, for each finite extension $E$ of $F$ in $\overline{F}$ let $k_E = R_E/(\boldsymbol{\pi}_E)$ be the residual field of $E$ and $q_E$ its cardinality. Put $\overline{k} = \cup_E k_E$ and $k = k_F$. Then

$$1 \to I_F \to \text{Gal}(\overline{F}/F) \to \text{Gal}(\overline{k}/k) \to 0,$$

where $I_F$ is the inertia subgroup, consisting of the $\sigma \in \mathrm{Gal}(\overline{F}/F)$ fixing $\overline{k}$. The Galois group $\mathrm{Gal}(\overline{k}/k)$ is the profinite group $\widehat{\mathbb{Z}} = \varprojlim \mathbb{Z}/n\mathbb{Z}$, topologically generated by $x \mapsto x^{q_F}$. Any element of $\mathrm{Gal}(\overline{F}/F)$ which maps to this generator is called *Frobenius* and denoted by $\mathrm{Fr}_F$. Then $W_F$ is the dense subgroup of $\mathrm{Gal}(\overline{F}/F)$ generated by the Frobenii. Thus the sequence $1 \to I_F \to W_F \to \mathbb{Z} \to 1$ is exact. The subgroup $I_F$ is a profinite subgroup of $\mathrm{Gal}(\overline{F}/F)$, and open in $W_F$, making $W_F$ a topological group. Then $\varphi : W_F \to \mathrm{Gal}(\overline{F}/F)$ is the inclusion and $r_E : E^\times \to W_E^{\mathrm{ab}}$ are the reciprocity law homomorphisms, $r_E(a)$ acts as $x \mapsto x^{|a|_E}$ on $\overline{k}$, the valuation being normalized by $|\boldsymbol{\pi}_E|_E = q_E^{-1}$.

When $F$ is a global function field the situation is similar to the previous case, with "residual field" replaced by "constant field", "inertia group $I_F$" by "geometric Galois group $\mathrm{Gal}(\overline{F}/F\overline{k})$", and the absolute value $|a|_E$ of the idèle class $a = (a_v) \in C_E$ is $\prod_v |a_v|_v$.

When $F$ is a number field Weil gave an abstract, cohomological construction of $W_F$, and asked for a natural construction. He showed that $\varphi : W_F \to \mathrm{Gal}(\overline{F}/F)$ is onto. Its kernel is the connected component of the identity in $W_F$.

The isomorphism $r_F : C_F \xrightarrow{\sim} W_F^{\mathrm{ab}}$ and the absolute value $C_F \to \mathbb{R}_{>0}^\times$, $x = (x_v) \mapsto |x|_F = \prod_v |x_v|_v$, define the norm $W_F \to \mathbb{R}_{>0}^\times$, $w \mapsto |w|$. Since $W_E^{\mathrm{ab}} \subset W_F^{\mathrm{ab}}$ corresponds via $r_E$ and $r_F$ to $N_{E/F} : C_E \to C_F$ and $|N_{E/F}a|_F = |a|_E$, the restriction of $W_F \to \mathbb{R}_{>0}^\times$ to $W_E$ coincides with the norm $W_E \to \mathbb{R}_{>0}^\times$, and we write simply $|w|$ instead of $|w|_F$. The kernel $W_F^1$ of $w \mapsto |w|$ is compact. The image of $w \mapsto |w|$ is $q_F^{\mathbb{Z}}$ and $W_F$ is $W_F^1 \rtimes \mathbb{Z}$ in the nonarchimedean and function field cases, while in the archimedean and number field cases the image is $\mathbb{R}_{>0}^\times$ and $W_F$ is $W_F^1 \times \mathbb{R}_{>0}^\times$.

Finally there are commutative squares of local-to-global maps, for each $v$,

$$\begin{array}{ccc} W_{F_v} & \longrightarrow & \mathrm{Gal}(\overline{F}_v/F_v) \\ \downarrow & & \downarrow \\ W_F & \longrightarrow & \mathrm{Gal}(\overline{F}/F). \end{array}$$

Class Field Theory, which asserts that $W_F^{\mathrm{ab}} \simeq C_F$, can then be phrased as an isomorphism between the set of continuous, complex valued characters of $W_F$, and the set of continuous, complex valued characters of $C_F$ ($= \mathbb{A}^\times/F^\times$ globally, $F^\times$ locally). One is interested in all finite dimensional (continuous, over $\mathbb{C}$) representations of the Weil group $W_F$, as by the Tannakian formalism these determine $W_F$ itself as the "motivic Galois group" of their category. The **hypothetical reciprocity law** would associate to an irreducible $n$-dimensional representation $\lambda : W_F \to \mathrm{GL}(n, \mathbb{C})$ a cuspidal representation $\pi$ of $\mathrm{GL}(n, \mathbb{A})$ if $F$ is global and of $\mathrm{GL}(n, F)$ if $F$ is local, and to $\oplus_{i=1}^r \lambda_i$ the representation $I_P(\pi_1, \ldots, \pi_r)$ normalizedly induced from the cuspidal representation $\pi_1 \otimes \cdots \otimes \pi_r$ of the parabolic $P$ (trivial on its unipotent radical) of type $(\dim \lambda_1, \ldots, \dim \lambda_r)$, where $\lambda_i \mapsto \pi_i$. We postpone the explanation of the new terms, but note that this new correspondence is defined similarly to the case of CFT, which is that of $n = 1$. The local analogue has recently been proven (by Harris-Taylor [HT], Henniart [He]) in the nonarchimedean case, and by Lafforgue [Lf] in the function field case.

Once the connection between $n$-dimensional representations of $W_F$ and admissible (locally) or automorphic (globally) representations is accepted, one would like to include all admissible and automorphic representations. For that the group $W_F$ has to be replaced by a bigger group, which is the *Weil-Deligne group* $W_F \times \mathrm{SU}(2, \mathbb{R})$, an extension of $W_F$ by a compact group (see [D2], [Tt], Kazhdan-Lusztig [KL], and Kottwitz [Ko2], §12) when $F$ is nonarchimedean (when $F$ is archimedean the group remains $W_F$). This is necessary for inclusion of the square integrable but noncuspidal representations of $\mathrm{GL}(n, F)$ in the reciprocity law. The representations $\lambda$ of $W_F \times \mathrm{SU}(2, \mathbb{R})$ of interest are analytic in the second variable, thus extend to $\mathrm{SL}(2, \mathbb{C})$. We embed $W_F$ in $W_F \times \mathrm{SL}(2, \mathbb{C})$ by $w \mapsto w \times \mathrm{diag}(|w|^{1/2}, |w|^{-1/2})$, where $|.| : W_F \to F^\times \to \mathbb{C}^\times$ is the composition of the usual absolute value with $W_F^{\mathrm{ab}} \simeq F^\times$.

For example, the Steinberg representation of $\mathrm{GL}(n, F)$ is parametrized by the homomorphism $\lambda$ which is trivial on $W_F$ while its restriction to $\mathrm{SL}(2, \mathbb{C})$ is the irreducible $n$-dimensional representation $\mathrm{Sym}^{n-1}$; it maps $\mathrm{diag}(a, a^{-1})$ to $\mathrm{diag}(a^{(n-1)/2}, a^{(n-3)/2}, \ldots, a^{-(n-1)/2})$ and $\left(\begin{smallmatrix} 1 & 1 \\ 0 & 1 \end{smallmatrix}\right)$ to the regular unipotent matrix $\exp((\delta_{(i,i+1)}))$. The nontempered trivial representation of $\mathrm{GL}(n, F)$ is the quotient of the normalizedly induced representation $I(\mu_1, \ldots, \mu_n)$ of $\mathrm{GL}(n, F)$, with $\mu_i = \nu^{\frac{n+1}{2} - i}$, $\nu(x) = |x|$, while the square integrable Steinberg is a subrepresentation. The quotient is parametrized by $\lambda$ trivial on the second factor, $\mathrm{SU}(2, \mathbb{R})$, and with $\lambda(w) = \mathrm{diag}(\mu_1(w), \ldots, \mu_n(w))$, $w \in W_F$.

If the reciprocity law holds, the category of the representations $\pi$ should be Tannakian, with addition $\pi_1 \boxplus \cdots \boxplus \pi_r$ being normalized induction $I_P(\pi_1, \ldots, \pi_r)$, and multiplication $\pi_1 \boxtimes \cdots \boxtimes \pi_r$, and fiber functor. At least when considering only those representations formed by twisting tempered representations, and assuming the *Ramanujan conjecture* ("cuspidal representations of $\mathrm{GL}(n, \mathbb{A})$ are tempered, i.e. all their components are tempered"), if the category is Tannakian, its motivic Galois group is expected to be the correct substitute for $W_F$, for which the reciprocity law holds. This hypothetical group is denoted $L_F$, named the "*Langlands group*". We often write $W_F$ below for what would one day be $L_F$. See Arthur [A5] for a proposed construction.

# 2. Reductive Groups

Since progress on the global reciprocity law for $\mathrm{GL}(n)$ is not expected soon, one looks for a **generalization to the context of any reductive connected $F$-group** $G$. This is not a generalization for its own sake, as it leads to two practical developments. The first is the theory of *liftings* of representations of one group to another. Reflecting simple relations of representations of Galois or Weil groups, one is led to deep relations of automorphic and admissible representations on different groups.

The second is the use of *Shimura varieties* (see [D5]) to actually prove parts of the global reciprocity law for groups which define Shimura varieties (symplectic, orthogonal and unitary groups, but not $\mathrm{GL}(n)$ and its inner forms if $n > 2$), and for "cohomological" representations, whose components at the archimedean places are discrete series or nontempered representations with cohomology $\neq 0$.

The reciprocity law for $G$ is stated in terms of the *Langlands dual group* ${}^LG = \widehat{G} \rtimes W_F$, where $\widehat{G}$ is the connected component of the identity of ${}^LG$, a complex group, and $W_F$ acts via its image in $\mathrm{Gal}(\overline{F}/F)$. The law relates homomorphisms $\lambda : W_F \to {}^LG$ whose composition with the projection to $W_F$ is the identity, with admissible and automorphic representations of $G(F)$ or $G(\mathbb{A})$, in fact with packets of such representations. It was proven by Langlands [L7] for archimedean local fields, as part of his classification of admissible representations of real reductive connected groups, and for tori in [L8]. Globally it is compatible with the theory of Eisenstein series [L3], [MW2]. For unramified representations of a $p$-adic group it coincides with the theory of the Satake transform, and for representations with a nonzero Iwahori fixed vector it was proven by Kazhdan and Lusztig [KL]. These results, and those on liftings and cohomology of Shimura varieties, in addition to the local and function field results for $\mathrm{GL}(n)$, give some hope that the reciprocity law is indeed valid. Of course, a final form of this law will be stated with $L_F$ replacing the Weil group $W_F$, once $L_F$ is defined.

We proceed to review the definition of the **connected dual group** $\widehat{G}$ and the $L$-group ${}^LG$, following Langlands [L1], Borel [Bo1], Kottwitz [Ko2], §1.

Books on linear algebraic groups include Borel [Bo2], Humphreys [Hu], Springer [Sp].

Associated with a torus $T$ defined over $F$ is the characters lattice $X^*(T) = \mathrm{Hom}(T, \mathbb{G}_m)$ and the lattice $X_*(T) = \mathrm{Hom}(\mathbb{G}_m, T)$ of 1-parameter subgroups, or cocharacters. These are free abelian groups, dual in the pairing

$$\langle .,. \rangle : X_*(T) \times X^*(T) \to \mathbb{Z} = \mathrm{Hom}(\mathbb{G}_m, \mathbb{G}_m).$$

The connected dual group of $T$ is the complex torus $\widehat{T} = \mathrm{Hom}(X_*(T), \mathbb{C}^\times)$. Then $X^*(\widehat{T}) = X_*(T)$, and by duality $X_*(\widehat{T}) = X^*(T)$. Thus $T \mapsto \widehat{T}$ interchanges $X_*$ and $X^*$. As $T$ is defined over $F$, $\mathrm{Gal}(\overline{F}/F)$ acts on $X_*(T)$, hence on $\widehat{T}$. An action of $\mathrm{Gal}(\overline{F}/F)$ on $\widehat{T}$, or ${}^LT = \widehat{T} \rtimes W_F$, determines $T$ as an $F$-torus (up to isomorphism), since the $F$-isomorphism class of $T$ is determined by the $\mathrm{Gal}(\overline{F}/F)$-module $X_*(T)(= X^*(\widehat{T}))$. The $\mathrm{Gal}(\overline{F}/F)$-action is trivial iff $T$ is an $F$-split torus.

Let $X$, $X^\vee$ be free $\mathbb{Z}$-modules of finite rank, dual in a $\mathbb{Z}$-valued pairing $< .,. >$. Suppose $\nabla \subset X$, $\nabla^\vee \subset X^\vee$ are finite subsets and $\alpha \mapsto \alpha^\vee$, $\nabla \to \nabla^\vee$, is a bijection with $< \alpha, \alpha^\vee > = 2$. The 4-tuple $(X, \nabla, X^\vee, \nabla^\vee)$ is a *root datum* if the reflection $s_\alpha(x) = x - \langle x, \alpha^\vee \rangle \alpha$ $(x \in X)$ maps $\nabla$ to itself, and $s_{\alpha^\vee}(y) = y - \langle \alpha, y \rangle \alpha^\vee$ $(y \in X^\vee)$ maps $\nabla^\vee$ to itself. Then $\nabla$ is the set of *roots* and $\nabla^\vee$ the set of *coroots*. The root datum is called *reduced* if $\alpha$ and $n\alpha$ in $\nabla$ $(n \in \mathbb{Z})$ implies that $n = \pm 1$. The set $\nabla$ defines a root system in a subspace of the the vector space $X \otimes \mathbb{R}$. Thus one has the notions of positive roots and simple roots. If $\Delta = \{\alpha\}$ is a set of simple roots, put $\Delta^\vee = \{\alpha^\vee\}$. The 4-tuple $\Psi = (X, \Delta, X^\vee, \Delta^\vee)$ is called a *based root datum* (it determines the root datum). The *dual* based root datum is $\Psi^\vee = (X^\vee, \Delta^\vee, X, \Delta)$, and the dual root datum is $(X^\vee, \nabla^\vee, X, \nabla)$.

A *Borel pair* $(B,T)$ of a reductive connected $F$-group $G$ is a maximal torus $T$ of $G$ and a Borel subgroup $B$ of $G$ containing $T$, both defined over $\overline{F}$. If $G$ has a Borel pair defined over $F$, it is called *quasisplit*. It is *split* if there is such a pair with $T$ split over $F$. Any pair $(B,T)$ defines a reduced root datum $\Psi(G,B,T) = (X^*(T), \Delta, X_*(T), \Delta^\vee)$. Here $\Delta = \Delta(B,T) \subset X^*(T)$ is the set of simple roots of $T$ in $B$, and $\Delta^\vee = \Delta^\vee(B,T) \subset X_*(T)$ is the set of coroots dual to $\Delta$. Any two Borel pairs are conjugate under the adjoint group $G^{\mathrm{ad}} = G/Z(G)$ of $G$ (here $Z(G)$ denotes the center of $G$). If $\mathrm{Int}(g)$ $(x \mapsto gxg^{-1})$ maps $(B,T)$ to $(B',T')$, it defines an isomorphism $\Psi(G,B,T) \xrightarrow{\sim} \Psi(G,B',T')$, independent of $g$. Using this, we identify the based root data, to get $\Psi(G)$. Then $\mathrm{Aut}(G)$ acts on $\Psi(G)$, with $G^{\mathrm{ad}}$ acting trivially.

A *connected dual group* for $G$ is a complex connected reductive group $\widehat{G}$ with an isomorphism $\Psi(\widehat{G}) \xrightarrow{\sim} \Psi(G)^\vee$.

The map $G \mapsto \Psi(G)$ defines a bijection from the set of $\overline{F}$-isomorphism classes of connected reductive groups $G$ to the set of isomorphism classes of reduced based root data $\Psi$. An isomorphism $G_1 \xrightarrow{\sim} G_2$ determines an isomorphism $\Psi(G_1) \xrightarrow{\sim} \Psi(G_2)$, which in turn determines $G_1 \xrightarrow{\sim} G_2$ up to an inner automorphism.

This classification theorem implies that a connected dual group $\widehat{G}$ of $G$ exists and is unique up to an inner automorphism. It depends only on the $\overline{F}$-isomorphism class of $G$.

If $(B,T)$ is a Borel pair for $G$ and $(\widehat{B}, \widehat{S})$ is a Borel pair for $\widehat{G}$, there exists a unique isomorphism $\widehat{T}$ (defined from $T$) $\to \widehat{S}$ inducing the chosen isomorphism

$$\Psi(\widehat{G}) = (X^*(\widehat{S}), \widehat{\Delta}, X_*(\widehat{S}), \widehat{\Delta}^\vee) \xrightarrow{\sim}$$

$$\Psi(G)^\vee = (X_*(T) = X^*(\widehat{T}), \Delta^\vee, X^*(T) = X_*(\widehat{T}), \Delta).$$

If $f : G \to G'$ is a *normal* morphism (its image is a normal subgroup), and $(B,T)$ is a Borel pair in $G$, there exists a Borel pair $(B',T')$ in $G'$ with $f(B) \subset B'$, $f(T) \subset T'$. Hence there is a map $\Psi(f) : \Psi(G) \to \Psi(G')$ and a dual map $\Psi^\vee(f) : \Psi(G')^\vee \to \Psi(G)^\vee$, and so a map $\widehat{f} : \widehat{G}' \to \widehat{G}$. Any other such map has the form $\mathrm{Int}(t) \cdot \widehat{f} \cdot \mathrm{Int}(t')$ $(t \in \widehat{T},\ t' \in \widehat{T}')$, mapping $\widehat{T}'$ to $\widehat{T}$, $\widehat{B}'$ to $\widehat{B}$.

The simplest **example** is that of $G = \mathrm{GL}(n)$. Then $X^*(T) = \mathbb{Z}^n$ has the standard basis $\{e_i; 1 \le i \le n\}$, and $X_*(T) = \mathbb{Z}^n$ the dual basis $\{e_i^\vee(= e_i)\}$. Also $\Delta = \{e_i - e_{i+1}; 1 \le i < n\}$ and $\Delta^\vee = \{e_i^\vee - e_{i+1}^\vee; 1 \le i < n\}$. Then $\Psi(G) = \Psi(G)^\vee$ and $\widehat{G} = \mathrm{GL}(n, \mathbb{C})$.

A more complicated example is $G = \mathrm{PGSp}(2) = \{g \in \mathrm{GL}(4),\ {}^tgJg = \lambda J\}/\mathbb{G}_m$, $J = \left(\begin{smallmatrix} 0 & w \\ -w & 0 \end{smallmatrix}\right)$, $w = \left(\begin{smallmatrix} 0 & 1 \\ 1 & 0 \end{smallmatrix}\right)$, the projective symplectic group of similitudes of rank 2. With the form $J$, a Borel subgroup $B$ is the upper triangular matrices, and a maximal torus $T$ is the diagonal subgroup. The simple roots in $X^*(T) = \mathbb{Z}^2$ are $\alpha = e_1 - e_2$, $\beta = 2e_2$, and the other positive roots are $\alpha + \beta = e_1 + e_2$, $2\alpha + \beta = 2e_1$. Then $\Delta^\vee = \{\alpha^\vee = e_1 - e_2,\ \beta^\vee = e_2\}$. The isomorphism from the lattice

$$X^*(\widehat{T}) = \{(x, y, z, t) \,\mathrm{mod}(n, m, m, n); x, y, z, t \in \mathbb{Z}\},$$

where $(x, y, z, t)$ takes $\mathrm{diag}(a, b, b^{-1}, a^{-1})$ in $\widehat{T}$ to $a^{x-t}b^{y-z}$, to the lattice

$$X_*(T) = \{(\alpha, \beta, \gamma, \delta) \,\mathrm{mod}(\epsilon, \epsilon, \epsilon, \epsilon); \alpha + \delta = \beta + \gamma,\ \alpha, \beta, \gamma, \delta, \epsilon \in \mathbb{Z}\},$$

is given by $(x, y, z, t) \mapsto (x + y, x + z, y + t, z + t)$, with inverse

$$\iota : (\alpha, \beta, \gamma, \delta) \mapsto (\alpha - \gamma, \alpha - \beta, 0, 0).$$

The isomorphism $\iota^* : X_*(\widehat{T}) \xrightarrow{\sim} X^*(T)$ dual to $\iota : X_*(T) \xrightarrow{\sim} X^*(\widehat{T})$ is defined by $\langle \iota(u), v \rangle = \langle u, \iota^*(v) \rangle$. Thus $v = (a, b, -b, -a) \in X_*(\widehat{T})$ maps to the character

$$\iota^*(v) : \mathrm{diag}(\alpha, \beta, \gamma, \delta) \mapsto (\alpha/\gamma)^a (\alpha/\beta)^b$$

of $T$. The character $\eta : \mathrm{diag}(\alpha, \beta, \gamma, \delta) \mapsto \mu_1(\alpha/\gamma)\mu_2(\alpha/\beta)$ of $T$ corresponds to the homomorphism

$$z (\in W_F^{\mathrm{ab}}) \mapsto \mathrm{diag}(\mu_1(z), \mu_2(z), \mu_2(z)^{-1}, \mu_1(z)^{-1}) \quad (\in \widehat{T}).$$

By an *isogeny* we mean a surjective homomorphism $f : G \to G'$ of algebraic groups whose kernel is finite and central (in $G$). The finite kernel is always central if char $F = 0$, and if char $F > 0$ our $f$ is usually named central isogeny.

A connected (linear algebraic) group is called *reductive* if its *unipotent radical* (the maximal connected unipotent normal subgroup) is trivial, and *semisimple* if its *radical* (replace "unipotent" by "solvable" in the definition of the unipotent radical) is trivial. A semisimple group $G$ is called *simply connected* if every isogeny $f : H \to G$, where $H$ is connected reductive (as is $G$), is an isomorphism, and *adjoint* if every such $f : G \to H$ is an isomorphism. The *adjoint group* of a reductive $G$ is $G^{\mathrm{ad}} = G/Z(G)$, where $Z(G)$ is the center of $G$. This $G^{\mathrm{ad}}$ is adjoint. The derived group $G^{\mathrm{der}}$ of a reductive $G$ (the closure of the subgroup generated by commutators $[x, y] = xyx^{-1}y^{-1}$) is semisimple, denoted also by $G^{\mathrm{ss}}$.

Let $G$ be semisimple and $\Psi(G) = (X, \Delta, X^\vee, \Delta^\vee)$, $\nabla$ the root system with basis $\Delta$ and $\nabla^\vee$ the root system with basis $\Delta^\vee$. Then $G$ is simply connected iff the lattice of weights $P(\nabla) \subset X \otimes \mathbb{Q}$ of $\nabla$ is $X$, and adjoint iff the group $Q(\nabla)$ generated by $\nabla$ in $X$ is $X$. Since

$$P(\nabla) = \{\Lambda \in X \otimes Q; \langle \Lambda, \nabla^\vee \rangle \subset \mathbb{Z}\}$$

and

$$P(\nabla^\vee) = \{\Lambda \in X^\vee \otimes \mathbb{Q}; \langle \Lambda, \nabla \rangle \in \mathbb{Z}\},$$

$G$ is simply connected iff $\widehat{G}$ is adjoint, $G$ is adjoint iff $\widehat{G}$ is simply connected.

A *simple group* $G$ (one which has no nontrivial connected normal subgroup) is characterized – up to isogeny – by its type $A_n, \dots, G_2$. The map $\Psi(G) \to \Psi(G)^\vee$ interchanges $B_n$ with $C_n$, and fixes all other types. Thus the connected dual of a simple group is a simple group of the same type unless $G$ is of type $B_n$ or $C_n$, and duality changes simply connected to adjoint. The classical simply connected simple groups and their duals are in type $A_n$ : $\mathrm{SL}(n)$, $\mathrm{PGL}(n, \mathbb{C})$; $B_n$ : $\mathrm{Spin}(2n+1)$, $\mathrm{PGSp}(2n, \mathbb{C})$; $C_n$ : $\mathrm{Sp}(2n)$, $\mathrm{SO}(2n+1, \mathbb{C})$; $D_n$ : $\mathrm{Spin}(2n)$, $\mathrm{PO}(2n, \mathbb{C})$.

The *dual group*, or $L$-group, ${}^L G = \widehat{G} \rtimes W_F$, is the semidirect product of the connected dual group $\widehat{G}$ with the Weil group $W_F$, which acts on $\widehat{G}$ via its image (by $\varphi$) in $\mathrm{Gal}(\overline{F}/F)$.

To explain how $\mathrm{Gal}(\overline{F}/F)$ acts, note that we have a split exact sequence

$$1 \to \mathrm{Inn}\, G \to \mathrm{Aut}\, G \to \mathrm{Out}\, G \to 1,$$

where $\mathrm{Inn}\, G = \mathrm{Int}\, G \simeq G^{\mathrm{ad}}$ is the subgroup of inner automorphisms of $G$, and the group $\mathrm{Out}\, G = \mathrm{Aut}\, G/\mathrm{Inn}\, G$ of outer automorphisms is isomorphic to the group $\mathrm{Aut}\, \Psi(G) = \mathrm{Aut}\, \Psi(\widehat{G})$ of automorphisms of $\Psi(G)$ (or $\Psi(\widehat{G})$).

A *splitting* for $\widehat{G}$ is a triple $\Sigma = (\widehat{B}, \widehat{T}, \{X_\alpha; \alpha \in \widehat{\Delta}\})$, where $X_\alpha$ is an $\alpha$-root vector in Lie $\widehat{G}$ for each simple root $\alpha$ of $\widehat{T}$ in $\widehat{B}$. The set of splittings is a principal homogeneous space for (the action of) $G^{\mathrm{ad}}$ (by conjugation). A choice of a splitting $\Sigma$ determines a splitting $\operatorname{Aut}\Psi(\widehat{G}) \to \operatorname{Aut} G$ of our exact sequence: an element of $\operatorname{Aut}\Psi(\widehat{G})$ maps to the unique automorphism of $G$ fixing $\Sigma$.

The action of $\mathrm{Gal}(\overline{F}/F)$ on $\Psi(G)^\vee = \Psi(\widehat{G})$ then lifts to an action on $\widehat{G}$ which fixes the fixed splitting $\Sigma$. The $L$-group ${}^LG = \widehat{G} \rtimes W_F$ depends on the choice of $\Sigma$, but a different choice gives rise to an isomorphic $L$-group.

If $v$ is a place of the number field $F$ there is a natural $W_F$-conjugacy class of embeddings $W_{F_v} \hookrightarrow W_F$, hence such a class of embeddings ${}^LG/F_v \hookrightarrow {}^LG/F$ which restrict to the identity $\widehat{G} \to \widehat{G}$.

# 3. Functoriality

The purpose of the principle of functoriality is to parametrize the admissible representations of $G(F)$ in the local case, and automorphic representations of $G(\mathbb{A})$ in the global case, in terms of $L$-**parameters**. These are the (continuous) homomorphisms $\lambda : L_F \to {}^LG = \widehat{G} \rtimes W_F$, where $L_F$ is $W_F$ if $F$ is archimedean and $W_F \times \mathrm{SU}(2,\mathbb{R})$ if $F$ is nonarchimedean, such that $\lambda$ followed by the projection to $W_F$ is the natural map $L_F \to W_F$, $\mathrm{pr}_{\widehat{G}} \circ \lambda$ is complex analytic if $F$ is archimedean, and $\mathrm{pr}_{\widehat{G}}(\lambda(w))$ is semisimple for all $w$ in $L_F$.

Two parameters $\lambda$, $\lambda'$ are called *equivalent* if $z \cdot \lambda' = \mathrm{Int}(g)\lambda$ for some $g$ in $\widehat{G}$ and $z : L_F \to Z(\widehat{G})$ such that the class of the 1-cocycle $z$ in $H^1(L_F, Z(\widehat{G}))$ is locally trivial.

If $\mathrm{Gal}(\overline{F}/F)$ acts trivially on the center $Z(\widehat{G})$ of $\widehat{G}$ then

$$H^1(L_F, Z(\widehat{G})) = \mathrm{Hom}(L_F, Z(\widehat{G})).$$

In this case Chebotarev density theorem for $L_F^{\mathrm{ab}} = W_F^{\mathrm{ab}}$ implies that any locally trivial element of $H^1(L_F, Z(\widehat{G}))$ is trivial. Thus $\lambda(w) = \phi(w) \times \delta(w)$ where $\delta$ denotes the projection $L_F \to W_F$ followed by $\varphi : W_F \to \mathrm{Gal}(\overline{F}/F)$, and $\phi$ is a (continuous) 1-cocycle of $L_F$ in $\widehat{G}$. The cocycle $\lambda'(w) = \phi'(w) \times \delta(w)$ is equivalent to $\lambda$ iff $\phi$ and $\phi'$ are cohomologous. Hence *the set of equivalence classes*, denoted $\Lambda(G/F)$, is the quotient of the group $H^1(L_F, \widehat{G})$ of (continuous) cohomology classes by $\ker[H^1(L_F, Z(\widehat{G})) \to \oplus_v H^1(L_{F_v}, Z(\widehat{G}))]$.

**Functoriality for tori** $T$ over $F$ concerns then (continuous) homomorphisms $\lambda : W_F \to {}^L T = \widehat{T} \rtimes W_F$ with $\mathrm{pr}_{W_F} \circ \lambda = \mathrm{id}_{W_F}$, thus $\lambda$ which factorize through the projection $L_F \to W_F$. Langlands [L8] shows that when $F$ is local, $H^1(W_F, \widehat{T})$ is canonically isomorphic to the group of characters of $T(F) = \mathrm{Hom}_{\mathrm{Gal}(E/F)}(X_*(T), E^\times)$, where $E$ is a finite Galois extension of $F$ over which $T$ splits. If $F$ is global the group of characters of $T(\mathbb{A}_F)/T(F)$ is the quotient of $H^1(W_F, \widehat{T})$ by the kernel of the localization maps

$$\ker[H^1(W_F, \widehat{T}) \to \oplus_v H^1(W_{F_v}, \widehat{T})].$$

Let $\pi : G(F) \to \mathrm{Aut}\, V$ be a representation (which simply means a homomorphism) of the group $G(F)$ of $F$-points of the connected reductive $F$-group $G$, on a complex vector space $V$. In other words, $V$ is a $G(F)$-module. If $F$ is a nonarchimedean local field, $\pi$ is called *algebraic* (Bernstein-Zelevinsky [BZ1]) or *smooth* if for each vector $v$ in $V$ there is an open subgroup $U$ of $G(F)$ which fixes $v$ (thus $\pi(U)v = v$). Such $\pi$ is called **admissible** if moreover, for every open subgroup $U$ of $G(F)$ the space $V^U$ of $U$-fixed vectors in $V$ is finite dimensional. Admissible representations $(\pi_1, V_1)$ and $(\pi_2, V_2)$ are *equivalent* if there exists a vector space isomorphism $A : V_1 \to V_2$ intertwining $\pi_1$ and $\pi_2$, thus $A(\pi_1(g)v) = \pi_2(g)Av$.

In the next few paragraphs we abbreviate $G$ for $G(F)$ (same for a parabolic subgroup $P$, its unipotent radical $N$, its Levi factor $M$), where $F$ is a local field. Put $\delta_P(p) = |\det(\mathrm{Ad}(p)|\,\mathrm{Lie}\, N)|$ for $p \in P$.

A useful construction in module theory is that of *induction*. Let $(\tau, W)$ be an admissible $M$-module. Denote by $\pi = I(\tau) = I(\tau; G, P)$ the space of all functions $f : G \to W$ with $f(nmg) = \delta_P^{1/2}(m)\tau(m)f(g)$ $(m \in M$, $n \in N$, $g \in G)$. It is viewed as a $G$-module by $(\pi(g)f)(h) = f(hg)$.

Another useful construction is that of the *module* $\pi_N$ *of* $N$*-coinvariants* of an admissible $G$-module $\pi$. Thus if $V$ denotes the space of $\pi$, put $'V_N$ for $V/\langle \pi(n)v - v; n \in N, v \in V\rangle$. Since the Levi factor $M = P/N$ of $P$ normalizes $N$, $'V_N$ is an $M$-module, with action $'\pi_N$. Put $\pi_N = \delta_P^{-1/2}\, '\pi_N$. The functor $\pi \mapsto \pi_N$ of $N$-coinvariants is exact and left-adjoint to the exact functor of induction. Indeed, this is the content of Frobenius reciprocity ([BZ1], 3.13): $\mathrm{Hom}_M(\pi_N, \tau) = \mathrm{Hom}_G(\pi, \mathrm{Ind}(\tau; G, P))$. Let $\overline{N}$ be the unipotent radical of the parabolic subgroup $\overline{P}$ opposite to $P$

(thus $P \cap \overline{P} = M$). Unpublished lecture notes of J. Bernstein show that the functor $\pi \mapsto \pi_{\overline{N}}$ is right adjoint to induction $\tau \mapsto I(\tau; G, P)$, namely $\mathrm{Hom}_G(\mathrm{Ind}(\tau; G, P), \pi) = \mathrm{Hom}_M(\tau, \pi_{\overline{N}})$.

An irreducible admissible representation $\pi$ is called *cuspidal* if $\pi_N$ is zero for all proper $F$-parabolic subgroups $P$ of $G$. Related notions are of square integrability and temperedness. Thus $\pi$ is *square integrable*, or *discrete series*, if its central exponents decay. A $\pi$ is *tempered* if its central exponents are bounded. A cuspidal $\pi$ is square integrable. A square integrable $\pi$ is tempered. Cuspidal $\pi$ exist only for $p$-adic $F$. Langlands classification parametrizes all irreducible $\pi$ as unique quotients of induced $I(\tau\nu^s; G, P)$ where $\tau$ is tempered on $M$ and $\nu^s$ is a character in the "positive cone" (see [L7], [BW], [Si]). As for *central exponents*, they are the central characters of the irreducibles in the $\pi_N$ for proper $P$. *Decay* means that these exponents are strictly less than 1 on the positive cone (defined by the positive roots being positive on the center of $M$), and *bounded* means that these exponents are $\leq 1$ there. All three definitions can be stated in terms of matrix coefficients of $\pi$.

Harish-Chandra used the term "supercuspidal" for what is termed in [BZ1] and above "cuspidal". He used the term "cuspidal" for what is currently named "square integrable" or "discrete series".

If $F$ is $\mathbb{R}$ or $\mathbb{C}$, let $K$ be a maximal compact subgroup of $G(F)$. By an "admissible representation of $G(F)$" we mean a $(\mathfrak{g}, K)$-**module** $V$, thus a complex vector space $V$ on which both $K$, and the Lie algebra $\mathfrak{g}$ of $G(F)$ act. The action is denoted $\pi$. The action of $\mathfrak{k}$ obtained from the differential of the action of $K$ coincides with the restriction to $\mathfrak{k}$ of the action of $\mathfrak{g}$, $\pi(\mathrm{Ad}(k)X) = \pi(k)\pi(X)\pi(k^{-1})$ ($k \in K$, $X \in \mathfrak{g}$). As a $K$-module, $V$ decomposes as a direct sum of irreducible representations of $K$, each occurring with finite multiplicities. A ($\mathfrak{g}$, $K$)-module ($\pi_1$, $V_1$) is equivalent to ($\pi_2$, $V_2$) if there is an isomorphism $V_1 \to V_2$ which intertwines the actions of both $K$ and $\mathfrak{g}$.

Denote by $\Pi(G(F))$ the *set of equivalence classes* of irreducible admissible representations of $G(F)$, namely $(\mathfrak{g}, K)$-modules when $F$ is $\mathbb{R}$ or $\mathbb{C}$.

The *local Langlands conjecture*, or the local **Principle of Functoriality**, predicts that there is a partition of the set $\Pi(G/F)$ of equivalence classes of irreducible admissible representations of $G(F)$ into finite sets, named ($L$-)packets, which are parametrized by the set $\Lambda(G/F)$ of admissible homomorphisms $\lambda$ of $L_F$ into ${}^LG$, the "$L$-parameters".

When $F$ is $\mathbb{R}$ or $\mathbb{C}$ the partition and parametrization were defined by Langlands [L7].

When $F$ is $p$-adic, a packet for $G = \mathrm{GL}(n, F)$ consists of a single irreducible, and the parametrization $\Pi(\mathrm{GL}(n)/F) = \Lambda(\mathrm{GL}(n)/F)$ is defined by means of (identity of) $L$- and $\varepsilon$- (or $\gamma$-) factors. The parametrization for $\mathrm{GL}(n, F)$ has recently been proven by Harris-Taylor [HT] and Henniart [He].

Packets for $G = \mathrm{SL}(n, F)$ can be defined to be the set of irreducibles in the restriction to $\mathrm{SL}(n, F)$ of an irreducible of $\mathrm{GL}(n, F)$. This is done for $G = \mathrm{SL}(2, F)$ in Labesse-Langlands [LL]. Alternatively, packets for $G(F) = \mathrm{SL}(n, F)$ can be defined to be the $G^{\mathrm{ad}}(F)$-orbit $\pi^g$ (where $\pi^g(h) = \pi(g^{-1}hg)$) of an irreducible $\pi$, as $g$ ranges over $G^{\mathrm{ad}}(F) = \mathrm{PGL}(n, F)$.

Other cases where packets were introduced are those of the unitary group $\mathrm{U}(3, E/F)$ in 3-variables ([F4]) and the projective symplectic group of similitudes of rank 2 ([F6]). Although the $G^{\mathrm{ad}}(F)$-orbit of an irreducible representation is contained in a packet, in both cases there are packets which consist of several orbits. In both cases the packets are defined by proving liftings to representations of $\mathrm{GL}(n, F)$ for a suitable $n$, by means of the trace formula and character relations. Such an intrinsic definition is given in [F3] for SL(2).

There are several **compatibility** requirements on the packets $\Pi_\lambda$ and their parameters $\lambda$. Some are:

(1) One element of $\Pi_\lambda$ is square integrable modulo the center $Z(G)(F)$ of $G(F)$ iff all elements of $\Pi_\lambda$ have this property, iff $\lambda(L_F)$ is not contained in any proper Levi subgroup of ${}^LG$.

(2) One element of $\Pi_\lambda$ is (essentially) tempered iff all elements are, iff $\lambda(L_F)$ is bounded (modulo the center $Z(\widehat{G})$ of $\widehat{G}$, resp.).

A representation $\pi$ is "essentially $*$" if its product with some character is $*$.

(3) A packet should contain at most one unramified irreducible, and be parametrized in this case by an unramified parameter (which is trivial on the factor $\mathrm{SU}(2, \mathbb{R})$ and the inertia subgroup $I_F$ of $W_F$), see below.

The parametrization is to be compatible with **central characters**. We proceed to explain this (for details see [L1]).

Given a parameter $\lambda : L_F \to {}^LG$, we define a character of the center $Z(G)(F)$ of $G(F)$ as follows. Suppose $Z$ is the maximal torus in $Z(G)$. The normal homomorphism $Z \hookrightarrow G$ defines a surjection ${}^LG \to {}^LZ$, hence a map $\Lambda(G/F) \to \Lambda(Z/F)$. Duality for tori associates to $\lambda \in \Lambda(G/F)$ a character $\omega_\lambda$ of $Z(F)$. If $Z(G)$ is a torus, this is the desired character. If not, choose a connected reductive $F$-group $G_1$ generated by $G$ and a central torus, whose center is a torus. The normal homomorphism $G \hookrightarrow G_1$ defines a surjection $\Lambda(G_1/F) \to \Lambda(G/F)$. We get a character of the center of $G_1(F)$, and by restriction one of the center of $G(F)$, independent of the choice of $G_1$.

Given a parameter $\zeta : L_F \to Z(\widehat{G}) \rtimes W_F$, equivalently $\zeta \in H^1(L_F, Z(\widehat{G}))$, where $Z(\widehat{G})$ is the center of $\widehat{G}$, we define a character $\xi_\zeta$ of $G(F)$ as follows. Let $H$ be a *z-extension* of $G$ (see [Ko1]), namely an extension $1 \to D \to H \to G \to 1$ of $G$ by a quasitrivial torus $D$ (product of tori $\mathrm{R}_{E/F}\mathbb{G}_m$, obtained by restriction of scalars from $\mathbb{G}_m$), $H$ and $D$ are defined over $F$ and the derived group of $H$ is simply connected, equal to $G^{\mathrm{sc}}$. Then the commutative diagram

$$\begin{array}{ccccccc} & & G^{\mathrm{sc}} & \xrightarrow{u} & G & & \\ & & \downarrow & & \| & & \\ 1 \to & D & \to & H & \longrightarrow & G & \to 1 \\ & \| & & \downarrow & & & \\ & D & \xrightarrow{v} & H/G^{\mathrm{sc}} & & & \end{array}$$

has as dual the commutative diagram

$$\begin{array}{ccccccc} & & (H/G^{\mathrm{sc}})^\wedge & \xrightarrow{\widehat{v}} & \widehat{D} & & \\ & & \downarrow & & \| & & \\ 1 \to & \widehat{G} & \longrightarrow & \widehat{H} & \longrightarrow & \widehat{D} & \longrightarrow 1 \\ & \| & & \downarrow & & & \\ & \widehat{G} & \xrightarrow{\widehat{u}} & (G^{\mathrm{sc}})^\wedge & & & \end{array}$$

As $(G^{\mathrm{sc}})^\wedge = \widehat{G}^{\mathrm{ad}}$, $Z(\widehat{G}) = \ker \widehat{u}$. A diagram chase implies that $Z(\widehat{G}) = \ker \widehat{v}$. Hence there is a map

$$H^1(L_F, Z(\widehat{G})) \to \ker[H^1(L_F, (H/G^{\mathrm{sc}})^\wedge) \to H^1(L_F, \widehat{D})].$$

Thus given $\zeta \in H^1(L_F, Z(\widehat{G}))$ there is a character $\sigma_\zeta$ of $(H/G^{\mathrm{sc}})(F)$ which is trivial on $D(F)$, hence a character $\xi_\zeta$ of $G(F) = H(F)/D(F)$. It can be shown that $\xi_\zeta$ is independent of the choice of $D$. We have

$$\omega_{\zeta\lambda} = \xi_\zeta \omega_\lambda, \qquad \zeta \in H^1(L_F, Z(\widehat{G})), \qquad \lambda \in \Lambda(G/F).$$

Further conditions on $\lambda \mapsto \Pi_\lambda$, $\Lambda(G/F) \to \Pi(G/F)$, are:
(1) The central character $\omega_\pi$ of $\pi \in \Pi_\lambda$ is $\omega_\lambda$.
(2) If $\lambda' = \zeta\lambda$ [$\lambda'$, $\lambda' \in \Lambda(G/F)$, $\zeta \in H^1(L_F, Z(\widehat{G}))$], then $\Pi_{\lambda'} = \{\xi_\zeta\pi; \pi \in \Pi_\lambda\}$.

Note that $(\xi_\zeta \cdot \pi)(g) = \xi_\zeta(g)\pi(g)$.

# 4. Unramified Case

Local functoriality for tori leads to **functoriality for unramified representations**. This is necessary for the global theory, as each irreducible admissible representation $\pi$ of $G(\mathbb{A})$ decomposes as the restricted product $\otimes\pi_v$ of representations $\pi_v$ of $G(F_v)$ over all places $v$ of $F$, where $\pi_v$ is unramified for almost all $v$. Thus assume that $F$ is local $p$-adic with residual field $\mathbb{F}_q$. Suppose $G$ is (connected reductive) *unramified* over $F$, namely $G$ is quasisplit over $F$ and split over an unramified extension of $F$. Then the inertia subgroup $I_F$ of $W_F$ acts trivially on $\widehat{G}$, hence $\widehat{G} \rtimes \langle \mathrm{Fr} \rangle$ is defined.

An $L$-parameter $\lambda$ is called *unramified* if it reduces to $\langle \mathrm{Fr} \rangle \to \widehat{G} \rtimes \langle \mathrm{Fr} \rangle$. It is determined by $\lambda(\mathrm{Fr}) = t \times \mathrm{Fr}$ where $t$ is semisimple in $\widehat{G}$. The set $\Lambda^{\mathrm{ur}}(G/F)$ of equivalence classes of unramified $L$-parameters is the set of $\widehat{G}$-conjugacy classes in ${}^LG$ of elements $t \rtimes \mathrm{Fr}$, where $t$ is semisimple. This set is naturally bijected with the set $\Pi^{\mathrm{ur}}(G/F)$ of equivalence classes of *unramified* representations $\pi$ of $G(F)$ (namely the irreducible admissible representations $(\pi, V)$ of $G(F)$ which have a nonzero $K$-fixed vector, where $K$ is a fixed hyperspecial ([Ti]) maximal compact subgroup $K$ of $G(F)$. Note that all such $K$ are conjugate under $G^{\mathrm{ad}}(F)$).

Let us explain the isomorphism $\Lambda^{\mathrm{ur}}(G/F) = \Pi^{\mathrm{ur}}(G/F)$ when $G$ is an $F$-torus $T$.

There is an isomorphism

$$u : T(F)/T(R) \to \mathrm{Hom}(X^*(T)_{\mathrm{Gal}(\overline{F}/F)}, \mathbb{Z}) = X_*(T)^{\mathrm{Gal}(\overline{F}/F)},$$

where $T(R)$ is the maximal compact subgroup of $T(F)$. The isomorphism is defined by $(u(t))(\chi) = \mathrm{ord}_F(\chi(t))$.

Here $\text{ord}_F$ is the map $F^\times \to \mathbb{Z}$, $\text{val}_F(x\boldsymbol{\pi}^n) = n$ if $x$ is in the group $R^\times$ of units ($|x| = 1$).

The surjectivity of $u$ follows on using an unramified splitting field $E$ of $T$ and descending using Hilbert's theorem 90, which implies

$$H^1(\text{Gal}(E/F), R_E^\times) = \{1\}, \quad \text{thus} \quad H^1(\text{Gal}(E/F), T(R)) = \{1\}.$$

Let $S$ denote the maximal $F$-split torus in $T$. Then $X_*(S) = X_*(T)^{\text{Gal}(\overline{F}/F)}$, so

$$\widehat{S} = \text{Hom}(X_*(T)^{\text{Gal}(\overline{F}/F)}, \mathbb{C}^\times) = \text{Hom}(T(F)/T(R), \mathbb{C}^\times) = \Pi^{\text{ur}}(T/F).$$

The inclusion $X_*(S) \to X_*(T)$ defines the exact sequence $1 \to \widehat{T}^{1-\text{Fr}} \to \widehat{T} \to \widehat{S} \to 1$. But $\widehat{S} = \widehat{T}/\widehat{T}^{1-\text{Fr}}$ is $\widehat{T} \rtimes \text{Fr} / \text{Int}(\widehat{T}) = \Lambda^{\text{ur}}(T/F)$.

When $G$ is an unramified reductive group, let $S$ be a maximal $F$-split torus in $G$, and $T$ a maximal $F$-torus containing $S$. There is a unique $\widehat{G}$-conjugacy class of embeddings of $\widehat{T}$ in $\widehat{G}$ compatible with $\iota : \Psi(\widehat{G}) \xrightarrow{\sim} \Psi(G)^\vee$. Choose such an embedding and a Borel $\widehat{B} \supset \widehat{T}$ such that $(\widehat{B}, \widehat{T})$ is fixed by the Galois action. Then we get ${}^LT \hookrightarrow {}^LG$ and a map $\widehat{S} = \Lambda^{\text{ur}}(T/F) \to \Lambda^{\text{ur}}(G/F)$. The Weyl group $W_F(T)$ (= normalizer of $T(F)$ in $G(F)$, quotient by $T(F)$) of $T(F)$ in $G(F)$ preserves $S$ and acts on $\widehat{S}$ by duality. The map factorizes to an isomorphism $\Lambda^{\text{ur}}(T/F)/W_F(T) = \Lambda^{\text{ur}}(G/F)$.

On the representation theoretic side there is a bijection

$$\Pi^{\text{ur}}(T/F)/W_F(T) \xrightarrow{\sim} \Pi^{\text{ur}}(G/F),$$

$\chi \mapsto \pi(\chi)$, constructed by means of the unramified principal series $I(\chi)$ as follows. Let $B$ be a Borel subgroup containing $T$, and $N$ its unipotent radical. Then $B(F) = T(F)N(F)$ and $G(F) = B(F)K$. Extend $\chi \in \Pi^{\text{ur}}(T/F)$ to a character of $B(F)$ trivial on $N(F)$. The induced representation $I(\chi)$ of $G(F)$ acts by right translation on the space of locally constant functions $f : G(F) \to \mathbb{C}$ with $f(nag) = \delta^{1/2}(a)\chi(a)f(g)$ for all $a \in T(F)$, $n \in N(F)$, $g \in G(F)$, where $\delta(a) = |\det(\text{Ad}(a)|\,\text{Lie}\,N)|$. Since $G(F) = B(F)K$, $I(\chi)$ is admissible and contains a unique (up to a scalar multiple) nonzero $K$-invariant vector. Hence $I(\chi)$ has a unique unramified irreducible constituent, denoted $\pi(\chi)$. Every unramified irreducible representation of $G(F)$ is of the form $\pi(\chi)$ for some unramified $\chi : T(F) \to \mathbb{C}^\times$, and $\pi(\chi) \simeq \pi(\chi')$ iff $\chi' = \chi \circ \mathrm{Int}(w)$, $w$ being a representative in $G(F)$ for $W_F(T)$.

The Hecke algebra $\mathbb{H}(G)$ of $G(F)$ with respect to $K$ is the convolution algebra of compactly supported $\mathbb{Z}$-valued $K$-biinvariant functions $f$ on $G(F)$. One has $\mathbb{H}_{\mathbb{C}}(G) = \mathbb{H}(G) \otimes \mathbb{C}$. The Satake transform $f \mapsto f^\vee$, $f^\vee(\pi) = \operatorname{tr} \pi(fdg)$ on $\Pi^{\mathrm{ur}}(G/F)$, is a map from $\mathbb{H}_{\mathbb{C}}(G)$ to the space of functions on the affine variety $\widehat{S}/W_F(T)$, whose coordinate ring is $\mathbb{C}[X_*(S)]^{W_F(T)}$. It is an algebra isomorphism.

Let $F$ be a **global** field, and $G$ a connected reductive group over $F$. A (smooth) representation $\pi$ of $G(\mathbb{A})$ is a vector space $V$ which is both a $(\mathfrak{g}_\infty, K_\infty)$-module ($K_\infty = \prod_{v\in\infty} K_v, G_\infty = \prod_{v\in\infty} G_v$, $\mathfrak{g}_\infty$ denotes the Lie algebra of $G_\infty$, $\infty$ signifies the set of archimedean places of $F$) and a (smooth) $G(\mathbb{A}_f)$-module (each vector of $V$ is fixed by some open subgroup of $G(\mathbb{A}_f)$), such that the action of $G(\mathbb{A}_f)$ commutes with that of $K_\infty$ and $\mathfrak{g}_\infty$. Let $K_v$ be a maximal compact subgroup of $G(F_v)$ at each place $v$ of $F$, which is hyperspecial ([Ti]) at almost all places, and put $K_f = \prod_{v\notin\infty} K_v$, $K = K_\infty K_f$.

A representation $\pi$ is called *admissible* if it is smooth and for each isomorphism class $\gamma$ of continuous irreducible representations of $K$, the $\gamma$-isotypic component of $V$ has finite dimension.

Every irreducible admissible representation $(\pi, V)$ of $G(\mathbb{A})$ is *factorizable* as the restricted tensor product of admissible irreducible representations $(\pi_v, V_v)$ of $G(F_v)$, over all $v$, where $\pi_v$ is unramified for almost all $v$. Thus we fix a nonzero $K_v$-fixed vector $\xi_v^0$ at each place $v$ where $\pi_v$ is unramified, and the space $V$ of $\pi$ is spanned by the products $\otimes_v \xi_v$, where $\xi_v \in \pi_v$ for all $v$ and $\xi_v = \xi_v^0$ for almost all $v$. We write $\pi = \otimes_v \pi_v$; the local components $\pi_v$ are uniquely determined by $\pi$ up to isomorphism.

Suppose $F_v$ is nonarchimedean, and $G(F_v)$ acts on a Hilbert space $H_v$ by a unitary representation $\pi_v$. The space $H_v^0$ of $K_v$-finite vectors is stable under the action of $G(F_v)$. If $H_v$ is irreducible, $H_v^0$ is admissible. Unitary $\pi_{1v}$, $\pi_{2v}$ are unitarily equivalent iff the admissible $\pi_{1v}^0$, $\pi_{2v}^0$ are equivalent.

If $\{H_v\}$ is a family of Hilbert spaces, fix a unit vector $x_v$ in $H_v$ for almost all $v$. The Hilbert restricted product $H = \widehat{\otimes}_{x_v} H_v$ is a Hilbert space with basis $\widehat{\otimes}_v h_v$, $h_v \in P_v$ for all $v$, $h_v = x_v$ for almost all $v$, where $P_v$ is an orthonormal basis of $H_v$, including $x_v$ for almost all $v$. If $\pi$ is a continuous irreducible unitary Hilbert space representation of $G(\mathbb{A})$ then there exist such representations $\pi_v$ of $G(F_v)$, unramified for almost all $v$, unique up to isomorphism, with $\pi \simeq \widehat{\otimes}\pi_v$. For each isomorphism class $\gamma$ of continuous irreducible representations of $K$, the $\gamma$-isotypic component of $\pi$ has finite dimension. The space $\pi^0$ of $K$-finite vectors in $\pi$ is an admissible irreducible $G(\mathbb{A})$-module. Then $\pi^0 = \otimes\pi_v^0$, and $\pi_v^0$ is isomorphic as an admissible $G(F_v)$-module to the space of $K_v$-finite vectors of $\pi_v$. For references and further comments see Flath [Fl].

By Schur's lemma ([BZ1]), an admissible irreducible representation $\pi_v$ has a central character, $\omega_v$. Thus if $Z(F_v)$ is the center of $G(F_v)$, $\pi_v(zg) = \omega_v(z)\pi_v(g)$ for all $z \in Z(F_v)$, $g \in G(F_v)$. Similarly, an admissible irreducible $\pi$ of $G(\mathbb{A})$ has central character, $\omega$.

# 5. Automorphic Representations

Very few of the admissible representations $\pi$ of $G(\mathbb{A})$ are of number theoretic significance. Those which are of interest are the **automorphic representations**. Let $Z$ denote the center of $G$, and let $\omega$ be a unitary character of $Z(\mathbb{A})/Z(F)$. Let $L = L^2_\omega(G(F)Z(\mathbb{A})\backslash G(\mathbb{A}))$ be the space of smooth functions $\phi$ on $G(F)\backslash G(\mathbb{A})$ with $\phi(zg) = \omega(z)\phi(g)$ $(z \in Z(\mathbb{A}))$ and $\int |\phi(g)|^2 dg < \infty$, where $dg$ is the unique up to scalar invariant measure on $G(F)Z(\mathbb{A})\backslash G(\mathbb{A})$. The completion of this space in the $L^2$-norm is a Hilbert space of the $\phi$ which are measurable (not smooth: right invariant under an open subgroup of $G(\mathbb{A}_f)$). The space $L$ is a $G(\mathbb{A})$-module under right translation: $(r(g)\phi)(h) = \phi(hg)$. Any irreducible constituent, or subquotient, of $(r, L)$, is called an *automorphic* representation.

The space $L$ decomposes as a direct sum of irreducible representations only when the homogeneous space $G(F)Z(\mathbb{A})\backslash G(\mathbb{A})$ is compact. In this case $G$ is called *anisotropic*, and all elements of $G(F)$ are semisimple.

In general Langlands theory of Eisenstein series [L3] decomposes $L$ as a direct sum of two invariant subspaces, the discrete spectrum $L_d$, and the continuous spectrum $L_c$. The *discrete spectrum* is the sum of all irreducible subspaces of $L$. Each irreducible summand, $\pi$, in $L_d$, occurs with finite multiplicity, $m(\pi)$. The continuous spectrum $L_c$ is the direct integral of families of representations induced from parabolic subgroups of $G(\mathbb{A})$.

The discrete spectrum splits as the direct sum of the cuspidal spectrum $L_0$, and the residual spectrum $L_r$. The *cuspidal spectrum* consists of the

$\phi$ in $L$ with $\int_{N(F)\backslash N(\mathbb{A})} \phi(ng)dn = 0$ for the unipotent radical $N$ of any proper $F$-parabolic subgroup $P$ of $G$, and any $g \in G(\mathbb{A})$. The residual spectrum is generated by residues of Eisenstein series associated with proper parabolic subgroups. The irreducible constituents in $L_r$, named *residual representations*, are quotients of properly induced representations. They are determined in Moeglin-Waldspurger [MW1] for $G = \mathrm{GL}(n)$, in terms of the divisors $d$ of $n$ and cuspidal representations of $\mathrm{GL}(d, \mathbb{A})$ (and the parabolic subgroup of the type $(d, \ldots, d)$). *Cuspidal representations* are the constituents of $L_0$. Langlands [L4] has shown that the constituents of an induced representation $I(\sigma)$ from a cuspidal representation $\sigma = \otimes\sigma_v$ of a parabolic subgroup $P(\mathbb{A})$ ($\sigma$ trivial on the unipotent radical $N(\mathbb{A})$) are the $\otimes_v\pi_v$, where $\pi_v$ is a constituent of $I(\sigma_v)$ for all $v$, and $\pi_v$ is the unique unramified constituent of $I(\sigma_v)$ for almost all $v$. Moreover, an admissible irreducible representation $\pi$ of $G(\mathbb{A})$ is automorphic iff $\pi$ is a constituent of $I(\sigma)$ for some $P$ and some $\sigma$.

The *global principle of functoriality* relates parameters $\lambda : L_F \to {}^LG$ with irreducible automorphic representations $\pi$ of $G(\mathbb{A})$. The relation is such that for almost all places, where the restriction $\lambda_v$ of $\lambda$ to $L_{F_v} \hookrightarrow L_F$ is unramified and the component $\pi_v$ of $\pi$ is unramified, the $\widehat{G}$-conjugacy class $\lambda_v(\mathrm{Fr}_v) = t(\lambda_v) \times \mathrm{Fr}_v$ in $\widehat{G} \times \langle \mathrm{Fr}_v \rangle$ corresponds to $\pi_v = \pi(\chi_v)$, $\chi_v$ in

$$\Pi^{\mathrm{ur}}(T/F_v)/W_{F_v}(T) = \Pi^{\mathrm{ur}}(G/F_v) = \Phi^{\mathrm{ur}}(G/F_v) = \Phi^{\mathrm{ur}}(T/F_v)/W_{F_v}(T).$$

In other words, the unramified components of $\lambda$ and $\pi$ correspond under the correspondence for unramified representations. For split groups, ${}^LG$ is a direct product, and the unramified $\lambda_v$ and $\pi_v$ are parametrized by semisimple conjugacy classes in $\widehat{G}$.

For the group $G = \mathrm{GL}(n)$ the principle can be stated as asserting that there is a bijection between the set of $n$-dimensional irreducible representations $\lambda : L_F \to \mathrm{GL}(n, \mathbb{C})$, and the set of cuspidal (irreducible) representations $\pi$ of $\mathrm{GL}(n, \mathbb{A})$. Here $\lambda$ is uniquely determined by $\lambda_v$ for almost all $v$ by the Chebotarev density theorem: the set of Frobenii at almost all $v$ is dense in $\mathrm{Gal}(\overline{F}/F)$. The cuspidal $\pi$ is uniquely determined by almost all of its unramified components, by the *rigidity theorem* for $\mathrm{GL}(n)$ ([JS]). When the global field $F$ is a function field, this principle was proven by Lafforgue [Lf].

This case has as an application the (Emil) *Artin conjecture*, which predicts that the $L$-function of an irreducible nontrivial representation $\lambda$ of

$\mathrm{Gal}(\overline{F}/F)$ is entire. Indeed, if $\lambda \leftrightarrow \pi$ then $L(s,\lambda) = L(s,\pi)$, and the $L$-function of a cuspidal $\pi$ is entire.

Note that if $\lambda_i \leftrightarrow \pi_i (1 \le i \le k)$ then $\oplus_i \lambda_i \leftrightarrow \boxplus_i \pi_i$, where $\boxplus_i \pi_i$ indicates the representation $I(\pi_1, \dots, \pi_k)$ normalizedly induced from $\pi_1 \otimes \cdots \otimes \pi_k$ on the parabolic subgroup $P(\mathbb{A})$ of $G(\mathbb{A})$ of type $(\dim \lambda, \dots, \dim \lambda_k)$ which is trivial on the unipotent radical $N$ of $P(\mathbb{A})$. The normalizing factor is $\delta^{1/2}$, where $\delta(m) = |\det(\mathrm{Ad}(m)|\,\mathrm{Lie}\, N)|$.

For general reductive connected group $G$ over a global field $F$, a weak form of the principle would assert the existence of an automorphic representation $\pi$ of $G(\mathbb{A})$ for each parameter $\lambda : L_F \to {}^L G$, such that $\lambda_v \leftrightarrow \pi_v$ for almost all $v$, and conversely, given such $\pi$ there is a $\lambda$. The last claim, that $\pi$ defines $\lambda$, is false even for $\mathrm{GL}(n)$, and the group $L_F$ has to be increased to $L_F \times \mathrm{SL}(2,\mathbb{R})$. Before we explain this, let us present a **strong form** of the conjectural principle of functoriality, in terms of all places.

Let $P_v$ be a packet of admissible irreducible representations of $G(F_v)$ for each place $v$ of the global field $F$, such that $P_v$ contains an unramified representation $\pi_v^0$ for almost all $v$. The *global packet* $P = P(\{P_v\}_v)$ consists of all $G(\mathbb{A})$-modules $\otimes_v \pi_v$ with $\pi_v \in P_v$ for all $v$ and $\pi_v = \pi_v^0$ for almost all $v$. It is the restricted product of the $P_v$ with respect to $\{\pi_v^0\}_v$. The global packet is called *automorphic* (discrete spectrum, cuspidal, ...) if it contains such a representation. The example of $\mathrm{SL}(2)$ shows that not all irreducibles in an automorphic packet need be automorphic.

A strong form of the principle would assert that there is a bijection between $\Lambda(G/F)$, the set of equivalence classes of parameters $\lambda : L_F \to {}^L G$, and the set of automorphic packets $P = \{\pi\} = \otimes\{\pi_v\}$, such that $\lambda_v \leftrightarrow \{\pi_v\}$ for all $v$. Moreover it would specify which members of $P = P_\lambda$ are automorphic.

# 6. Residual Case

As noted above, the group $L_F$ does not carry sufficiently many parameters $\lambda : L_F \to {}^L G$ to account for all discrete spectrum, or even cuspidal, automorphic representations. These $\lambda$ correspond, by the Ramanujan conjecture, to those discrete spectrum representations whose local components are all tempered. A bigger group than $L_F$ has to be introduced to account for the discrete spectrum, including cuspidal, representations of $G(\mathbb{A})$ which are not tempered, in fact at almost all places. To present it, we consider first the **case of** $\mathrm{GL}(n)$.

The discrete spectrum representations of $\mathrm{GL}(n,\mathbb{A})$ have been determined by Moeglin and Waldspurger [MW1] in terms of the divisors $d$ of $n$, and the cuspidal representations $\tau$ of $\mathrm{GL}(d,\mathbb{A})$. Denote by $P$ the standard parabolic subgroup of $\mathrm{GL}(n)$ of type $\mathbf{d}=(d,d,\ldots,d)$, and by $N$ its unipotent radical. Put $\delta_{P_v}(p)=|\det(\mathrm{Ad}(p)|\,\mathrm{Lie}\,N)|_v$ for $p\in P(F_v)$. Thus

$$\delta_{P_v}(\mathrm{diag}(g_1,\ldots,g_m))=\prod_{1\le i\le m}|\det g_i|_v^{(m+1-2i)/2},$$

where $md=n$, for $g_i\in\mathrm{GL}(d,F_v)$. Thus

$$\delta_{P_v}=\nu_v^{m-1}\times\nu_v^{m-3}\times\cdots\times\nu_v^{-(m-1)},$$

where $\nu_v(g)=|\det g|_v$. The normalizedly induced representation

$$I(\delta_{P_v}^{1/2}\tau_v^{\mathbf{d}})=I(\nu_v^{\frac{m-1}{2}}\tau_v,\nu_v^{\frac{m-3}{2}}\tau_v,\ldots,\nu_v^{-\frac{m-1}{2}}\tau_v)$$

is realized in the space of smooth functions $f:G(F_v)\to V_v\otimes\cdots\otimes V_v$ ($V_v$ is the space of $\tau_v$) with

$$f(pg)=\delta_{P_v}(p)[\tau_v(g_1)\otimes\cdots\otimes\tau_v(g_m)]f(g)\qquad(g\in\mathrm{GL}(n,F_v)),$$

where $\mathrm{diag}(g_1,\ldots,g_m)$ is the Levi component of $p$. It has a unique quotient $J(\delta_{P_v}^{1/2}\tau_v^{\mathbf{d}})=J(\nu_v^{\frac{m-1}{2}}\tau_v,\ldots,\nu_v^{-\frac{m-1}{2}}\tau_v)$ when $\tau_v$ is generic (or tempered), by [Z]. The discrete spectrum representations of $\mathrm{GL}(n,\mathbb{A})$ are precisely the

$$J(\delta_P^{1/2}\tau^{\mathbf{d}})=J(\nu^{\frac{m-1}{2}}\tau,\nu^{\frac{m-3}{2}}\tau,\ldots,\nu^{-\frac{m-1}{2}}\tau)$$

$$=\otimes_v J(\nu_v^{\frac{m-1}{2}}\tau_v,\nu_v^{\frac{m-3}{2}}\tau_v,\ldots,\nu_v^{-\frac{m-1}{2}}\tau_v)=\otimes_v J(\delta_{P_v}^{1/2}\tau_v^{\mathbf{d}})$$

as $d$ ranges over the divisors of $n$, $m=n/d$, and $\tau$ range over the cuspidal representations of $\mathrm{GL}(d,\mathbb{A})$.

If the cuspidal representations $\pi$ of $\mathrm{GL}(n,\mathbb{A})$ are parametrized by the $\lambda: L_F\to {}^LG=\mathrm{GL}(n,\mathbb{C})\times W_F$, namely $n$-dimensional representations $\lambda:L_F\to\mathrm{GL}(n,\mathbb{C})$, the discrete spectrum representations can be parametrized by the equivalence classes of the irreducible complex representations

$$\alpha : L_F \times \mathrm{SL}(2,\mathbb{C}) \to \mathrm{GL}(n,\mathbb{C}),$$

where $\alpha$ is the tensor product $\alpha_{\mathrm{ss}} \otimes \alpha_{\mathrm{unip}}$. Here $\alpha_{\mathrm{ss}} : L_F \to \mathrm{GL}(d,\mathbb{C})$ and $\alpha_{\mathrm{unip}} : \mathrm{SL}(2,\mathbb{C}) \to \mathrm{GL}(m,\mathbb{C})$ are irreducible representations with $n = dm$. In particular $\alpha_{\mathrm{unip}}\left(\left(\begin{smallmatrix} 1 & 1 \\ 0 & 1 \end{smallmatrix}\right)\right)$ is a regular unipotent element in $\mathrm{GL}(m,\mathbb{C})$ (single Jordan block).

The cuspidal representations can then be viewed as the semisimple ones, while the unipotent representations are those with $\alpha_{\mathrm{ss}} = 1$. The associated discrete spectrum representation $J$ is the trivial representation of $\mathrm{GL}(n,\mathbb{A})$. Further, the map

$$\lambda_\alpha(w) = \alpha\left(w, \left(\begin{smallmatrix} |w|^{1/2} & 0 \\ 0 & |w|^{-1/2} \end{smallmatrix}\right)\right), \quad \lambda_\alpha : L_F \to \mathrm{GL}(n,\mathbb{C}),$$

is the $n$-dimensional representation of $L_F$ which parametrizes $J(\delta_P^{1/2}\tau^{\mathbf{d}})$, $\tau = \tau(\alpha_{ss})$, by the principle of functoriality. Here $|.|$ is the composition of $L_F \to W_F \to W_F^{\mathrm{ab}} \simeq C_F$ and the absolute value on $C_F$.

The group $\mathrm{GL}(n)$ has the special property that the decomposition of its discrete spectrum into the cuspidal and residual parts is conjecturally the same as its decomposition into tempered and nontempered representations. Indeed, the Ramanujan conjecture predicts that all local components of any cuspidal representation of $\mathrm{GL}(n,\mathbb{A})$ are tempered. From the explicit description given above of the residual spectrum it is clear that each component of a residual representation of $\mathrm{GL}(n,\mathbb{A})$ is nontempered. Such partition, cuspidal equals temperedness and residual equals nontemperedness, does not hold for groups which are not closely related to $\mathrm{GL}(n)$, such as inner forms or $\mathrm{SL}(n)$.

To describe a conjectural picture of the automorphic representations of $G(\mathbb{A})$ for a reductive connected group $G$ over a global field $F$, Arthur ([A2], [A3], [A4]) introduced the notion of what we call $A$-**parameter**. It is a homomorphism

$$\alpha : L_F \times \mathrm{SL}(2,\mathbb{C}) \to {}^L G$$

whose restriction to $L_F$ is an essentially tempered $L$-parameter (the projection to $\widehat{G}$ of $\alpha(L_F)$ is bounded modulo $Z(\widehat{G})$, the composition of $\alpha|L_F$ with the projection ${}^L G \to W_F$ is the natural map $L_F \to W_F$, $\mathrm{pr}_{\widehat{G}} \circ \alpha(w)$ is semisimple for every $w \in L_F$) and whose restriction to the factor $\mathrm{SL}(2,\mathbb{C})$ is a homomorphism $\mathrm{SL}(2,\mathbb{C}) \to \widehat{G}$ of complex algebraic groups. Moreover, $\alpha$ is *globally relevant*: if $\mathrm{pr}_{\widehat{G}}$ of its image lies in a parabolic subgroup of $\widehat{G}$ the corresponding parabolic subgroup of $G$ has to be defined over $F$. Thus a tempered $L$-parameter $\lambda$ is an $A$-parameter; an $A$-parameter $\alpha$ whose restriction to the second factor $\mathrm{SL}(2,\mathbb{C})$ is trivial (thus $\alpha$ is also an $L$-parameter) is tempered; and the restriction of $\alpha$ to $L_{F_v} \times \mathrm{SL}(2,\mathbb{C})$ defines a local parameter $\alpha_v$ up to equivalence, for each $v$.

Two $A$-parameters $\alpha_1$ and $\alpha_2$ are called *equivalent* if there exist $g$ in $\widehat{G}$ and a 1-cocycle $z$ of $L_F$ in $Z(\widehat{G})$ with $\mathrm{Int}(g)\alpha_1 = z\alpha_2$ such that the class of $z$ in $H^1(L_F, Z(\widehat{G}))$ is locally trivial (lies in $\ker[H^1(L_F, Z(\widehat{G})) \to \prod_v H^1(L_{F_v}, Z(\widehat{G}))]$). If $\mathrm{Gal}(\overline{F}/F)$ (hence $L_F$, $W_F$) acts trivially on $Z(\widehat{G})$ then $H^1(L_F, Z(\widehat{G})) = \mathrm{Hom}(L_F, Z(\widehat{G}))$ and Chebotarev density theorem for $L_F^{\mathrm{ab}} = W_F^{\mathrm{ab}}$ implies that $z$ is trivial.

Denote by $\aleph(G/F)$ the *set of equivalence classes* of $A$-parameters for $G$ over $F$.

For any $\alpha$ the parameter

$$\lambda_\alpha(w) = \alpha\left(w, \begin{pmatrix} |w|^{1/2} & 0 \\ 0 & |w|^{-1/2} \end{pmatrix}\right)$$

lies in $\Lambda(G/F)$. Here $w \in L_F$, and $L_F \to W_F \to W_F^{\mathrm{ab}} \simeq C_F$ together with the absolute value on $C_F$ defines $w \mapsto |w|$. The map $\alpha \mapsto \lambda_\alpha$ injects $\aleph(G/F)$ in $\Lambda(G/F)$, $\Lambda(G/F)$ is the subset of $\aleph(G/F)$ of $\alpha$ with $\alpha = 1$.

Locally, to each $\alpha \in \aleph(G/F_v)$ there should be associated a finite set $\prod_\alpha$ of irreducibles, containing $\prod_{\lambda_\alpha}$. The set $\prod_\alpha$, named *A-packet* or *quasipacket*, does not partition the set of representations. Examples of $\mathrm{U}(3, E/F)$ ([F4]) and $\mathrm{PGSp}(2, F)$ ([F6]) show that a quasipacket has nontrivial intersection with a packet of cuspidal representations. Quasipackets come up in character relations which define liftings, by means of the trace formula. They do however define a global partition of the discrete spectrum.

We define a global quasipacket as the restricted product over all $v$ of a family of local quasipackets for all $v$ which contain a fixed unramified irreducible $\pi_v^0$ for almost all $v$. In fact $\pi_v^0$ is $\prod_{\lambda_{\alpha_v}}$ for $v$ where $\alpha_v = \alpha|(L_{F_v} \times \mathrm{SL}(2,\mathbb{C}))$ is unramified.

However, not every irreducible in a quasipacket is discrete spectrum, or automorphic.

Let $S_\alpha = S_\alpha(G)$ be the set of $s \in \widehat{G}$ such that $s\alpha(w')s^{-1} = z(w')\alpha(w')$ for all $w'$ in $L_F \times \mathrm{SL}(2,\mathbb{C})$, where $z(w') \in Z(\widehat{G})$ depends only on the $L_F$-factor $w$ of $w'$, and the class of the cocycle $z$ in $H^1(L_F, Z(\widehat{G}))$ is locally trivial, namely in the kernel $\ker[H^1(L_F, Z(\widehat{G})) \to \prod_v H^1(L_{F_v}, Z(\widehat{G}))]$ of all localization maps. Put $\overline{S}_\alpha = S_\alpha / S_\alpha^0 \cdot Z(\widehat{G}) = \pi_0(S_\alpha/Z(\widehat{G}))$. Then $\overline{S}_\alpha \to \overline{S}_{\lambda_\alpha}$ is surjective, where $\overline{S}_\lambda = \pi_0(S_\lambda/Z(\widehat{G}))$ and $S_\lambda$ is the group of $s \in \widehat{G}$ with $s\lambda(w)s^{-1} = z(w)\lambda(w)$ $(w \in L_F)$, where $z(w) \in Z(\widehat{G})$ defines a locally trivial element in $H^1(L_F, Z(\widehat{G}))$.

The composition of the map $L_{F_v} \times \mathrm{SL}(2,\mathbb{C}) \to L_F \times \mathrm{SL}(2,\mathbb{C})$ with $\alpha$ defines a parameter $\alpha_v \in \aleph(G/F_v)$. There are natural maps $S_\alpha \to S_{\alpha_v}$ and $\overline{S}_\alpha \to \overline{S}_{\alpha_v}$. Arthur ([A2], 1.3.3) then expects to have a finite set, $\prod_{\alpha_v}$, of irreducible representations of $G(F_v)$, containing $\prod_{\lambda_{\alpha_v}}$, and a function $\varepsilon_{\alpha_v} : \prod_{\alpha_v} \to \{\pm 1\}$ which is 1 on $\prod_{\lambda_{\alpha_v}}$, and which is 1 if $\alpha_v$ is tempered, and a pairing $\langle .,. \rangle_v : \overline{S}_{\alpha_v} \times \prod_{\alpha_v} \to \mathbb{C}^1$, with various properties, including:
(i) $\pi \in \prod_{\lambda_{\alpha_v}} (\subset \prod_{\alpha_v})$ iff $\langle ., \pi \rangle_v$ is a character of $\overline{S}_{\alpha_v}$ pulled via $\overline{S}_{\alpha_v} \twoheadrightarrow \overline{S}_{\lambda_{\alpha_v}}$ from a character of $\overline{S}_{\lambda_{\alpha_v}}$.
(ii) The invariant distribution $\sum_{\pi \in \prod_{\alpha_v}} \varepsilon_{\alpha_v}(\pi)\langle 1, \pi \rangle \operatorname{tr} \pi$ is *stable* (depends only on the stable orbital integrals of the test measure $fdg$).
(iii) $\prod_{\lambda_{\alpha_v}}$ contains an unramified irreducible $\pi_v^0$ whenever $\alpha_v$ is unramified (trivial on the inertia subgroup of $W_{F_v}$) and $G$ is unramified over $F_v$.

There should also be a function $c_v : S_{\alpha_v}/Z(\widehat{G}) \to \{\pm 1\}$ which is conjugacy invariant, such that the map $\pi \mapsto c_v(s)\langle \overline{s}, \pi \rangle_v$ on $\prod_{\alpha_v}$ is independent of the pairing $\langle ., \rangle_v$. Here $\overline{s}$ is the projection of $s$ to $\overline{S}_{\alpha_v}$. It is used in endoscopy.

We name the $\prod_{\alpha_v}$ quasipackets. When $\alpha_v$ is trivial on the factor $\mathrm{SL}(2,\mathbb{C})$ the quasipacket $\prod_{\alpha_v}$ is simply a packet. The quasipackets do not partition the set of (equivalence classes of) irreducible admissible representations. The examples of $\mathrm{U}(3, E/F)$ ([F4]) and $\mathrm{PGSp}(2)$ ([F6]) show that often a quasipacket consists of a nontempered irreducible together with a cuspidal representation, and the cuspidal lies in a packet of cuspidals. These examples show that quasipackets naturally occur in character relations describing liftings, and are necessary to describe the discrete spectrum automorphic representations.

Given $\alpha \in \aleph(G/F)$ we define the quasipacket $\prod_\alpha$ as the restricted tensor product of the local quasipackets $\prod_{\alpha_v}$ with respect to the unramified

$\pi_v^0 \in \prod_{\lambda_{\alpha_v}}$ for almost all $v$. There should be a global pairing

$$\langle .,. \rangle : \overline{S}_\alpha \times \Pi_\alpha \to \mathbb{C}^1, \qquad \langle \overline{s}, \pi \rangle = \prod_v \langle \overline{s}_v, \pi_v \rangle_v$$

where $\overline{s}_v$ is the image of $\overline{s}$ in $\overline{S}_{\alpha_v}$. Further there should be a function

$$\varepsilon_\alpha : \Pi_\alpha \to \{\pm 1\}, \qquad \varepsilon_\alpha(\pi) = \prod_v \varepsilon_{\alpha_v}(\pi_v), \qquad \pi = \otimes \pi_v.$$

Almost all $\varepsilon_{\alpha_v}(\pi_v)$ should be 1, and $\langle \overline{s}_v, \pi_v \rangle_v = 1$ for almost all $v$. Further one expects that for $s \in S_\alpha / Z(\widehat{G})$ the product $\prod_v c_v(s_v)$ is 1, where $s_v$ is the image of $s$ in $S_{\alpha_v}/Z(\widehat{G})$.

It is expected of the quasipackets, parametrized by $\alpha \in \aleph(G/F)$, to partition the automorphic representations of $G(\mathbb{A})$. The automorphic $\pi$ in $\prod_\alpha$ occur in the discrete spectrum iff $S_\alpha$ is finite. If $S_\alpha$ is finite there should exist an integer $d_\alpha > 0$ and a homomorphism $\xi_\alpha : \overline{S}_\alpha \to \{\pm 1\}$ such that the **multiplicity** $m(\pi)$ with which $\pi \in \prod_\alpha$ occurs in the discrete spectrum of $L^2(G(F)/G(\mathbb{A}))$ is

$$\frac{d_\alpha}{|\overline{S}_\alpha|} \sum_{\overline{s} \in \overline{S}_\alpha} \langle \overline{s}, \pi \rangle \xi_\alpha(\overline{s}).$$

In particular, if $\overline{S}_\alpha$ and each $\overline{S}_{\alpha_v}$ are abelian then the multiplicity of $\pi$ is $d_\alpha$ if $\langle ., \pi \rangle = \xi_\alpha$, and 0 otherwise.

If $\overline{S}_\alpha$ consists of a single element then the multiplicity $m(\pi)$ is constant on $\prod_\alpha$, and we say that $\prod_\alpha$ is *stable.*

In case the quasipackets have nonzero intersection, the multiplicity $m(\pi)$ will be the sum of the expressions displayed above over all $\alpha$ such that $\pi \in \prod_\alpha$.

## 7. Endoscopy

An auxiliary notion is that of an **endoscopic group** $H$ of $G$. It comes up on stabilizing the trace formula, which permits lifting representations from $H$ to $G$. We recall its definition following Kottwitz [Ko2].

Let $G$ be a connected reductive group over a local or global field $F$. An *endoscopic datum* for $G$ is a pair $(s, \rho)$. The $s$ is a semisimple element of $\widehat{G}/Z(\widehat{G})$. Put $\widehat{H}$ for the connected centralizer $Z_{\widehat{G}}(s)^0$ of $s$ in $\widehat{G}$. The $\rho : W_F \to \mathrm{Out}(\widehat{H})$ is a homomorphism (which factorizes via $W_F \to \mathrm{Gal}(\overline{F}/F)$. We may work with $\mathrm{Gal}(\overline{F}/F)$ instead of $W_F$). For each $w$ in $W_F$ the element $\rho(w)$ is required to have the form $n \times w \in \widehat{G} \times w$ and it normalizes $\widehat{H}$. In particular $\rho$ induces an action of $W_F$ on $Z(\widehat{H})$. Of course $W_F$ acts on $\widehat{G}$ and on its subgroup $Z(\widehat{G})$. The map $Z(\widehat{G}) \hookrightarrow Z(\widehat{H})$ is a $W_F$-map. The exact sequence $1 \to Z(\widehat{G}) \to Z(\widehat{H}) \to Z(\widehat{H})/Z(\widehat{G}) \to 1$ gives a long exact sequence ([Ko2], Cor. 2.3)

$$\cdots \to \pi_0(Z(\widehat{H})^{W_F}) \to \pi_0([Z(\widehat{H})/Z(\widehat{G})]^{W_F}) \to H^1(F, Z(\widehat{G})) \to \ldots .$$

The element $s \in Z(\widehat{H})/Z(\widehat{G})$ is required to be fixed by $W_F$, and its image in

$$\pi_0([Z(\widehat{H})/Z(\widehat{G})]^{W_F})$$

is in the subgroup $\mathfrak{K}(s, \rho)$, consisting of the elements whose image in $H^1(F, Z(\widehat{G}))$ is trivial if $F$ is local, and locally trivial if $F$ is global.

An *isomorphism* of endoscopic data $(s_1, \rho_1)$ and $(s_2, \rho_2)$ is $g \in \widehat{G}$ with

$$\mathrm{Int}(g)\widehat{H}_1 = \widehat{H}_2; \quad \rho_2 = (\mathrm{Int}\, g)^0 \circ \rho_1$$

($(\mathrm{Int}\, g)^0$ is the isomorphism $\mathrm{Out}(\widehat{H}_1) \to \mathrm{Out}(\widehat{H}_2)$ induced by $\mathrm{Int}\, g$; $\mathrm{Int}(g)s_1$ and $s_2$ have the same image in $\mathfrak{K}(s, \rho)$.

Write $\mathrm{Aut}(s, \rho)$ for the group of automorphisms of $(s, \rho)$. It is an algebraic subgroup of $\widehat{G}$ with identity component $\widehat{H}$. Put

$$\Lambda(s, \rho) = \mathrm{Aut}(s, \rho)/\widehat{H}.$$

An endoscopic datum $(s, \rho)$ is *elliptic* if $(Z(\widehat{H})^{W_F})^0 \subset Z(\widehat{G})$. Then the 3rd condition in the definition of an isomorphism can be replaced by $\mathrm{Int}(g)s_1 = s_2$.

An *endoscopic group* $H$ of $G$ is in fact a triple $(H, s, \eta)$, where $H$ is a quasisplit connected reductive $F$-group, $s \in Z(\widehat{H})$, and $\eta : \widehat{H} \to \widehat{G}$ is an embedding of complex groups. It is required that

(1) $\eta(\widehat{H})$ is the connected centralizer $Z_{\widehat{G}}(\eta(s))^0$ of $\eta(s)$ in $\widehat{G}$, and that

(2) the $\widehat{G}$-conjugacy class of $\eta$ is fixed by $W_F$ (that is, by $\varphi(W_F) \subset \mathrm{Gal}(\overline{F}/F)$).

We regard $Z(\widehat{G})$ as a subgroup of $Z(\widehat{H})$. By (2), the $W_F$-actions on $Z(\widehat{G})$ and $Z(\widehat{H})$ are compatible. Define a subgroup $\mathfrak{K}(H/F)$ of

$$\pi_0([Z(\widehat{H})/Z(\widehat{G})]^{W_F})$$

analogously to $\mathfrak{K}(s,\rho)$ above. It is further required that

(3) the image of $s$ in $Z(\widehat{H})/Z(\widehat{G})$ is fixed by $W_F$ and its image in $\pi_0([Z(\widehat{H})/Z(\widehat{G})]^{W_F})$ lies in $\mathfrak{K}(H/F)$.

An *isomorphism* of endoscopic groups $(H_1, s_1, \eta_1)$ and $(H_2, s_2, \eta_2)$ is an $F$-isomorphism $\alpha : H_1 \to H_2$ satisfying:

(1) $\eta_1 \circ \widehat{\alpha}$ and $\eta_2$ are $\widehat{G}$-conjugate. ($\widehat{\alpha}$ is defined up to $\widehat{H}_1$-conjugacy; it induces a canonical isomorphism $\mathfrak{K}(H_2/F) \xrightarrow{\sim} \mathfrak{K}(H_1/F)$).

(2) The elements of $\mathfrak{K}(H_i/F)$ defined by $s_i$ correspond under

$$\mathfrak{K}(H_2/F) \xrightarrow{\sim} \mathfrak{K}(H_1/F).$$

The group $\mathrm{Aut}(H,s,\eta)$ of automorphisms of $(H,s,\eta)$ contains $H^{\mathrm{ad}}(F)(= (\mathrm{Int}\, H)(F))$ as a normal subgroup. Put

$$\Lambda(H,s,\eta) = \mathrm{Aut}(H,s,\eta)/H^{\mathrm{ad}}(F).$$

An endoscopic group $(H,s,\eta)$ determines an endoscopic datum $(\eta(s),\rho)$, where $\rho$ is the composition

$$W_F \to \mathrm{Aut}(\widehat{H}) \xrightarrow{\sim} \mathrm{Aut}(Z_{\widehat{G}}(\eta(s))^0) \to \mathrm{Out}(Z_{\widehat{G}}(\eta(s))^0).$$

Every endoscopic datum arises from some endoscopic group. There is a canonical bijection from the set of isomorphisms from an endoscopic group $(H_1, s_1, \eta_1)$ to another, $(H_2, s_2, \eta_2)$, taken modulo $\mathrm{Int}(H_2)$, to the set of isomorphisms from the corresponding endoscopic datum $(\eta(s_1), \rho_1)$ to $(\eta(s_2), \rho_2)$, taken modulo $Z_{\widehat{G}}(\eta(s_2))^0$. Thus there is a bijection from the set of isomorphism classes of endoscopic groups to the set of isomorphism classes of endoscopic data. Moreover, there is a canonical isomorphism $\Lambda(H,s,\eta) \xrightarrow{\sim} \Lambda(\eta(s),\rho)$.

We say that $(H,s,\eta)$ is *elliptic* if $(\eta(s),\rho)$ is elliptic.

Twisted endoscopic groups are defined, discussed and used to stabilize the twisted trace formula in [KS].

For further discussion of parameters and (quasi) packets see [A4].

Let $f : G^* \to G$ be an $\overline{F}$-isomorphism of $F$-groups. It defines a map $\overline{f} : \Psi(G^*) \to \Psi(G)$. It is called an *inner twist* if for every $\sigma$ in $\mathrm{Gal}(\overline{F}/F)$ there is $g_\sigma$ in $G(\overline{F})$ with $f(\sigma(g)) = \mathrm{Int}(g_\sigma)(\sigma(f(g)))$. In this case $G^*$ is called an *inner form* of $G$. The $L$-group ${}^LG$ depends only on the class of inner forms of $G$. In each such class there exists a unique quasisplit form. The $L$-group determines the $F$-isomorphism class of the quasisplit form. The Galois action on $\widehat{G}$ is trivial iff $G$ is an inner form of a split group. The $L$-parameters of $G$ are only those which factorize through ${}^LP$ for an $F$-parabolic subgroup $P$ of the quasisplit form $G^*$ of $G$, provided $P$ is *relevant*, namely is an $F$-parabolic subgroup of $G$ itself.

The group $G$ is defined over a field $F$, and the theory for $G$ depends on the choice of $F$. What would happen if we **replace the base field** by a finite extension $E$ of $F$? For this it is convenient to recall the theory of *induced groups*. Let $A'$ be a subgroup of finite index in a group $A$. The example of interest to us will later be $A = \mathrm{Gal}(\overline{F}/F)$ and $A' = \mathrm{Gal}(\overline{F}/E)$. Suppose $A'$ acts on a group $G$. The induced group $I_{A'}^A(G) = \mathrm{Ind}_{A'}^A(G)$ is defined to consist of all $f : A \to G$ with $f(a'a) = a'f(a)$ ($a \in A$, $a' \in A'$). The group structure is $(ff')(a) = f(a)f'(a)$. The group $A$ acts by $(r(a)f)(x) = f(xa)$ ($a, x \in A$). For a coset $s$ in $A'\backslash A$ put

$$G_s = \{f \in I_{A'}^A(G); f(a) = 0 \text{ if } a \notin s\}.$$

It is a group and $I_{A'}^A(G)$ is $\prod_{s\in A'\backslash A} G_s$. The groups $G_s$ are permuted by $A$. The subgroup $G_{\overline{e}}$ is stable under $A'$, and $f \mapsto f(e)$, $G_{\overline{e}} \to G$, is an $A'$-module isomorphism. Shapiro's lemma asserts $H^1(A, I_{A'}^A(G)) = H^1(A', G)$.

Let $B$ be a group, $\mu : B \to A$ a homomorphism, put $B' = \mu^{-1}(A')$, and suppose $\mu$ induces a bijection $B'\backslash B \xrightarrow{\sim} A'\backslash A$. Then $B'$ acts on $G$ via $\mu : b' \cdot g = \mu(b')g$. The map $f \mapsto \mu \circ f$ is a $\mu$-equivarient isomorphism $\mu' : I_{A'}^A(G) \xrightarrow{\sim} I_{B'}^B(G)$. We have $r(\mu(a))(\mu \circ f)(x) = \mu \circ f(xa)$.

If $E/F$ is a finite field extension, we have

$$W_E\backslash W_F = \mathrm{Gal}(\overline{F}/E)\backslash \mathrm{Gal}(\overline{F}/F) = \mathrm{Hom}_F(E, \overline{F}).$$

If $G$ is an $E$-group, its restriction of scalars $G' = \mathrm{R}_{E/F}G$ is the $F$-group $IG$ where $I = I_{\mathrm{Gal}(\overline{F}/E)}^{\mathrm{Gal}(\overline{F}/F)}$. Thus $G'(\overline{F}) = I(G(\overline{F})) = \prod_\sigma G(\overline{F})_\sigma$ where $\sigma$ ranges over $\mathrm{Hom}_F(E,\overline{F})$. Let $\sigma_i \in W_F$ $(1 \leq i \leq [E:F])$ be a set of representatives for $W_E\backslash W_F$. Define an action of $\tau \in W_F$ on $(i; 1 \leq i \leq [E:F])$ by $W_E\sigma_i\tau^{-1} = W_E\sigma_{\tau(i)}$. Put $\tau_i = \sigma_{\tau(i)}\tau\sigma_i^{-1} \in W_E$. Then

$$(\gamma\tau)_i = \sigma_{(\gamma\tau)(i)}\gamma\tau\sigma_i^{-1} = \sigma_{(\gamma\tau)(i)}\gamma\sigma_{\tau(i)}^{-1} \cdot \sigma_{\tau(i)}\tau\sigma_i^{-1} = \gamma_{\tau(i)}\tau_i \in W_E.$$

The group $W_F$ acts on $G'(\overline{F}) = \prod_i G(\overline{F})$ by $(g_i)\gamma = (\gamma_i^{-1}(g_{\gamma(i)}))$. Indeed,

$$\begin{aligned}(g_i)(\gamma\tau) &= ((\gamma\tau)_i^{-1}(g_{(\gamma\tau)(i)})) = ((\gamma_{\tau(i)}\tau_i)^{-1}(g_{\gamma(\tau(i))}))\\ &= (\gamma_i^{-1}(g_{\gamma(i)}))\tau = ((g_i)\gamma)\tau.\end{aligned}$$

In particular $G'(F) = G(E)$ and if $E/F$ is Galois then $G'(E) = \prod_\sigma G(E)$.

Suppose $G$ is reductive connected. Then $\Psi(G') = (X', \nabla', X'^*, \nabla'^\vee)$ is related to $\Psi(G) = (X, \nabla, X^*, \nabla^\vee)$ by $X' = IX$, $\nabla = \cup_\sigma \nabla\sigma$ $(\sigma \in \mathrm{Gal}(\overline{F}/E)\backslash \mathrm{Gal}(\overline{F}/F))$. Similarly, bases $\Delta'$ and $\Delta$ of $\nabla'$ and $\nabla$ are related by $\Delta' = \cup_\sigma \Delta\sigma$. In particular we have a natural isomorphism $\widehat{G}' \xrightarrow{\sim} I(\widehat{G})$, thus $\widehat{G}' \simeq \widehat{G}^{[E:F]}$.

The map $P \mapsto \mathrm{R}_{E/F}P$ induces a bijection from the set of $E$-parabolic subgroups of $G$ to the set of $F$-parabolic subgroups of $G'$, $P$ is a Borel subgroup of $G$ iff $\mathrm{R}_{E/F}P$ is one of $G'$. Hence $G$ is quasisplit over $E$ iff $G'$ is quasisplit over $F$.

If $\alpha : L_E \times \mathrm{SL}(2,\mathbb{C}) \to {}^LG$ is an $A$-parameter for $G$, then the corresponding parameter $\alpha' : L_F \times \mathrm{SL}(2,\mathbb{C}) \to {}^LG'$ is defined by

$$\alpha'(\tau \times s) = (\alpha(\tau_1 \times s), \ldots, \alpha(\tau_{[E:F]} \times s)) \times \tau.$$

The diagonal embedding $\widehat{G} \hookrightarrow \widehat{G}'$ induces $S_\alpha \to S_{\alpha'}$, and by Shapiro's lemma gives $\ker^1(L_E, Z(\widehat{G})) \to \ker^1(L_F, Z(\widehat{G}'))$, where $\ker^1$ denote the set of classes in $H^1$ which are locally trivial. We have a commutative square

$$\begin{array}{ccc} S_\alpha & \longrightarrow & S_{\alpha'} \\ \downarrow & & \downarrow \\ \ker^1(L_E, Z(\widehat{G})) & \to & \ker^1(L_F, Z(\widehat{G}')). \end{array}$$

Hence $S_{\alpha'} = Z(\widehat{G}') \cdot \mathrm{Im}(S_\alpha)$, and the diagonal map yields an isomorphism $\overline{S}_\alpha = \overline{S}_{\alpha'}$. In other words, the representation theory of $G(E)$ is the same as that of $G'(F)$.

# 8. Basechange

As an example, let us consider the case of **basechange** lifting. It concerns an $F$-group $G$, and "lifting" admissible representations of $G(F)$ to such representations of $G(E)$ if $F$ is local and $E/F$ is a finite extension of fields, or automorphic representations of $G(\mathbb{A}_F)$ to such representations of $G(\mathbb{A}_E)$ if $E/F$ is an extension of global fields. We need to view $G(E)$ (or $G(\mathbb{A}_E)$) as the group of points of an $F$-group in order to compare $L$-parameters of the $F$-group $G$ with those of what should describe $G(E)$. Such a group is given by $G' = \mathrm{R}_{E/F}G$, which is an $F$-group with $G'(F) = G(E)$. As for $L$-parameters, we have that the composition of $\lambda : W_F \to {}^LG$ with the diagonal embedding

$$\mathrm{bc}_{E/F} : {}^LG = \widehat{G} \rtimes W_F \to {}^LG' = \widehat{G}' \rtimes W_F = (\widehat{G} \times \cdots \times \widehat{G}) \rtimes W_F$$

gives an $L$-parameter $\lambda' = \mathrm{bc}_{E/F}(\lambda) : W_F \to {}^LG'$. In particular, the group $W_F$ permutes the factors $\widehat{G}$ in $\widehat{G}'$. The parameter $\lambda'$ can be viewed as the restriction $\lambda_E : W_E \to {}^LG = \widehat{G} \rtimes W_E$ of $\lambda$ from $W_F$ to $W_E$.

As a special case, suppose $G$ is split, thus the group $W_F$ acts trivially on $\widehat{G}$, but it permutes the factors $\widehat{G}$ in $\widehat{G}'$. Suppose $E/F$ is an unramified local fields extension. Then an unramified representation $\pi$ of $G(F)$ is determined by the image $t(\pi) = \lambda(\mathrm{Fr})$ of the Frobenius in $\widehat{G}$. This image is determined up to conjugacy. The image of $t(\pi)$ in ${}^LG'$ is the conjugacy class of $t(\pi') = (t(\pi) \times \cdots \times t(\pi)) \rtimes \mathrm{Fr} = (t(\pi)^{[E:F]}, 1 \ldots, 1) \rtimes \mathrm{Fr}$, which is the conjugacy class of $t(\pi)^{[E:F]}$ in the $L$-group ${}^L(G/E)$ of $G$ over $E$.

For example, the unramified irreducible constituent $\pi$ in the normalizedly induced representation $I(\mu_1, \ldots, \mu_n)$ of $\mathrm{GL}(n, F)$, where $\mu_i : F^\times \to \mathbb{C}^\times$ are unramified characters, lifts to the unramified irreducible constituent $\pi_E$ in the normalizedly induced representation $I(\mu_1 \circ N_{E/F}, \ldots, \mu_n \circ N_{E/F})$, $N_{E/F} : E^\times \to F^\times$ being the norm.

If $v$ is a place of a global field $F$ which splits in $E$, thus $E_v = E \otimes_F F_v = F_v \oplus \cdots \oplus F_v$, then $\mathrm{bc}_{E/F}(\pi_v) = \pi_v \times \cdots \times \pi_v$ is a representation of $G(E_v) = G(F_v) \times \cdots \times G(F_v)$.

The problem of basechange is to show, given an automorphic $\pi$ of $G(\mathbb{A}_F)$, the existence of an automorphic $\pi_E$ of $G(\mathbb{A}_E) = G'(\mathbb{A}_F)$ with $t(\pi_{E,v}) = \mathrm{bc}_{E/F}(t(\pi_v))$ for almost all $v$. For $G = \mathrm{GL}(n)$, if $\pi_E$ exists it is unique by rigidity theorem for $\mathrm{GL}(n)$.

A related question is to define and prove the existence of the local lifting. In any case, basic properties of basechange are, suitably interpreted:

• transitivity: if $F \subset E \subset L$ then $\mathrm{bc}_{L/E}(\mathrm{bc}_{E/F}(\pi)) = \mathrm{bc}_{L/F}(\pi)$ for $\pi$ on $G(F)$.

• twists: $\mathrm{bc}_{E/F}(\pi \otimes \chi) = \mathrm{bc}_{E/F}(\pi) \otimes \chi_E$, $\chi_E = \chi \circ N_{E/F}$ (if $G = \mathrm{GL}(n)$, $\chi : \mathbb{A}_F^\times / F^\times \to \mathbb{C}^\times$).

• parameters compatibility: $\mathrm{bc}_{E/F}(\pi(\lambda)) = \pi(\lambda_E)$.

For $G = \mathrm{GL}(n)$, cyclic basechange (thus when $E/F$ is a cyclic, in particular Galois, extension of number fields) was proven by Arthur-Clozel [AC]. A simple proof, but only for $\pi$ with a cuspidal component, is given in [F2;II], where the trace formula simplifies on using a regular-Iwahori component of the test function. The case of $n = 2$ had been done by Langlands [L6], using ideas of Saito and Shintani (twisted trace formula, character relations). A simple proof of basechange for GL(2), with no restrictions, is given in [F2;I], again using regular-Iwahori component to simplify the trace formula. Basechange for $\mathrm{GL}(n)$ asserts (see [AC]): Let $E/F$ be a cyclic extension of prime degree $\ell$.

• Given a cuspidal automorphic representation of $\mathrm{GL}(n, \mathbb{A}_F)$ there exists a unique automorphic representation $\pi_E = \mathrm{bc}_{E/F}(\pi)$ of $\mathrm{GL}(n, \mathbb{A}_E)$ which is the basechange lift of $\pi$. It is cuspidal unless $\ell$ divides $n$ and $\pi\omega = \pi$ for some character $\omega \neq 1$ of $\mathbb{A}_F^\times / F^\times N_{E/F} \mathbb{A}_E^\times$.

• If $\pi$ and $\pi'$ are cuspidal then $\mathrm{bc}_{E/F}(\pi) = \mathrm{bc}_{E/F}(\pi')$ iff $\pi' = \pi\omega$ for some character $\omega$ of $\mathbb{A}_F^\times / F^\times N_{E/F} \mathbb{A}_E^\times$

• A cuspidal representation $\pi_E$ of $\mathrm{GL}(n, \mathbb{A}_E)$ is the basechange $\mathrm{bc}_{E/F}(\pi)$ of a cuspidal $\pi$ of $\mathrm{GL}(n, \mathbb{A}_F)$ iff ${}^\sigma\pi_E = \pi_E$ for all $\sigma \in \mathrm{Gal}(E/F)$. Here ${}^\sigma\pi_E(g) = \pi_E(\sigma g)$.

• If $n = \ell m$ and $\pi$ is a cuspidal representation of $\mathrm{GL}(n, \mathbb{A}_F)$ with $\pi\omega = \pi$, $\omega \neq 1$ on $\mathbb{A}_F^\times / F^\times N_{E/F} \mathbb{A}_E^\times$ (thus $\omega$ has order $\ell = [E : F]$), then there is a cuspidal representation $\tau$ of $\mathrm{GL}(m, \mathbb{A}_E)$ with ${}^\sigma\tau \neq \tau$ for all $\sigma \neq 1$ in $\mathrm{Gal}(E/F)$ such that $\mathrm{bc}_{E/F}(\pi)$ is the representation $I(\tau, {}^\sigma\tau, {}^{\sigma^2}\tau, \ldots, {}^{\sigma^{\ell-1}}\tau)$ normalizedly induced from $\tau \otimes {}^\sigma\tau \otimes \cdots \otimes {}^{\sigma^{\ell-1}}\tau$ on the parabolic of type $(m, \ldots, m)$.

The last statement can also be stated as $\tau \mapsto \pi$, as follows.

Let $E/F$ be a cyclic extension of prime degree $\ell$. Let $\tau$ be a cuspidal representation of $\mathrm{GL}(m, \mathbb{A}_E)$ with ${}^\sigma\tau \neq \tau$ for all $\sigma \neq 1$ in $\mathrm{Gal}(E/F)$. Conjecturally this $\tau$ is parametrized by an $L$-parameter $\lambda^E : W_E \to \mathrm{GL}(m, \mathbb{C})$.

Consider $\lambda = \mathrm{Ind}_E^F \lambda^E$. It is a representation of $W_F$ in $\mathrm{GL}(n, \mathbb{C})$, $n = m\ell$. The group $\mathrm{GL}(m, E)$, or $\mathrm{R}_{E/F}\,\mathrm{GL}(m)$, can be viewed as an $\omega$-twisted endoscopic group of $\mathrm{GL}(n)$ over $F$, where $\omega$ is a primitive character on $\mathbb{A}_F^\times/F^\times N_{E/F}\mathbb{A}_E^\times$. At a place $v$ of $F$ which stays prime in $E$, an unramified representation $I(\mu_i; 1 \leq i \leq m)$ of $\mathrm{GL}(m, E_v)$ would correspond to $I(\zeta^j \mu_i^{1/\ell}; 0 \leq j \leq \ell, 1 \leq i \leq m)$ on $\mathrm{GL}(n, F_v)$. Here $\zeta$ is a primitive $\ell$th root of 1. At a place $v$ of $F$ which splits in $E$, $\lambda_v = \oplus_{w|v}\lambda_w^E$ and $\tau_v$ of $\mathrm{GL}(m, E_v)$, which is $\otimes_{w|v}\tau_w$ of $\prod_{w|v}\mathrm{GL}(m, F_w)$ corresponds to $I(\otimes_{w|v}\tau_w)$. The last result stated above, as part of basechange for $\mathrm{GL}(n)$, asserts that **endoscopic lifting** for $\mathrm{GL}(n)$ exists. Denote it be $\mathrm{end}_{E/F}(\tau)$. Namely

- Let $E/F$ be a cyclic extension of prime degree $\ell$, and $\tau$ and a cuspidal representation of $\mathrm{GL}(m, \mathbb{A}_E)$. Then $\pi = \mathrm{end}_{E/F}(\tau)$ exists as an automorphic representation of $\mathrm{GL}(n, \mathbb{A}_F)$, which is cuspidal when ${}^\sigma\tau \neq \tau$ for all $\sigma \neq 1$ in $\mathrm{Gal}(E/F)$. Moreover $\pi\omega = \pi$ for any character $\omega$ of $\mathbb{A}_F^\times/F^\times N_{E/F}\mathbb{A}_E^\times$. Any cuspidal $\pi$ of $\mathrm{GL}(n, \mathbb{A}_F)$ with $\omega\pi = \pi$ for such $\omega \neq 1$ is $\pi = \mathrm{end}_{E/F}(\tau)$ for a cuspidal $\tau$ of $\mathrm{GL}(m, \mathbb{A}_E)$ with ${}^\sigma\tau \neq \tau$ for all $\sigma$ in $\mathrm{Gal}(E/F)$. Further, $\mathrm{end}_{E/F}(\tau') = \mathrm{end}_{E/F}(\tau)$ iff $\tau' = {}^\sigma\tau$.

This result was first proven for $m = 1$, thus $\ell = n$, by Kazhdan [K1] for $\pi$ with a cuspidal component, and by [F1;I] without such restriction, and by Waldspurger [W3] and [F1;I] for all $m$. This technique, of endoscopic lifting (twisted by $\omega$, into $\mathrm{GL}(n, F)$), has the advantage of giving (local) character relations which are useful in the study of the metaplectic correspondence ([FK1]). The theory of basechange gives other character relations, and lifts $\pi$ with $\pi\omega = \pi$ to $I(\tau, {}^\sigma\tau, \dots)$. The endoscopic case of $n = 2 = \ell$ had been done by Labesse-Langlands [LL]. See also [F6].

The basechange and endoscopic liftings described above were proven using the trace formula, and they apply only to cyclic (Galois) extensions $E/F$. By means of the converse theorem, Jacquet, Piatetski-Shapiro, Shalika ([JPS]) showed

- Let $E/F$ be a non-Galois extension of degree 3 of number fields.

If $\pi$ is a cuspidal representation of $\mathrm{GL}(2, \mathbb{A}_F)$ then the basechange lift $\mathrm{bc}_{E/F}(\pi)$ exists and is a cuspidal representation of $\mathrm{GL}(2, \mathbb{A}_E)$.

Again, the lifting is defined by means of almost all components $\pi_v$, and $\mathrm{bc}_{E/F}(\pi)$ is unique – if it exists – by rigidity theorem for $\mathrm{GL}(2)$.

# II. ON ARTIN'S CONJECTURE

Let $F$ be a number field, $\overline{F}$ an algebraic closure, and $\lambda : \mathrm{Gal}(\overline{F}/F) \to \mathrm{Aut}\, V$, $\dim_{\mathbb{C}} V < \infty$, an irreducible representation. Define $L(s, \lambda)$ to be the product over all finite places $v$ of $F$ of the local factors $L(s, \lambda_v) = \det[1 - q_v^{-s} \cdot (\lambda_v | V^{I_v})(\mathrm{Fr}_v)]^{-1}$, where $V^{I_v}$ is the space of vectors in $V$ fixed by the inertia group $I_v$ at $v$, and $\lambda_v$ is the restriction of $\lambda$ to the decomposition group $D_v$ at $v$. Artin's conjecture asserts that the $L$-function $L(s, \lambda)$ is entire unless $\lambda$ is trivial ($= 1$). Langlands proposed approach to it is to show that there exists a cuspidal representation $\pi(\lambda)$ of $\mathrm{GL}(\dim V, \mathbb{A}_F)$ with $L(s, \lambda_v) = L(s, \pi(\lambda)_v)$ for almost all $v$. In this case, the holomorphy follows from the fact that $L(s, \pi) = \prod_v L(s, \pi_v)$ is entire for a cuspidal $\pi = \otimes \pi_v \neq 1$. Thus $\pi = \pi(\lambda)$ is related to $\lambda$ by the identity $t(\pi_v) = \lambda(\mathrm{Fr}_v)$ of semisimple conjugacy classes in $\mathrm{GL}(n, \mathbb{C})$ for almost all $v$. If this relation holds, $\lambda$ is uniquely determined by Chebotarev's density theorem, and $\pi$ is uniquely determined by the rigidity theorem for cuspidal representations of $\mathrm{GL}(n, \mathbb{A}_F)$. The case of $\dim_{\mathbb{C}} V = 1$ is that of Class Field Theory, which asserts that $\pi(\lambda)$ exists as a character of $\mathbb{A}_F^\times / F^\times$.

Suppose $\dim \lambda$ (i.e., $\dim V$) is two. Denote by $\mathrm{Sym}^2 : \mathrm{GL}(2, \mathbb{C}) \to \mathrm{GL}(3, \mathbb{C})$ the irreducible 3 dimensional representation of $\mathrm{GL}(2, \mathbb{C})$ which maps $g$ to $\mathrm{Int}(g)$ on $\mathrm{Lie}\, \mathrm{SL}(2)$. Its image is $\mathrm{SO}(3, \mathbb{C})$ and its kernel is the center of $\mathrm{GL}(2, \mathbb{C})$ (thus it gives

$$\mathrm{PGL}(2, \mathbb{C}) \xrightarrow{\sim} \mathrm{SO}(3, \mathbb{C}) \subset \mathrm{SL}(3, \mathbb{C})).$$

The finite subgroups of $\mathrm{SO}(3, \mathbb{C})$ are cyclic, dihedral, the alternating groups $A_4$ or $A_5$, or the symmetric group $S_4$ on four letters; see, e.g., Artin [A], Ch. 5, Theorem 9.1 (p. 184). If $\mathrm{Im}(\mathrm{Sym}^2 \circ \lambda)$ is cyclic then $\mathrm{Im}(\lambda)$ is contained in a torus of $\mathrm{GL}(2, \mathbb{C})$ and $\lambda$ is reducible, the sum of two characters. This case reduces to the case of CFT.

Let $\lambda : G \to \mathrm{GL}(2, \mathbb{C})$ be an irreducible two dimensional representation of a finite group.

1. PROPOSITION. $\mathrm{Im}(\mathrm{Sym}^2 \circ \lambda)$ *is dihedral iff* $\lambda = \mathrm{Ind}_H^G \chi$ *is induced from a character* $\chi$ *of an index two subgroup* $H$ *of* $G$*, and* ${}^g\chi \neq \chi$ *for all* $g \in G - H$.

PROOF. Assume $\lambda$ is faithful by replacing $G$ with $G/\ker\lambda$. Let $T$ be the cyclic subgroup of $\mathrm{Im}(\mathrm{Sym}^2 \circ \lambda)$ of index two. Since the kernel of $\mathrm{Sym}^2$ is central in $\mathrm{GL}(2,\mathbb{C})$ and $\lambda$ is faithful, the inverse image $H$ of $T$ in $G$ is abelian. Hence the restriction of $\lambda$ to $H$ is the sum of two one dimensional representations, $\chi$ and $\chi'$. If $\chi = \chi'$, Clifford's theory implies that $\chi$ extends to $G$ in two different ways (differing by the sign character on $G/H$). But $\lambda$ is irreducible two-dimensional, hence $\chi' \neq \chi$, $\chi' = {}^g\chi$ for any $g \in G - H$, and $\lambda = \mathrm{Ind}_H^G \chi$. □

2. COROLLARY. *Suppose* $\mathrm{Im}(\mathrm{Sym}^2 \circ \lambda)$ *is dihedral, where* $\lambda : \mathrm{Gal}(\overline{F}/F) \to \mathrm{GL}(2,\mathbb{C})$ *is two-dimensional. Then* $\pi(\lambda)$ *exists as a cuspidal representation of* $\mathrm{GL}(2,\mathbb{A}_F)$.

PROOF. By Proposition 1 there is a quadratic extension $E$ of $F$ and a character $\chi$ of $\mathrm{Gal}(\overline{F}/E)$ such that $\lambda = \mathrm{Ind}_E^F \chi$, $\chi \neq {}^\sigma\chi$ for all $\sigma \in \mathrm{Gal}(\overline{F}/F) - \mathrm{Gal}(\overline{F}/E)$. The existence of $\pi(\mathrm{Ind}_E^F \chi)$ is proven in [JL], [LL], [F3]. □

The irreducible representations of the symmetric group $S_n$ are parametrized by the partitions of $n$, and the associated Young tableaux. The representation $\lambda'$ associated to the dual Young tableaux is $\lambda \cdot \mathrm{sgn}$, where $\lambda$ is associated with the original Young tableaux, and sgn is the nontrivial character of $S_n/A_n$. The representation $\lambda$ of $S_n$ becomes reducible when restricted to $A_n$ precisely when the Young tableaux is selfdual. The dimension of $\lambda$ is the number of removal chains, by which we means a chain of operations of deleting a spot of a Young diagram at the right end of a row under which there is no spot. For example, $S_4$ has the representations listed in the table on the next page.

There the partitions (4) and (1,1,1,1) are dual. They parametrize the trivial and sgn one dimensional representations of $S_4$. The partitions (3,1) and (2,1,1) are dual (obtained from each other by transposition), and parametrize 3-dimensional representations whose restrictions $\lambda_3$ to $A_4$ remain irreducible and equal to one another. The selfdual partition (2,2) parametrizes the 2-dimensional irreducible representation of $S_4$ whose restriction to $A_4$ is reducible, equal to the sum of the two quadratic characters of $A_4$ (trivial on the 3-Sylow subgroup).

| partition | Young T | chains | dim |
|---|---|---|---|
| $(4)$ | $xxxx$ | $(xxx, xx, x)$ | 1 |
| $(3,1)$ | $\begin{matrix} xxx \\ x \end{matrix}$ | $(xxx, xx, x), \begin{pmatrix} xx & xx, & x \\ x\;, & & \end{pmatrix}, \begin{pmatrix} xx & x & x \\ x & , & x, \end{pmatrix}$ | 3 |
| $(2,2)$ | $\begin{matrix} xx \\ xx \end{matrix}$ | $\begin{pmatrix} xx & xx, & x \\ x\;, & & \end{pmatrix}, \begin{pmatrix} xx & x & x \\ x & , & x, \end{pmatrix}$ | 2 |
| $(2,1,1)$ | $\begin{matrix} xx \\ x \\ x \end{matrix}$ | $\begin{pmatrix} x & x & x \\ x & x, & \\ x, & & \end{pmatrix} \begin{pmatrix} xx & xx, & x \\ x\;, & & \end{pmatrix}, \begin{pmatrix} xx & x & x \\ x & , & x, \end{pmatrix}$ | 3 |
| $(1,1,1,1)$ | $\begin{matrix} x \\ x \\ x \\ x \end{matrix}$ | $\begin{pmatrix} x & x & x \\ x & x, & \\ x, & & \end{pmatrix}$ | 1 |

The 3-dimensional representation $\lambda_3$ of $A_4$ is induced. Indeed, consider the 2-Sylow subgroup $A_4'$ of $A_4$. It is generated by (12)(34), (13)(24), (14)(23), and is isomorphic to $\mathbb{Z}/2\oplus\mathbb{Z}/2$. The quotient $A_4/A_4'$ is $\mathbb{Z}/3$. The restriction to the abelian $A_4'$ of the irreducible 3-dimensional representation $\lambda_3$ of $A_4$ is the sum of 3 characters permuted by the quotient $\mathbb{Z}/3$ of $A_4$, hence $\lambda_3$ is induced $\mathrm{Ind}_{A_4'}^{A_4}\chi$, $\chi^2 = 1 \neq \chi$.

3. THEOREM. *There exists a cuspidal representation* $\pi(\lambda)$ *of* $\mathrm{GL}(2, \mathbb{A}_F)$ *where* $\lambda : \mathrm{Gal}(\overline{F}/F) \to \mathrm{GL}(2, \mathbb{C})$ *is an irreducible representation such that* $\mathrm{Im}(\mathrm{Sym}^2 \circ\lambda) = A_4$.

We record Langlands' proof ([L6]).

4. LEMMA. *There exists a cuspidal representation* $\pi(\mathrm{Sym}^2 \circ\lambda)$ *of* $\mathrm{GL}(3, \mathbb{A}_F)$.

PROOF. The composition of $\mathrm{Sym}^2 \circ\lambda$ with the projection $A_4 \to \mathbb{Z}/3$ is a surjective map $\mathrm{Gal}(\overline{F}/F) \to \mathbb{Z}/3$. Its kernel has the form $\mathrm{Gal}(\overline{F}/E)$, where $E/F$ is a cubic extension. As noted before the lemma, $\mathrm{Sym}^2 \circ\lambda = \mathrm{Ind}_E^F \chi$, where $\chi : \mathrm{Gal}(\overline{F}/E) \twoheadrightarrow \{\pm 1\}$, and ${}^\sigma\chi \neq \chi$ for $\sigma \neq 1$ in $\mathrm{Gal}(E/F) = \mathbb{Z}/3$. The existence of $\pi(\mathrm{Ind}_E^F \chi)$ now follows from the theory of (cubic) basechange for GL(3) [AC] or the endoscopic lifting for SL(3) of [K1] and [F1;I]. □

Put $\lambda_E$ for $\lambda | \mathrm{Gal}(\overline{F}/E)$.

5. LEMMA. *There exists a cuspidal representation* $\pi(\lambda_E)$ *of* $\mathrm{GL}(2, \mathbb{A}_E)$.

PROOF. We claim that $\lambda_E$ is irreducible. If not, it would be the direct sum of two characters, permuted by $\mathrm{Gal}(\overline{F}/F)/\mathrm{Gal}(\overline{F}/E) = \mathrm{Gal}(E/F) = \mathbb{Z}/3$. This action would then be trivial and $\lambda$ be reducible. But $\lambda$ is irreducible, hence so is $\lambda_E$. Now $\mathrm{Sym}^2 \circ \lambda_E$ has as image the order 4 dihedral group, hence $\pi(\lambda_E)$ exists. □

6. PROPOSITION. *Suppose* $\pi$ *is a cuspidal representation of* $\mathrm{GL}(2, \mathbb{A}_F)$ *whose basechange* $\mathrm{bc}_{E/F}(\pi)$ *to* $E$ *is* $\pi(\lambda_E)$*, whose central character* $\omega_\pi$ *is* $\det \lambda$*, and such that its symmetric square lift* $\mathrm{Sym}^2(\pi)$ *is* $\pi(\mathrm{Sym}^2 \circ \lambda)$*. Then* $\pi = \pi(\lambda)$.

PROOF. Denote by $[a, b]$ the conjugacy class of $\mathrm{diag}(a, b)$ in $\mathrm{GL}(2, \mathbb{C})$. At any place $v$ where $\pi_v$ and $\lambda_v = \lambda | D_v$ are unramified ($D_v \simeq \mathrm{Gal}(\overline{F}_v/F_v)$ is the decomposition group of $v$ in $\mathrm{Gal}(\overline{F}/F)$), put $t(\pi_v) = [a, b]$ and $\lambda(\mathrm{Fr}_v) = [\alpha, \beta]$. If $v$ splits in $E$ then $\mathrm{bc}_{E/F}(\pi) = \pi(\lambda_E)$ implies that $[a, b] = [\alpha, \beta]$. We need to show this also when $E_v = E \otimes_F F_v$ is a field, to conclude that $\pi = \pi(\lambda)$ by rigidity theorem for GL(2). When $E_v$ is a field, from $\mathrm{bc}_{E/F}(\pi) = \pi(\lambda_E)$ we conclude that $[a^3, b^3] = [\alpha^3, \beta^3]$, and from $\omega_\pi = \det \lambda$ that $ab = \alpha\beta$. Hence $a = \zeta\alpha$ and $b = \zeta^2\beta$ for some $\zeta \in \mathbb{C}^\times$ with $\zeta^3 = 1$. As $\mathrm{Sym}^2(\pi) = \pi(\mathrm{Sym}^3 \circ \lambda)$, we have $[a/b, 1, b/a] = [\alpha/\beta, 1, \beta/\alpha]$. From $t(\pi_v) = [\zeta\alpha, \zeta^2\beta]$, $\zeta \neq 1$, we then conclude that $\zeta^{-1}\alpha/\beta = \beta/\alpha$, hence that $\alpha/\beta = \pm\zeta^2$. If $\alpha/\beta = \zeta^2$, then $a = \zeta\alpha = \beta$, $b = \zeta^2\beta = \alpha$ and $[a, b] = [\alpha, \beta]$. If $\alpha/\beta = -\zeta^2$ then $\mathrm{Sym}^2 \circ \lambda(\mathrm{Fr}_v) = [-\zeta^2, 1, -\zeta]$, but $A_4$ has no element of order 6. □

It remains to show that $\pi$ as in Proposition 6 exists. Since ${}^\sigma\lambda_E = \lambda_E$ for all $\sigma$ in $\mathrm{Gal}(\overline{F}/F)$, we have ${}^\sigma\pi(\lambda_E) = \pi(\lambda_E)$. Hence there exists a cuspidal $\pi$ of $\mathrm{GL}(2, \mathbb{A}_F)$ with $\mathrm{bc}_{E/F}(\pi) = \pi(\lambda_E)$. This $\pi$ is unique only up to a twist by a character of $\mathbb{A}_F^\times / F^\times N_{E/F} \mathbb{A}_E^\times = \mathbb{Z}/3$. From $\mathrm{bc}_{E/F}(\pi) = \pi(\lambda_E)$ we get $\omega_\pi \circ N_{E/F} = \det \lambda \circ N_{E/F}$, hence $\omega_\pi \omega = \det \lambda$ for some character $\omega$ of $\mathbb{A}_F^\times / F^\times N_{E/F} \mathbb{A}_E^\times$. As $\omega_{\pi \otimes \omega^2} = \omega_\pi \omega^4 = \omega_\pi \omega$, we may and do choose $\pi$ with

$\omega_\pi = \det\lambda$. It remains to show that $\pi_1 = \mathrm{Sym}^2(\pi)$ and $\pi_2 = \pi(\mathrm{Sym}^2 \circ\lambda)$ are equal, namely that the classes $t(\pi_{1v})$ and $t(\pi_{2v})$ are equal for almost all $v$. For this we use the following theorem of Jacquet and Shalika [JS].

7. LEMMA. *Let $\pi_1$, $\pi_2$ be automorphic representations of* $\mathrm{GL}(n, \mathbb{A}_F)$ *with $\pi_2$ cuspidal, such that $t(\pi_{1v}) \otimes t(\check{\pi}_{2v}) = t(\pi_{2v}) \otimes t(\check{\pi}_{2v})$ for almost all $v$, where $\check{\pi}_{2v}$ denotes the representation contragredient to $\pi_{2v}$. Then $\pi_1 = \pi_2$.*

We take $n = 3$, and note that $t(\pi_{1v}) = t(\pi_{2v})$ when $v$ is split in $E$. It remains to verify the requirement of the Lemma when $v$ stays prime in $E$. In this case the image of $\mathrm{Fr}_v \in \mathrm{Gal}(\overline{F}/F)$ in $A_4$ has order 3, namely $t(\pi_{2v}) = \mathrm{Sym}^2(\lambda(\mathrm{Fr}_v)) = [1, \zeta, \zeta^2]$ for some $\zeta \neq 1 = \zeta^3$. Hence $\lambda(\mathrm{Fr}_v) = [\alpha, \zeta\alpha]$ for some $\alpha \in \mathbb{C}^\times$. Since $\pi_2$ is self-contragredient, we have $t(\check{\pi}_{2v}) = t(\pi_{2v})$. But $t(\pi_v)^3 = \lambda(\mathrm{Fr}_v)^3$ and $\det(t(\pi_v)) = \det(\lambda(\mathrm{Fr}_v))$. Hence $t(\pi_v) = [a, b]$ with $a^3 = b^3 = \alpha^3$ and $ab = \zeta\alpha^2$. So $t(\pi_v) = [\alpha, \zeta\alpha]$ or $[\zeta^2\alpha, \zeta^2\alpha]$. Consequently $t(\pi_{1v})$ is $[1, \zeta, \zeta^2]$ or $[1, 1, 1]$, and we have $t(\pi_{1v}) \otimes t(\check{\pi}_{2v}) = t(\pi_{2v}) \otimes t(\check{\pi}_{2v})$ in both cases.

This completes the proof of the existence of a cuspidal representation $\pi(\lambda)$ of $\mathrm{GL}(2, \mathbb{A}_F)$ where $\lambda : \mathrm{Gal}(\overline{F}/F) \to \mathrm{GL}(2, \mathbb{C})$ is irreducible with $\mathrm{Im}(\mathrm{Sym}^2 \circ\lambda) = A_4$. □

The next case, completed by Tunnell [Tu] after some work of Langlands, is that of

8. THEOREM. *Let $\lambda : \mathrm{Gal}(\overline{F}/F) \to \mathrm{GL}(2, \mathbb{C})$ be an irreducible representation with*

$$\mathrm{Im}(\mathrm{Sym}^2 \circ\lambda) = S_4 \quad (\simeq \mathrm{PGL}(2, \mathbb{F}_3)).$$

*Then $\pi(\lambda)$ exists as a cuspidal representation of* $\mathrm{GL}(2, \mathbb{A}_F)$.

Suppose $\ker(\mathrm{Sym}^2 \circ\lambda) = \mathrm{Gal}(\overline{F}/N)$, thus $N/F$ is an $S_4$-Galois extension. The subgroup $S_0$ of $S_4$, generated by $(12)(34)$, $(13)(24)$, $(14)(23)$, is normal in $S_4$, isomorphic to $\mathbb{Z}/2 \oplus \mathbb{Z}/2$, and there is an exact sequence

$$1 \to S_0 \to S_4 \to S_3 \to 1.$$

As $S_0$ is normal in $S_4$, its fixed field, $M$, is a Galois extension of $F$ of type $S_3$. The sgn character on $S_3$ defines a character of $S_4$; let $E$ be the quadratic extension of $F$ defined by its kernel. Let $K$ be the nonGalois cubic extension of $F$ fixed by a fixed 2-Sylow subgroup containing $S_0$. Since $\mathrm{Im}(\mathrm{Sym}^2 \circ \lambda_K)$ is a dihedral group ($\lambda_K = \lambda | \mathrm{Gal}(\overline{F}/K)$), $\pi(\lambda_K)$ exists as a cuspidal representation of $\mathrm{GL}(2, \mathbb{A}_K)$. Since $\mathrm{Im}(\mathrm{Sym}^2 \circ \lambda_E)$ is $A_4$, $\pi(\lambda_E)$ exists as a cuspidal representation of $\mathrm{GL}(2, \mathbb{A}_E)$. As usual, by $\mathrm{bc}_{A/F}(\pi)$ we mean the basechange of $\pi$ from $\mathrm{GL}(2, \mathbb{A}_F)$ to $\mathrm{GL}(2, \mathbb{A}_A)$.

We have the following diagram of fields

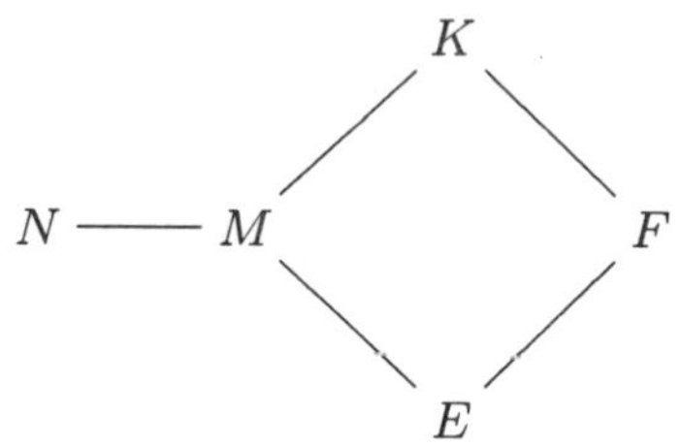


9. LEMMA. *Let $\pi$ be a cuspidal representation of $\mathrm{GL}(2, \mathbb{A}_F)$ such that $\mathrm{bc}_{E/F}(\pi) = \pi(\lambda_E)$ and $\mathrm{bc}_{K/F}(\pi) = \pi(\lambda_K)$. Then $\pi = \pi(\lambda)$.*

PROOF. At a place $v$ of $F$ where both $\pi$ and $\lambda$ are unramified, put $t(\pi_v) = [a, b]$ and $\lambda(\mathrm{Fr}_v) = [\alpha, \beta]$. If $v$ splits in $E$ or if $K_v = K \otimes_F F_v$ has $F_v$ as a direct summand, we have $t(\pi_v) = \lambda(\mathrm{Fr}_v)$ (equality of conjugacy classes in $\mathrm{GL}(2, \mathbb{C})$). If not, we get $t(\pi_v)^2 = \lambda(\mathrm{Fr}_v)^2$ and $t(\pi_v)^3 = \lambda(\mathrm{Fr}_v)^3$. If $t(\pi_v)$ and $\lambda(\mathrm{Fr}_v)$ share an eigenvalue, say $a = \alpha$, then $b^2 = \beta^2$ and $b^3 = \beta^3$ imply $b = \beta$ and $t(\pi_v) = \lambda(\mathrm{Fr}_v)$. If $t(\pi_v) \neq \lambda(\mathrm{Fr}_v)$ then they do not share an eigenvalue, and we may assume that $a = -\alpha$. As $t(\pi_v)^3 = \lambda(\mathrm{Fr}_v)^3$, we have $\beta = \zeta a$, $\zeta^3 = 1 \neq \zeta$. Hence $\lambda(\mathrm{Fr}_v) = [-a, \zeta a]$ and $(\mathrm{Sym}^2 \circ \lambda)(\mathrm{Fr}_v) = [-\zeta, 1, -\zeta^2]$, an element of order 6, which does not exist in $S_4$. Hence $t(\pi_v) = \lambda(\mathrm{Fr}_v)$. □

It remains to manufacture $\pi$ as in Lemma 9. Since $\lambda_E$ extends to $\lambda$ we have ${}^\sigma\lambda_E = \lambda_E$ for $\sigma \in \mathrm{Gal}(\overline{F}/F)$, $\sigma | E \neq 1$. Hence ${}^\sigma\pi(\lambda_E) = \pi(\lambda_E)$, and basechange theorem for $\mathrm{GL}(2, \mathbb{A}_F)$ implies that there exist precisely two cuspidal representations $\pi_1$ and $\pi_2$ of $\mathrm{GL}(2, \mathbb{A}_F)$ with $\mathrm{bc}_{E/F}(\pi_i) = \pi(\lambda_E)$, and $\pi_2 = \pi_1 \otimes \chi_{E/F}$, where $\chi_{E/F}(g)$ is 1 iff $\det g \in \mathbb{A}^\times / F^\times N_{E/F} \mathbb{A}_E^\times$.

Since $\mathrm{Im}(\mathrm{Sym}^2 \circ \lambda_M)$ is dihedral, and $\lambda_M$ is irreducible (see Lemma 5 in proof of Theorem 3), the cuspidal $\pi(\lambda_M)$ of $\mathrm{GL}(2, \mathbb{A}_M)$ exists. Hence $\pi(\lambda_K)$ lifts to $\pi(\lambda_M)$. But $\pi(\lambda_M) = \mathrm{bc}_{M/K}(\pi')$ for precisely two cuspidal representations $\pi'$ of $\mathrm{GL}(2, \mathbb{A}_K)$, and these two differ by a twist with $\chi_{M/K}$. Hence $\pi'$ are $\pi(\lambda_K)$ and $\pi(\lambda_K) \otimes \chi_{M/K}$.

At this stage we require a theorem of Jacquet, Piatetski-Shapiro and Shalika [JPS].

10. PROPOSITION. *Let $K/F$ be a field extension of degree 3 which is nonGalois. Then the basechange $\mathrm{bc}_{K/F}(\pi)$ of a cuspidal representation $\pi$ of $\mathrm{GL}(2, \mathbb{A}_F)$ exists and is a cuspidal representation of $\mathrm{GL}(2, \mathbb{A}_K)$.* □

This is proven by means of the converse theorem.

In particular $\mathrm{bc}_{K/F}(\pi_1)$ and $\mathrm{bc}_{K/F}(\pi_2)$ exist. They lift to $\pi(\lambda_M)$. Indeed, basechange is transitive, and is compatible with the Langlands correspondence $\lambda \mapsto \pi(\lambda)$. Hence

$$\mathrm{bc}_{M/K}(\mathrm{bc}_{K/F}(\pi_i)) = \mathrm{bc}_{M/F}(\pi_i)$$

$$= \mathrm{bc}_{M/E}(\mathrm{bc}_{E/F}(\pi_i)) = \mathrm{bc}_{M/E}(\pi(\lambda_E)) = \pi(\lambda_M).$$

But $\pi_1 = \pi_2 \otimes \chi_{E/F}$, and $\chi_{M/K} = \chi_{E/F} \circ N_{K/F}$. By the compatibility of basechange with twisting,

$$\mathrm{bc}_{K/F}(\pi_1) = \mathrm{bc}_{K/F}(\pi_2 \otimes \chi_{E/F}) = \mathrm{bc}_{K/F}(\pi_2) \otimes \chi_{M/K}.$$

Hence $\mathrm{bc}_{K/F}(\pi_i) = \pi(\lambda_K)$ for either $i = 1$ or $i = 2$. This $\pi_i$ has the properties required by Lemma 9, hence theorem 8 follows. □

# REFERENCES


[A1] J. Arthur, On a family of distributions obtained from Eisenstein series II: Explicit formulas, *Amer. J. Math.* 104 (1982), 1289-1336.

[A2] J. Arthur, On some problems suggested by the trace formula, *Lie group representations* II, Springer Lecture Notes 1041 (1984), 1-49.

[A3] J. Arthur, Unipotent automorphic representations: conjectures, in: Orbites unipotentes et representations, II, *Asterisque* 171-172 (1989), 13-71.

[A4] J. Arthur, Unipotent automorphic representations: global motivation, in *Automorphic forms, Shimura varieties, and L-functions*, Vol. I (Ann Arbor, MI, 1988), 1-75, Perspect. Math., 10, Academic Press, Boston MA 1990.

[A5] J. Arthur, A note on the automorphic Langlands group, *Canad. Math. Bull.* 45 (2002), 466-482.

[AC] J. Arthur, L. Clozel, *Simple algebras, base change, and the advanced theory of the trace formula*, Annals Math. Studies 120, Princeton, NJ, 1989. xiv+230 pp.

[A] M. Artin, *Algebra*, Prentice Hall, Englewood Cliffs, N.J., 1991.

[BZ1] J. Bernstein, A. Zelevinskii, Representations of the group $\mathrm{GL}(n, F)$ where $F$ is a nonarchimedean local field, *Russian Math. Surveys* 31 (1976), 1-68.

[BZ2] J. Bernstein, A. Zelevinsky, Induced representations of reductive $p$-adic groups, *Ann. Sci. École Norm. Sup.* 10 (1977), 441-472.

[Bla] D. Blasius, On multiplicities for $\mathrm{SL}(n)$, *Israel J. Math.* 88 (1994), 237-251.

[Bo1] A. Borel, Automorphic $L$-functions, in *Automorphic forms, representations and L-functions*, Proc. Sympos. Pure Math. 33 II, Amer. Math. Soc., Providence, R.I., (1979), 27-63.

[Bo2] A. Borel, *Linear algebraic groups*, 2nd enl. ed., Graduate texts in math. 126, Springer-Verlag, 1991.

[BC] A. Borel, W. Casselman, $L_2$-cohomology of locally symmetric manifolds of finite volume, *Duke Math. J.* 50 (1983), 625-647.

[BJ] A. Borel, H. Jacquet, Automorphic forms and automorphic representations, in *Automorphic forms, representations and L-functions*, Proc. Sympos. Pure Math. 33 I, Amer. Math. Soc., Providence, R.I., (1979), 189-202.

[BW] A. Borel, N. Wallach, *Continuous cohomology, discrete subgroups, and representations of reductive groups*, 2nd Ed., AMS, Providence, R.I., (2000).

[C] W. Casselman, Characters and Jacquet modules, *Math. Ann.* 230 (1977), 101-105.

[Ch] P.-S. Chan, *Invariant representations of GSp(2)*, thesis, the Ohio State University, 2005.

[Cl] L. Clozel, Characters of nonconnected, reductive $p$-adic groups, *Canad. J. Math.* 39 (1987), 149-167.

[CLL] L. Clozel, J. P. Labesse, R. Langlands, Seminar on the trace formula, Lecture Notes, Institute for Advanced Study, Princeton.

[D1] P. Deligne, Travaux de Shimura, in *Séminaire Bourbaki*, 23ème année (1970/71), Exp. No. 389, 123-165, Lecture Notes in Math. 244, Springer, Berlin, 1971.

[D2] P. Deligne, Formes modulaires et représentations de GL(2), in *Modular functions of one variable*, Lecture Notes in Math. 349 (1973), 55-105.

[D3] P. Deligne, Les constantes des equations fonctionnelles des fonctions L, *Modular Functions of One Variable* II, Springer Lecture Notes 349 (1973), 501-597.

[D4] P. Deligne, Le support du caractére d'une represéntation supercuspidale, *C. R. Acad. Sci. Paris* 283 (1976), 155-157.

[D5] P. Deligne, Variétés de Shimura: interprtation modulaire, et techniques de construction de modèles canoniques, in *Automorphic forms, representations and L-functions*, Proc. Sympos. Pure Math. 33 II, 247-289, Amer. Math. Soc., Providence, R.I., 1979.

[D6] P. Deligne, La conjecture de Weil II, *Publ. Math. IHES* 52 (1980), 137-252.

[Fl] D. Flath, Decomposition of representations into tensor products, in *Automorphic forms, representations and L-functions*, Proc. Sympos. Pure Math. 33(1) (1977), 179-183, Amer. Math. Soc., Providence, R.I., 1979.

[F1] Y. Flicker, I. On twisted liftings, *Trans. Amer. Math. Soc.* 290 (1985), 161-178;
II. Rigidity for Automorphic forms, *J. Analyse Math.* 49 (1987), 135-202.

[F2] Y. Flicker, I. Regular trace formula and base-change lifting, *Amer. J. Math.* 110 (1988), 739-764;
II. Regular trace formula and base-change for GL($n$), *Ann. Inst. Fourier* 40 (1990), 1-30.

[F3] Y. Flicker, *On the symmetric square lifting*, research monograph, 2003;
On the symmetric square. IV. Applications of a trace formula, *Trans. Amer. Math. Soc.* 330 (1992), 125-152;
V. Unit elements, *Pacific J. Math.* 175 (1996), 507-526;
VI. Total global comparison; *J. Funct. Analyse* 122 (1994), 255-278.

[F4] Y. Flicker, *Automorphic representations of the unitary group* U($3, E/F$), research monograph, 2003;
I. Elementary proof of a fundamental lemma for a unitary group, *Canad. J. Math.* 50 (1998), 74-98;
II. Packets and liftings for U(3), *J. Analyse Math.* 50 (1988), 19-63;
III. Base change trace identity for U(3), *J. Analyse Math.* 52 (1989), 39-52;
IV. Characters, genericity, and multiplicity one for U(3), *J. Analyse Math.* 94 (2004), 1-15.

[F5] Y. Flicker, Matching of orbital integrals on GL(4) and GSp(2), *Mem. Amer. Math. Soc.* no. 655, vol. 137 (1999), 1-114.

[F6] Y. Flicker, Lifting automorphic forms of PGSp(2) and SO(4) to PGL(4), part 1 of this book;
Automorphic forms on PGSp(2), Arbeitstagung 2003, Max-

Planck-Institut für Mathematik, Bonn, June 2003, 1-14; Automorphic forms on PGSp(2), *Elect. Res. Announc. Amer. Math. Soc.* 10 (2004), 39-50, http://www.ams.org/era; Automorphic forms on SO(4), *Proc. Japan Academy* 80 (2004), 100-104.

[F7] Y. Flicker, Zeta functions of Shimura varieties of PGSp(2), part 2 of this book.

[FK1] Y. Flicker, D. Kazhdan, Metaplectic correspondence, *Publ. Math. IHES* 64 (1987), 53-110

[FK2] Y. Flicker, D. Kazhdan, A simple trace formula, *J. Analyse Math.* 50 (1988), 189-200.

[FK3] Y. Flicker, D. Kazhdan, Geometric Ramanujan conjecture and Drinfeld reciprocity law, in *Number theory, trace formulas and discrete groups* (Oslo, 1987), 201-218, Academic Press, Boston MA, 1989.

[FZ] Y. Flicker, D. Zinoviev, Twisted character of a small representation of PGL(4), *Moscow Math. J.* 4 (2004), 333-368.

[Fu] K. Fujiwara, Rigid geometry, Lefschetz-Verdier trace formula and Deligne's conjecture, *Invent. Math.* 127 (1997), 489-533.

[FH] W. Fulton, J. Harris, *Representation Theory, A First Course*, Graduate Text in Mathematics 129, Springer-Verlag (1991).

[G] S. Gelbart, An elementary introduction to the Langlands program, *Bull. Amer. Math. Soc.* 10 (1984), 177-219.

[GRS] D. Ginzburg, S. Rallis, D. Soudry, On generic forms on SO($2n+$ 1); functorial lift to GL($2n$), endoscopy and base change, *Internat. Math. Res. Notices* 14 (2001), 729-763.

[Ha] G. Harder, *Eisensteinkohomologie und die Konstruktion gemischter Motive*, Springer Lecture Notes 1562 (1993).

[H] Harish-Chandra, Admissible invariant distributions on reductive $p$-adic groups, notes by S. DeBacker and P. Sally, *AMS Univ. Lecture Series* 16 (1999); see also: *Queen's Papers in Pure and Appl. Math.* 48 (1978), 281-346.

[HT] M. Harris, R. Taylor, *The geometry and cohomology of some simple Shimura varieties*, Annals of Mathematics Studies, 151, Princeton University Press, Princeton, NJ, 2001.

[He] G. Henniart, Progrès récents en fonctorialité de Langlands,

Séminaire Bourbaki, Vol. 2000/2001. Astérisque 282 (2002), Exp. No. 890, ix, 301-322.

[Hu] J. Humphreys, *Linear algebraic groups*, Graduate texts in math. 21, Springer-Verlag, 1975.

[JL] H. Jacquet, R. Langlands, *Automorphic forms on* GL(2), Springer Lecture Notes 114 (1970).

[JS] H. Jacquet, J. Shalika, On Euler products and the classification of automorphic forms (I and) II, *Amer. J. Math.* 103 (1981), (499-558 and) 777-815.

[JPS] H. Jacquet, I. Piatetski-Shapiro, J. Shalika, Relèvement cubique non normal, *C. R. Acad. Sci. Paris Sér. I Math.* 292 (1981), no. 12, 567-571.

[K1] D. Kazhdan, On lifting, in *Lie groups representations II*, Springer Lecture Notes 1041 (1984), 209-249.

[K2] D. Kazhdan, Cuspidal geometry of $p$-adic groups, *J. Analyse Math.* 47 (1986), 1-36.

[KL] D. Kazhdan, G. Lusztig, Proof of the Deligne-Langlands conjecture for Hecke algebras, *Invent. Math.* 87 (1987), 153-215.

[Kim] H. Kim, Residual spectrum of odd orthogonal groups, *Internat. Math. Res. Notices* 17 (2001), 873-906.

[Kn] A. Knapp, *Representation Theory of Semisimple Groups*, Princeton University Press (1986).

[Ko1] R. Kottwitz, Rational conjugacy classes in reductive groups, *Duke Math. J.* 49 (1982), 785-806.

[Ko2] R. Kottwitz, Stable trace formula: cuspidal tempered terms, *Duke Math. J.* 51 (1984), 611-650.

[Ko3] R. Kottwitz, Shimura varieties and twisted orbital integrals, *Math. Ann.* 269 (1984), 287-300.

[Ko4] R. Kottwitz, Shimura varieties and $\lambda$-adic representations, *Automorphic forms, Shimura varieties, and L-functions*, I (Ann Arbor, MI, 1988), 161-209, Perspect. Math. 10, Academic Press, Boston MA, 1990.

[Ko5] R. Kottwitz, On the $\lambda$-adic representations associated to some simple Shimura varieties, *Invent. Math.* 108 (1992), 653-665.

[Ko6] R. Kottwitz, Points on some Shimura varieties over finite fields, *J. Amer. Math. Soc.* 5 (1992), 373-444.

[KS] R. Kottwitz, D. Shelstad, Foundations of twisted endoscopy, *Asterisque* 255 (1999), vi+190 pp.

[KRS] S. Kudla, S. Rallis, D. Soudry, On the degree 5 $L$-function for Sp(2), *Invent. Math.* 107 (1992), 483-541.

[Lf] L. Lafforgue, Chtoucas de Drinfeld, formule des traces d'Arthur Selberg et correspondance de Langlands, *Proceedings of the International Congress of Mathematicians*, Vol. I (Beijing, 2002), 383-400, Higher Ed. Press, Beijing, 2002.

[LL] J.-P. Labesse, R. Langlands, $L$-indistinguishability for SL(2), *Canad. J. Math.* 31 (1979), 726-785.

[L1] R. Langlands, Problems in the theory of automorphic forms, in *Lectures in modern analysis and applications*, III, 18-61, Lecture Notes in Math. 170, Springer, Berlin, 1970.

[L2] R. Langlands, *Euler products*, Yale Mathematical Monographs, 1. Yale University Press, New Haven, Conn.-London, 1971. v+53 pp.

[L3] R. Langlands, *On the functional equations satisfied by Eisenstein series*, Lecture Notes in Math. 544. Springer-Verlag, 1976.

[L4] R. Langlands, On the notion of an automorphic representation, in *Automorphic forms, representations and L-functions*, Proc. Sympos. Pure Math. 33 I, Amer. Math. Soc., Providence, R.I., (1979), 203-207.

[L5] R. Langlands, I. Qn the zeta functions of some simple Shimura varieties, *Canad. J. Math.* 31 (1979), 1121-1216;
II. Shimura varieties and the Selberg trace formula, *Canad. J. Math.* 29 (1977), 1292-1299;
III. Automorphic representations, Shimura varieties, and motives, in *Automorphic forms, representations and L-functions*, Proc. Sympos. Pure Math. 33 II, (1979), 205-246.

[L6] R. Langlands, On the classification of irreducible representations of real algebraic groups, in *Representation theory and harmonic analysis on semisimple Lie groups*, 101-170, Math. Surveys Monogr., 31, Amer. Math. Soc., Providence, RI, 1989.

[L7] R. Langlands, *Base change for* GL(2), Annals of Math. Studies, 96. Princeton University Press, Princeton, N.J., 1980.

[L8] R. Langlands, Representations of abelian algebraic groups, *Pacific J. Math.* (1997), special issue, 231-250.

[Lar] M. Larsen, On the conjugacy of element-conjugate homomorphisms, *Israel J. Math.* 88 (1994), 253-277; II, *Quart. J. Math. Oxford Ser.* (2) 47 (1996), 73-85.

[Lau] G. Laumon, Sur la cohomologie à supports compacts des variétés de Shimura pour GSp(4, $\mathbb{Q}$), *Compos. Math.* 105 (1997), 267-359.

[MW1] C. Moeglin, J.L. Waldspurger, Le spectre résiduel de GL($n$), *Ann. Sci. Ecole Norm. Sup.* 22 (1989), 605-674.

[MW2] C. Möeglin, J.-L. Waldspurger, *Spectral decomposition and Eisenstein series. Une paraphrase de l'écriture*, Cambridge Tracts in Math. 113, Cambridge University Press, Cambridge, 1995.

[PS1] I. Piatetski-Shapiro, On the Saito-Kurokawa lifting, *Invent. Math.* 71 (1983), 309-338; A remark on my paper: "On the Saito-Kurokawa lifting", *Invent. Math.* 76 (1984), 75-76.

[PS2] I. Piatetski-Shapiro, Work of Waldspurger, *Lie group representations* II, Springer Lecture Notes 1041 (1984), 280-302.

[PR] V. Platonov, A. Rapinchuk, *Algebraic groups and number theory*, Pure and Applied Math. 139, Academic Press, Boston, MA, 1994.

[P] R. Pink, On the calculation of local terms in the Lefschetz-Verdier trace formula and its application to a conjecture of Deligne, *Ann. of Math.* 135 (1992), 483-525.

[Rb] B. Roberts, Global L-packets for GSp(2) and theta lifts, *Doc. Math.* 6 (2001), 247-314.

[Ro1] F. Rodier, Modèle de Whittaker et caractères de représentations, *Non-commutative harmonic analysis*, Springer Lecture Notes 466 (1975), 151-171.

[Ro2] F. Rodier, Sur les répresentations nonramifiées des groupes réductifs $p$-adiques; l'example de GSp(4), *Bull. Soc. Math. France* 116 (1988), 15-42;
Errata: http://xxx.lanl.gov/pdf/math.RT/0410273.

[ST] P. Sally, M. Tadic, Induced representations and classifications for GSp(2, $F$) and Sp(2, $F$), *Mem. Soc. Math. France* 52

(1993), 75-133.

[Sch] R. Schmidt, Generalized Saito-Kurokawa liftings, Habilitation 2002.

[Sh] E. Shpiz, Deligne's conjecture in the constant coefficients case, Ph.D. thesis, Harvard 1990.

[S] J.-P. Serre, *Cohomologie Galoisienne*, Springer Lecture Notes 5, 5$^e$ éd. (1994).

[Sh1] F. Shahidi, On certain L-functions, *Amer. J. Math.* 103 (1981), 297-355.

[Sh2] F. Shahidi, A proof of Langlands' conjecture on Plancherel measures; complementary series for $p$-adic groups, *Ann. of Math.* 132 (1990), 273-330.

[Sh3] F. Shahidi, Langlands' conjecture on Plancherel measures for $p$-adic groups, in *Harmonic analysis on reductive groups*, Progr. Math., 101, pp. 277-295, Birkhäuser, Boston, MA, 1991.

[Shal] F. Shalika, A theorem on semisimple $p$-adic groups, *Ann. of Math.* 95 (1972), 226-242.

[Si] A. Silberger, The Langlands quotient theorem for $p$-adic groups, *Math. Ann* 236 (1978), 95-104.

[So] D. Soudry, A uniqueness theorem for representations of GSO(6) and the strong multiplicity one theorem for generic representations of GSp(4), *Israel J. Math.* 58 (1987), 257-287.

[Sp] T. A. Springer, *Linear algebraic groups*, 2nd ed., Progress in math. 9, 1998.

[Tt] J. Tate, Number theoretic background, in *Automorphic forms, representations and L-functions*, Proc. Sympos. Pure Math. 33 II, Amer. Math. Soc., Providence, R.I., (1979).

[Ty] R. Taylor, On the $\ell$-adic cohomology of Siegel threefolds, *Invent. Math.* 114 (1993), 289-310.

[Ti] J. Tits, Reductive groups over local fields, in *Automorphic forms, representations and L-functions*, Proc. Sympos. Pure Math. 33(1), 29-69, Amer. Math. Soc., Providence, R.I., (1979).

[Tu] J. Tunnell, Artin's conjecture for representations of octahedral type, *Bull. Amer. Math. Soc.* (N.S.) 5 (1981), 173-175.

[Va] Y. Varshavsky, Lefschetz-Verdier trace formula and a generalization of a theorem of Fujiwara, http://arXiv.org/abs/math/0505564, 2005.

[Vo] D. Vogan, The Kazhdan-Lusztig conjecture for real reductive groups, in *Representation Theory of Reductive Groups*, P. Trombi, Ed., Birhauser (1983), 223-264.

[VZ] D. Vogan, G. Zuckerman, Unitary representations with nonzero cohomology, *Compos. Math.* 53 (1984), 51-90.

[W1] J.-L. Waldspurger, Un exercise sur GSp$(4, F)$ et les représentations de Weil, *Bull. Soc. Math. France* 115 (1987), 35-69.

[W2] J.-L. Waldspurger, Correspondances de Shimura et quaternions, *Forum Math.* 3 (1991), 219-307.

[W3] J.-L. Waldspurger, Le lemme fondamental implique le transfert, *Compositio Math.* 105 (1997), 153-236.

[W] R. Weissauer, http://www.mathi.uni-heidelberg.de/~weissaue

[Z] A. Zelevinsky, Induced representations of reductive $p$-adic groups II, *Ann. Sci. Ecole Norm. Sup.* 13 (1980), 165-210.

[Zi] Th. Zink, The Lefschetz trace formula for an open algebraic surface, in *Automorphic forms, Shimura varieties, and L-functions*, Vol. II (Ann Arbor, MI, 1988), 337-376, Perspect. Math. 11, Academic Press, Boston MA, 1990.

[Zu] S. Zucker, $L_2$-cohomology of Shimura varieties, in *Automorphic forms, Shimura varieties, and L- functions*, Vol. II (Ann Arbor, MI, 1988), 377-391, Perspect. Math., 11, Academic Press, Boston, MA, 1990.

# INDEX

# AUTOMORPHIC FORMS AND SHIMURA VARIETIES OF PGSp(2)

**by Yuval Z. Flicker (The Ohio State University, USA)**

The area of automorphic representations is a natural continuation of the 19th and 20th centuries studies in number theory and modular forms. A guiding principle is a reciprocity law relating the infinite dimensional automorphic representations, with finite dimensional Galois representations. Simple relations on the Galois side reflect deep relations on the automorphic side, called "liftings". This monograph concentrates on an initial example of the lifting, from a rank 2 symplectic group PGSp(2) to PGL(4), reflecting the natural embedding of $\mathrm{Sp}(2,\mathbb{C})$ in $\mathrm{SL}(4,\mathbb{C})$. It develops the technique of comparison of twisted and stabilized trace formulae. Main results include:

- A detailed classification of the representations of PGSp(2).
- A definition of the notions of "packets" and "quasi-packets".
- A statement and proof of the "lifting" by means of character relations.
- Proof of multiplicity one and rigidity theorems for the discrete spectrum.

These results are then used to study the decomposition of the cohomology of an associated Shimura variety, thereby linking Galois representations to geometric automorphic representations.

To put these results in a general context, the book ends with a technical introduction to Langlands' program in the area of automorphic representations. It includes a proof of known cases of Artin's conjecture.

This research monograph will benefit an audience of graduate students and researchers in number theory, algebra and representation theory.